European Conference on Neutron Scattering 2023 (ECNS 2023)

EPJ Web of Conferences Volume 286 (2023)

Garching, Germany
20 - 23 March 2023

Editors:

Ina Lommatzsch
Olaf Holderer
Henrich Frielinghaus
Yixi Su
Bjorn Pedersen
Martin Meven
Sebastian Busch
Christian Stieghorst
Anatoliy Senyshyn
Robert Georgii
Bastian Markisch
Stefano Pasini

ISBN: 978-1-7138-8043-1

Printed from e-media with permission by:

Curran Associates, Inc.
57 Morehouse Lane
Red Hook, NY 12571

Some format issues inherent in the e-media version may also appear in this print version.

For additional information, please contact EDP Sciences – Web of Conferences at the address below.

EDP Sciences – Web of Conferences
17, Avenue du Hoggar
Parc d'Activité de Courtabœuf
BP 112
F-91944 Les Ulis Cedex A
France

Phone: +33 (0) 1 69 18 75 75
Fax: +33 (0) 1 69 28 84 91

contact-edps@webofconferences.org

Additional copies of this publication are available from:

Curran Associates, Inc.
57 Morehouse Lane
Red Hook, NY 12571 USA
Phone: 845-758-0400
Fax: 845-758-2633
Email: curran@proceedings.com
Web: www.proceedings.com

TABLE OF CONTENTS

edp sciences

Web of Conferences

Statement of Peer review

In submitting conference proceedings to *Web of Conferences*, the editors of the proceedings certify to the Publisher that

1. They adhere to its **Policy on Publishing Integrity** in order to safeguard good scientific practice in publishing.
2. All articles have been subjected to peer review administered by the proceedings editors.
3. Reviews have been conducted by expert referees, who have been requested to provide unbiased and constructive comments aimed, whenever possible, at improving the work.
4. Proceedings editors have taken all reasonable steps to ensure the quality of the materials they publish and their decision to accept or reject a paper for publication has been based only on the merits of the work and the relevance to the journal.

Title, date and place of the conference

European Conference on Neutron Scattering 2023 (ECNS 2023), 20.-23.3.2023, Garching, Germany

Proceedings editor(s):

Ina Lommatzsch, Olaf Holderer, Henrich Frielinghaus, Yixi Su, Björn Pedersen, Martin Meven, Sebastian Busch, Christian Stieghorst, Anatoliy Senyshyn, Robert Georgii, Bastian Märkisch, Stefano Pasini

Date and editor's signature

2023-09-26, on behalf of all co-editors

17, avenue du Hoggar - PA de Courtabœuf – BP 112 - 91944 Les Ulis Cedex A (France)
Tél. : 33 (0)1 69 18 75 75 - Fax : 33(0)1 69 07 45 17 – www.edpsciences.org

Editorial: European Conference on Neutron Scattering 2023 in Garching

Ina Lommatzsch[1,*], *Olaf* Holderer[1,**], *Henrich* Frielinghaus[1], *Yixi* Su[1], *Björn* Pedersen[2], *Martin* Meven[3], *Sebastian* Busch[4], *Christian* Stieghorst[2], *Anatoliy* Senyshyn[2], *Robert* Georgii[2], *Bastian* Märkisch[5], *Stefano* Pasini[1], *Stephan* Förster[6], *Martin* Müller[4], and *Peter* Müller-Buschbaum[2]

[1]Jülich Centre for Neutron Science (JCNS) at Heinz Maier-Leibnitz Zentrum (MLZ), Forschungszentrum Jülich GmbH, Garching, Germany
[2]Heinz Maier-Leibnitz Zentrum (MLZ), Technical University of Munich, Lichtenbergstr. 1, 85748 Garching, Germany
[3]RWTH at Heinz Maier-Leibnitz Zentrum (MLZ), RWTH University Aachen, Garching, Germany
[4]Hereon at Heinz Maier-Leibnitz Zentrum (MLZ), Hereon Forschungszentrum, Garching, Germany
[5]TUM School of Natural Sciences, Technical University Munich, James-Franck-Str. 1, 85748 Garching
[6]Jülich Centre for Neutron Science (JCNS), Forschungszentrum Jülich GmbH, Jülich, Germany

Abstract. This edition of the European Conference on Neutron Scattering, which is held every four years, took place in Garching, Germany and gathered over 500 people from all areas of neutron science. From new source or instrumental concepts to fundamental physics, and chemistry, all areas of this community were represented. Lively discussions in the lecture halls, poster areas,and social events around this conference contributed to a very fruitful exchange of new ideas.

Hosted by the Heinz Maier-Leibnitz Zentrum (MLZ), the 8th European Conference on Neutron Scattering (ECNS) took place on the site of the TU Munich and in the vicinity of the Research Neutron Source Heinz Maier-Leibnitz (FRM II) in Garching/ Germany on March 20-23, 2023. The ECNS is organised by the European Neutron Scattering Association (https: //ensa.tudelft.nl/) and thus, every four years, neutron scatterers and instrument developers from all over the world gather at another beautiful place for the ECNS to discuss new thoughts and ideas. From its first appearance on the calendar of the neutron scattering community in 1996 to the 8th edition in Garching the ECNS made a tour of Europe with stops in Budapest (1999), Montpellier (2003), Lund (2007), Prag (2011), Zaragoza (2015), and St. Petersburg (2019).

Being the first big neutron conference in person after the pandemic, the invitation to Garching was very well received: 538 scientists not only from 20 European countries but also from America, Asia, and Australia participated! They gave and listened to 165 talks, and presented and visited 277 posters during this four-days event.

The program of this conference, with sessions on scientific topics such as magnetism, soft matter, biology, including biological membranes and proteins, glasses and liquids, or

*e-mail: i.lommatzsch@fz-juelich.de
**e-mail: o.holderer@fz-juelich.de

functional materials, reflected the broad field of applications of neutron science. Two microsymposia were dedicated to Compact Accelerator-driven Neutron Sources (CANS) and deuteration services (organised by DEUNET). Neutron instrumentation and neutron source developments were presented with a variety of talks, posters, and many animated discussions in all areas of the conference. New instrument concepts for the European Spallation Source were not only discussed at the ECNS but also at the SAGA satellite workshop dedicated especially to a new GISANS instrument at the ESS. Contacts were further intensified at the conference dinner held at the famous Munich Hofbräuhaus. The participants enjoyed traditional Bavarian beer, food, and dances very much!

The conference was opened by a welcome adress of the representative of German Ministry of Education and Research (BMBF) Dr. Renate Feldmann, of the Senior Vice President Research and Innovation of the Technical University of Munich, Prof. Gerhard Kramer as well as of the Scientific Director of the FRM II and MLZ Prof. Dr. Peter Mü ller-Buschbaum. Prof. Mirijam Zobel spoke on behalf of the German Committee Research with Neutrons (KFN) while the welcome speech from ENSA was held by Prof. Henrik Rønnow. They all pointed out the importance of neutron research as a tool to solve future challenges for science and society.

The conference location on the campus of the TUM and next to the site of FRM II offered the ideal infrastructure to accommodate the scientific program with its many parallel and poster sessions. The location was not accidental but chosen also for its vicinity to the FRM II. So after the closing speech of the Scientific Director of the FRM II and MLZ all the participants was given the possibility of a guided tours at the neutron source FRM II. Many took advantage of the offer. A novelty of this ECNS event was a new conference app. This could be easily downloaded form the conference web page and allowed to browse the sessions, read the abstracts, and by adding them to your agenda, to create your very own program for each event on your mobile. It was also used by the conference office to send messages and inform about changes in the program.

The final session of the last day chaired by the head of the ENSA, Prof. Henrik Rønnow celebrated the winners of the following prices: The Neutron Instrumentation and Innovation Award was awarded to Dr. Alessandro Tengattini (ILL), "in recognition of his outstanding contributions to the development of Neutron Imaging, in particular for realising the NeXT-Grenoble instrument at the ILL and perfecting all its capabilities to the highest level that is achievable" (https://www.ill.eu/about-the-ill/documentation/a-bit-of-history/prizes-and-honours).

The Erwin Félix Lewy Bertaut Prize was awarded to Dr. Matilde Saura-Múzquiz (Universidad Complutense de Madrid) "for her impressive early carrier contributions to the field of crystallography of functional materials, in particular those related to the studies of magnetic and non-magnetic complex oxide structures (scheelites, fergusonites, perovskites, pyrochlores,) and studies on nanoparticle morphology, nuclear and magnetic texture and magnetic performance of sintered materials, while combining neutron and X-ray diffraction methods". (https://ensa.tudelft.nl/laureate-of-the-2023-erwin-felix-lewy-bertaut-prize-is-dr-matilde-saura-muzquiz/).

The Walter Hälg Prize was awarded to Prof. Toby Perring (ISIS Neutron and Muon Facility) "for his exceptional contributions to neutron science through his work on time-of-flight single crystal spectroscopy". (https://www.isis.stfc.ac.uk/Pages/Congratulations-Toby-Perring!.aspx).

Congratulations to all award winners!

Supporters of the ECNS 2023 could choose between four options: platinum, gold, silver and bronze. Eleven neutron facilities, companies and associations contributed to the event:

1. The Institut Laue-Langevin
2. The Europan Spallation Source
3. AVS Neutron
4. The League of Advanced Neutrons Sources (LENS)
5. Paul Scherrer Institue
6. S-DH Neutron Guides
7. SwissNeutronics
8. Amsterdam Scientific Instruments
9. LND Detectors
10. Mirrotron
11. W-IE-NE-R Power electronics GmbH

Acknowledgement

We thank all authors and reviewers for their effort in contributing to these proceedings. The organisers of the conference would like to thank as well all participants for contributing to the success of the event.

Figure 1. The family of neutron scientists met in Garching. Foto: Bernhard Ludewig, FRM II / TUM

In Memoriam

Our much-loved, highly respected long-standing colleague, Dr Marie-Sousai Appavou, passed away suddenly on 13 September 2023 at the age of 46.

We will remember him for his passionate dedication to his research as well as his ready willingness to help others, always with kindness, modesty, and generosity – qualities that were very much his own.

After graduating from Cergy-Pontoise University in 2001 with a degree in biochemistry, Marie-Sousai moved to the Laboratoire Léon Brillouin (LLB) for his PhD. He brought together two very distinct areas of physics – high pressure and neutron scattering – to elucidate questions rooted in biochemistry in his thesis entitled "Influence de la température et de la pression sur la structure et la dynamique de l'inhibiteur de la trypsine pancréatique bovine. Une étude par diffusion de neutrons." His PhD supervisor, Dr Marie-Claire Bellissent-Funel, recalls Marie-Sousai showing great perseverance as he carried out "a remarkable piece of research."

In 2006, Marie-Sousai moved to Germany to take up a postdoctoral position with Dr Wolfgang Doster in the Physics Department of the Technical University of Munich, where he studied the effects of hydration, secondary structure, temperature, and pressure on the dynamics of proteins. To this end, he designed and oversaw the construction of a high-pressure cell, tirelessly fighting leaks as he pushed materials to their limits.

From 2007, Marie-Sousai worked at the newly established outstation of Forschungszentrum Jülich, the Jülich Centre for Neutron Science (JCNS) at the FRM II research reactor in Garching. As instrument scientist for small-angle neutron scattering, he always supported users and students with his exceptional dedication, abundance of patience, and outstanding expertise. At the same time, he furthered his own research on the structure and dynamics of biomolecules using a wide range of neutron scattering methods.

Shortly after beginning at JCNS, Marie-Sousai once again broadened his competence by diving into transmission electron microscopy. After spending months on end with experts to master this technique, he established a completely new, JCNS-operated transmission electron microscopy laboratory for soft matter. Subsequently, he initiated numerous collaborations with institutes and universities all over Europe as well as in the United States and Japan. In 2021, he was instrumental in acquiring a scanning electron microscope for the jointly operated Hereon and JCNS laboratory at MLZ. All this activity did not distract him from neutron research and he often participated in measuring campaigns. Most recently, he was preparing an experiment at ILL scheduled for November 2023. He successfully managed to bridge the gap between real and reciprocal space and was keen to help others do the same.

Marie-Sousai's colleagues across the scientific community are deeply saddened by their loss. Wherever he went, he left a lasting impression with his smile and human touch. Sometimes, in the middle of a technical discussion, he would start searching for the necessary reference in his archive: an imposing ensemble of paper towering around him on his ever-too-narrow desk. We will all miss Marie-Sousai's warm and open-hearted nature, which made him approachable for scientific as well as private discussions, whatever the hour. His passing leaves an irreplaceable void within our community – not only do we lose an excellent scientist, we also lose a friend.

Our thoughts go to Marie-Sousai's family, and in particular to his wife and son. We wish them strength and fortitude at this sad and difficult time.

On behalf of his colleagues at FZJ, at MLZ – JCNS, GEMS, FRM II – as well as his former colleagues at LLB.

September 2023, Garching bei München

Endurance – Modernisation of the instrumentation suite at the Institut Laue-Langevin

Charles Dewhurst[1], and *Andreas* Meyer[1]

[1]Institut Laue-Langevin, 71 avenue des Martyrs, 38042 Grenoble, France

Abstract. Endurance encompasses more than 30 new or upgraded instrument and infrastructure projects, rolled out over 8 years between 2016 and 2023. Many new or upgraded instruments have already been deployed and are in user-operation including: the fission-fragment gamma ray spectrometer, FIPPS; the upgraded cold-neutron TOF spectrometer IN5; new thermal TOF spectrometer PANTHER; and a second protein crystallography station, DALI. The D3 hot-neutron diffractometer and IN20 thermal triple-axis spectrometer have been upgraded while new and additional detectors for the SANS instruments D11 and D22 have been installed. The D16 cold-neutron diffractometer has been fully modernised and new cold-neutron imaging instrument, NeXT, has been installed. Delivering a full suite of modernised instrumentation is dependent on the renewed in-pile beam extraction, H1-H2, new H24 (thermal) and H15 (cold) neutron guides and the in-house development and manufacture of critical technologies such as neutron detectors, monochromator optics and polarisation components. The new H24 guide provides dedicated end-of-guide positions to the upgraded D10+ single crystal diffractometer, IN13 backscattering instrument and the new XtremeD powder and single-crystal diffractometer. H15 will accommodate a substantially upgraded D(00)7 polarised diffuse scattering and spectroscopy instrument while D11 will be rebuilt and relocated with an optically cleaner collimation. Two additional end-of-guide positions are available for new instrumentation: The SHARP+ cold-neutron TOF spectrometer and a 4th SANS instrument, SAM. The success of Endurance will provide users with a fully modernised suite of world-class instruments.

1 Introduction

The ILL has maintained its position as international leader in neutron science by paying constant attention to its capacity for innovative engineering and through the construction of ever more powerful instrumentation. To this end, the Millennium Programme [1] has been succeeded by Endurance [2]. The first phase of Endurance began in 2016 with a particular emphasis on the replacement of the H24 thermal-neutron guide and associated instrumentation. In 2019 the ILL Associates agreed to fund a second phase of Endurance in full, thus allowing the complex and interdependent H15 cold-neutron guide and instrument projects to go ahead. For each phase of Endurance an external and independent dedicated instrument subcommittee was convened in order to advise the ILL management and ILL's scientific council as to the prioritisation of projects. Many Endurance projects that are independent of large infrastructure modifications such as neutron guides have already been delivered. With the replacement of key neutron guide infrastructure during the recent long shutdown we have delivered the large interdependent suite of instrumentation on the new H24 neutron guide and are installing the new H15 guide and instruments.

Endurance is an ambitious programme that has required careful planning to ensure that the construction schedules for the new instruments and critical infrastructure remain coherent with the reactor maintenance programme and operating cycles. In 2020, the ILL launched the multiannual ILL20-23 programme, which combines a major upgrade of the instrument suite (Endurance), safety implementations and maintenance operations to the reactor. The ILL20-23 project involved a master resource-loaded schedule for coordinating instrumentation work and reactor activities whilst securing a maximum of reactor cycles for our users.

During the H1-H2 long shutdown from October 2021 to February 2023 we have rolled out the build of our modernised infrastructure and instrument suite under the Endurance programme and in parallel with major maintenance, security and safety work on the reactor. The new H24 guide and instrument suite has been completed and has begun commissioning with the reactor restart on 27th February 2023. The H15 guide and associated instrumentation is on time for delivery in the second half of 2024. The first neutrons of 2023 have allowed us to perform important radioprotection validation of the new installations and the scientific commissioning of new or upgraded instrumentation. Gold-foil activation measurements and the first commissioning data from the new instruments have validated the performance of the new guide installations.

We are now ready to welcome the return of user experiments.

2 Endurance

2.1 Endurance phase 1 (2016 – 2018)

The first phase of Endurance has already seen the delivery and entry into user operation of the new fission-fragment gamma ray spectrometer FIPPS [3], the upgraded IN5 cold time-of-flight (TOF) spectrometer [4], upgraded IN20 thermal triple-axis spectrometer (TAS), new thermal TOF spectrometer PANTHER and an upgraded D3 hot-neutron diffractometer to allow for efficient measurement of hydrogen-containing liquid samples [5]. Meanwhile, after a long and complex cryogenic project the SuperSUN source of ultra-cold-neutrons [6] has begun commissioning with the aim to produce the highest densities of UCN's and allow science to begin on the PanEDM experiment. The BASTILLE and NESSE infrastructure projects have continuously delivered software for data treatment and analysis and new sample environment capabilities throughout Endurance.

The 'backbone' of Endurance 1 is the Chartreuse project involving the renewal of the H24 thermal neutron guide and the upgrade of the D10+ single crystal diffractometer, IN13 (CRG – collaborative research group) backscattering spectrometer [7] and a new extreme conditions powder and single crystal diffractometer, XtremeD (CRG). The 'Chartreuse' project has now finally been delivered with the completion of the H1-H2 shutdown in February 2023 and is described in Section 3.1.

2.2 Endurance phase 2 (2019 – 2023)

The second phase of Endurance had its tentative scope and budget set during 2017 with review and recommendations by a dedicated Instrument Subcommittee and endorsement by the Scientific Council. In a first round of funding for Endurance 2 we were able to rapidly construct detectors for the SANS instruments D22 and D11. A second protein crystallography station, DALI, was installed on a modified H141 guide (ex IN11) while the secondary spectrometer of the D16 cold-neutron diffractometer has also been upgraded including a new wide-angle detector [8]. The neutron imaging facility, NeXT, entered the user programme in 2020 [9] and has now seen a full re-build of the imaging instrument backbone allowing for state-of-the-art imaging and a second (monochromatic) imaging station, MOTO to be operational in 2024.

The 'lion's share' of Endurance 2 is the Vercors project and involves the renewal of the H15 cold-neutron guide and associated suite of upgraded instruments D(00)7 [10], D11, and T3 and new CRG instruments, SAM (SANS) and SHARP+ (cold TOF spectrometer), each of which receiving their own high-performance and dedicated end-of-guide beam position. The H15 project is vast in terms of the scope as well as both the engineering and optical complexity and began installation during the recent H1-H2 shutdown.

A summary of all completed and ongoing Endurance projects is presented in Section 4 while the layout of the ILL instrument suite and neutron guides referred to in this article is presented in Figure 7.

3 H1-H2 shutdown (2021 – 2023)

The most critical project during the past 16 months was the replacement of the H1-H2 beam tube and in-pile neutron guides that allow the extraction and transport of thermal- and cold-neutrons to instrumentation in the ILL7 guide hall (Figure 1). We have also taken the opportunity to modernise and renew infrastructure such as the H1-H2 swimming pool providing biological protection to the beam extraction, and the C5 third-barrier guide feeds through the outer reactor wall to ensure the highest safety standards and longevity of our key neutron delivery infrastructure.

Fig. 1. (top) Installation of the H1-H2 beam tube. (bottom) 'Pink'-housing with the assembly of guide plateaus with the H1 (cold neutron) guides on the top and H2 (thermal neutron) guides on the bottom.

By the beginning of 2022 the ILL 7 guide hall had been excavated down to the river bed of the Grenoble valley (Figure 2). Reconstruction began promptly with extensive concrete landscaping ready to accept the new guide and instrument suites for both the H24 and H15 projects. Civil engineering work was phased beginning with the new D10+ zone and the H24 guide and working roughly in instrument order though the Chartreuse side of the guide hall. A similar situation was seen on the Vercors side of the guide hall with the dismounting of the old H1 casemate, H15 guide and instruments in preparation for civil engineering works. The staggered work plan for the H24 and H15 suites allowed for an optimisation of ILL staff, construction subcontractors and material resources.

Fig. 2. The point of maximum destruction in the ILL 7 guide hall showing the excavation works for the new H24 (left) and H15 (right) guides and instrumentation suites.

Fig. 3. (top) The D10+ thermal single crystal diffractometer installed on the new H24 guide. (middle) H15 works showing the D11 and SHARP+ flight chambers relocated into their new positions. (bottom) Guide housings for the new H15 guide under preparation in front of the new H1 casemat in ILL7.

Concrete landscaping on the H24 side was completed in the summer of 2022 and marble laid allowing installation of the first instrument components for D10+ and XtremeD. Civil engineering work on the H15 side was also well advanced with work on the new H1 steel casemate and foundation pillars for the relocated D11 detector tank. By the end of the summer, the last sections of H24 were installed up to the new FIPPS position along with the H24 steel shielding tunnel. The IN13 and CT2 instruments began re-installation while the new floors for FIPPS and SHARP+ were prepared at the ends of the H24 and H15 beamlines respectively.

The new H24 guide was completed by autumn 2022 while work on the instruments D10+, IN13, XtremeD and CT2 progressed (Figure 3 top). On the H15 side the D11 detector tube was moved into its new position and the SHARP+ flight chamber was delicately 'air-lifted' though the roof of ILL7 into its new position at the end of the guide hall (Figure 3 middle). Meanwhile, in the reactor building and following the successful replacement of the H1-H2 beam tube, 'pink'-carter and plateau of in-pile neutron guides, connection of the new guide system to new swimming pool and H5 casemate guides progressed ready to meet up with the ILL7 guides.

Fig. 4. (top) New cold-neutron and x-ray imaging instrument NeXT. (bottom) New secondary spectrometer of the D16 cold-neutron diffractometer with in-house manufactured trench multi-wire proportional counter (TMWPC) detector.

We have completed all 'skeleton' tasks and much more during the H1-H2 shutdown concerning both the instrument suite under Endurance and reactor maintenance. The H1-H2 shutdown and Endurance has delivered a fully modernised H24 guide and instrument suite (D10+, IN13, XtremeD and CT2) as well as a number of independent instrument projects such as NeXT (Figure 4 top). The cold neutron diffractometer and wide-angle scattering instrument D16 has received a fully renewed secondary spectrometer including a new position sensitive detector covering four times the solid angle of the previous detector allowing for more rapid measurements over an extended q-range (Figure 4 bottom, Figure 6). Although already commissioned in 2020, the thermal TOF spectrometer PANTHER further increases its performance thanks to its new cascade of five background choppers. SuperSUN has finally produced its first ultra-cold neutrons after an extended and complex instrument project and commissioning. The renewed suite of H15 instruments (D007, D11, T3, Sharp+ and Sam) is on schedule for delivery and commissioning during 2024.

3.1 H24 guide and instruments D10+, IN13, XtremeD, CT2, FIPPS

The new H24 guide brings dedicated thermal neutron guides to upgraded D10+, IN13 and the new XtremeD (CRG) powder and single-crystal diffractometer as well as providing neutrons to the test instrument CT2 and a repositioned FIPPS. Each instrument benefits from a dedicated end-of-guide position allowing for optimised beam-shaping and monochromator optics. The new guide has a high critical angle coating (m=3) and exploits the two radii of curvature of the H241 (R=14000m) and H242 (R=8000m) down-stream sub-branches to naturally curve and expand the guide over a distance of 22 m. Gold-foil activation give a neutron flux of 2.5 x 10^9 n / s / cm^2 at the position of the D10+ instrument compared with a value of 0.7 x 10^9 n / s / cm^2 for the old guide system measured at a similar distance from the source just before the renovation. We have measured a total gain in count rate of 11 times on the new D10+ single crystal diffractometer (Cu monochromator at 1.26Å) with respect to the old D10 instrument due to the efficiency of the new detector (x1.6) and neutron flux (x6.6) due to the increased divergence, guide and monochromator size. Similar gains in intensity have been measured on IN13, primarily due to the performance of the new H24 guide coupled with a new temperature-gradient monochromator which is in the process of being optimised (Figure 5 right). New capabilities and capacity is available with the new CRG powder and single crystal diffractometer XtremeD having a performance to rival that of D20c. The renewed CT2 instrument for the testing of detectors also forms part of the H24 instrument suite while the end of the new H24 guide will make way for the future relocation of the fission fragment spectrometer FIPPS giving the instrument space, reduced background and most importantly a stable and intense neutron beam.

3.2 H15 guide and instruments T3, D007, D11, SAM, SHARP+

The H15 guide has a rather complex opposing-curved expanding section, referred to as 'the trumpet'. The trumpet serves two important purposes: The first is to spatially expand the neutron guide, thus converting the up-stream high divergence to a lower divergence but extended over a greater area. This allows the guide to be split into multiple individual guide branches allowing five dedicated end-of-guide beam positions for instrumentation. The second is that the opposing curve leaves a 'fingerprint' correlation between the divergence profile and spatial position at the end of the trumpet. In other words, neutrons on the left of the guide have a divergence distribution on average pointing to the left and conversely for the right side of the guide. Importantly, this allows guide branches to be more widely separated in angle, therefore allowing space for substantially more instrumentation downstream.

Civil engineering works for H15 are now complete allowing the cold-neutron instruments on the Vercors side of ILL7, excluding the H15 instruments, to restart with the ILL reactor. The H15 works have been scheduled intentionally so as to allow continued construction of the out-of-casemat H15 guide and instruments during reactor operation in 2023. The next long shutdown at the end 2023 will allow the final in-casemat installation of the new H15 guides, finally connecting the new guide to its new instrumentation suite ready for commissioning and user operation during the second half of 2024. We expect significant gains in performance from instruments such as D(00)7 and SHARP+ (CRG) compared to their predecessors D7 and IN6 due to the new high performance H15 guide, dedicated focussing guide and beam optics and improved or larger detectors. In the case of D(00)7 the expected order of magnitude increase in flux will make polarisation and spectroscopic analysis measurements realistic as well as massive gains for the more usual diffuse scattering measurements. On the cold TOF spectrometer SHARP+ we expect more than ten times the count rate of its predecessor IN6. D11 will be rebuilt and relocated with an optically cleaner collimation and should re-gain its theoretical brilliance to match that of D22 and D33 as should that of the new CRG SANS instrument SAM which includes a MIEZE option. The test instrument T3, essential for the testing of polarised supermirrors produced in-house, will be rebuilt and upgraded and completes the H15 instrument suite offering increased flux and beam polarisation.

Fig. 5. (top-left) Double-focussing copper monochromator for PANTHER. (bottom-left) Single-focussing back-to-back copper and graphite (HOPG) monochromators for D10+. (right) Final alignment of the new temperature gradient monochromator for IN13 on T13c with the first neutrons of 2023.

https://doi.org/10.1051/epjconf/202328601001

3.3 In-house technologies

In-house developed technologies have been key to successful and efficient instrument upgrades [8,11,12,13,14]. Both single- and double-focussing monochromators have been produced for various instruments, some of which are show in Figure 5. The detector technology used for XtremeD (Figure 6) is an excellent example of the high-quality research, development and manufacturing carried out in-house by ILL's services. The so-called Trench Multi-Wire Proportional Counter (TMWPC) [8] is a novel concept in detector technology allowing high count-rate and spatial resolutions down to 1.5 mm using ^{3}He detection gas. This same technology is being used to provide detectors for D16 (Figure 4 bottom, Figure 6) and D20c.

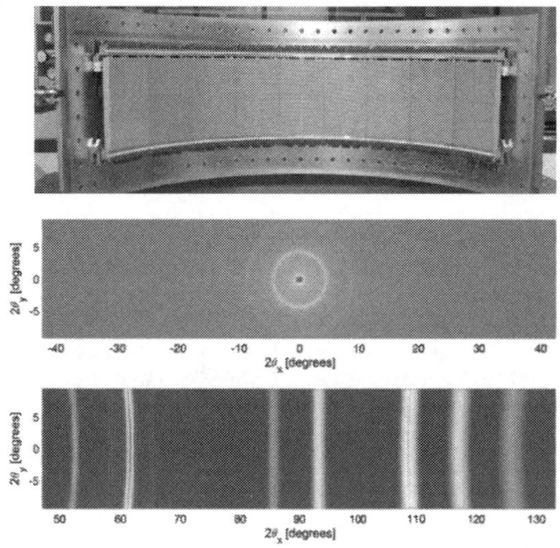

Fig. 6. (top) ILL designed and manufactured Trench Multi-Wire Proportional Counter (TMWPC) detector technology for XtremeD, D16 and D20 [9]. D16's new 'banana' detector has a radius of 1.15 m and 1152 x 192 pixels of resolution 1.5 mm and 2 mm [x,y]. (middle) Small-angle scattering from silver behenate. (bottom) Wide-angle diffraction from Yttrium-Iron-Garnet powder.

4 Endurance projects

Endurance encompass more than 30 new or upgraded instrument and infrastructure projects. We have already delivered many new or upgraded instruments and infrastructure packages that are already in user operation.

4.1 Completed projects

• FIPPS: Fission fragment gamma ray spectrometer giving new capabilities in the identification of prompt short-lived fission products and entirely complimentary to the existing Lohengrin mass spectrometer [3]. (2016)

• RAINBOWS: A re-scoped project to upgrade the D17 reflectometer with a new focussing guide and chopper system [15] to pursue an alternative high-flux mode of operation using the so-called 'coherent summing' method [16]. (2018)

• IN5 / H16: Cold-neutron TOF spectrometer. The new elliptically focussing H16 guide boasts huge gains in intensity, in particular at shorter wavelengths while focussing onto much smaller samples [4]. Access to shorter wavelengths on IN5 improves the overlap with the new thermal-neutron TOF spectrometer PANTHER. (2019)

• NESSE 1 & 2: Stimulus in sample environment capabilities covering all scientific domains such as low- and high-temperatures, high pressure, magnetic fields, humidity environments, and equipment for soft-matter and biological sciences. (2019, 2023)

• BASTILLE 1 & 2: Stimulus in scientific computing - Modern data reduction and analysis software tools via the Mantid project. Mantid is now well developed and fully deployed over our suite of TOF spectrometers and begins adoption on our powder diffraction, reflectometry and SANS instruments (solution scattering). (2019, 2023)

• PANTHER: New thermal-neutron TOF spectrometer replacing IN4 and with performance (signal / noise) approximately 60 times that if its predecessor. (2020). A cascade of 5 background choppers have been installed during the H1-H2 shutdown to further improve the performance of PANTHER. (2022)

• IN20: Upgraded thermal triple-axis spectrometer. A velocity selector for wavelength filtering and allowing much greater flexibility in instrument use and accessible energies. (2020). New graphite monochromator and analyser and a new Heussler monochromator [11]. (2023)

• Orient Express: Strategic move of instrument to H23 to allow uninterrupted access for sample alignment during the H24 works and to free up end-of-guide position for future relocation of FIPPS. (2020)

• DALI: Second protein crystallography instrument. Doubles the capacity and throughput of protein crystallography measurements with an increased neutron flux and flexibility due to the use of a velocity selector. (2020)

• D3 Liquids: The hot-neutron diffractometer is back in user operation following its beam tube replacement and upgrade with a new position-sensitive area detector and polarisation components, allowing for efficient measurement of hydrogen-containing liquid samples with precise discrimination of incoherent scattering [5]. (2021)

• D11 Detector: Replaced the aging multidetector with a modern, increased area, efficiency and count-rate detector. (2021)

• D22 Detector: An additional high-angle detector, massively extending the instrument's dynamic q-range and making for more rapid measurements with a reduced number of instrument configurations. (2021)

• D16: New secondary spectrometer, including a new wide-angle detector bank [8] with approximately four times the angular coverage (85 degrees) of the previous detector. (2023)

• NeXT: A full re-build of the neutron imaging instrument allowing for state-of-the-art imaging with advanced contrast techniques, high spatial resolution, intense neutron flux and combined X-ray imaging [9]. A second (monochromatic) imaging station, MOTO, for measurements such as Bragg-edge and dark-field imaging as well as technique development is under installation. (2023)

• H24 guide: New thermal-neutron guide providing dedicated end-of-guide positions to D10+, IN13, the new powder and single-crystal diffractometer XtremeD (CRG) as well as providing neutrons to the test instrument CT2 and a relocated FIPPS. (2023)

• D10+: Renewed thermal single-crystal diffractometer with count-rates more than ten times that of the previous D10 instrument due to the new H24 guide, new and larger focussing monochromators and improved efficiency detector. (2023)

• IN13: New primary spectrometer, including new temperature gradient monochromator, optimised for the dedicated end-of-guide position on the new H24 guide [7]. (2023)

• XtremeD: New CRG thermal neutron powder and single crystal diffractometer with performance to rival that of D20. (2023)

• CT2: Renewed test instrument for the characterisation of neutron detectors. (2023)

• SuperSUN: The new ultra-cold neutron (UCN) source SuperSUN is in the commissioning phase with the successful fabrication and installation of the converter guide and cool-down of the superfluid He converter volume. This long and complex project has presented a great number of technical and operational challenges along the way but SuperSUN has begun to produce high densities of UCNs and will allow science to begin on the PanEDM experiment [6]. (2023)

4.2 Ongoing projects

A number of Endurance projects remain in execution such as the H15 guide and associated instrument suite. We have also been able to re-launch several new instrument upgrade projects in a last wave of the independent Endurance projects. Ongoing Endurance projects are:

• H15 guide: New cold-neutron guide providing dedicated end-of-guide positions to T3, D(00)7, D11, new SANS instrument SAM (CRG) and upgraded SHARP+ (CRG) cold TOF spectrometer. (2024)

• T3: Renewed reflectometry test instrument for the characterisation of supermirrors. (2024)

• D(00)7: Upgraded diffuse scattering diffractometer and spectrometer. The expected increased flux will make polarisation and analysis spectroscopic studies realistic [10]. (2024)

• D11: SANS instrument relocated onto the new H15 guide and upgraded with reliable and optically clean beam collimation. (2024)

• SAM: New SANS CRG instrument with MIEZE option. (2024)

• SHARP+: Relocation of the SHARP instrument (previously IN6) onto the new H15 guide and upgrade of the primary spectrometer with focussing guide, monochromator and chopper cascade. (2024)

• D20c Detector: Replace the existing aging microstrip detector [17] with a new banana detector of TMWPC technology [8] (Figure 6). (2024)

• MARMOT: Multiplexed energy and angle analysis on the cold-neutron TAS instrument Thales. Detector elements and bent silicon analyser crystals for the spectrometer will use in-house technologies and fabrication. (2024 / 2025)

• WASP: The wide-angle spin-echo instrument [18] will receive its full complement of detectors with completion of the first additional detector bank due in 2024. (2024 / 2025)

5 Post-Endurance instrument upgrades and opportunities

We intend to continue upgrading our instrumentation after the completion of Endurance 2 wherever we can achieve the highest impact and cost-efficient upgrades that are independent of large infrastructure. We will capitalise on the high-quality and unique expertise of our in-house technical services in the provision of neutron technologies such as detectors, optical components, electronics, computing and sample environment.

In order to strengthen our research capabilities with high societal impact, e.g. materials research, process development, and health, potential smaller upgrade projects that are already identified include:

• Move FIPPS to a H24 end position to provide more intense and cleaner beam, more space for experimentation, and to allow an upgrade of SALSA (autumn 2023).

• Additional (Cu 200) monochromator for SALSA providing increased flux (up to x5) and instrument capabilities at longer wavelength (e.g. access to titanium-based materials and carbides).

- Additional thermal neutron imaging station providing additional capabilities and capacity.
- Velocity selector upgrade to LADI allowing increased flexibility in choice of wavelength and increasing the flux by up three times compared to the current wavelength filter.
- Camera-based detector system for LADI and DALI allowing quicker and continuous readout and higher detection efficiency.
- Use of machine learning and AI techniques for optimised instrument control, data acquisition and analysis.
- Continuous development of sample environments (e.g. for levitation techniques and in-operando additive manufacturing set-ups).
- Improved shielding and background reduction – overall improvement of signal-to-noise on instruments.
- New choppers, velocity selectors, monochromators, analysers, detectors.

The ILL scientific community will be consulted for upgrade proposals in due course and when we have greater visibility of the financial landscape for post-Endurance upgrades.

6 Summary

The present article gives an overview of the modernisation of the instrument suite and critical infrastructure at the Institut Laue-Langevin. The Endurance programme (2016 – 2023) has seen the upgrade, modernisation or complete renewal of more than 30 instrument and infrastructure projects, many of which have now been delivered and are available to users in the science programme. The remaining ongoing Endurance projects, in particular the H15 neutron guide and associated suite of instruments, will be completed and enter the user programme in 2024. We intend to continue to upgrade our instrument suite after the completion of Endurance and have identified a number of high-impact projects. The Millennium and Endurance programmes over the last 20 years have ensured ILL has a fully modernised suite of world-class instruments ready to operate for the decade to come.

We would like to acknowledge and thank all ILL staff for their outstanding contributions in the execution of the Endurance projects and the massive undertaking of works during the H1-H2 shutdown. The goodwill, mobility and competence of our staff has ensured these vast works remained on schedule. In particular we would like to mention the project managers and instrument teams involved in all Endurance, infrastructure and reactor projects over recent years. Endurance could not have succeeded without the broad technical oversight of Benjamin Giroud in the implantation of the H24 and H15 neutron guide projects and their associated suites of instruments. Our fully modernised instrument suite would not be possible without the expertise of the ILL technical services for detectors, optics, electronics and instrument control, design, engineering and mechanics, and construction.

References

1. https://www.ill.eu/users/instruments/modernisation-programmes/millennium-programme
2. https://www.ill.eu/users/instruments/modernisation-programmes/endurance
3. C. Michelagnoli, A. Blanc, E. Ruez-Martinez, A. Chebboubi, H. Faust, E. Froidefond, G. Kessedjian, M. Jentschel, U. Köster, P. Mutti, G. Simpson, EPJ Web Conf. 193, 04009 (2018)
4. J. Ollivier, B. Giroud, M. Kreuz, E. Farhi, J. Beaucour, C. Dewhurst, J. Neutron Res. **20**, 123 (2019)
5. A. Stunault, S. Vial, D. Jullien, G. Cuello, Physica B **551**, 373 (2018)
6. E. Chanel, S. Baudoin, MH Baurand, N. Belkhier, E. Bourgeat-Lami, S. Degenkolb, M. van der Grinten, M. Jentschel, V. Joyet, M. Kreuz, E. Lelièvre-Berna, J. Lucas, X. Tonon, O. Zimmer, J. Neutron Res. **24**, 111 (2022)
7. A. J. C. Dennison, F. Natali, E. Farhi, G. Manzin, P. Courtois, J. Peters, Nucl. Instrum. Methods Phys. Res. A **908**, 182 (2018)
8. J. C. Buffet, J. F. Clergeau, S. Cuccaro, B. Guerard, N. Mandaroux, J. Marchal, J. Pantenero, M. Platz, P. Van Esch, JINST **12**, C12009 (2017)
9. A. Tengattini, N. Lenoir, E. Ando, B. Giroud, D. Atkins, J. Beaucour, G. Viggiani, Nucl. Instrum. Methods Phys. Res. A **968**, 163939 (2020)
10. G. J. Nilsen, K. L. Brown, L. Mangin-Thro, W. Clancey, B. Giroud, C. D. Dewhurst, A. R. Wildes, Nucl. Instrum. Methods Phys. Res. A **951**, 162990 (2020)
11. P. Courtois, B. Hamelin, K. H. Andersen, Nucl. Instrum. Methods Phys. Res. A **529**, 158 (2004)
12. P. Courtois, J. Phys.: Conf. Ser. **746** 012005 (2016)
13. P. Courtois, T. Bigault, K. H. Andersen, J. Baudin-Cavallo, K. Ben Saidane, M. Berneron, A. El-Aazzouzzi, D. Gorny, W. Graff, T. Guiblain, R. Hehn, E. Hetzler, C. Menthonnex, B. Mestrallet, C. Dewhurst, Physica B **385**, 1271 (2006)
14. T. Bigault, G. Delphin, A. Vittoz, V. Gaignon, P. Courtois, J. Phys.: Conf. Ser. **528** 012017 (2014).
15. T. Saerbeck, R. Cubitt, A. Wildes, G. Manzin, K. H. Andersen, P. Gutfreund, J. Appl. Cryst. **51**, 249 (2018)
16. R. Cubitt, T. Saerbeck, R. A. Campbell, R. Barker, P. Gutfreund, J. Appl. Cryst. 48, 2006 (2015)
17. J. F. Clergeau, P. Convert, D. Feltin, H. E. Fischer, B. Guerard, T. Hansen, G. Manzin, A. Oed, P. Palleau, Nucl. Instrum. Methods Phys. Res. A **471**, 60 (2001)
18. R. Holzel, P. M. Bentley, P. Fouquet, Rev. Sci. Inst. **77**, 105107 (2006)

 https://doi.org/10.1051/epjconf/202328601001

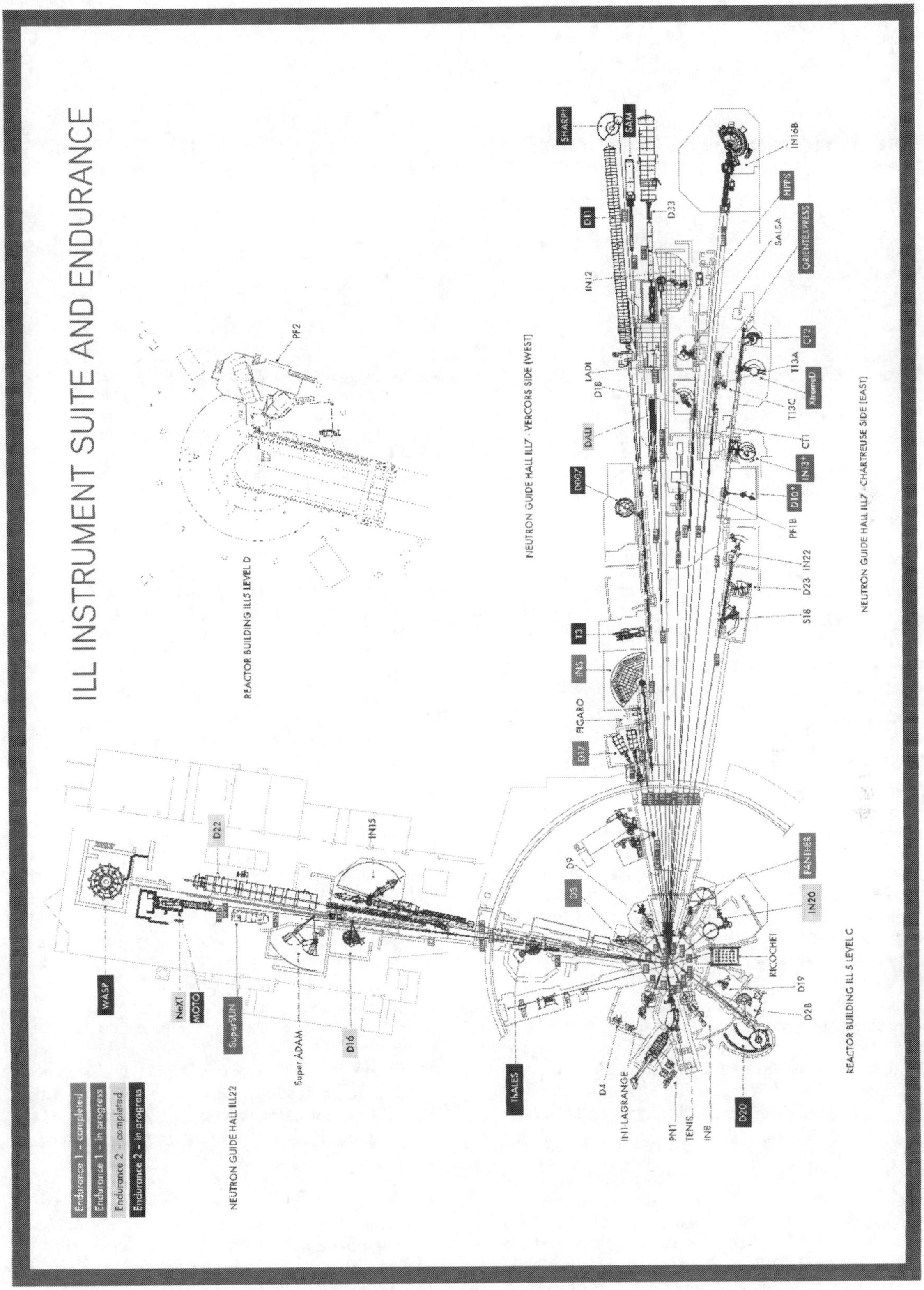

Fig. 7. Layout of the ILL instrument suite over the ILL 5 reactor building and the ILL 7 and ILL 22 guide halls. Instruments completed or in-progress as part of the Endurance programme are highlighted according to the legend while neutron guides are also indicated.

The future of ESS is bright

Werner Schweika[1,2*], *Mats* Lindroos[1], *Luca* Zanini[1] and *Tord* Ekelöf[3]

[1]European Spallation Source, SE-22100 Lund, Sweden
[2]Forschungszentrum Jülich, 52425 Jülich, Germany
[3]Uppsala University, 751 20 Uppsala, Sweden

Abstract. The European Spallation Source ESS has still a huge upgrade potential by using an accelerator ring structure for proton pulse compression that can change the long pulse to a medium pulse structure. Therefore, we consider the performance of a medium pulse structure on the existing ESS target, moderator and neutron instrumentation. A medium proton pulse will enhance the neutron peak brightness of thermal and cold neutrons by about one order of magnitude and even up to two orders of magnitude at shortest wavelengths used at ESS, largely increasing the performance of the ESS instruments for neutron scattering. The arguments for a medium pulse length are that it is best adapted to the resolution requirements of the ESS instruments, the coupled moderator, the typical long instruments; furthermore, that it seems feasible to realise a medium pulse length by advanced, slow extraction from an accelerator ring, and feasible for the rotating tungsten target to take the high-power load. We discuss the implications of a medium proton pulse length and its specific choice for the instrument resolution, and for moderator and target. The proposed upgrade is stimulated by the new project ESSnuSB for a neutrino super beam at ESS and motivated by the synergy effect of using a common pulse compressor ring. The upgrade will be a most economical and efficient path for the ESS to a next higher level from an already world leading long-pulse source, transforming with a medium pulse structure into an even much more powerful, future next generation neutron source.

1 Introduction

The European Spallation Source ESS [1] is approaching operation with beam on target in 2025. Yet it is timely to look ahead into future perspectives of ESS. The present work is motivated further by the synergies of a major ESS upgrade and a new project, the ESS neutrino super-beam project ESSnuSB, which is detailed in a recently completed conceptual design report [2]. For any explanation of that the universe has survived from the matter-antimatter annihilation after the Big Bang, there must be enough charge-parity violation (CPV) in nature. Such a CPV has till now not been observed. However, an observation of a difference in the flavour oscillation of neutrinos and antineutrinos, respectively, would be a sign of CPV. It is therefore now a paramount interest to try to discover such a difference and in particular also to measure its value. With the world's most powerful proton LINAC at ESS, it has been concluded that a more precise value of the CPV can be measured with the ESSnuSB large neutrino detector located at a distance from the ESS neutrino source in Lund where the neutrino flavour oscillation has its second maximum, as compared to the other long baseline neutrino experiments in the world, which have their neutrino detectors located at the closer first oscillation maximum. Therefore, it is proposed to raise the power of the ESS LINAC from 5 MW to 10 MW by increasing the frequency of the 2.86 ms long proton pulses accelerated in the LINAC from 14 Hz to 28 Hz, extract the additional 14 proton pulses per second sideways from the LINAC, inject them into a 400 m circumference compressor ring during many turns, and finally extract the stored protons in one turn, thereby creating 1.3 μs short pulses. Such a ring could also be used for pulse compression for neutron spallation at ESS, offering a tremendous possible increase in neutron peak brightness. In this context, it is most important to point out the appropriate best use of compressed pulses for neutron instrumentation at ESS, and next how to meet the challenge of an unprecedented peak power of protons hitting the target. Taking the proton beam with a compressed pulse of 1.3 μs on the neutron spallation target would increase the instantaneous power load on the ESS target to a peak power of 275 GW. As proposed earlier, one mitigation to achieve an acceptable peak power, is a slow extraction of protons in multiple turns from the ring [3]. Clearly, to optimise the structures for generating spallation neutrons and neutrinos is an overarching endeavour, combining considerations of the best benefit for neutron instrumentation with possible technical realisations for target and accelerator. Hence, we will aim for a new flexible accelerator ring that can deliver short 1.3 μs proton pulses for neutrino production and medium proton pulses of about 50 μs, - as will be discussed here -, for spallation neutrons, in

* Corresponding author: werner.schweika@ess.eu

each case with 14 pulses per second and each with final 5 MW power.

ESS is based on the long-pulse spallation source concept [4,5]. The ESS long pulse structure is directly generated by a powerful 5 MW LINAC without the typical use of a compressor ring at short-pulse spallation sources. This choice has been taken as a best cost-benefit compromise [5]. There are certainly benefits of the long pulse structure as discussed in [6]. Perhaps most importantly, it can make consequently use of fully coupled moderators providing highest average flux. Only one moderator system is required to serve all instruments. Any instrument specific resolution can be generated by pulse shaping choppers, which can favourably tailor sharper pulses with even symmetric pulse shape. With such flexibility in setting resolution requirements, typically an ESS instrument can cover more experimental needs than single instruments at other facilities. The ESS average flux is comparable to the existing most intense reactor sources ILL at Grenoble, FRM-2 at Munich and HFIR at Oak Ridge, albeit with a peak brightness that at least for cold neutrons is superior to the best current short pulse spallation sources, JPARC and SNS. The first construction goal at ESS will be to achieve 2 MW at 0.8 GeV. With further optimization of the moderator [7], it will already yield about the same neutron flux that has been predicted by the originally design for 5 MW [1].

A compressor ring at ESS producing short pulses, which is required by ESSnuB to achieve a sufficient suppression of the background from cosmic radiation in the large underground neutrino detector, also creates new opportunities for neutron production and its time structure. The conventional approach for an upgrade by using short pulses would be to build a new target station and new instrumentation. Such an approach however would take neutrons from the existing neutron target and diminish its value. In this paper, we want to explore an exiting alternative, the potential of using compressed pulses for the existing target and the coupled moderator, and the existing neutron instrumentation [6]. Most characteristic features of the ESS are the long proton pulse of 2.86 ms generated with 14 Hz repetition rate by a most powerful 5 MW superconducting LINAC, a He-cooled rotating tungsten target, and a new developed coupled moderator of high brightness, which offers cold and thermal neutrons for each neutron beam port [7]. To achieve adequate time-wavelength resolution, beamlines and instruments at ESS, shown in Figure 1, tend to be comparatively long, up to 160 m, nonetheless, pulse-shaping choppers are required for most of the instruments, essentially cutting intensity for resolution. Now a first simple goal is to preserve the full flux by generating a sufficiently short pulse for all instruments.

The potential of a medium pulse source to reach higher neutron peak brightness has been recognised earlier [3,8]. In Ref [8] the authors envisaged to build a new target station with dedicated neutron instruments, which would require a much longer lead time for realisation. In contrast, the upgrade concept presented here can take a short-cut with realisation within the next decade.

In the following, we first briefly discuss the aspects for accelerator and target, and next the time-wavelength dependent response of the coupled ESS moderator. The consequences for the various neutron instruments will be discussed within a preliminary analysis to guide a realistic and most promising upgrade path of ESS.

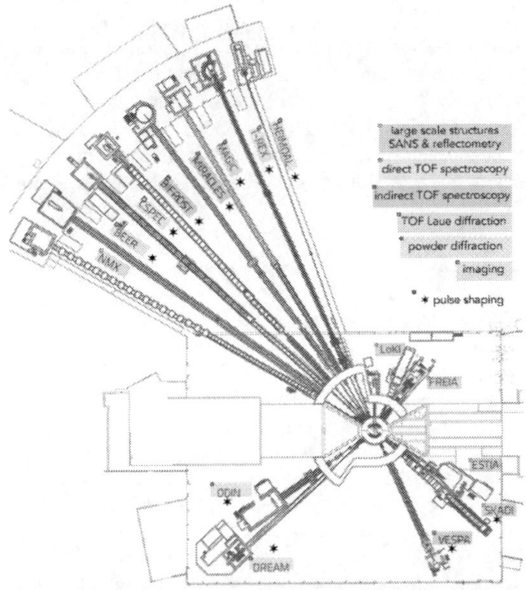

Fig. 1. Layout of the ESS neutron scattering instruments. A major fraction of the instruments, marked with an asterisk, uses pulse shaping choppers.

2 Accelerator, target and moderator

2.1 Accelerator

The plan for the ESS neutrino Super Beam project ESSnuSB is to operate the ESS LINAC at 28 Hz, where the additional pulses are used for neutrino production. As already described, this implies a doubling of the power from 5 to 10 MW and an accumulator ring, which requires the addition of an H^- source, with pulse compression from 2.86 ms to 1.3 μs. Within this upgrade, the energy of the ESS LINAC will increase from 2 GeV to the originally designed 2.5 GeV. The short pulses are required for detecting neutrino oscillations with a sufficient signal-to-background ratio. In fact, these pulses have a sub-structure of four separated pulses, and to decrease the instantaneous power load for the target, the beam will be split and delivered to four separate targets.

With advances of accelerator technology, it appears feasible to extract alternatively longer pulses to adjust to the acceptable peak power for the neutron spallation target and to better match to the moderation times of thermal and cold neutrons. A proposal in this direction has been made already in 2013 [3]. This scheme is proposing to use a large stacked (4 rings) storage ring with a circumference of 1100 m and fast strip line kickers to enable low loss bunch-by-bunch extraction generating a bunch-train of a total length of 50-100 microseconds. Another very interesting proposal has been made by S. Machida [9,10] to use a cyclotron like extraction scheme from an accumulation ring. Back-of-the-envelope calculations indicate that the scheme would have low losses in the extraction region and that the proposed extraction mechanism is feasible. Finally, slow extraction from the ESSnuSB accumulation ring using a fast integer resonance scheme has also been proposed by M. Olvegaard [9,10]. The challenges to operate a high-power H^- ion LINAC are substantial, which can be met with recent technical advances by direct H^+ injection to a ring as presented by C. Prior [9,10].

2.2 Target

The ESS target [11] is a rotating tungsten wheel, which is cooled by He-gas flow, and seems to be a most favourable choice in view of high-power load. The original design stems from the SNQ design study by G. Bauer *et al.* in 1981 [12], which already included a compressor ring as an upgrade option. With the choice of tungsten, the SNQ target should have been capable to take the compressed pulses from the 5 MW source operating at 100 Hz, corresponding to a peak power of 71 GW, for compressed pulses of 0.7 μs [12]. The rotating tungsten target has been taken also as the preferred choice for the SNS second target station STS [13]. At final average power at SNS of 2.8 MW at 60 Hz repetition rate and 0.85 μs proton pulse duration, the peak power will reach 55 GW. One may further compare to the peak power of targets at other short pulse spallation sources, 40 GW at JPARC using a liquid mercury target like the current SNS target, and 50 GW on a stationary tungsten target for the ISIS-II upgrade project [14]. This comparison suggests that without severe modifications, the ESS target should be capable for taking similarly a peak power near to 50 GW, which would correspond to an acceptable minimal pulse length of about 7 μs.

Note that the pulse compression has no impact on slow processes like target cooling, although radiation embrittlement may have an impact for the required elastic properties. The relevant fast process is due to thermo-mechanical shock waves. The dynamical stress limit of the tungsten material is 90 N/mm². The dynamical stress/strain due to shock waves can be estimated by $\sigma = E\,\alpha\,\Delta T$, where $E\,\alpha$ is the product of the elasticity module and thermal expansion coefficient, and ΔT is the temperature gradient emerging in a tungsten block upon the proton pulse impact. A rough, conservative estimate yields $\Delta T{\sim}150$ K, with 357 kJ/pulse, the specific heat of tungsten, and heat distribution profile [10], indicating that the stress/strain can be unacceptably high, $\sigma{\sim}270$ N/mm².

There are two mitigations. One is stress release for $t_p > t_0$, *i.e.* if the proton pulse duration t_p is larger than the time t_0 of sound waves passing through the heated region. Estimating $t_0 \sim 20$ μs for a characteristic length of $\Delta l \sim 10$ cm yields that a pulse of $t_p > \sim 60$ μs is acceptable. The second, and relative straightforward option seems to be a modification of the currently large tungsten block structure towards a more granular target. Filling the rotating target wheel with smaller blocks or grains of a few cm size will reduce temperature gradients to less than 50 K to stay below the dynamical stress/strain limit.

Clearly, a thorough analysis and simulation of the ESS target and possible modifications are of very high interest, since the above considerations suggest a possible stable solution avoiding destructive shock waves even for short proton pulses. There are more material aspects to consider in detail, like the internal steel structure in the wheel, its enclosure material with the proton beam window. With respect to the proton beam spread, the current rastering scheme could be replaced by an appropriate defocusing of the proton beam, however, in comparison to the LINAC, the ring will generate already much a larger beam size.

2.3 Moderator

Optimising for high neutron flux and long pulses the ESS has chosen a coupled, 3 cm flat moderator with a "butterfly" shape, where a thermal water moderator is placed next to a para-hydrogen cold source, with open view to all instrument beamlines [7]. It is straightforward to evaluate the performance of the ESS moderator for shorter pulses, since the response of reflector and moderator to generated spallation neutrons is unchanged and the neutron pulse is determined by a convolution of the proton pulse with this response function. The corresponding wavelength-time dependent neutron flux distribution for a 50 μs short pulse as calculated by Vitess [15] is shown in Fig. 2.

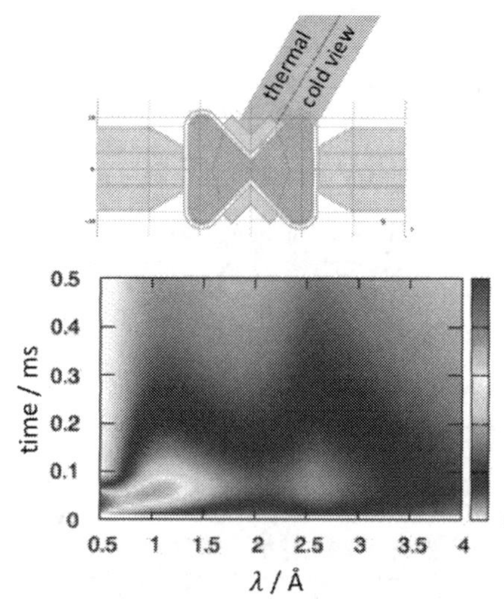

Fig. 2. (top) The ESS butterfly moderator and (bottom) the wavelength-time dependent brightness for 50 μs proton pulses in an approximate Vitess model.

There are two peaks related to the spectra from the thermal and cold moderators. The time dependence displays a characteristic asymmetry, whose asymptotic behaviour is independent of the proton pulse duration. Upon entering the epithermal region near 0.5 Å, one may note relatively sharp neutron pulses, since the neutron energies essentially decouple from the moderator.

The effect of varying proton pulse duration is shown in Fig. 3. While the integrated flux remains unchanged, a large increase in peak brightness is found for a compressed shorter pulse length: for a 50 μs pulse, the gain is approximately an order of magnitude at 4 Å, a factor 18 at 1Å, and a factor 57 at 0.5 Å. For cold neutrons, as shown in Fig. 3 (*top*), the neutron pulse width is essentially the same for proton pulses shorter than ~100 μs. In the MCNPX [16] simulations, the cold moderator is viewed under an angle of 30 degrees, and thus viewing the full depth of the cold para-hydrogen moderator resulting in a pulse width of ~300 μs. For larger viewing angles, the width is expected to be shorter down to ~200 μs.

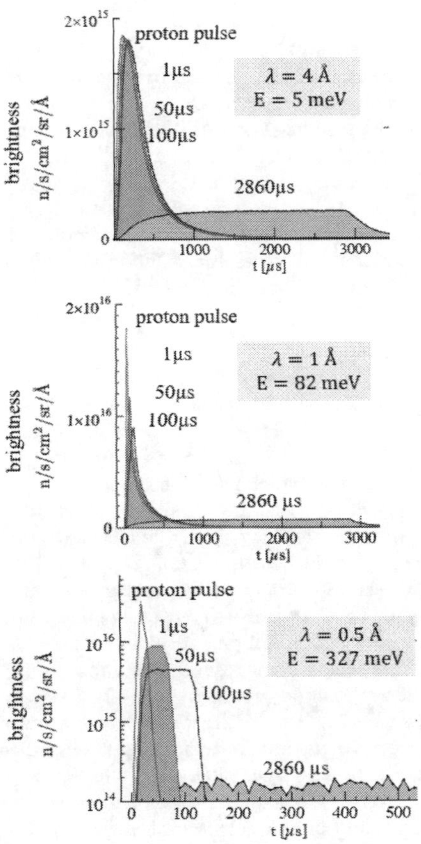

Fig. 3. Neutron brightness of the ESS moderator obtained by MCNPX simulations for various proton pulse lengths, from short pulse 1μs to medium pulse lengths of 50 and 100 μs, and the long pulse of 2860 μs, for neutrons at 4 Å (*top*), 1Å (*middle*) and 0.5Å (*bottom, note log-scale*).

For thermal neutrons, as shown in Fig.3 (*middle*) for 1 Å, the neutron pulse width varies significantly with the proton pulse length, which leads to significant further gain. For a 50μs proton pulse the gain factor is 18. Particularly, upon entering the epithermal region, the widths of neutron pulses are essentially determined by only by the proton pulse width for proton pulses, if longer than ~20 μs. For 50 μs proton pulse duration, we observe a gain factor in peak brightness of 57 at 0.5Å, see also Fig. 3 (*bottom*) on log-scale.

3 Neutron instruments

A first evaluation of the possible performance gain for the neutron scattering instruments at ESS [6] can be obtained by comparing the gain in Fig. 4 for the wavelength band used by each instrument, resulting in a gain varying from almost one order of magnitude at long wavelengths up to two orders of magnitude at short wavelength below 1 Å. However, the latter fact also reflects that the peak brightness of the long pulse is comparatively low to short pulse sources. The shortest wavelengths used at ESS instruments are about 0.5 Å for competitiveness, increasing difficulties to use choppers and efficient transport by neutron guides. A more accurate estimate of the actual gain for the instruments needs to consider the instrument specific requirements for time-wavelength resolution and the wavelength band of operation.

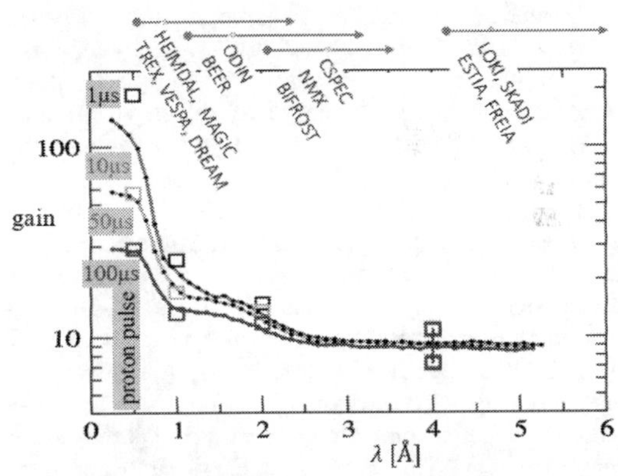

Fig. 4. The relative gain in peak brightness for ESS instruments as a function of wavelength for medium proton pulse lengths with respect to the case of the long 2860 μs proton pulse. Lines show results of gains from Vitess simulations for 10 μs (*green*), 50 μs (*orange*), and 100 μs (*blue*), while squares represent estimates from MCNPX for 1 μs (*black*), 50 μs (*orange*), and 100 μs (*blue*).
Red arrows indicate the approximate the wavelength band of operation for four groups of instruments, see text for more details.

The class of instruments for **large scale structures** appears to be best suited for the long pulse. Nonetheless, in all cases, for the SANS instruments LOKI and SKADI, for reflectometers ESTIA and FREIA, as well as for NMX the time-of-flight Laue diffractometer for macromolecular crystallography, improved resolution is generally desirable, certainly not a disadvantage.

Depending on the resolution required for the science application, one may expect modest gains by factors varying from one to four.

There are two **direct time of flight spectrometers**, the cold and thermal instruments CSPEC and TREX, respectively. The possible gains are an order of magnitude in the cold range, and even higher for thermal neutrons at wavelengths below 1Å, see Fig. 4, achieving a resolution, which otherwise requires pulse shaping. The pulse-shaping for multiple discrete wavelengths focuses on the central part of the neutron pulse. Hence, the existing choppers can still be used to improve the resolution further at a much higher peak brightness or can be applied to cut-off the asymptotic tail of the pulse without loss in signal or band width.

All other instruments at ESS use pulse shaping choppers near to the monolith at ~6.2 m distance from the source to generate, -depending on the instrument length-, one or more continuous wavelength spectra with flexible resolution. Reducing the neutron pulse width at the moderator, will shrink each wavelength band. While this will be useful and possible to work with, the preference should be a pulse that is short enough that instruments normally can operate without pulse-shaping.

There are three **indirect time of flight spectrometers**, the backscattering spectrometer MIRACLES, the extreme environment spectrometer BIFROST, and the vibrational spectrometer VESPA, which use a wavelength band near ~6.3 Å for backscattering, and larger than approximately 1.5 Å and 0.5 Å, for BIFROST and VESPA respectively. At large wavelength, see Fig.2, the impact of the proton pulse duration on the neutron pulse width is negligible for $t_p <$ 100 μs. For the backscattering instrument, viewing the source at an angle of 54°, the expected neutron pulse width is ~200 μs, which results in a total resolution of 3.5 μeV. Pulse shaping can still yield an ultimate ~2 μeV resolution within 260 μeV energy transfer near the elastic line. A time resolution of ~200 μs at 20 meV incoming energy is a typical goal for the BIFROST instrument as specified in the instrument proposal. It is also achievable for medium proton pulses of less than ~100 μs. A better time resolution would be preferred for the instrument VESPA, covering energy transfers from ~250 meV down to 4 meV. A pulse width of 70 μs results in 1% resolution for 250 meV, suggesting to preferably aim for a proton pulse of ~50 μs. The performance gain for these instruments is up to one order of magnitude, with even possibly higher gain at short wavelengths depending on the achievable proton pulse, see Figure 4. Essentially there are no needs anymore to employ pulse shaping choppers.

In the diffraction class of instruments, the **powder diffractometers** DREAM and HEIMDAL use rather flexible pulse shaping to adapt resolution. Choosing a typical good time-wavelength resolution of ~0.3% can be achieved with a 50 μs proton pulse. The existing pulse shaping chopper systems can still be used for much finer resolution, to detect subtle symmetry breaking and peak splitting, albeit using the much higher peak brightness on a smaller wavelength band. Modifying the slit pattern of the chopper disks can generate multiple wavelength frames to recover the full wavelength band. These instruments using wavelengths $\lambda > 0.5$ Å strongly benefit from shorter pulses and higher peak brightness. Similar estimates can be made for the **engineering diffractometer** BEER and the polarised **single crystal diffractometer** MAGIC, starting with $\lambda > 1$ Å and $\lambda > 0.6$ Å, respectively. For the specific engineering case of strain analysis, a symmetric peak shape would be preferable. Again, this can be done with the existing pulse-shaping choppers for a smaller wavelength band, or with many wavelength frames for the full wavelength band using a modified slit pattern.

While in many cases for **imaging** the time structure is not used, - so called 'white beam' imaging -, there is an enormous information potential in the wavelength dependent transmission signal that can be measured at pulsed sources with sufficient time resolution. Examples include diffraction contrast at Bragg edges and inelastic scattering contrast, whereas epithermal short pulses also allow for resonance absorption contrast [17]. The resolution requirements given for the imaging instrument ODIN vary from 0.3% to 10% and these are realized by pulse shaping choppers.

In summary, we note that a ~50 μs proton pulse will cover the most demanding cases. It is further worthwhile to note that shorter pulses will improve the signal-to-background ratio for all instruments.

It seems fortunate that many of the ESS instruments will strongly benefit from a proton pulse compression. In view of future instruments at ESS, this opens new opportunities for further optimisation particularly for less costly shorter instruments using a larger bandwidth. Hence, a shorter pulse raises the value of many still unused beamlines for shorter instruments.

4 Conclusion

With respect to the upgrade potential of the ESS, there is an obvious large possible gain in neutron peak brightness with proton pulse compression using a storage ring structure, which has been proposed as part of the ESSnuSB project. Most importantly, such an upgrade is of interest for the existing target, moderator, and neutron scattering instrumentation at the ESS. Therefore, the specific choice of optimal proton pulse duration should be that the resulting neutron pulse fits to the requirements for resolution for all instruments, at least for their standard modes of operation. In this respect, the resolution requirements for powder diffraction as well as for spectroscopy near 200 meV define the most demanding cases that could be fulfilled with a proton pulse length near 50 μs or of shorter times.

As a side remark, would it be even better to aim for a typical 1 μs short pulse? For the instruments using a flux optimised coupled moderator, there is practically no further gain compared to 10 or 50 μs. Furthermore, it does not seem to be attractive for future instrumentation to transform the ESS into a short pulse spallation source, since these usually operate more efficiently at higher frequency.

The current review of instrument performance aims to guide the common optimization for accelerator and target to realize a medium pulse at ESS. A more detailed consideration should be based on simulations of virtual experiments for all instruments. With beam on target and the following ramp-up of power, there will be a great opportunity to test a new time structure with shorter pulses at the instruments. While the average peak brightness of ESS will be already superior for a long pulse, a medium pulse length will considerably boost its peak brightness, which will preserve a world leading role of ESS in future among other existing neutron sources and upgrade projects [13-14] as shown in Fig. 5.

It is good to open this perspective now, as it assures the sustainable value of the current project and investment by a continuous upgrade path for the ESS, and it is timely to explore the detailed technical design that benefits to science with neutrons and neutrinos at ESS.

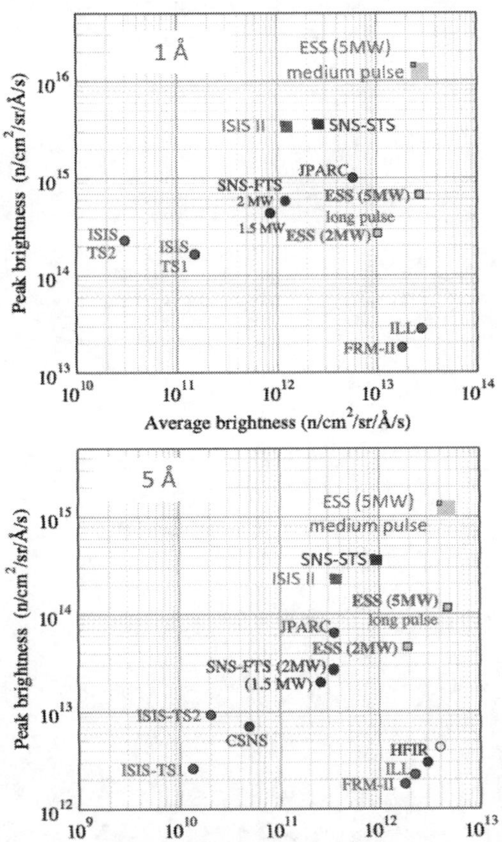

Fig. 5. Average brightness and peak brightness for the ESS with long and medium pulses in comparison to other neutron sources and upgrade projects. Top: thermal (1Å) and bottom: cold (5Å) neutrons.

Acknowledgements

Klaus Lieutenant is gratefully acknowledged for supporting the VITESS simulations. We gratefully acknowledge Masatoshi Arai, Kevin Jones, Ferenc Mezei, Pascale Deen, Mikhail Feygenson, Esko Oksanen, Manuel Morgano and Robin Woracek for useful comments and critical reading of the manuscript.

References

1. European Spallation Source Technical Design Report, DOI: 10.13140/RG.2.1.2040.6483/1, ISBN: 978-91-980173-2-8 (2013)
2. A. Alekou *et al.*, *The European Spallation Source Neutrino Super Beam Conceptual Design Report,* Eur. Phys. J. Spec. Top. **231**, 3779–3955 (2022); arXiv :2206.01208v1 (2022)
3. D. McGinnis, M. Lindroos, R. Miyamoto, *European spallation source afterburner concept,* Proceedings of IPAC2013, Shanghai, China, ISBN 978-3-95450-122-9 3926, (2013)
4. F. Mezei, *Comparison of neutron efficiency of reactor and pulsed source instruments,* Proceedings of ICANS-XII (1993)
5. F. Mezei, *The raison d'être of long pulse spallation sources*, J. Neutron Research **6**, 3-32 (1997); F. Mezei, P. Tindemans and K. Bongardt, *The 5 MW LP ESS; best price-performance*, 2008.
6. K.H. Andersen *et. al.*, *The instrument suite of the European Spallation Source,* Nuclear Inst. and Methods in Physics Research, A **957** 163402 (2020) https://doi.org/10.1016/j.nima.2020.163402
7. L. Zanini *et al.*, *Design of the cold and thermal neutron moderators for the European Spallation Source*, Nuclear Inst. And Methods in Physics Research, A **925** 33–52 (2019)
8. K.H. Andersen and C.J. Carlile, *A Proposal for a Next Generation European Neutron Source,* J. Phys.: Conf. Ser. **746**, 012030 (2016)
9. *Mini-workshop on Medium pulse length at ESS,* https://indico.esss.lu.se/event/3179/ (2023)
10. M. Lindroos *et al.*, *Future ESS upgrade to medium pulse length – what are the technical challenges for the accelerator and target?* Proceeding of IPAC'23, Venice, Italy (2023)
11. A. Aguilara *et al.*, *Design specification for the European Spallation Source neutron generating target element*, Nuclear Inst. And Methods in Physics Research A **856**, 99-108 (2017)
12. G. Bauer, H. Sebening, J.E. Vetter, H. Willax, *Realisierungsstudie zur Spallations-Neutronenquelle,* Jül-Spez-**113**, KfK 3175 (1981). [p.156 for target design]
13. *SNS Second Target Station STS: Technical design report* (2105), https://doi.org/10.2172/1185891
14. J.-B. Lagrange *et al.*, *Progress on design studies for the ISIS II upgrade*, IPAC2019, 10th Int. Particle Accelerator Conf., JACoW Publishing, doi:10.18429/JACoW-IPAC2019-TUPTS068,

2075-2078 (2019) ; *ISIS II: the future development of the ISIS Neutron and Muon Source*, S. Langridge, ECNS 2023; https://www.isis.stfc.ac.uk/Pages/isis-ii-working-group-report16266.pdf

15. G. Zsigmond *et al.*, *A survey of simulations of complex neutronic systems by VITESS,* Nuclear Inst. And Methods in Physics Research A **529**, 218-222 (2004)
16. L.S. Waters *et al.*, *The MCNPX Monte Carlo radiation transport code,* AIP Conf. Proc. **896** 81-90 (2007)
17. K.V. Tran *et al.*, *Spectral neutron tomography*, Materials Today Advances **9,** 100132 (2021). https://doi.org/10.1016/j.mtadv.2021.100132

Expression of ^{2}H, ^{13}C, ^{15}N-labeled NIST-Fab fragment in the methylotrophic yeast *Komagataella phaffii* for nuclear magnetic resonance studies

Kinlin L. Chao[1,3*], *William B.* O'Dell[1,2,3], *Tsega L.* Solomon[1,2], *Robert G.* Brinson[1,2], *John P.* Marino[1,2] and *Zvi* Kelman[1,2,3]

[1]Institute for Biosciences and Biotechnology Research, [2]National Institute of Standards and Technology and the University of Maryland, [3]Biomolecular Labeling Laboratory, Rockville, MD 20850, USA.

Abstract. Labeling of proteins with deuterium is an essential tool in overcoming size limitations in the application of nuclear magnetic resonance (NMR) spectroscopy to proteins larger than 30 kilodaltons (kDa). A non-originator antigen-binding fragment (Fab) of NIST RM 8671 NISTmAb, so called yNIST-Fab, is a ~ 50 kDa protein, with 5 native disulfide linkages, that can be expressed in properly folded form in methylotrophic *Komagataella phaffii* (formerly *Pichia pastoris*). Further, the *K. phaffii* host can support the production of perdeuterated yNIST-Fab which is necessary to obtain well-resolved TROSY-based triple-resonance NMR spectra for chemical shift assignment of the peptide backbone resonances. Here, we examined growth conditions and effects of media composition to maximize biomass generation and expression yield of the ^{2}H, ^{13}C, ^{15}N-enriched NIST-Fab fragment. Triple-labeled yNIST-Fab with ~93% deuteration reduced the $^{1}H_{N}$, ^{15}N and ^{13}C-linewidths in the NMR spectra, allowing sequential NMR assignment of backbone resonance - a key step toward sequence-specific structural and dynamic studies of Fab fragments and intact antibodies.

1 Introduction

Monoclonal antibodies (mAbs) are a vital therapeutic platform for the treatment of cancers and various diseases with 162 antibody therapies currently approved by drug regulatory agencies in the US, Europe, Japan, and China [1]. The cluster of differentiation 3 (CD3) protein complex and T cell co-receptor appears the most common target with 10 approved antibodies for organ transplant rejection (The Antibody Therapies Database, Umabs-DB, https://umabs.com). Yet the anti-CD3 antibody development pipeline continues with 6 in Phase-3, 48 in Phase-2, 87 in Phase-1, and 319 in Preclinical trials. It is critical to evaluate the stability and efficacy of these pharmaceutical proteins during manufacture, formulation, and storage. The majority of therapeutic mAbs are of the immunoglobulin G1 (IgG1) subclass. IgG1s are 150 kDa tetrameric proteins consisting of two heavy chains (HC) and two light chains (LC) covalently linked by disulfide bonds. The heavy chain is divided into variable (V_H) and constant (CH_1, CH_2 and CH_3) domains and light chains into V_L, and C_L domains. Antibodies are comprised of functional antigen binding (Fab) and crystallizable (Fc) regions. The Fab portion, comprising V_H, CH_1, V_L and C_L domains, binds specifically to the antigenic epitope. The Fc region, disulfide linked CH_2 and CH_3 domains with N-linked glycosylation at asparagine residue 297, interacts with antibody receptors on the cell surface.

Fingerprinting methods using nuclear magnetic resonance (NMR) spectroscopy at natural isotopic abundance is gaining increased acceptance for assessment of the critical quality attributes (CQA) of higher order structure (HOS) for protein-based therapeutics [2-6]. Spectral changes, in the form of shifted cross peaks or the appearance and disappearance of signals, reflect HOS perturbations, can arise from misfolding, aggregation, degradation, and/or chemical modifications of proteins. NMR fingerprinting methods have been applied to many different protein modalities, including peptides, small proteins, and intact mAbs [7-10]. Despite the success of NMR fingerprinting methods for industrial applications, any observed spectral perturbations for intact mAbs cannot be directly traced to specific structural elements without the availability of spectral assignments.

NMR resonance assignment opens the possibility for detailed structural analysis including sequence specific assessment of HOS perturbations of biologics. Although there are relatively few backbone assignments for proteins over 50 kDa, the HOS of intact mAbs can be considered modular. Previous work has established that the constituent Fc and Fab domains retain the HOS of the parent molecule with only minor chemical shift perturbations observed for residues near in the hinge region [11]. Until our recent publication of NMR assignments of a Fab domain [12], only assignments of the Fc domain were available including from

* Corresponding author: kchao@umd.edu

glycosylated and aglycosylated versions from CHO (Chinese Hamster Ovary) and *Escherichia coli* expression systems [11, 13]. The first step in chemical shift assignment is to conduct a suite of triple-resonance NMR experiments to correlate the chemical shifts of the amide proton (H_N) with the chemical shifts of Cα and Cβ [14]. While these experiments require uniformly ^{13}C and ^{15}N isotope labeling, proteins larger than 30 kDa also need deuterium labeling to eliminate efficient proton relaxation pathways and allow for optimal implementation of the transverse relaxation optimized spectroscopy (TROSY) element within pulse sequences [15], leading to sensitivity enhancements for high resolution NMR spectra. Despite these technical requirements for application of NMR to Fabs, prior to this study no expression systems have been described for the production of a properly folded, uniformly labeled ^2H, ^{13}C, ^{15}N Fab domain.

To address the need for a ^2H, ^{13}C, ^{15}N Fab expression system, here we report the expression of uniformly, triply-labeled Fab domain in *Komagataella phaffii* (formerly *Pichia pastoris*). This Fab domain was derived from the publicly available NISTmAb Reference Material (RM 8671) and is referred to as yNIST-Fab due to its yeast production. The NISTmAb is an IgG1κ monoclonal antibody developed as a Reference Material (RM) by the National Institute of Standards and Technology (NIST) [16, 17]. The proper fold and disulfide bonds are formed in *K. phaffii* for the yNIST-Fab, which is then secreted into the culture media under the control of Alcohol oxidase 1 (AOX1) promoter [18]. Previous reports demonstrate that some degree of process optimization is necessary to maximize yields of cell biomass and isotopically enriched proteins in yeast from a simple minimal defined ^2H_2O-media [19-21]. We therefore further investigated *K. phaffii* culture growth conditions and purification protocols to achieve production of ~6.6 mg/L of triple-labeled yNIST-Fab with 93% deuteration. The TROSY-based spectra of this deuterated yNIST-Fab were acquired to demonstrate the suitability of the material for NMR experiments and enabled the first report of backbone assignments for a Fab domain [12].

2 Materials and methods

2.1 Media

Table 1. Components for supplemented minimal growth media.

Buffered minimal (BM) Media without supplements
0.1 M Potassium phosphate, pH 6.0
0.34% (w/v) Yeast nitrogen base without amino acids and ammonium sulfate (Research Product International Y20060)
1% (w/v) ^{15}N-Ammonium sulfate (MilliporeSigma 299286)
0.00004% (w/v) Biotin (Thermo Fisher Scientific BP-232-1)
0.0002% (v/v) Antifoam 204 (MilliporeSigma A6426)
98% (v/v) ^2H_2O (Cambridge Isotope Laboratories DLM-4-99.8-1000)
1% (v/v) glycerol
Supplements added to above BM media during the biomass generation and protein expression phases.
1 x Minimum Essential Medium (MEM) essential amino acids (Corning 25-030-CI)
1 x MEM nonessential amino acids (Corning 25-025-CI)
1 x MEM vitamins (Corning 25-020-CI)
1 x Pichia Trace Mineral 1 salts (VWR 241)
0.2 - 1% (w/v) ^2H, ^{13}C, ^{15}N-ISOGRO® (MilliporeSigma DCN-IG 608297-125)
Carbon source during protein expression
0.5% (v/v) ^{13}C-methanol (MilliporeSigma 277177), ^{13}C, $^2H_4$-methanol (MilliporeSigma 293865), or $^2H_4$-methanol (CIL DLM-24RG)

2.2 Engineering of *K. phaffii* for improved production of yNIST-Fab

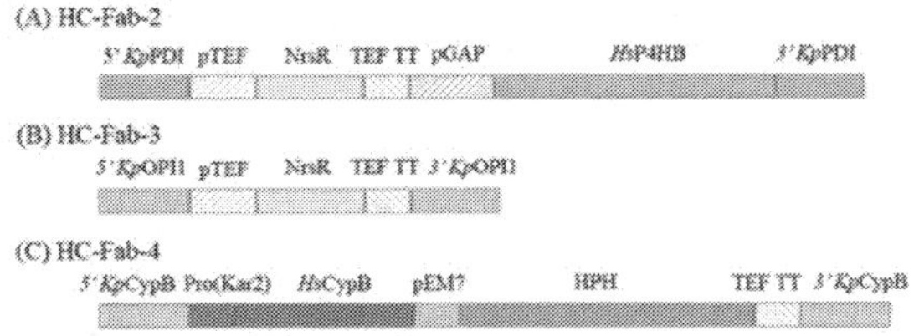

Fig. 1. (A) Disruption cassette to generate HC-Fab-2 clone by replacement of *K. phaffii* protein disulfide isomerase (PDI, NCBI Gene symbol PAS_chr4_0844) gene with human prolyl 4-hydroxylase subunit β (*Hs* P4HB) to enhance disulfide bond formation of human Fab fragment. The 500 base pairs (bp) of 5' upstream and 3'downstream flanking regions of *Kp* PDI gene are colored in blue, the NrsR gene (nourseothricin resistance, ClonNat) under the control of Translation Elongation Factor promoter (pTEF) and transcription terminator (TEF TT) in yellow, residues 17 to 508 of human (*Hs* P4HB) under *K. phaffii* glyceraldehyde-3-phosphate dehydrogenase constitutive promoter (pGAP), and the *Saccharomyces cerevisiae* α mating factor pre-peptide in orange. **(B)** Disruption cassette to generate HC-Fab-3 clone by deletion of *K. phaffii* OPI1 (PAS_chr1-1_0033) gene to enlarge endoplasmic reticulum luminal volume and to increase in antibody/fragment maturation capacity. The 500 bp of 5' and 3' flanking regions of *Kp* OPI1 are colored in green with NrsR gene, pTEF, and TEF TT in yellow. **(C)** Disruption cassette to generate HC-Fab-4 clone by replacement of *K. phaffii* cyclophilin B (PAS_chr1-1_0267) gene with human cyclophilin B (*Hs* CypB) to increase the rate limiting trans-cis isomerization of $Proline_{32}$ of CH_1 domain of Fab fragment. The 500 bp of 5' and 3' flanking regions of *Kp* CypB gene are colored in pink, 281 bp of 5' promoter region of protein folding chaperone Kar2 (PAS_chr2-1_0140) in magenta, *Hs* CypB gene in purple, the HPH gene for hygromycin resistance under the pEM7 promoter in grey, and TEF TT in yellow. HC-Fab-5 was modified with both cassettes used for the construction of HC-Fab-3 and HC-Fab-4 clones.

The original clone of *K. phaffii* X33 expressing yNIST-Fab (HC-Fab-1 clone) was kindly provided by Dr. Yves Aubin (Health Canada, Ottawa, ON, Canada) and was first reported by Brinson *et al.* [18]. The recombinant yNIST-Fab is a disulfide-linked heterodimer with 98% sequence identity to Fab fragment generated with papain cleavage of NISTmAb [18]. yNIST-Fab comprises residues $Glutamine_1$ to $Histidine_{228}$ of heavy chain and residue Aspartic $acid_1$ to $Cysteine_{213}$ of light chain. Because the dipeptidyl aminopeptidase A protease (STE13 gene product) failed to cleave the α-factor mating signal sequence of *Saccharomyces cerevisiae* at its recognition site, the heavy and light chain of yNIST-Fab contain vector-derived Glu-Ala-Glu-Ala residues on the N-termini of the heavy and light chains [18].

HC-Fab-1 was modified by double-crossover homology-directed recombination gene replacement using synthetic cassettes to generate HC-Fab clones 2-5 (Fig. 1). A culture of HC-Fab-1 was made electrocompetent and transformed by electroporation according to the Thermo Fisher Scientific EasySelect *Pichia* Expression manual. Transformed colonies were selected by ClonNAT or hygromycin resistance. Genomic modifications were confirmed by gel electrophoresis and Sanger sequencing of genomic DNA PCR product amplified using primer pairs unique to specific *K. phaffii* sequences adjacent to the 5′ and 3′ homology sequences.

2.3 Expression of ^{2}H, ^{13}C, ^{15}N-yNIST-Fab

All cultivations were performed in baffled shake flasks at 29 °C, 260 RPM. Cell biomass was monitored by measuring optical density at 600 nm (OD_{600}). Briefly, a loopful of freshly streaked cells from a minimal dextrose agar plate was inoculated into 10 mL of 90% ^2H_2O-buffered minimal (BM) glycerol media and was grown to an OD_{600} of 5 (~ 24 h; Table 1). Approximately 0.4 mL of this culture was transferred into 100 mL of 98% ^2H_2O-BM glycerol and was incubated until OD_{600} of 7.5 (~72 h). The 98% ^2H_2O-adapted culture was expanded by inoculating ~16 mL to 500 mL of 98% ^2H_2O-BM media containing 0.5% (w/v) $^{13}C_6$-glucose, and was grown to OD_{600} ~ 4.4 (~48 h). For the supplemented culture, ~17 mL of 98% ^2H_2O-adapted cells were inoculated into 500 mL of 98% ^2H_2O-BM with 0.4% (w/v) DCN-ISOGRO® (herein referred to as DCN-algae extract) and 0.5% (v/v) $^2H_8$-glycerol, and incubated until OD_{600} ~ 9-12 (~48 h). Typically, HC-Fab cultures reached saturation growth (OD_{600}~20-25) in 90% ^2H_2O-media with 2% (v/v) glycerol, whereas the cultures in 98% ^2H_2O-minimal media grew slower, requiring longer incubation time to reach $OD_{600} \geq 10$.

The cultures were centrifuged in sterile bottles for 20 min at 4650 x g, 15 °C to collect the biomass. The cell pellets were suspended in 150 mL BM media containing 0.01% (v/v) glycerol, and the centrifugation was repeated to remove traces of $^{13}C_6$-glucose. The washed biomass was suspended in fresh 500 mL 98% ^2H_2O-BM media with or without DCN-algae extract supplement. The cultures were starved for ~2 h before the addition of 0.5% (v/v) methanol to induce protein expression. The cultures were fed every 24 hours with 5 mL of 50% (v/v) methanol in ^2H_2O (^{13}C-, $^2H_4$-, or ^{13}C, $^2H_4$-methanol). After 3 days, the culture supplemented with 0.4% (w/v) DCN-algae extract grew to slightly higher density (OD_{600} ~13.5) than the unsupplemented BM culture (OD_{600} ~11). The media was harvested by centrifugation as above.

2.4 Purification of secreted ^{2}H, ^{13}C, ^{15}N-yNIST-Fab

Clarified media was adjusted to pH ~3.7 with 12 M hydrochloric acid and then applied onto a 3 mL DEAE Sephacel (Cytiva 17-0500-01) column, equilibrated with 25 mM sodium acetate, pH 4.0. The anion exchange column removed cell debris from the media and bound the majority of brown pigment from the media supplement. The flow-through from the DEAE Sephacel column was directly loaded onto a 5 mL SP Sepharose (Cytiva-17-0729-01) column, equilibrated with 25 mM sodium acetate, pH 4.0. The cation exchange column was washed with 15 mL of the same buffer, followed by 10 mL of 25 mM sodium acetate pH 4.0, 0.1 M sodium chloride. Bound yNIST-Fab, disulfide-linked light chain dimer, and monomeric light chains were eluted stepwise with 25 mM sodium acetate pH 4.0, 0.2 M sodium chloride, and then with 35 mL of 25 mM sodium acetate, pH 5.5, 0.5 M sodium chloride buffer. The separation of yNIST-Fab from the light chains shifted depending on the total protein concentration of the media. Fractions were analyzed by SDS-PAGE, and those containing yNIST-Fab were concentrated to ≥ 2 mg/mL using 10 kDa molecular weight cut-off (MWCO) centrifugal concentrators at 2100 x g at 12 °C. The concentrated ^{2}H, ^{13}C, ^{15}N-protein was dialyzed in 1 L phosphate-buffered saline (PBS; 0.137 M sodium chloride, 2.7 mM potassium chloride, 10 mM dibasic sodium phosphate, 1.8 mM monobasic potassium phosphate, pH 7.3), overnight at 4 °C, prior to application onto a 5 mL IgG-CH1 affinity (Thermo Fisher Scientific 1943200) column equilibrated in PBS. The affinity column was washed with 25 mL of PBS to remove the light chain-only species, and bound yNIST-Fab was eluted with 35 mL of 0.1 M glycine, pH 3.5. These yNIST-Fab fractions were concentrated using 10 kDa MWCO concentrators to ~1 mL before applying onto a Superdex 75 10/300 GL (Cytiva 17-5174) column equilibrated with 25 mM bis-tris, pH 6.0, 0.1 M sodium chloride buffer as a polishing step. The extinction coefficient of 69,955 M^{-1} cm^{-1} (1.44 mL/mg) at 280 nm wavelength was used to determine yNIST-Fab concentration by spectrophotometry.

2.5 Western immunoblotting

Western blot analysis was used to monitor the time course of yNIST-Fab expression with the CaptureSelect Biotin CH1-XL conjugate antibody (Thermo Fisher Scientific 7103462100), Streptavidin-Horseradish peroxidase secondary antibody (Pierce-21130), and luminol chemiluminescent detection system. Relative signals from 2 to 20 µL of clarified media or fractions from chromatographic columns, spotted as dot blots on PVDF membranes, were quantified using BioRad Image Lab software with 10 to 150 ng of yNIST-Fab spotted on each blot as standard to normalize the signals.

2.6 MALDI-TOF mass spectroscopy

The intact molecular mass of yNIST-Fab was determined by the matrix-assisted laser desorption time-of-flight (MALDI-TOF) mass spectrometry with sinapinic acid (4-hydroxy-3,5-dimethyloxycinnamic acid, MilliporeSigma D-7927) as a matrix. The calculated molecular masses of unlabeled (48,436 Da) yNIST-Fab, triple-labeled yNIST-Fab (53,650 Da), and the expected masses gained from isotope incorporations were calculated from the Protein Calculator [22] (https://www.gmclore.org). The calculated molecular mass increases for 100% incorporation of ^{15}N, ^{13}C and ^{2}H into yNIST-Fab are 561 Da, 2123 Da and 2560 Da, respectively. The intact mass for yNIST-Fab was determined to be 48,446 Da by LC-MS/MS, whereas its molecular mass by lower resolution MALDI-TOF MS was 48,784 ± 49 Da, ~348 Da higher due to peak broadening. This molecular mass of yNIST-Fab was used to estimate the isotope enrichment by MALDI-TOF mass spectrometry as follows:

$$\frac{(MW_{labeled} - MW_{unlabeled})}{(\text{expected mass increase for complete labeling})} \times 100 = \% \text{ isotope enrichment} \quad (1)$$

2.7 NMR spectroscopy

NMR spectroscopic experiments were conducted as previously described [12]. Partial refolding of ^{2}H, ^{13}C, ^{15}N-yNIST-Fab was necessary to promote a full deuterium-to-hydrogen back exchange of backbone amides to observe the corresponding resonances [12]. For estimation of ^{2}H-incorporation, the peak area from the methyl region (0.5 to 2.0 ppm) of 1D ^{1}H-spectra was integrated for the unlabeled and ^{2}H-labeled yNIST-Fab samples. The reference sodium 3-(tri-methyl-silyl)propane-1-sulfonate-D6 (DSS-d6, 150 µM) in each sample was used to normalize the spectra differences from data collections. All spectra were normalized to a spectrum from a 200 µM unlabeled yNIST-Fab sample with 100% protonation. ^{2}H isotope enrichment was estimated by subtracting the peak area of labeled sample from corresponding peak from unlabeled yNIST-Fab with a correction for concentration difference.

3 Results and discussion

3.1 Expression of yNIST-Fab

Temperature, pH, and carbon source are parameters for optimal growth and yield of secreted recombinant protein that vary among different *K. phaffii* strains in defined minimal media [19, 20, 23, 24]. Experiments to determine the optimal growth conditions for yNIST-Fab expression were conducted in H_2O-minimal media with methanol as the carbon source during the induction phase. The amount of yNIST-Fab present in the culture media was determined by western dot blot analysis using a CH_1-domain primary antibody that specifically recognizes Fab. Samples were collected every 24 hours after induction (Fig. 2A, top). The time course of expression from HC-Fab-1 clone shows that a detectable amount of yNIST-Fab appeared in the media after 24 hours post-induction with slow accumulation up to 72 hours (Fig. 2A, bottom). Extending the expression from 72 hours to 120 hours did not significantly increase the secreted yNIST-Fab levels. Unlike A_2AR GPCR [20] or *Viteroscilla* hemoglobin [25], the expression of yNIST-Fab at 23 °C during the induction phase did not change the protein yield relative to cultures grown at 29 °C (data not shown).

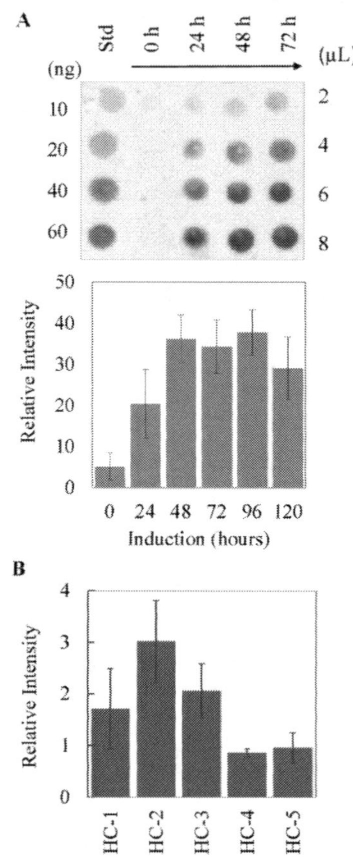

Fig. 2. (**A**) Top panel - Western dot blot to monitor the expression of secreted yNIST-Fab as a function of induction time from HC-Fab-1 clone. Bottom panel - Quantitation of yNIST-Fab signal from 2-8 µL media on the dot blot, using the 10-150 ng yNIST-Fab standards and normalized to 1 µL media. (**B**) Secreted yNIST-Fab from the HC-Fab-1 and four modified clones (HC-Fab-2 to HC-Fab-5 as described in *Materials and methods*): 72 hours post-induction in H_2O-miminal methanol media. The error bars represent standard deviations from three experiments.

The HC-Fab-1 clone was engineered to improve the production of secreted yNIST-Fab by modulating the levels of endoplasmic reticulum (ER) proteins involved in protein folding and secretion [26]. Additional HC-Fab-2 to HC-Fab-5 clones were constructed to test co-expression of chaperones and modulation of the ER (Fig. 1). Of these, the HC-Fab-2 clone with constitutively overexpressed human prolyl 4-hydroxylase subunit β (P4HB), a protein disulfide isomerase, had consistently ~1.5 to 2-fold higher yield of yNIST-Fab than the original HC-Fab-1 clone (Fig. 1A & 2B). Other modifications of HC-Fab-1 clone provided limited benefit or were even detrimental to the yield of secreted yNIST-Fab (Fig. 1B-1C, 2B). An HC-

Fab-3 clone was constructed to test the anticipated effect of ER expansion by deleting *K. phaffii* OPI1 gene, a known regulator of phospholipid synthesis in *S. cerevisiae* [26-28]. HC-Fab-4 clone was further constructed to examine the effect of humanizing *K. phaffii* cyclophilin B by replacement with human cyclophilin B under the control of an ER stress-inducible promoter (Fig. 1C) [29]. In addition, an HC-Fab-5 clone was constructed, which combined the engineering of HC-Fab-3 and HC-Fab-4 (Fig. 1B-1C). It should be noted that neither overexpression of *Hs* P4HB protein in HC-Fab-2 nor overexpression of *Hs* CypB protein in HC-Fab-4 and HC-Fab-5 were confirmed. It was also beyond the scope of this work to assess ER expansion in HC-Fab-3 and HC-Fab-5 clones.

Glucose and glycerol are typically used as the carbon sources for the generation of yeast biomass [30, 31]. The majority of isotopic enrichment in overexpressed proteins is reported to come from the yeast biomass. A small percentage of isotope enrichment also arises from methanol during the induction phase [20, 21, 23, 24, 32]. Since yeast growth is limited in defined minimal media with methanol as the sole carbon source, a preferred procedure is to induce protein overexpression at high cell density [20]. For yNIST-Fab production, HC-Fab-1 and HC-Fab-2 cultures were cultivated in 98% ^2H_2O-minimal media with either glucose or glycerol to determine a preferred carbon source for biomass accumulation. After 24 h in ^{15}N, 90% ^2H_2O-media, both HC-Fab-1 and HC-Fab-2 cultures grew to an OD_{600} of 9 with glucose, while those in glycerol media grew to a slightly higher OD_{600} of 14. This trend continued in the next adaptation step to ^{15}N, 98% ^2H_2O-media. From the initial OD_{600} of ~0.03, the HC-Fab-1 and HC-Fab-2 cultures in glucose media grew to OD_{600} of 4.4 after 2 days, whereas the cultures in ^{15}N, 98% ^2H_2O-media with glycerol grew to OD_{600} of 5.5 and 7.4, respectively. These data demonstrate a preference for glycerol by these yeast clones during biomass accumulation. The cell biomasses were washed to remove traces of glucose before resuspension in fresh media, and protein expression was induced with methanol for 3 days. The secreted proteins in clarified media were concentrated on an SP sepharose column, and all bound yNIST-Fab and light chain species were eluted with 25 mM acetate, pH 4.0, 0.2 M sodium chloride buffer. SDS-PAGE analysis of fractions showed the light chain dimer as the predominant product from biomass growth on glucose (Fig. 3A, Glucose). By contrast, the amount of yNIST-Fab is equal to or greater than that of the light chain species from the glycerol-derived biomasses of HC-Fab-1 and HC-Fab-2 clones (Fig. 3A, Glycerol). It is unknown why the yNIST-Fab expression is lower from the biomasses generated from glucose-media.

K. phaffii cultures can tolerate a pH range from 3 to 7 with an optimal growth rate at pH 6 [19]. Without continuous buffering, the media pH can drop from neutral to acidic range upon release of acetic acid if yeast cultures continue to grow during protein expression phase. For secreted expression of single chain Fv (scFv) antibody fragments in *K. phaffii,* an optimal yield of A33scFv antibody fragment was found to be at pH 3 to 4, whereas Anti-Serpin-scFv production was best at pH ≥ 6 [19, 33]. In addition, the effect of pH on yNIST-Fab expression for cultures grown in H_2O- or ^2H_2O-media was examined. After 24 hours expression, the higher expression of yNIST-Fab corresponded with increasing pH for both H_2O- or ^2H_2O-media types (Fig. 3B).

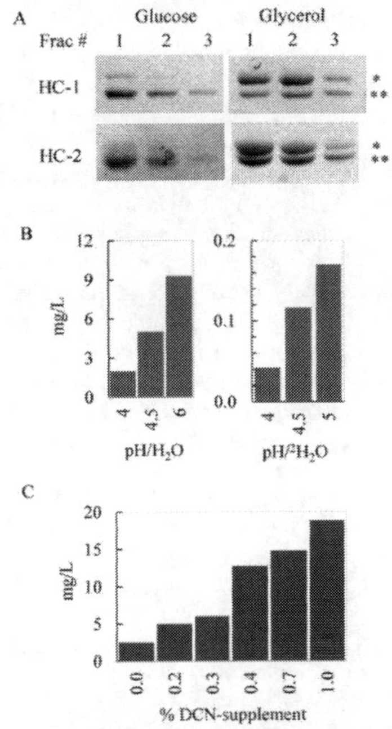

Fig. 3. (**A**) Expression of yNIST-Fab is higher in 1% (v/v) glycerol media than in 1% (w/v) glucose as the carbon source in ^{15}N, ^2H_2O-cultures. 12% SDS-PAGE under non-reducing condition of fractions from 1 mL SP sepharose column eluted with 25 mM sodium acetate, pH 4.0, 0.2M sodium chloride. HC-Fab-1 and HC-Fab-2 clones labeled as HC-1 and HC-2 with 48.8 kDa Fab and 46.2 kDa Light chain-dimer as *and **, respectively. (**B**) Yield of yNIST-Fab from H_2O-media at pH 4-6 (left) and ^2H, ^{15}N-yNIST-Fab from ^{15}N, ^2H_2O-media at pH 4-5 (right). (**C**) Yield of ^2H, ^{13}C, ^{15}N-yNIST-Fab in ^{15}N, ^2H_2O-media supplemented with (w/v) DCN-algae extract.

This result also demonstrates very low yield of ^2H, ^{15}N-yNIST-Fab from ^{15}N, 98% ^2H_2O-media (< 0.5 mg/L from pH 6 media). This effect is expected since the slower *K. phaffii* growth rate in high [^2H_2O]-minimal media leads to reduction in the accumulated biomass for the protein expression on methanol [21, 34]. For *K. phaffii* expressing *Plasmodium falciparum* merozoite surface protein-1, doubling time increased from 2.5 hours in H_2O media to 8 to 10 hours in 95% to 99.8% ^2H_2O-media [21]. The subsequent yield of U-^2H, ^{15}N-labeled small protein was approximately fourfold lower than that of unlabelled version [21]. Massou *et al.* reported that 140 g/L of *K. phaffii* biomass was obtained from 0 to 75% ^2H_2O-media, but the amount of biomass dropped to 16 g/L in 97.5% ^2H_2O-media [34]. A similar growth pattern with *K. phaffii* HC-Fab-2 clone was observed. In H_2O-media, the HC-Fab-2 cultures reached saturation growth (OD_{600} ~25), while cell density remained well below OD_{600} of 10 in 98% ^2H_2O-minimal media. The final cell densities of these cultures ranged from OD_{600} of 4 to 10 which generated lower amounts of biomass for protein production.

3.2 Impact of DCN-algae extract on yNIST-Fab expression

To produce fully deuterated protein for NMR experiments, the yield of ^{2}H, ^{13}C, ^{15}N-yNIST-Fab from ^{15}N, 98% $^{2}H_2O$ media supplemented with DCN-labeled algae extract was investigated. The labeled amino acids and peptides, comprising ~65% of algae extract by mass, were expected to lessen the metabolic burden of *K. phaffii* Fab cultures grown in $^{2}H_2O$-minimal media, especially during the protein expression phase with methanol. The presence of algae supplements (0.4%, 0.7%, 1% (w/v) DCN-algae extract) increased the cell biomass slightly (OD_{600} ~ 9 to 10) over that from the minimal media (OD_{600} ~ 6). During the induction phase with methanol, the unsupplemented control culture in 98% $^{2}H_2O$-minimal media did not accumulate biomass, while the algae supplement cultures in 98% $^{2}H_2O$-media grew after an initial 1 to 2 day delay. The yield of ^{2}H, ^{13}C, ^{15}N-yNIST-Fab increased from ~2 to 8 fold (2.5 mg/L to ~18.8 mg/L at 1% algae extract) in supplemented ^{15}N, 98%-$^{2}H_2O$ media (Fig. 3C). However, above 0.4% (w/v) algae extract concentration, the ^{2}H, ^{13}C, ^{15}N-yNIST-Fab expression level only improved by < 1.5 fold in ^{15}N, 98% $^{2}H_2O$-media. For comparison, ~17 to 22 mg/L of ^{15}N, 15% ^{13}C-yNIST-Fab were obtained from H_2O-minimal media. Therefore, it is possible that the secretion pathway in *K. phaffii* was limiting the production of secreted yNIST-Fab rather than the availability of supplemented nutrients in ^{15}N, 98% $^{2}H_2O$-media. In addition, ^{2}H, ^{13}C, ^{15}N-yNIST-Fab purified from ^{15}N, 98% $^{2}H_2O$-media supplemented with greater than 0.5% (w/v) isotope-labeled algae extract was irreversibly associated with brown pigment from the extract that could not be removed by dialysis or preparative size exclusion chromatography.

3.3 Purification of ^{2}H, ^{13}C, ^{15}N-yNIST-Fab

In contrast to *E. coli* expression systems, in which many eukaryotic proteins with multiple disulfide bonds are difficult to produce, the alternative methylotrophic *K. phaffii* host offers a suitable expression system to produce full length, glycosylated antibodies, and is amendable to grow in deuterium minimal media for isotope-labeling of proteins [21, 35, 36]. However, protein expression in *K. phaffii* used twice as much ^{15}N, $^{2}H_2O$-medium compared against *E. coli* systems, incurring higher costs. The yeast biomass must be transferred into fresh ^{15}N, $^{2}H_2O$-medium prior to the induction of protein expression with methanol (see *Materials and methods*), highlighting the importance of capturing as much ^{2}H, ^{13}C, ^{15}N-yNIST-Fab from the culture media and to minimize protein loss during purification. Two steps in our purification protocol affected the final yield of yNIST-Fab. Initially, pressure-driven tangential flow filters or stirred cell devices were used to concentrate liters of culture supernatant prior to purification. However, a significant amount of yNIST-Fab passed through 10 kDa MWCO regenerated cellulose or polyethersulfone (PES) membrane filters as detected by western immunoblots. As an alternative, the SP sepharose cation ion exchange resin could reliably capture all yNIST-Fab and light chain species from the supernatant while concentrating the secreted proteins greater than 100 fold. (Fig. 4A). The IgG-CH1 affinity matrix provided critical intermediate purification by removing unwanted light chain homodimer and free light chain species (Fig. 4B). Despite a highly specific association, residual yNIST-Fab was observed in the flow through of IgG-CH1 affinity column even after overnight batch binding. This resulted in multiple IgG-CH1 column experiments to capture a majority of yNIST-Fab eluted from the S Sepharose column. The amount of bound yNIST-Fab to 3 mL IgG-CHI column increased proportionally to the amount of loaded protein (Fig. 4C). This unexpectedly low capture capacity was resolved by concentrating the input material to ≥ 2 mg/mL (Fig. 4C), suggesting that protein concentration was a factor in the effective dynamic binding capacity of the IgG-CH1 affinity matrix for yNIST-Fab.

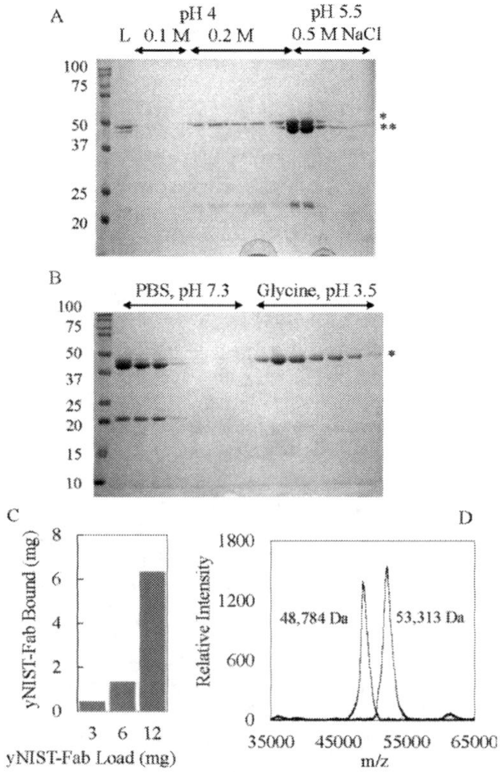

Fig. 4. 12% SDS-PAGE of ^{2}H, ^{13}C, ^{15}N-yNIST-Fab fractions from (**A**) SP sepharose chromatography and (**B**) CaptureSelect IgG-CH1 chromatography, under non-denaturing condition with * (48.8 kDa Fab), ** (46.2 kDa Light chain-dimer) and L (load). (**C**) Binding capacity determination for 3 mL CaptureSelect IgG-CH1 column loaded with 3 mg, 6 mg, or 12 mg yNIST-Fab. Bound yNIST-Fab eluted with 0.1 M glycine, pH 3.5. (**D**) MALDI-TOF MS overlay of unlabeled (left peak) and ^{2}H, ^{13}C, ^{15}N-yNIST-Fab (right peak).

3.4 Estimation of stable isotope incorporation by MALDI-TOF Mass-spectroscopy

To estimate the stable isotope incorporation, the intact masses for a series of stable-isotope-labeled yNIST-Fab were measured using MALDI-TOF MS (Table 2). This procedure for stable-isotope incorporation was previously established for the EGF domain of thrombomodulin with >98% ^{15}N-enrichment [24]. For ovine interferon-tau proteins, the combined ^{13}C, ^{15}N incorporation varied with labeling protocols: 68 to 83% for one protocol and 99% for another protocol [23]. Uniform ^{15}N- and 90% ^{13}C-labeling has been reported for eukaryotic seven-transmembrane rhodopsin expressed in *K. phaffii* for solid state NMR studies [30]. This approach allowed the evaluation of the contribution of ^2H_2O media and methanol to overall isotope enrichment of yNIST-Fab. The intact masses of unlabeled and ^{2}H, ^{13}C, ^{15}N-yNIST-Fab from MALDI-TOF analyses are 48,784 ± 49 Da (Fig. 4D, Table 2, sample-1) and 53,313 ± 52 Da, respectively (Fig. 4D, Table 2, sample-14). The 1D ^{1}H spectra of protonated ^{13}C, ^{15}N- and ^{2}H, ^{13}C, ^{15}N-yNIST-Fab with ~93% deuteration show reduction in ^{1}H signals (Fig. 5A). ^{2}H-yNIST-Fab from 98% ^2H_2O-media with methanol as the carbon source has a molecular mass of 51,189 ± 125 Da (Table 2, sample-2), corresponding to an average substitution of ^{2}H at 94% of non-exchangeable positions in yNIST-Fab. Deuteration of human μ-opiate receptor in *K. phaffii* showed that 80 ± 5% deuteration of amino acids came from ^2H_2O-media with methanol, while 20 ± 10% of deuteration of amino acids was derived from $^2H_4$-methanol as the carbon source in H_2O-media [34]. Similarly, the presence of $^2H_4$-methanol in ^2H_2O-media contributed an additional 9% deuteration of the C-terminal fragment of the *P. falciparum* merozoite surface protein-1 from 72% to 81% enrichment in ^2H_2O-media with protonated methanol as carbon source [21]. The deuterated levels of ^{2}H, ^{15}N-labeled yNIST-Fab (Table 2, samples 3-4) from either methanol or $^2H_4$-methanol were the same (~85%) within experimental error, suggesting minimal contribution from $^2H_4$-methanol. Likewise, 1D ^{1}H-NMR experiments of ^{2}H, ^{13}C, ^{15}N-yNIST-Fab from ^{15}N, 98% ^2H_2O-media with 0.2 to 0.4% (w/v) DCN-algae extract supplement and $^2H_4$-methanol (Table 2, samples 5-7) showed ~85 to 87% deuteration with no apparent contribution from those added carbon sources. By comparison, ^{2}H, ^{13}C, ^{15}N-yNIST-Fab from ^{15}N, 98% ^2H_2O media with 100% ^{13}C-methanol gave a deuteration level of 58% from 1D ^{1}H NMR analysis (Table 2, sample 13). ^{13}C, $^2H_4$-methanol was used to produce ^{2}H, ^{13}C, ^{15}N-yNIST-Fab (Table 2, sample 14) with as high a level of deuteration as possible from ^{15}N, 98% ^2H_2O media, 0.4% (w/v) DCN-algae extract. The 1D ^{1}H NMR analysis revealed this ^{2}H, ^{13}C, ^{15}N-yNIST-Fab (sample-14) has 93% deuteration. With one exception, the 1D ^{1}H NMR and MALDI-TOF MS analyses showed ~85% to 94% ^{2}H-enrichment for most deuterated yNIST-Fab samples.

In addition to optimizing ^{2}H labeling, incorporation of ^{13}C was also investigated to determine the level of contribution from cell biomass and from methanol. Wood and Komvies [24] reported that 70% of ^{13}C-enrichment in thrombomodulin's EGF domain comes from cell biomass grown in ^{13}C-glucose and the remaining 30% from ^{13}C-methanol during the induction phase. In contrast to ^{2}H-enrichment, the ^{13}C-methanol and ^{13}C-labeled peptides and amino acids of algae extract contribute equally to the ^{13}C-incorporation into yNIST-Fab (Table 2, samples 8-13). When expressed in ^{15}N, H_2O-minimal media with 20%^{13}C/80%-methanol as carbon source, mass spectrometry analysis revealed ~14% to 16% ^{13}C-enrichment for ^{13}C, ^{15}N yNIST-Fab (Table 2, sample 12). The ^{13}C-incorporation increased to ~36% in ^{13}C, ^{15}N-yNIST-Fab purified from cultures grown with 100% ^{13}C-methanol (Table 2, sample 13). When the ^{13}C-source was solely from the *K. phaffii* biomass grown in ^2H_2O-media supplemented with CN- or DCN-algae extract, the ^{13}C-enrichments in ^{2}H, ^{13}C, ^{15}N-yNIST-Fab (Table 2, samples 6 and 7) are estimated to be in 24% to 28% range. In the presence of both ^{13}C-methanol and CN/DCN-algae extracts, the ^{13}C-enrichments were greater than 50% for ^{2}H, ^{13}C, ^{15}N-yNIST-Fab samples 8-11 (Table 2). The TROSY-based backbone NMR experiments for resonance assignment were conducted on the ~85% ^{13}C-enriched ^{2}H, ^{13}C, ^{15}N-yNIST-Fab sample-14 [12].

Table 2. Molecular mass of unlabeled and isotope-labeled yNIST-Fab by MALDI-TOF MS for the estimation of isotope-enrichment.

Sample #	Methanol	MALDI-TOF		NMR
		Daltons	% Isotope	(^{2}H)
1) -	MeOH	48,784±49	-	nd
2) ^{2}H	MeOH	51,189±125	94% ^{2}H	nd
3) ^{2}H^{15}N	MeOH	51,513±47	85% ^{2}H	nd
4) ^{2}H^{15}N	d_4-MeOD	51,476±39	83% ^{2}H	nd
5) ^{2}H^{15}N, 0.2% DCN[a]	d_4-MeOD	51,555±73	1% ^{13}C[a]	85%
6) ^{2}H^{15}N, 0.3% DCN[a]	d_4-MeOD	52,035±44	24% ^{13}C[a]	86%
7) ^{2}H^{15}N, 0.4% DCN[a]	d_4-MeOD	52,110±52	28% ^{13}C[a]	87%
8) ^{2}H^{15}N, 0.4% DCN[a]	^{13}C-MeOH	52,573±123	50% ^{13}C[a]	nd
9) ^{2}H^{15}N, 0.7% DCN[a]	^{13}C-MeOH	52,585±79	50% ^{13}C[a]	nd
10) ^{2}H^{15}N, 1.0% DCN[a]	^{13}C-MeOH	52,787±69	58% ^{13}C[a]	nd
11) ^{15}N^{13}C, 0.5% CN[b]	^{13}C-MeOH	50,516±39	55% ^{13}C	nd
12) ^{15}N, 20%^{13}C	20% ^{13}C-MeOH	49,626±56	14-16% ^{13}C	nd
13) ^{2}H^{15}N, No sup.	^{13}C-MeOH	52,287±81	36% ^{13}C[a]	58%
14) ^{2}H^{15}N, 0.4% DCN[a]	^{13}C,d_4-MeOD	53,313±52	85% ^{2}H/ 85% ^{13}C[c]	93%

nd: not determined, insufficient sample; [a]DCN: ^{2}H,^{13}C,^{15}N-ISOGRO®, [b]CN:^{13}C, ^{15}N-ISOGRO®, 85% deuteration and 100% ^{15}N-incorporation assumed for the ^{13}C-enrichment calculation

3.5 Quality of amide backbone spectra from triple-labeled yNIST-Fab

A total of 422 non-proline amide ^{1}H, ^{15}N cross peaks are expected from 48.4 kDa U-^{15}N, 20% ^{13}C-yNIST-Fab, and the corresponding backbone fingerprint is congested with overlapping cross peaks [18]. The reduction of the amide proton and nitrogen linewidths is critical in obtaining well-resolved TROSY spectra for the sequence specific assignment [14]. The overlay of 2D ^{1}H,^{15}N spectra of ^{13}C, ^{15}N-yNIST-Fab (sample 11) and ^{2}H, ^{13}C, ^{15}N-yNIST-Fab (sample 14) illustrate a high-quality backbone fingerprint of the deuterated sample resulting from narrow linewidths with well dispersed ^{1}H and ^{15}N dimensions (Fig. 5B). The backbone chemical shift assignments of NIST-Fab were made using this triple-labeled yNIST-Fab sample, representing a step toward in-depth molecular assessment of the HOS of NIST-Fab fragment.

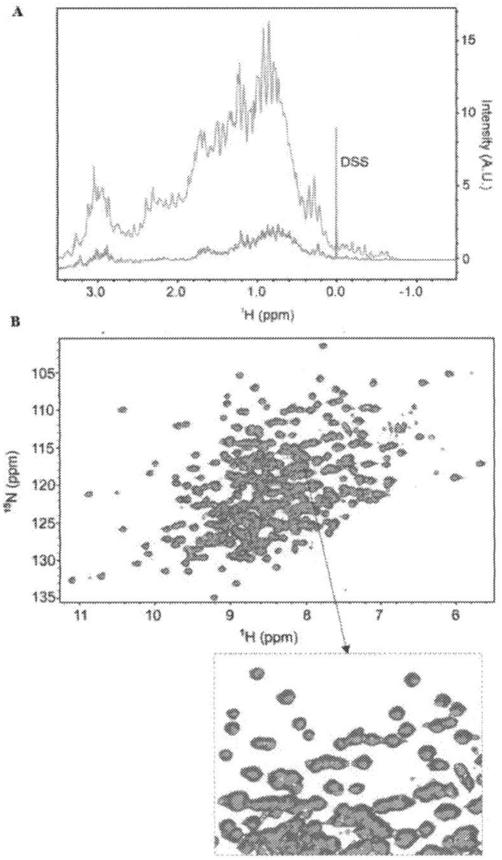

Fig. 5. (**A**) 1D ^{1}H-NMR spectra overlay of 490 μM U-^{13}C, ^{15}N-yNIST-Fab (cyan), 500 μM ^{2}H, ^{13}C, ^{15}N-yNIST-Fab (gold), and DSS (sodium 3-(trimethylsilyl)propane-1-sulfonate-D6 reference) at 37 °C and 900 MHz. (**B**) 2D ^{1}H-^{15}N-HSQC of U-^{13}C, ^{15}N-yNIST-Fab (cyan) and TROSY-HSQC of ^{2}H, ^{13}C, ^{15}N-yNIST-Fab (gold). Inset shows an expanded region of the spectrum from 105 to 120 ppm in the nitrogen dimension and 6.5 to 9.2 ppm in the proton dimension.

4 Conclusion

Here we report a protocol for optimized expression and purification of highly deuterated, ^{13}C, ^{15}N-enriched Fab fragment of the NISTmAb RM 8671 reference IgG1κ with native disulfide bonds in the *K. phaffii* expression system. U-^{15}N, 93% ^{2}H, 85% ^{13}C-yNIST-Fab enabled application of TROSY-based NMR experiments for the assignment of the Fab backbone chemical shifts, opening the door to detailed, sequence-specific structural analysis of Fab fragments as well as intact mAbs. This simple protocol can be easily implemented for the isotope labeling of proteins in *K. phaffii* using commonly available lab equipment. We showed that yeast host cell engineering and the addition of algae supplement to the 98% $^{2}H_2O$-minimal media contribute to obtaining a high expression of ^{2}H, ^{13}C, ^{15}N-yNISTFab sample for NMR studies. Finally, our results demonstrate MALDI-TOF mass spectrometry as a reliable method in estimating isotope-incorporation levels..

We are indebted to Genevieve Gingras and Yves Aubin (Centre for Biologics Evaluation, Biologics and Genetic Therapies Directorate, Health Canada, Ottawa, ON, Canada) for sharing the original clone of *K. phaffii* X33 expressing yNIST-Fab.

5 Disclaimer

Certain commercial equipment, instruments or materials are identified in this article in order to specify the experimental procedure adequately. Such identification is not intended to imply recommendation or endorsement by the National Institute of Standards and Technology, nor is it intended to imply that the equipment, instruments or materials identified are necessarily the best available for the purpose.

References

1. Lyu, Q. Zhao, J. Hui, T. Wang, M. Lin, K. Wang, J. Zhang, J. Shentu, P.A. Dalby, H. Zhang, B. Liu, Antib Ther. **5**, 233-257 (2022)
2. F. Cantini, E. Andreano, I. Paciello, V. Ghini, F. Berti, R. Rappuoli, L. Banci, Pharmaceutics. **14**, 1981 (2022)
3. G.W. Haxholm, B.O. Petersen, J. Malmstrom, J Pharm Sci. **108**, 3029-3035 (2019)
4. T.L. Hwang, D. Batabyal, N. Knutson, M. Wikstrom, Molecules. **26**, 2714 (2021)
5. S. Majumder, D.S. Bhattacharya, A. Langford, A.A. Ignatius, Pharm Res. **39**, 529-539 (2022)
6. Y. Tokunaga, K. Takeuchi, Int J Mol Sci. **22**, 46 (2020)
7. Y. Aubin, G. Gingras, S. Sauve, Anal Chem. **80**, 2623-2627 (2008)
8. B. Japelj, G. Ilc, J. Marusic, J. Sencar, D. Kuzman, J. Plavec, Sci Rep. **6**, 32201 (2016)
9. S. Majumder, A. Saati, S. Philip, L.L. Liu, E.

Stephens, J.C. Rouse, A.A. Ignatius, Pharm Res. **36**, 130 (2019)

10. D.D. Shah, S.M. Singh, K.M.G. Mallela, Pharm Res. **35**, 232 (2018)
11. D. Liu, D. Ren, H. Huang, J. Dankberg, R. Rosenfeld, M.J. Cocco, L. Li, D.N. Brems, R.L. Remmele, Jr., Biochemistry. **47**, 5088-5100 (2008)
12. T.L. Solomon, K. Chao, G. Gingras, Y. Aubin, W.B. O'Dell, J.P. Marino, R.G. Brinson, Biomol NMR Assign. **17**, 75-81 (2023)
13. H. Yagi, Y. Zhang, M. Yagi-Utsumi, T. Yamaguchi, S. Iida, Y. Yamaguchi, K. Kato, Biomol NMR Assign. **9**, 257-260 (2015)
14. M. Sattler, S.W. Fesik, Structure. **4**, 1245-1249 (1996)
15. M. Salzmann, K. Pervushin, G. Wider, H. Senn, K. Wuthrich, Proc Natl Acad Sci U S A. **95**, 13585-13590 (1998)
16. J.E. Schiel, A. Turner, Anal Bioanal Chem. **410**, 2067-2078 (2018)
17. J.E. Schiel, A. Turner, T. Mouchahoir, K. Yandrofski, S. Telikepalli, J. King, P. DeRose, D. Ripple, K. Phinney, Anal Bioanal Chem. **410**, 2127-2139 (2018)
18. R.G. Brinson, J.P. Marino, F. Delaglio, L.W. Arbogast, R.M. Evans, A. Kearsley, G. Gingras, H. Ghasriani, Y. Aubin, G.K. Pierens, X. Jia, M. Mobli, H.G. Grant, D.W. Keizer, K. Schweimer, J. Stahle, G. Widmalm, E.R. Zartler, C.W. Lawrence, P.N. Reardon, J.R. Cort, P. Xu, F. Ni, S. Yanaka, K. Kato, S.R. Parnham, D. Tsao, A. Blomgren, T. Rundlof, N. Trieloff, P. Schmieder, A. Ross, K. Skidmore, K. Chen, D. Keire, D.I. Freedberg, T. Suter-Stahel, G. Wider, G. Ilc, J. Plavec, S.A. Bradley, D.M. Baldisseri, M.L. Sforca, A.C.M. Zeri, J.Y. Wei, C.M. Szabo, C.A. Amezcua, J.B. Jordan, M. Wikstrom, mAbs. **11**, 94-105 (2019)
19. L.M. Damasceno, I. Pla, H.J. Chang, L. Cohen, G. Ritter, L.J. Old, C.A. Batt, Protein Expr Purif. **37**, 18-26 (2004)
20. I. Dikiy, L.D. Clark, K.H. Gardner, D.M. Rosenbaum, Methods Enzymol. **614**, 37-65 (2019)
21. W.D. Morgan, A. Kragt, J. Feeney, J Biomol NMR. **17**, 337-347 (2000)
22. N.J. Anthis, G.M. Clore, Protein Sci. **22**, 851-858 (2013)
23. E. Rodriguez, N.R. Krishna, J Biochem. **130**, 19-22 (2001)
24. M.J. Wood, E.A. Komives, J Biomol NMR. **13**, 149-159 (1999)
25. J.M. Wu, S.Y. Wang, W.C. Fu, Int J Mol Sci. **13**, 13212-13226 (2012)
26. J.C. de Ruijter, E.V. Koskela, A.D. Frey, Microb Cell Fact. **15**, 87 (2016)
27. M. Chen, L.C. Hancock, J.M. Lopes, Biochim Biophys Acta. **1771**, 310-321 (2007)
28. S. Schuck, W.A. Prinz, K.S. Thorn, C. Voss, P. Walter, J Cell Biol. **187**, 525-536 (2009)
29. M.J. Feige, S. Groscurth, M. Marcinowski, Y. Shimizu, H. Kessler, L.M. Hendershot, J. Buchner, Mol Cell. **34**, 569-579 (2009)
30. J. Liu, C. Liu, Y. Fan, R.A. Munro, V. Ladizhansky, L.S. Brown, S. Wang, J Biomol NMR. **65**, 7-13 (2016)
31. A. Sola, H. Maaheimo, K. Ylonen, P. Ferrer, T. Szyperski, Eur J Biochem. **271**, 2462-2470 (2004)
32. Y. Laroche, V. Storme, J. De Meutter, J. Messens, M. Lauwereys, Biotechnology (N Y). **12**, 1119-1124 (1994)
33. X. Shi, T. Karkut, M. Chamankhah, M. Alting-Mees, S.M. Hemmingsen, D. Hegedus, Protein Expr Purif. **28**, 321-330 (2003)
34. S. Massou, V. Puech, F. Talmont, P. Demange, N.D. Lindley, M. Tropis, A. Milon, J Biomol NMR. **14**, 231-239 (1999)
35. C.P. Liu, T.I. Tsai, T. Cheng, V.S. Shivatare, C.Y. Wu, C.Y. Wu, C.H. Wong, Proc Natl Acad Sci U S A. **115**, 720-725 (2018)
36. A. Nylen, M.T. Chen, Methods Mol Biol. **1674**, 37-48 (2018)

EPJ Web of Conferences **286**, 01004 (2023) https://doi.org/10.1051/epjconf/202328601004
ECNS 2023

Effective Synthesis of Deuterated n-Octylamine and Its Analogues

Kazuhiro Akutsu-Suyama[1*], *Misaki* Ueda[1], *Mitsuhiro* Shibayama[1], *Kosuke* Ishii[2], and *Naoya* Nishi[2]

[1] Neutron Science and Technology Center, Comprehensive Research Organization for Science and Society, Tokai, Ibaraki 319-1106, Japan
[2]Department of Energy and Hydrocarbon Chemistry, Graduate School of Engineering, Kyoto University, Kyoto 615-8510, Japan

Abstract. In neutron scattering studies, deuterium-labelled compounds play a key role in controlling the contrast of organic samples and reducing the incoherent scattering background from the samples. As amine compounds play a vital role as functional molecules, we have developed a new synthesis method using an amide compound as a starting material to synthesize deuterated amines and their analogues. We determined the deuteration ratio of the obtained deuterated 1-octylamine by mass spectrometry, nuclear magnetic resonance (NMR), and neutron reflectometry techniques. As a result, the deuteration ratio was estimated to be ~60 %. The deuteration ratio of the synthesized 1-octylamine was not high because the method used did not deuterate its α-protons and NH_2 group. However, this synthesis method is suitable for the large-scale synthesis of deuterated amine compounds for neutron research because it is easy to increase the synthetic scale.

1 Introduction

Deuterium (^2H) is a stable isotope of hydrogen and ^2H-labelled compounds have been extensively used in a variety of research fields, beginning with its isolation in 1932 [1]. For example, various deuterated compounds have been used in the field of nuclear magnetic resonance (NMR), infrared spectrometry, mass spectrometry (MS), and neutron research [2–7].

Various compounds, such as carboxylic acids, aromatic compounds, and ionic liquids can be effectively deuterated using platinum group metal-catalyzed direct deuterium-labelling methods [8–12]. In contrast, even though amines are important functional molecules, there is no effective direct multi-deuteration method for amine compounds. Several direct deuteration methods (Scheme 1) that are selective for the α-position or α,β-position of alkylamines have been developed [13–15], with a recent advancement presented to a multi-deuteration via heterogeneous catalysts for ammonium/amine compounds [16]. Deuterium labelling methods are important in neutron scattering to control the contrast of the organic samples, and thus the prepared compounds reduce the incoherent scattering. However, the synthesis of deuterated amines with high deuteration ratios is preferable for neutron studies.

Herein, to synthesize amine compounds with a higher deuteration ratio, we have developed a new method for the synthesis of deuterated amines (Scheme 2). In this approach, we have used an amide compound as a starting material for deuterated amine synthesis because amide compounds can be deuterated directly using a Pd/C and Rh/C mixed catalyst system [17, 18] and amines can be easily synthesized by reducing amide compounds. Additionally, carboxylic acid compounds can also be synthesized by the hydrolysis of amide compounds in the presence of acids or bases. Therefore, we synthesized three deuterated compounds, namely amide, amine, and carboxylic acid, using the method shown in the Scheme 2. The deuteration ratio of these deuterated compounds was analyzed by electrospray ionization mass spectrometry (ESI-MS), NMR, and neutron reflectometry (NR) techniques.

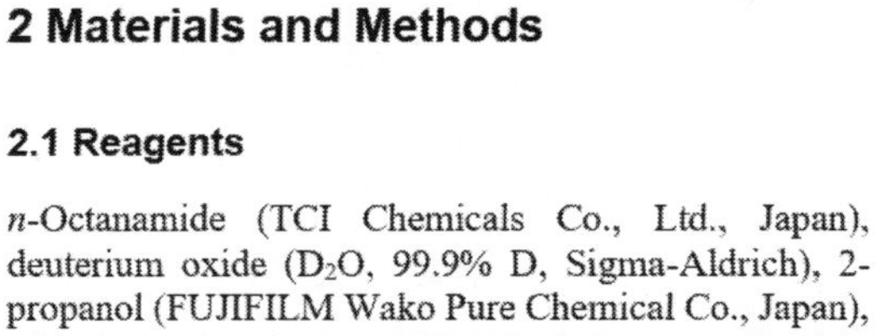

Scheme 1. Direct deuteration method for amine compounds.

2 Materials and Methods

2.1 Reagents

n-Octanamide (TCI Chemicals Co., Ltd., Japan), deuterium oxide (D_2O, 99.9% D, Sigma-Aldrich), 2-propanol (FUJIFILM Wako Pure Chemical Co., Japan), Pd/C (10 wt% Pd, N.E. CHEMCAT Co., Japan), Rh/C (5 wt%, FUJIFILM Wako Pure Chemical Co., Japan), lithium aluminium hydride ($LiAlH_4$, Sigma-Aldrich),

* Corresponding author: k_akutsu@cross.or.jp

hydrochloric acid (FUJIFILM Wako Pure Chemical Co., Japan), magnesium sulfate (FUJIFILM Wako Pure Chemical Co., Japan), methanol (FUJIFILM Wako Pure Chemical Co., Japan), ethanol (FUJIFILM Wako Pure Chemical Co., Japan), anhydrous tetrahydrofuran (FUJIFILM Wako Pure Chemical Co., Japan), diethyl ether (FUJIFILM Wako Pure Chemical Co., Japan), 1,4-dioxane (FUJIFILM Wako Pure Chemical Co., Japan), $CDCl_3$ (FUJIFILM Wako Pure Chemical Co., Japan), and CD_2Cl_2 (FUJIFILM Wako Pure Chemical Co., Japan) were used without further purification. Silicon wafer (diameter = 5.08 cm, thickness = 2.0 mm) was supplied by Crystal Base Co. Ltd. (Osaka, Japan). Ultrapure water (18.2 MΩ.cm) was produced with a deionized purified water production system (RFU424TA system, ADVANTEC, Japan) and used throughout this study.

2.2 Instruments

ESI-MS spectra were recorded on a mass spectrometer EXTREM-MS-100P (JASCO Corporation). The percentage of deuteration of the molecules was calculated using isotope distribution analysis of different isotopologues [8, 12]. This was achieved by analysing the area under each MS peak which corresponds to a defined number of deuterium atoms. Based on the relative amount of carbon-13 (^{13}C) natural abundance found or estimated in the protonated version, the ^{13}C contribution to the value of the area under each MS signal was subtracted.

^1H NMR (400 MHz) and ^2H NMR (61.4 MHz) spectra were recorded on a 400 MHz NMR spectrometer (JMTC-400/54/JJ/YH, JEOL Ltd., Tokyo, Japan).

In principle, the measurement was performed three times for each sample and the results were shown as mean deuteration ratios (with standard deviations).

2.3 Synthesis of deuterated *n*-octanamide

Deuteration reactions (Scheme 2) were carried out in a stainless-steel reactor (TSSR, TPR1-VSI-300, SUS316, Taiatsu Techno Corp., Tokyo, Japan) [19].

A mixture of *n*-octanamide (1.95 g, 13.6 mmol), Pd/C (10 wt% Pd, 1.60 g, 1.50 mmol (11 mol%)), and Rh/C (5 wt% Rh, 3.20 g, 1.55 mmol (11 mol%)) in 2-PrOH (40 mL)/D_2O (80 mL) mixed solvent was loaded into the TSSR. The mixture was vacuum degassed for 10 min to remove oxygen. The reactor was purged with Ar for 10 s and then sealed. The mixture was heated to 180 °C and stirred continuously for 48 hours. After cooling to 20 °C, the mixture was filtered through a short plug of celite to remove the catalyst and washed a second time with EtOH (150 mL). The filtrate was evaporated to dryness under reduced pressure to yield deuterated *n*-octanamide (1.31 g, 64.3 %). ^1H NMR (400 MHz, $CDCl_3$) δ 0.79–0.84 (m), 1.21–1.25 (m), 1.58 (residual signal), 2.12 (residual signal), 5.69 (s), 6.09 (s). ^2H NMR (61.4 MHz, $CDCl_3$) δ 0.78 (s), 1.21 (s), 1.54 (s), 2.13 (s), 5.74 (s), 6.11 (s).

2.4 Synthesis of deuterated 1-octylamine

Deuterated *n*-octanamide (0.48 g, 3.2 mmol) was dissolved in 5 mL of tetrahydrofuran, and 1.0 M $LiAlH_4$/THF solution (8 mL, 8 mmol (2.5 eq)) was slowly added dropwise to the deuterated *n*-octanamide solution under moderate stirring. The mixture was then heated at 70 °C for 3 hours. After cooling the mixture in an ice bath, cold water was added slowly until the reaction was complete. The mixture was then extracted with diethyl ether (3×30 mL). The combined extracts were dried with anhydrous magnesium sulfate, filtered, and evaporated to dryness under reduced pressure to yield deuterated 1-octylamine (0.40 g, 91.7 %). ^1H NMR (400 MHz, CD_2Cl_2) δ 0.79–0.86 (m), 1.20–1.44 (m), 2.57 (s, 2H). ^2H NMR (61.4 MHz, CD_2Cl_2) δ 0.82 (s), 1.23 (s), 1.35 (s), 1.48 (s).

2.5 Synthesis of deuterated 1-ocanoic acid

Deuterated *n*-octanamide (0.23 g, 1.5 mmol) was dissolved in 50 mL of 6 M HCl aq (0.3 mol, 200 eq), and the mixture was heated at 100 °C for 6 hours. After cooling to 20 °C, the mixture was extracted with diethyl ether (3×20 mL). The combined extracts were dried with anhydrous magnesium sulfate, filtered, and evaporated to dryness under reduced pressure to yield deuterated 1-octanoic acid (0.18 g, 80.2 %). ^1H NMR (400 MHz, CD_2Cl_2) δ 0.86 (residual signal), 1.28 (residual signal), 1.56 (residual signal), 2.29 (residual signal). ^2H NMR (61.4 MHz, CD_2Cl_2) δ 0.79 (s), 1.20 (s), 1.52 (s), 2.25 (s).

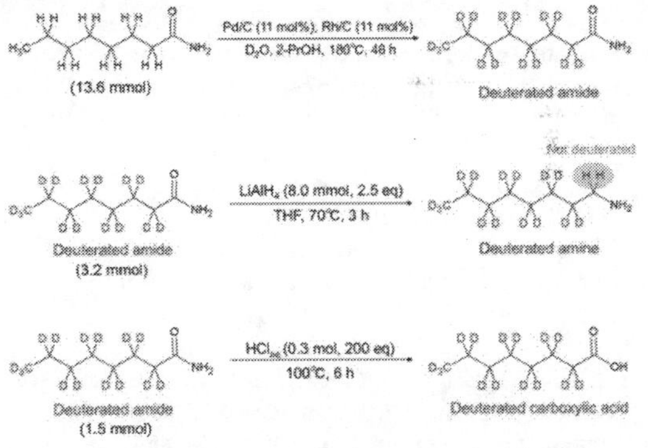

Scheme 2. Synthesis of deuterated amide, amine, and carboxylic acid in this study.

2.6 Neutron reflectivity measurement

NR measurements were performed at a polarized neutron reflectometer (BL17 SHARAKU, Japan) with a horizontal scattering geometry installed at the Materials and Life Science Experimental Facility (MLF) at J-PARC [20, 21]. The incident beam power of the proton accelerator was 800 kW (peak neutron flux at 10 meV [n/eV·s·cm^2] was about 4.8×10^{12}) for all measurements. Pulsed neutron beams were generated in a mercury target at 25 Hz, and the NR data were measured using the time-of-flight (TOF) technique. The wavelength (λ) range of the incident neutron beam was tuned to approximately 2.2–8.8 Å for the unpolarized neutron mode by disk choppers. The covered scattering vector (Q_z) range was 0.008–0.09 Å^{-1}, where $Q_z = (4\pi/\lambda)\sin\theta$ (θ represents the angle of incidence). A 10 mm beam footprint was maintained on the sample surface using six different types of collimating slits. TOF neutron data were collected using a ^{3}He gas tube detector without spatial resolution. All measurements were taken at ambient temperature. The MLF uses the event recording method as a standard data acquisition system [22]. Data reduction, normalization, and subtraction procedures were performed using a program installed at BL17 SHARAKU. Motofit software [23] was used to fit the NR profiles using the least-squares approach to minimize the deviation of the fit. The scattering length density (SLD) value was evaluated using Motofit. The molar composition and physical density of the sample can be used to determine the SLD value, which is a nuclear property of individual atoms. Therefore, the deuteration ratio of the sample can be calculated using the SLD value.

A sample cell for NR experiments developed at BL17 SHARAKU was used for this experiment. The Si-substrate was placed in the sample cell and deuterated *n*-octylamine was added. Then, the sample cell was mounted on the reflectometer sample stage.

3 Results and Discussions

3.1 Synthesis of deuterated compounds

The deuteration reaction of *n*-octanamide was carried out as shown in the Scheme 2. This reaction gave the deuterated *n*-octanamide in a moderate yield (64.3 %). The pressure of the TSSR reactor reached 0.2 MPa at 20 °C after the reaction, indicating that some of the *n*-octanamide decomposed during the reaction. On the other hand, the other reactions in the Scheme 2 proceeded quantitatively.

Figures 1–3 show the spectra of ^{1}H NMR, ^{2}H NMR, and ESI-MS of deuterated *n*-octanamide, 1-octylamine, and 1-octanoic acid, respectively. The obtained deuteration ratios of these deuterated materials by the ESI-MS analysis were 73 % (±0.2 %) for *n*-octanamide, 60 % (±0.1 %) for 1-octylamine, and 75 % (±0.7 %) for 1-octanoic acid, respectively. Since the α-position of the 1-octylamine was not deuterated, the deuteration ratio of deuterated 1-octylamine is lower than that of other deuterated compounds. In addition, as mentioned in the following section, the deuteration ratio of 1-octylamine was calculated to be lower than that of other compounds because the NH_2 group in 1-octylamine contains two protons.

Moreover, the deuteration ratios of the alkyl-chain moiety of these deuterated materials were determined by ^{1}H NMR analysis using 1,4-dioxane as an internal standard. The obtained deuteration ratios of these deuterated materials by the ^{1}H NMR analysis were 73 % (±0.2 %) for *n*-octanamide, 60 % (±0.2 %) for 1-octylamine, and 69 % (±0.3 %) for 1-octanoic acid. The difference in the obtained deuteration ratio between the ESI-MS and ^{1}H NMR analysis was within 4.3 %. Therefore, the ^{1}H NMR and ESI-MS analysis results are consistent.

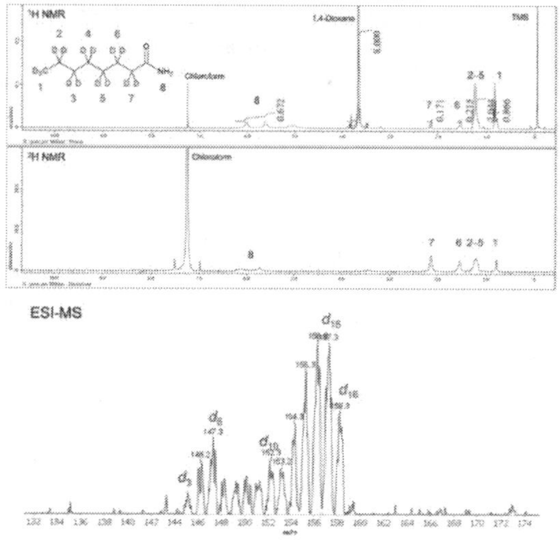

Figure 1. ^{1}H NMR ($CDCl_3$, 400 MHz), ^{2}H NMR ($CDCl_3$, 61.4 MHz), and ESI-MS data of deuterated *n*-octanamide. ESI-MS in negative mode of *n*-octanamide showing the mass distribution of the different isotopologues, which ranges from d_3–d_{17}. The distribution of the isotopologues is as follows [M–H]$^-$: 1.6 %, d_3; 4.0 %, d_4; 7.1 %, d_5; 2.9 %, d_6; 3.1 %, d_7; 3.6 %, d_8; 3.8 %, d_9; 4.6 %, d_{10}; 4.5 %, d_{11}; 6.6 %, d_{12}; 12.3 %, d_{13}; 17.6 %, d_{14}; 17.6 %, d_{15}; 10.0 %, d_{16}; 0.7 %, d_{17}.

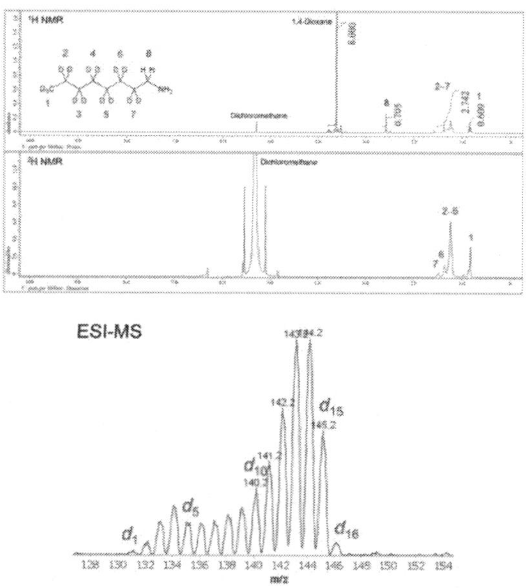

Figure 2. ^{1}H NMR (CD_2Cl_2, 400 MHz), ^{2}H NMR (CD_2Cl_2, 61.4 MHz), and ESI-MS data of deuterated 1-octylamine. ESI-MS in positive mode of 1-octylamine showing the mass distribution of the different isotopologues, which ranges from d_1–d_{15}. The distribution of the isotopologues is as follows [M+H]$^+$: 0.2 %, d_1; 0.8 %, d_2; 2.7 %, d_3; 4.0 %, d_4; 2.7 %, d_5; 2.6 %, d_6; 2.8 %, d_7; 3.4 %, d_8; 3.8 %, d_9; 5.2 %, d_{10}; 7.6 %, d_{11}; 12.8 %, d_{12}; 19.7 %, d_{13}; 19.7 %, d_{14}; 10.9 %, d_{15}; ; 1.0 %, d_{16}.

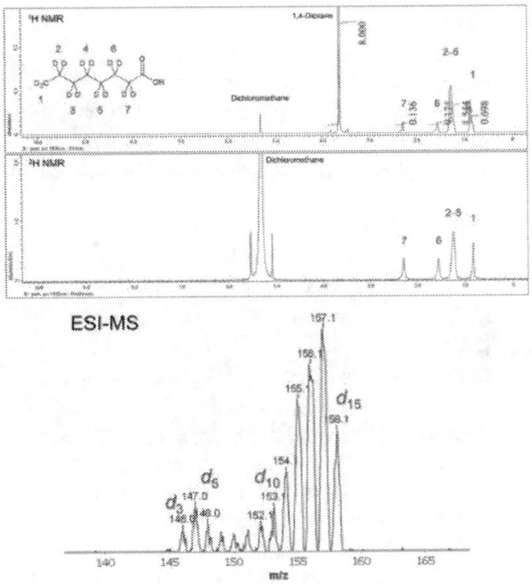

Figure 3. ^{1}H NMR (CD_2Cl_2, 400 MHz), ^{2}H NMR (CD_2Cl_2, 61.4 MHz), and ESI-MS data of deuterated 1-octanoic acid. ESI-MS in negative mode of 1-octanoic acid showing the mass distribution of the different isotopologues, which ranges from d_3–d_{15}. The distribution of the isotopologues is as follows [M–H]$^-$: 1.8 %, d_3; 6.4 %, d_4; 2.2 %, d_5; 1.1 %, d_6; 1.2 %, d_7; 2.2 %, d_8; 2.1 %, d_9; 2.9 %, d_{10}; 7.5 %, d_{11}; 14.5 %, d_{12}; 21.7 %, d_{13}; 25.2 %, d_{14}; 11.5 %, d_{15}.

As stated above, compounds with higher deuteration ratio are preferable for neutron experiments. In the case of the deuteration of malonamide compounds using the same catalyst system, the deuteration ratio increased with increasing of the reaction temperature. However, this also accelerates the decomposition reaction of malonamides [18]. It is inferred that *n*-octanamide probably has similar properties, and the deuteration ratio of *n*-octanamide will increase in exchange for the increased decomposition rate of *n*-octanamide. For enhancing the deuteration ratio of *n*-octanamide, it is recommended to optimize the reaction temperature. It is worth noting that reusing the deuterated *n*-octanamide for a second cycle with new D_2O and catalyst would lead to a higher deuteration ratio of *n*-octanamide. It is considered that this method can obtain *n*-octanamine with a high deuteration ratio without worrying about the decomposition of *n*-octanamine.

3.2 Neutron reflectivity analysis

In neutron studies, the SLD profile is the main factor for analyzing the fine structure of nano-materials. However, it is difficult to accurately determine the SLD value of 1-octylamine by ^{1}H NMR and ESI-MS methods because the NH_2 protons are easily exchanged with the surrounding solvent (D_2O and/or H_2O). However, neutron scattering techniques can be successfully applied to determine the SLD value (and deuteration ratio) of organic samples [24, 25]. Therefore, in this study, we determined the SLD value of deuterated 1-octylamine using NR.

The NR data were obtained using a BL17 SHARAKU neutron reflectometer. Figure 4 shows the NR profiles obtained from the deuterated 1-octylamine/Si substrate sample and the fitting curve of the sample obtained using Motofit software. Although a few experimental points at around $Q_z = 0.03$ Å^{-1} which is a data stitching zone deviated significantly from the fitting curve, data analysis didn't have any issues. The SLD value of deuterated 1-octylamine was estimated to be 3.63×10^{-6} Å^{-2}. Since the calculated SLD values for protonated (NH_2) species and deuterated (ND_2) species are 3.67×10^{-6} Å^{-2} and 4.36×10^{-6} Å^{-2}, respectively, it can be suggested that the NH_2 protons of deuterated 1-octylamine are not deuterated. Therefore, we have concluded that two protons of the NH_2 group in 1-octylamine were protonated during the synthesis procedure, and as a result, the obtained SLD values of deuterated 1-octylamine appeared to be smaller.

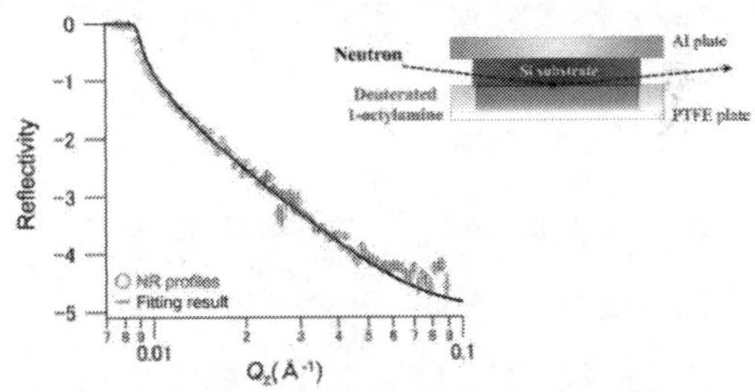

Figure 4. NR profiles obtained from the deuterated 1-octylamine/Si substrate sample and its fitting curve. The inset shows an image of the NR sample system.

4 Summary

In this study, we have efficiently synthesized three deuterated compounds with high deuteration ratios, *n*-octanamide, 1-octylamine, and 1-octanoic acid, using the synthesis method shown in Scheme 2. In particular, deuterated 1-octylamine with alkyl chains having a high deuteration ratio. If the $LiAlD_4$ catalyst system [26, 27] was used in the reduction reaction, the SLD value and deuteration ratio of deuterated 1-octylamine can be improved to 5.16×10^{-6} Å^{-2} and 77.1 %, respectively. Therefore, fully deuterated amine synthesis using the $LiAlD_4$ catalyst system seems to be an attractive option for the use of deuterated amines in neutron studies.

This work was supported by JSPS KAKENHI (Grant Number) 21K05121. Direct deuteration reactions were conducted at the User Experiment Preparation LabIII (CROSS). ^{1}H NMR, ^{2}H NMR, and ESI-MS spectroscopy experiments were conducted at the Deuteration Laboratory in J-PARC MLF. Neutron experiments were conducted at the BL17 SHARAKU apparatus in J-PARC, Tokai, Japan (Proposal No. 2022A0001).

References

1. H. C. Urey, F. G. Brickwedde, and G. M. Murphy, Phys. Rev. **40**, 1 (1932)

2. Y.H. Lee, K. Tamura, M. Maeda, M. Hoshino, K. Sakurai, S. Takahashi, T. Ikegami, T. Hase, Y. Goto, J. Biol. Chem. **282**, 5959 (2007)
3. T. Tenno, K. Fujiwara, H. Tochio, K. Iwai, E. H. Morita, H. Hayashi, S. Murata, H. Hiroaki, M. Sato, K. Tanaka, M. Shirakawa, Genes Cells. **8**, 865 (2004)
4. A. Ge, Q. Peng, L. Qiao, N. R. Yepuri, T. A. Darwish, M. Matsusaki, M. Akashi, S. Ye, Phys. Chem. Chem. Phys. **17**, 18072 (2015)
5. J. A. McEwan, A. J. Clulow, P. E. Shaw, A. Nelson, T. A. Darwish, P. L. Burn, I. R. Gentle, Adv. Mater. Interfaces **3**, 1600184 (2016)
6. M. Kofu, R. Watanuki, T. Sakakibara, S. Ohira-Kawamura, K. Nakajima, M. Matsuura, T. Ueki, K. Akutsu, O. Yamamuro, Sci. Rep. 11, 12098 (2021)
7. K. Akutsu, M. Cagnes, T. Niizeki, Y. Hasegawa, T. A. Darwish, Physica B **551**, 262 (2018)
8. K. Akutsu-Suyama, M. Cagnes, K. Tamura, T. Kanaya, T.A. Darwish, Phys. Chem. Chem. Phys. **21**, 17512 (2019)
9. H. Sajiki, K. Hattori, F. Aoki, K. Yasunaga, K. Hirota, Synlett. **7**, 1149 (2002)
10. K. Park, T. Matsuda, T. Yamada, Y. Monguchi, Y. Sawama, N. Doi, Y. Sasai, S. Kondo, Y. Sawama, H. Sajiki, Adv. Synth. Catal. **360**, 2303 (2018)
11. K. Akutsu-Suyama, H. Sajiki, M. Ueda, M. Asamoto, Y. Tsutsumi, RSC Adv. **12**, 24821 (2022)
12. N. R. Yepuri, S A. Holt, G. Moraes, P. J. Holden, K. R. Hossain, S. M. Valenzuela, M. James, T. A. Darwish, Chem. Phys. Lipids **183**, 22 (2014)
13. B. Chatterjee, V. Krishnakumar, C. Gunanathan, Org. Lett. **18**, 5892 (2016)
14. L. Neubert, D. Michalik, S. Bähn, S. Imm, H. Neumann, J. Atzrodt, V. Derdau, W. Holla, M. Beller. J. Am. Chem. Soc. **134**, 12239 (2012)
15. Y. Chang, A. Yesilcimen, M. Cao, Y. Zhang, B. Zhang, J. Z. Chan, M. Wasa. J. Am. Chem. Soc. **141**, 14570 (2019)
16. Y. Sawama, T. Matsuda, S. Moriyama, K. Ban, H. Fujioka, M. Kamiya, J. Shou, Y. Ozeki, S. Akai, H. Sajiki, Asian J. Org. Chem. **12**, e202200710 (2023)
17. N. Modutlwa, T. Maegawa, Y. Monguchi, H. Sajiki, J. Labelled Compd. Radiopharm. **53**, 686 (2010)
18. C. Micheau, Y. Ueda, K. Akutsu-Suyama, D. Bourgeois, R. Motokawa, Sol. Extr. Ion Exch. **41**, 221 (2023)
19. K. Akutsu-Suyama, K. Park, R. Takakura, K. Tamura, M. Cagnes, T. A. Darwish, T. Yamada, Y. Sawama, H. Sajiki, JPS Conf. Proc. **33**, 011150 (2021)
20. M. Takeda, M. Arai, J. Suzuki, D. Yamazaki, K. Soyama, R. Maruyama, H. Hayashida, H. Asaoka, T. Yamazaki, M. Kubota et al, Chin. J. Phys. **50**, 161 (2012)
21. K. Akutsu-Suyama, H. Kira, N. Miyata, T. Hanashima, T. Miyazaki, S. Kasai, D. Yamazaki, K. Soyama, H. Aoki, Polymers **12**, 2180 (2020)
22. K. Sakasai, S. Satoh, T. Seya, T. Nakamura, K. Toh, H. Yamagishi, K. Soyama, D. Yamazaki, R. Maruyama, T. Oku et al, Quantum Beam Sci. **1**, 10 (2017)
23. A. Nelson, J. Appl. Crystallogr. **39**, 273 (2006)
24. M. A. Klenner, M. Cagnes, K. Wood, K. Mita, M. Kishimoto, T. A. Darwish, Polym. Chem. **11**, 4986 (2020)
25. K. Akutsu-Suyama, N. L Yamada, Y. Ueda, R. Motokawa, H. Narita, Appl. Sci. **12**, 1215 (2022)
26. T. Murphy, S. K. Callear, N. Yepuri, K. Shimizu, M. Watanabe, J. N. C. Lopes, T. Darwish, G. G. Warr, R. Atkin, Phys. Chem. Chem. Phys. **18**, 17224 (2016)
27. M. Moir, N. R. Yepuri, D. L. Marshall, S. J. Blanksby, T. A. Darwish, Adv. Synth. Catal. **364**, 3670 (2022)

EPJ Web of Conferences **286**, 02001 (2023) https://doi.org/10.1051/epjconf/202328602001
ECNS 2023

ICONE – Towards a French HiCANS Neutron Source for materials science and industry

Frédéric Ott[1*], *Jacques* Darpentigny[1], *Burkhard* Annighöfer[1], *Mariano Andrés* Paulin[1], *Jean-Louis* Meuriot[1], *Alain* Menelle[1], *Nadia* Sellami[2] and *Jérôme* Schwindling[2]

[1] IRAMIS, Laboratoire Léon Brillouin CEA-CNRS, Univ. Paris-Saclay, 91190 Gif sur Yvette, FRANCE
[2] IRFU, CEA, Univ. Paris-Saclay, 91190 Gif sur Yvette, FRANCE

Abstract. We present the ICONE project which proposes to build a HiCANS source in France. The aim of the ICONE project is to be able to provide the French neutron user community sufficient instrumental capacity to continue performing neutron scattering experiments for their research programs. The baseline goal is to offer performances equivalent to a medium power research reactor or spallation source (such as Orphée or ISIS). We consider that such a machine would fulfil the needs of at least two-thirds of the users which not require ultimate performances but simply beam-time to perform their experiments. We also describe the experimental work ongoing at Saclay around the various technologies necessary to build a HiCANS.

1 Introduction

The decade 2000 was a golden age for neutron scattering in Europe as more than 30000 instruments-days were available across Europe for neutron scatterers. In the future however, the neutron provision is expected to steadily decrease due to the closure of aging facilities and the available beam-days in Europe is currently foreseen to decrease by 40% by the middle of the next decade [1]. If one focusses on the specific situation of France, the situation is far worse. Following the closure of the Orphée reactor in 2019, the number of French neutron scattering instruments, measured in Full Time Equivalent (FTE), dropped from 29 to 9, the current available instruments corresponding to the French share at the ILL and the French shares in the different CRGs around the ILL[†]. Should the Grenoble High Flux reactor shut down, the French access to neutron scattering would be reduced to 2 FTE instruments at ESS[‡] at the beginning of the next decade. France has been at the forefront of neutron science as the second most publishing country, on par with Germany, of scientific publications using neutron scattering techniques. This was made possible because French users had access to a share of 13% of the world neutron scattering instruments[§]. This share is currently foreseen to drop to 1% in the next decade.

Neutron scattering has been identified as a key scientific tool to address current technical and social challenges: Europe is engaged in the construction of the highest performance spallation source, the ESS, the United States are building a second target station around SNS and are upgrading the HFIR reactor at Oak Ridge and the NIST reactor in Washington, China has recently opened two new sources, CSNS and CARR, the United Kingdom is considering the upgrade of the ISIS facility… What could be done so that France remains competitive in this environment and avoids losing its expertise in the field?

Until recently, it was considered that only nuclear research reactors or spallation sources could provide sufficient neutron flux to perform neutron scattering experiments and welcome user programs. However, in the last decade the situation has been reconsidered owing to the progress in several fields: (i) high current proton accelerators, (ii) neutron moderation techniques, (iii) neutron instrumentation.

It is nowadays possible to build and operate proton accelerators with proton currents of the order of 100 mA or even higher [2]. The concept of low dimensional moderators is expected to boost the brilliance of neutrons sources by factors of 3 to 10 [3]. Neutron instrumentation adapted to exploit long neutron pulses has been developed for ESS [4].

Based on these new technical capabilities, the concept of HiCANS, High Current Accelerator-driven Neutron Source has been proposed. The underlying idea is to use low energy proton accelerators (tens of MeV)

* Corresponding author: Frederic.Ott@cea.fr
† France has access to 25% of 27.7 FTE neutron scattering instruments at the ILL (= 7 FTE instruments) and its shares in the different CRGs correspond to 2 FTE instruments.
‡ The currently foreseen French share of 14% in ESS would correspond to 2.1 FTE instruments (14% * 15 instruments)
§ *ICONE, une nouvelle source de diffusion neutronique française (to be published)*

and low energy nuclear processes to produce neutrons. The low neutron yield of such processes is compensated by the use of a high proton current, a very strong coupling between the target (which can be very small) and the moderator, and an optimized time structure which allows to use a very large fraction of the produced neutrons. Monte-Carlo simulations show that if all components of such a source are optimized, the performance of such sources for neutron scattering could be on par with medium power nuclear research reactors or spallation sources [5-6].

This has led to the proposition from various institutes to build HiCANS sources to provide neutrons to their user community. We can mention the HBS project in Germany, the ARGITU project in Spain, the SARAF project in Israel, the PC-CANS in Canada or the DARIA project in Russia. In Europe, the institutes working in the field of HiCANS have gathered into ELENA, the European Low Energy accelerator-driven Neutron facilities Association [7]. HiCANS has been considered by the LENS, the League of Advanced European Neutron Sources, as a possible component in the future neutron scattering landscape [1].

France is also considering the HiCANS solution to mitigate the foreseen neutron drought in France. In the following, we describe the ICONE source which aims to provide the French user community with a new facility in the coming decade. We describe the scope of the project and the current technical developments under way.

2 The ICONE project

2.1 Scientific scope

During the last four years, more than 300 different French laboratories have been performing neutron scattering experiments, either within the 2FDN, the Fédération Française de Diffusion Neutronique [8], which is gathering the French laboratories proposing neutron beam time or at the ILL Institut Laue Langevin.

The aim of the ICONE project is to be able to provide the French neutron user community with sufficient instrument capacities to continue performing standard neutron scattering experiments for their research programs. The baseline goal is to offer performances equivalent to a medium power research reactor or spallation source (such as Orphée or ISIS). We consider that such a device would fulfil the needs of at least two-thirds of the users who do not require ultimate performances but simply beam-time to perform their experiments. This does not preclude the fact that the facility would conduct a high-level scientific program. Statistics show that the quality of the scientific production of facilities is not connected to the power of the source [9].

In parallel, the French users will have access to ESS to perform experiments which require ultimate performances and cannot be performed on ICONE.

The Figure 1 shows a sketch of the ICONE facility. The aim is to offer a suite of instruments covering most of the neutron scattering techniques, mainly diffraction, spectroscopy, SANS, reflectivity and radiography. In order to maximize the efficiency of the source, the instruments will operate in time-of-flight mode as has been adopted on all recent spallation sources. In addition, it is considered to provide different neutrons pulses lengths and time structures by using several target stations, thus being able to fully optimize the need for lower resolution instruments and higher resolution instruments.

2.2 Scientific capabilities

Monte-Carlo modelling of HiCANS has been performed at various institutes across Europe in the past years. The results suggest that the use of a pulsed proton accelerator with an energy of 20 or more MeV and with peak proton currents of 50 mA or more, coupled to an optimized Target – Moderator – Reflector (TMR) system could provide performances on par with reactors [5-6]. The neutron brilliance which can be extracted from the TMR allows to simulate the performances of neutron scattering instruments which would be installed around a HiCANS.

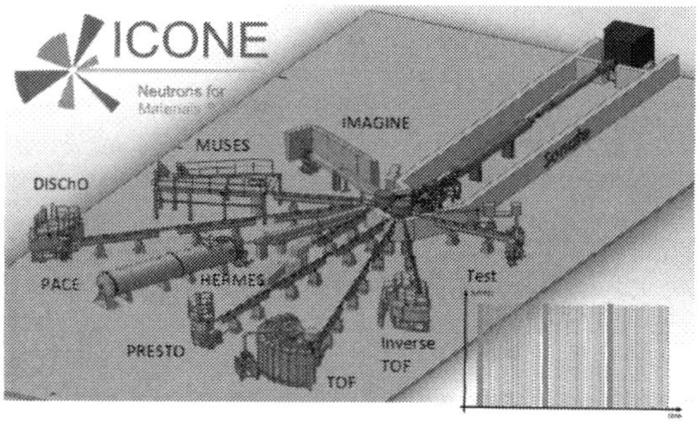

Fig. 1. Sketch of the ICONE facility. The actual facility will require extensive concrete shielding around the accelerator, the target stations.

At the CEA Saclay we consider the reference design described in Table 1. The proton accelerator energy is maintained in the lower energy range (25 MeV) in order to avoid significant activation of the machine. It also reduces the costs of construction and operation. Note that in order to maximize the efficiency of the source, it is necessary to be able to provide different time structures to serve best both high resolution and low resolution instruments. The current configuration which is considered consists of two interlaced time structures (see Fig. 1, bottom –right): (a) 2 ms – 20 Hz and (b) 200 μs – 100 Hz illuminating two different target stations TS1 and TS2. On TS1, the time structure (a) is reminiscent of "long pulse" spallation sources such as ESS and is ideally suited to low wavelength resolution instruments where resolution is traded for flux and bandwidth such as SANS, reflectometry or spin-echo. On TS2, the second time structure (b) is more suited to higher resolution experiments. The repetition rate is increased to maximize the flux at the expense of the wavelength bandwidth. The proposed time structures allow to build very short instruments in the range of 10 to 20 meters with up to 40 m for the longest instruments.

Table 1. The ICONE design parameters.

Accelerated particles	Protons
Protons energy	25 MeV
Peak current	80-100mA
Duty cycle	6% (4% + 2%)
Accelerator	Ion source RFQ 4 vanes (3.6 MeV) Hot DTL (25MeV)
Time structure (2 targets)	2ms – 20 Hz 200μs – 100 Hz
Target	Beryllium
Moderators	Thermal (H2O) Cold (methane) Cold (para-hydrogen)
Instrumentation	5 on Target Station 1 (low res.) 5 on Target Station 2 (high res.)

Using these design parameters, we performed calculations of the flux at the sample position for different techniques by transposing existing instrument designs. The Figure 2 shows the comparison of the flux at the sample position at Orphée (green) and ICONE (yellow). We should underline that one can boost the performances of instruments by going beyond conservative designs and by implementing innovative techniques: focussing techniques for reflectometry [10], focussing SANS for small Q [11], larger area Position Sensitive Detectors for powder diffraction [12], Micro-channel plate detectors for radiography [13] and large solid angle analysis for spin-echo. All these upgrades are either readily available or currently being developed or implemented at various institutes. The conclusion is that it is likely that new instruments on ICONE would probably outperform former instruments around the Orphée reactor. This conclusion is not too surprising as past experience shows that neutron sources have steadily improved the efficiency of their instruments over the decades (sometimes by an order of magnitude per decade) while the source itself remained unchanged. The ILL is a perfect illustration of this process.

Hence, we are confident that the ICONE source would satisfy the needs of a vast community of users.

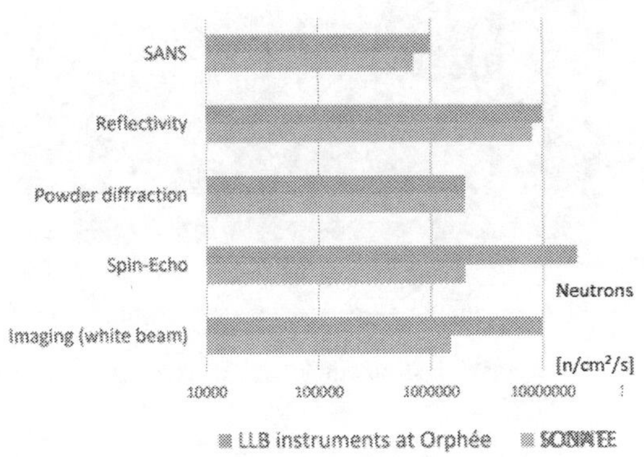

Fig. 2. Comparison of the performances of various types of neutron instruments: (green) typical instruments around Orphée, (yellow) similar instruments around ICONE.

Table 2. Foreseen instrumental suite and projection of the number of experiments, which could be done on the facility.

Instrument	***Typical run duration***	***Nb. Run / year***
SANS 1 (soft matter)	2.5 days	64
SANS 2 (hard matter)	5 days	32
Diffractometer 1	5 days	32
Diffractometer 2	2.5 days	64
Diffractometer 3	5 days	32
Reflectometry	5 days	32
Radiography	5 days	32
Direct ToF spectroscopy	5 days	32
Inverse ToF spectroscopy	15 days	10
Backscattering spectroscopy	10 days	16
Spin-Echo spectroscopy	15 days	10
TOTAL		**356**

Table 2 proposes a typical instrumental suite, which would meet the needs of the major part of the French user community. Assuming typical run durations extrapolated from earlier experience at Orphée-LLB, we estimate that it should potentially be possible to perform more than 300 experimental runs per year. This should lead to about 100-150 scientific publications per year.

2.3 Development plan

Owing to the fact that there are no HiCANS currently in operation, the construction of a demonstrator machine is a necessity to validate the technical concepts and the actual performances. This strategy is explicitly proposed by LENS [1].

HiCANS machines are especially suited to the construction of demonstrators as such machines are very modular. The proton energy can be upgraded over the lifetime of the machine by adding DTL modules. Provided that there is enough reserve power, the number of illuminated targets can be increased by distributing the proton beam on several targets. The target and moderator monoliths are rather modest in size and can be designed to be rather easily modified.

Hence, we propose to decompose the project ICONE into two phases: PRELUDE-ICONE which would be a demonstrator using an accelerator with an energy of about 10 MeV to produce sufficient neutrons. A single target would be illuminated and a limited number of instruments would be built (e.g. one diffractometer, SANS, reflectometer and radiography station). Such a limited scope machine would nevertheless allow to demonstrate all key technologies in actual operation and validate the operation and performances of typical neutron scattering instruments. The construction of a full-fledge user facility (25-30 MeV, 10 instruments) could be started with virtually no pending technological risks. Besides, the final machine ICONE could reuse most of the equipment of the demonstrator machine.

3 State of the art of the technologies

Over the period 2018-2022, the IRAMIS and IRFU Institutes at the CEA were supported by a SESAME programme of the region Île-de-France to develop the *IPHI – Neutron* platform. The goal was to setup a testbed for several technologies essential to the construction of HiCANS. This included in particular the operation of a proton accelerator at a high proton current (of the order of 100 mA), the construction of a target able to cope with a total incident proton power of 50 kW and the demonstration of the operation of a neutron scattering instrument around a CANS.

3.1 Accelerator

The *IPHI-Neutron* platform is built around the Injecteur de Proton à Haute Intensité (IPHI) accelerator. This accelerator was built to demonstrate the capability of operating a proton accelerator at a current up to 100 mA with a high current Electron Cyclotron Resonance (ECR) ion source and a 4-vanes Radio-Frequency Quadrupole (RFQ). The IPHI technology has been subsequently used for the design and the construction of the ion source of the IFMIF Prototype Accelerator (LIPAc), currently under commissioning at Rokkasho, Japan [14], and of the ESS RFQ, currently under commissioning at Lund, Sweden [15].

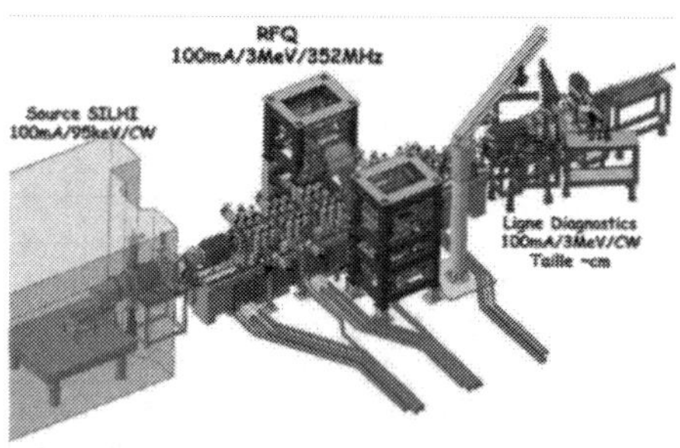

Fig. 3. Sketch of the IPHI accelerator. The machine is using an ECR proton source, a 4-vanes RFQ and a conical beam dump (able to withstand 300 kW). A deviated beam line is available to perform experiments.

In the case of the accelerator of the ICONE facility, we are considering using components derived from the ESS accelerator front-end, namely an ECR source, a 4-vane RFQ with a final energy of 3.6 MeV and a hot DTL section. There should be no major technical hurdle to boost the peak current from 62 mA (nominal value at ESS) to higher values (80 – 100 mA).

Fig. 4. (a) Front-end of the ESS accelerator: ion source, RFQ and first section of the DTL. (b) Ion source of the IFMIF Prototype Accelerator capable to produced currents up to 150 mA in continuous mode.

3.2 Target

As we aim to operate an accelerator with a proton energy as low as possible to minimize the costs of construction and operation and also to minimize activation issues, we have opted for a beryllium target which offers the best neutron yield in the range 10-30 MeV [16].

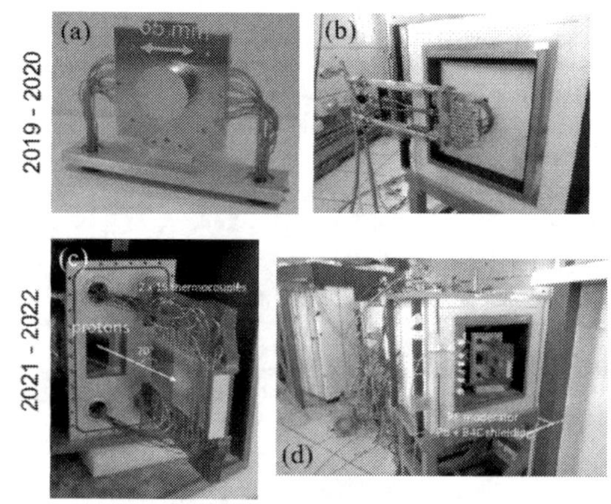

Fig. 5. (a) Beryllium targets mounted on a cooling plate. (b) Target in the moderator box. (c) 50 kW target prototype operating at a 20° incidence angle. (d) Target integrated in the moderator box.

The motto of the design was to build a system as sturdy as possible as we are aiming for very high reliability. It was thus decided to physically separate the target and the cooling system so that a failure of the target would have no impact on the accelerator as could be the case for a water leak in the accelerator circuit. Hence it was decided to attach beryllium blocks with screws on a copper cooling support (see Fig.5b). We started with a « lower power » prototype of a few kW. The power density on the target can go up to 500 W/cm^2. For the current full scale 50 kW prototype, we used an inclined geometry at 20° with respect to the incident proton beam. The target is composed of several 20 mm sub-bricks which can be individually replaced in case of a failure. The proton beam can be redirected in order to evenly spread the power over the whole target surface.

The high power target was tested over a period of 100 hours in a 2 weeks run to evaluate its durability [17].

3.3 Moderators

The TMR assembly is currently only comprised of a polyethylene moderator with one extraction hole (see Fig. 5b). The TMR is installed on trolleys so that it can be easily opened for maintenance and operation.

We are currently developing the technologies necessary to build a directional moderator using para-hydrogen. We recently achieved a para-hydrogen polarisation of 99.7% (without neutron irradiation). The system will be using a thermosiphon to fill the hydrogen finger moderator. Liquid hydrogen will serve as the cooling fluid and avoids an extra helium cooling circuit. The cooling can be achieved with an entry level cryogenerator as the radiation heating in a HiCANS TMR is rather limited. This facilitates the design of the cold source and makes its integration in the TMR easier. This is especially important as the use of finger moderators will require that each instrument using cold neutrons on ICONE has its own dedicated moderator.

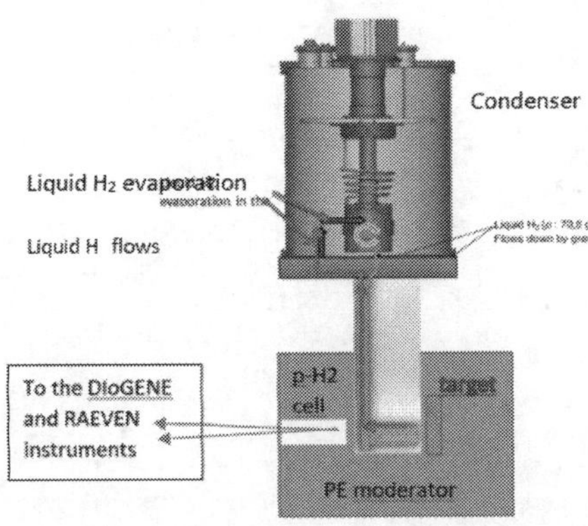

Fig. 6. Sketch of a cold finger moderator using para-hydrogen. The setup is using the principle of the thermo-siphon where the hydrogen is itself the cooling fluid. This configuration has been used in the MINOS and COCOTIER nuclear physics experiments to build liquid hydrogen targets [18].

3.4 Instrumentation

The DIoGENE instrument, DIffusion GEnérique de NEutrons, has recently been installed around the IPHI source. The instrument is a general-purpose neutron scattering instrument aimed at investigating the performance of the neutron TMR, the issues related to background noise due to fast neutrons and gamma rays productions and more generally the ToF data acquisition protocols and processing in event mode. The picture below shows the instrument in the *IPHI – Neutron* casemate. One limitation of the instrument concerning its resolution is its short flight path (L = 6.6 m).

Fig. 7. The IPHI – Neutron facility: the accelerator in the back, the TMR + shielding box (white) with one small extraction hole and the DIoGENE instrument.

During a recent operation of the source, it was nevertheless possible to observe the diffraction of an austenite steel rod when operating the machine in pulsed mode at a power of 200 watts [19].

Due to the low power used for the measurement, the counting time was 60 minutes. However, we think that a number of rather simple improvements are possible which could boost its flux by almost two orders of magnitude. This would reduce the acquisition times from hours to minute.

The long-term goal would be to increase the proton energy from 3 MeV to 25 MeV which would multiply the neutron flux by a factor of 300 bringing the instrument at a performance level comparable to the high flux powder diffractometer G41 around Orphée.

In order to acquire as much experience as possible, we are also collaborating with the JCNS Jülich Centre for Neutron Science. The HERMES reflectometer was recently transferred from the Orphée reactor to the JULIC test platform at the Forschungszentrum Jülich [20-21].

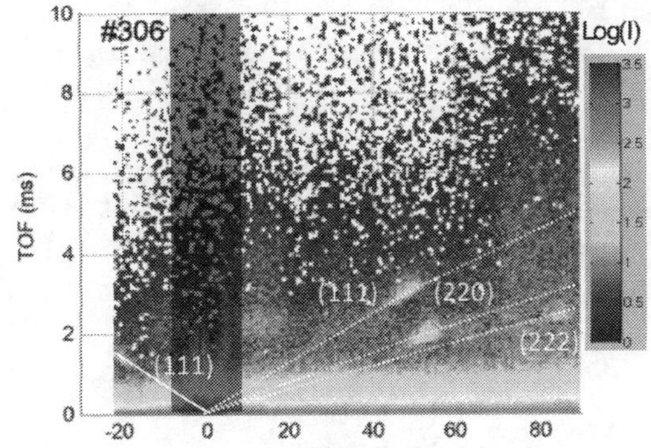

Fig. 8. Representation of the diffraction on a steel rod in the ToF Vs diffraction angle plane. The diffraction peaks have been indexed as (111), (220), (222) in the austenite phase.

4 Conclusion and outlook

We are convinced that the performance of a compact source is potentially equivalent to a medium scale research reactor or spallation source while offering low construction and operation costs.

There are no identified technological hurdles. The accelerator technology, while still quite ambitious, is available. Various technological solutions for the target are being considered and developed by several institutes within the ELENA network. The moderator technology is rather simple and does not require complex cryogenic systems, making the construction and handling of cold sources much easier than on fission or spallation sources. A number of new instrumentation concepts have been developed in the past decade during the R&D effort for ESS and are currently being implemented.

From a more specific French perspective, it is possible to benefit from a rich ecosystem: (i) scientific and technical expertise at Saclay and Grenoble, (ii) a wide user base, (iii) the possibility to reuse the R&D efforts injected into ESS and the acquired expertise.

The last point is that all developments for the ESS in the field of data reduction and data processing could be reused and would not require extra efforts.

We are thus proposing the construction of the HiCANS source ICONE in France [22] to support the French users in materials science research. This source would conduct its own research programs and does not aim at being in competition with ESS. ICONE would be integrated into a future European neutron ecosystem consisting of a network of local CANS or HiCANS national facilities and a few high-end national facilities, while ESS would be the reference source if ultimate capabilities are required.

This work has been funded by the "CANS Inflexion" program at the CEA and the "IPHI-Neutron" SESAME project of the Île-de-France region.

It is also part of the collaboration within ELENA and LENS on the development of HiCANS.

We are thankful to many colleagues who are contributing to the IPHI – Neutron and ICONE projects : H.N. Tran, L. Thulliez, A. Marchix, A. Letourneau for Monte-Carlo simulations, G. Gigante and S. Come for radiation protection issues, J. Schwindling, B. Bolzon, N. Chauvin, F. Sénée, B. Pottin, M. Oublaid, G. Perreu, A. Dubois, D. Chirpaz, Y. Gauthier, Y. Sauce, O. Kuster, C. Deberles for the construction of the test station and the operation of the IPHI accelerator, X. Fabrèges, A. Menelle, F. Porcher, F. Damay for aspects related to neutron instrumentation, K. Jiguet, W. Josse, F. Gibert, B. Homatter, P. Lavie, J.-L. Meuriot, F. Prunes, G. Exil, R. Lautié, E. Jorgji, P. Permingeat for the technical support and C. Alba-Simionesco, R. Duperrier, A. Leservot for strategic aspects.

References

1. Neutron Science in Europe, LENS Vision document (2022).
2. LIPAC Highlights – April 2023. https://www.ifmif.org/lipac-highlights-april-2023/ *Extracted currents of 155mA D+ at 100% DC was reached.*
3. L. Zanini et al, EPJ Web Conf. **231**, 04006 (2020), https://doi.org/10.1051/epjconf/202023104006
4. K. Andersen et al, NIM A **957**, 163402 (2020). https://doi.org/10.1016/j.nima.2020.163402
5. F. Ott, H. N. Tran, X. Fabrèges, A. Menelle, N. Chauvin, J. Schwindling, A. Letourneau, A. Marchix, C. Alba-Simionesco, J. Phys.: Conf. Ser. **1021**, 012007 (2018).
6. T. Brückel, T. Gutberlet (Eds.) Conceptual Design Report Jülich High Brilliance Neutron Source (HBS), Schriften des Forschungszentrums Jülich, General, Vol.8 (2020).
7. ELENA, European Low Energy accelerator-based Neutron facilities Association
8. 2FDN, Fédération Française de Diffusion Neutronique
9. T. Gutberlet et al, Neutron News **29**(2), 18-24 (2018). Do neutrons publish? A neutron publication survey, 2005–2015. Figure 9. DOI: 10.1080/10448632.2018.1514199 arXiv:1804.07185v1
10. ESTIA, Focusing Reflectometer at ESS. https://europeanspallationsource.se/instruments/estia
11. H. Wu et al, NIM A 972 (2020) 164072. https://doi.org/10.1016/j.nima.2020.164072
12. A. Houben et al, J. Appl. Cryst. 56 (2023) 633. POWTEX visits POWGEN. https://doi.org/10.1107/S1600576723002819
13. *Superior Neutron Imaging with the Neutronic [i].* https://ilphotonics.com/superior-neutron-imaging-with-the-neutronic-i/
14. K. Masuda et al., "Commissioning of IFMIF Prototype Accelerator Towards CW Operation", in Proc. LINAC'22, Liverpool, UK, Aug.-Sep. 2022, pp. 319-323. doi:10.18429/JACoW-LINAC2022-TU2AA04
15. D. Noll et al., "First Beam Matching and Transmission Studies on the ESS RFQ", in Proc. LINAC'22, Liverpool, UK, Aug.-Sep. 2022, pp. 414-417. doi:10.18429/JACoW-LINAC2022-TUPOPA04
16. I.S. Anderson et al, Physics Report **654** (2016) 1-58. https://doi.org/10.1016/j.physrep.2016.07.007
17. J. Schwindling et al, Journal of Neutron Research, vol. **24**, no. 3-4, pp. 289-298 (2022). DOI: 10.3233/JNR-220024
18. A. Obertelli et al, The European Physical Journal A **50**, 8 (2014). https://doi.org/10.1140/epja/i2014-14008-y
19. J. Darpentigny and F. Ott, Journal of Neutron Research **24**, pp. 385–393 (2022). DOI 10.3233/JNR-220018
20. M. Paulin et al, these proceedings
21. T. Brückel et al, these proceedings
22. *ICONE, une nouvelle source de diffusion neutronique française.* Avant-Projet Sommaire. To be published (2023).

Towards a conceptual design for the moderators of ARGITU: a preliminary neutronics study

Octavio G. del Moral, *Miguel* Magán, *Fernando* Sordo, *Félix* J. Villacorta* *and Mario* Pérez

ESS-Bilbao, Bizkaia Technology Park, Laida Bidea, Building 207 B, Ground Floor, 48160 Derio (Spain)

Abstract. This work reports on preliminary calculations of potential low-dimensional moderators for the high-current accelerator-driven neutron source (HiCANS) ARGITU. Simulations start with the selection of materials, depending on the neutron energy range to be used for each instrument. Results evaluate the performance of water for thermal moderators to deliver neutrons for thermal instruments, whereas para-H_2 and solid methane have been analyzed to deliver cold neutrons. Additionally, alternative concepts like using hybrid moderators (water/liquid methane) or reduced-size cold moderators (reduced-size para-H_2) have been scrutinized for their use in bispectral instruments that require a wide range of intermediate neutron energies. The dimensions of the moderators have been refined to improve the neutron yield for the neutron wavelength range selected in each case. After that, a space-efficient layout has been proposed to implement four moderators next to the Be target, linked to a preliminary suite of instruments that these moderators would serve. Although the eventual selection shall consider both the final instrument suite and the phase space volume required for such neutron instruments (that defines their neutron optics features), the results presented here represent a qualitative step towards the conceptual development of the ARGITU neutron source.

1 Introduction

High-Current Accelerator-driven Neutron Sources (HiCANS) represent a strategic move towards the development the next-generation neutron research facilities, oriented to mitigate the sudden decrease of available beamtime for neutron experiments. The HiCANS facility concept, by which neutrons are generated using a medium power accelerator (20-100 kW) that boosts a proton or deuterium beam towards a metal target through low energy nuclear reactions, are a new and efficient approach to produce beams with competitive neutron brilliance of thermal and cold neutrons, with lower construction and operation costs. Coordinated efforts to develop such kind of neutron sources are taking shape in several countries, like the HBS project[1] in Germany and ICONE (formerly SONATE) in France,[2] with recent developments and conceptual studies are taking place in Spain, Hungary, Italy and Sweden, all of them framed within the European Low-Energy accelerator-based Neutron facilities Association (ELENA).[3]

ARGITU ("Light up" in Basque language) is the proposal for a neutron source in Spain.[4] Located in the Basque Country, this facility belongs to the pioneer group of HiCANS proposed to keep Europe at their leading position in neutron science. Its main function will be to serve the local scientific community and enable everyday science aimed to address current major societal challenges related to new materials for computing and digital world (e.g., quantum computing, new sensors), energy and climate (e.g. new types of Li and Na batteries, fuel cells, hydrogen storage), and health (e.g., drug delivery, molecular dynamics in biosystems). Simultaneously, the facilityis intended to train the next generation of neutron scientists, since maintaining a broad community is essential to promote scientific discussion and address such new challenges through neutron science.

The conceptual design of ARGITU consists of a proton accelerator that delivers pulsed beams of duration 1.5 ms at a frequency of 30 Hz.[5] The proton beam (peak current 32 mA), reaching a final energy up to 31.5 MeV, by means of a 4-segment radiofrequency quadrupole and a 2-tank drift-tube linear accelerator (linac) hits a cooled beryllium target producing neutrons through stripping $^{9}Be(p,n)$ reactions. Similar concepts have been validated in existing target designs for other compact sources,[6,7] and can be scaled for HiCANS, which delivers a qualitative higher beam power. Neutrons generated are reflected by a beryllium reflector. Thermal and cold neutrons are usually obtained by means of thermalization processes occurring within the moderators.[8,9] A limited number of instruments at one target station are allowed, due to the worsening of the efficiency of the reflector with the increasing number of holes in such reflector to accommodate the neutron beam ports to integrate neutron scientific instruments. In ARGITU, every target station can deliver neutrons up to 4 instruments. This

* Corresponding author: fjimenez@essbilbao.org

work intends to provide a preliminary picture of the moderators that can be implemented on the first target station of ARGITU.

2 Methods

The neutron yield of beryllium thick targets was simulated using MCNP6.2 code[10] utilizing the nuclear data ENDF-B/VIII[11] for neutrons (S(α,β) for thermal scattering) and ENDF-B/VII[12] for protons. Tallies are placed to count neutrons going through the viewed surface of the moderator. At striping reactions, incident particles (i.e., protons) combine with the target nucleus (^{9}Be) and neutrons are generated as subproduct of the original particle with almost the same momentum and direction. SuperMC software is used to convert CAD geometries to MCNP format and built up the model.[13,14]

Beryllium is a very suitable option as a target material due to its good heat conduction, high melting point and high neutron yield at energy region up to 50 MeV.[15] Moreover, although beryllium is the chosen material for the ARGITU target, additional simulations including ^{7}Li(p,n) stripping reactions were carried out as a benchmark for validation of the results. In the calculations, thicknesses of 2.23 cm for Li (not in the literature) and 0.7 cm for Be are sufficient to stop a 31.5 MeV proton beam.

Results confirm a good agreement between nuclear data cross sections and experimental results at the same incident proton energy (Figure 1).[16,17,18] A Be target hit by a proton beam of 31.5 MeV energy leads to an unmoderated neutron yield of 2.1×10^{11} neutrons/μC (1 Coulomb = 6.24×10^{18} protons) with a wide energy distribution from 1 keV to 31.5 MeV with an average energy of 215 keV. Results for 3 MeV protons also match previous results in literature (in this case, 0.3 cm of Li is enough to stop protons of 3 MeV), which would lead to get 1.65×10^{9} neutrons/μC for lithium and $\sim5\times10^{8}$ neutrons/μC for beryllium. In summary, results are harnessed to validate that the total neutron yield at these energy levels can be reproduced by MCNP simulations with an appropriate accuracy using nuclear data cross sections without nuclear models.

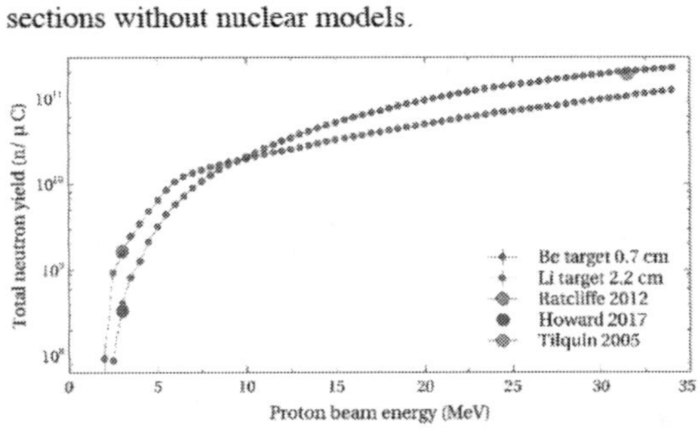

Fig. 1. Neutron yield for low-energy proton beam reactions (p, n) using Li and Be targets, and comparison with referenced literature [16,17,18]

Due to stripping reaction and target shape, the neutron spectra are not isotropic. Most of the neutron flux follows the original proton beam path, which is the privileged direction with the most energetic neutron spectra. This marked angle-dependence warn against placing neutron beam extraction, NBEX, for thermal and cold neutron applications at low angles.

3 Results and Discussion

3.1 Moderators: materials and neutron yield

The scalability of HiCANS in order to maximize the performance of neutron scientific instruments allows the possibility of a dedicated and customized moderator for each one of them. Moreover, one of the advantages of neutron sources with moderate power (below 100 kW) is that the neutron moderators and the neutron optics can be located very close to the target (the damage by irradiation estimated makes this affordable).[19] This allows the integration of moderators attached to the neutron beam extraction, NBEX, and this moderator-NBEX assembly to be close to the target.

The neutron moderation process is based on the collision of neutrons with a material (moderator) that brings the average energy of neutrons stemming from the target to an appropriate energy spectrum useful for neutron scattering experiments. At this stage of the ARGITU development, four possible thermal and cold moderating materials are being proposed: water at room temperature, liquid methane at 100K, solid methane at 22K and parahydrogen.

- Water (300 K). It is a common thermal moderator: it is a very effective thermal moderator due to its high hydrogen density (111 g/dm^{3}) which have a good neutron cross section.
- Methane (solid at 22 K and liquid at 100 K). it has a high hydrogen density (106 g/dm^{3}), due to its 4 hydrogen atoms per molecule. It has a low absorption rate for neutrons and many low-energy excitation states to scatter them with a short free path. Depending on the temperature, it is a fairly good candidate for cold or bispectral instruments.
- Para-H_2 (~20 K): Despite being pure hydrogen, the hydrogen population (70.8 g/dm^{3}) is lower than the hydrogenous materials mentioned above. Moreover, it has quite long neutron free path. However, it is ideal to enhance the population of cold neutrons well beyond the thermal energies.

Initially, a first approach will consist of simulating one neutron beamline with a single cylindrical moderator (radius 2.5 cm) normal to the Be target surface, i.e., at an angle of 135° with respect to the proton beam (see inset of Fig. 2).

The use of a thin water pre-moderator has been rendered necessary to slow down neutrons before they reach into the moderator and the NBEX, increasing in turn the moderator efficiency for all cases under study. This element, wrapping the moderator but slightly detached from it, consists of a room temperature water layer. An optimum thickness for the pre-moderator is between 1 cm to 2 cm (see Fig. 2). Thicker pre-moderators visibly reduce the outgoing neutron flux.

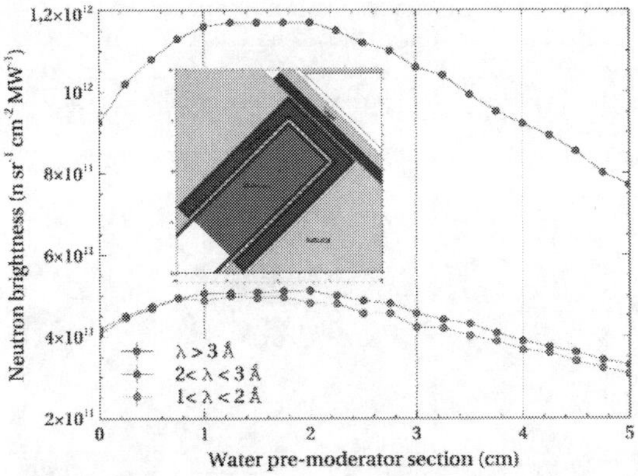

Fig. 2. Neutron brightness vs pre-moderator thickness for several energy ranges (Inset: MCNP moderator model, showing the configuration of the water pre-moderator).

Figure 3 shows the dependence of neutron flux on the moderator radius at certain neutron energy ranges for the four moderating configurations. Water and para-H_2 provide an optimized flux with a radius around 1 cm. On the other hand, methane moderators display a maximum between 1.5 and 2.5 cm (diameter: 3-5 cm). In this preliminary study, the section of the moderator has been fixed to 5 cm diameter, although a future redesign of the neutron instrument suite is envisioned as soon as the conceptual design of the instrument suite takes shape.

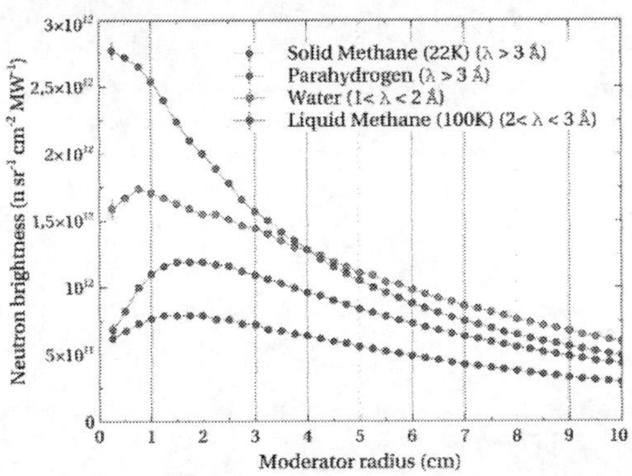

Fig. 3. Neutron brightness vs moderator section (radius).

The next step is to analyze the optimum thickness (height of the cylinder or length in the direction normal to the target surface) for the moderator materials considering the previous layout of one single neutron line at 135°. Every moderator requires to find a maximum of performance. A too thin moderator does not clash enough with neutrons to slow them down and a too thick one would increase the portion of absorbed neutrons reducing the effective flux.

Room temperature water is the suitable candidate for thermal neutron moderator and just requires 3 cm to get an unmatched performance, as shown in Figure 4a for neutrons with wavelengths between λ = 1 Å and λ = 2 Å.

On the other hand, parahydrogen has been demonstrated to be an ideal cold moderator to increase the flux of cold neutrons. Considering the yield of cold neutrons with λ > 3 Å displayed on Figure 4c, the best performance is achieved with a long moderator ($\approx$ 13 cm), very similar to the concept of "finger moderator" proposed for cold instruments at the HBS.[8]

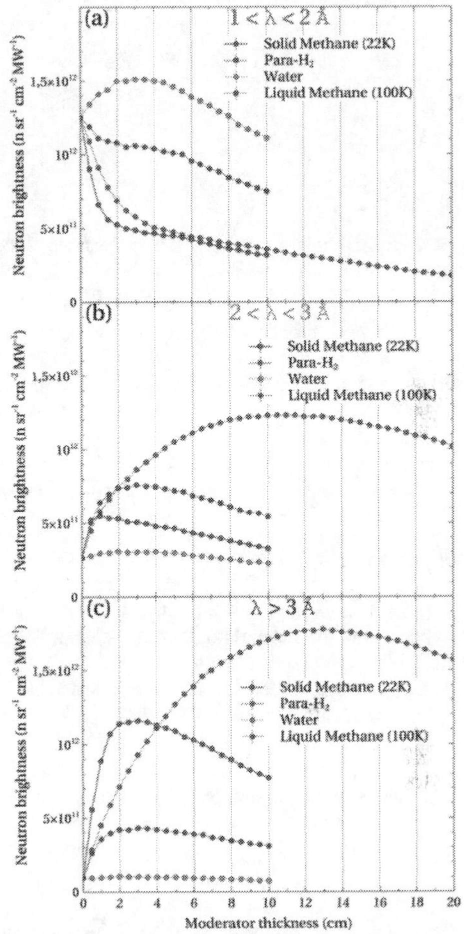

Fig. 4. Neutron flux as a function of moderator length (thickness) for the 4 moderators proposed, at different neutron wavelength ranges: (a) 1-2 Å; (b) 2-3 Å; and (c) >3 Å. The moderator section was fixed to 5 cm diameter in all cases.

Solid methane shows a good behavior to deliver cold neutrons with a thickness of 2-3 cm (better than para-H_2 with the same thickness up to 4 cm, see Fig. 4c). This must be taken into consideration in cases where there are dimensional restrictions.

Finally, liquid methane shows an overall good performance in the thermal-bispectral range (between λ = 1 Å and λ = 3 Å, see Figures 4a and 4b).

Figure 5 shows the spectra in the thermal and cold range delivered by these moderators at their best performance, i.e. with their optimized thickness. Results confirm the suitability of 3 cm-thick water (1 cm pre-moderator plus 2 cm of moderator) as thermal moderator and 13 cm-thick para-H_2 as the chosen cold moderator, keeping 3 cm-thick solid methane as an adequate alternative in case of geometrical/dimensional restrictions.

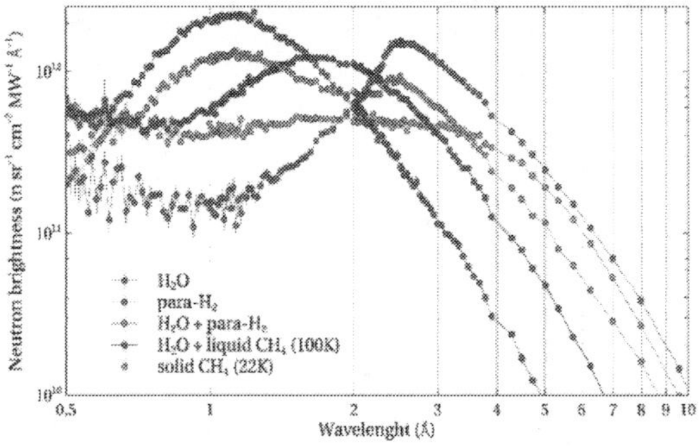

Fig. 5. Neutron spectra for several moderator materials at their optimized thickness. All moderator materials include a pre-moderator consisting of a 1 cm thick water layer.

Furthermore, additional calculations have been carried out using a combination of two moderating materials to achieve a good performance at intermediate wavelengths. The purpose is to find a suitable moderator for bispectral capabilities. Two options have been considered (the final ratio between the two moderating materials can be tuned up regarding the relative viewed areas of both moderators):

i. Reducing the diameter of the cold para-H_2 moderator to 60% of the neutron line (3 cm), a bi-spectral moderator can be generated with a good flux yield at thermal and cold ranges, slightly peaking at 1.1 Å and at 2.4 Å.
ii. Using a water moderator (3 cm thick) and adding liquid methane (0.8 cm thick) yields good neutron spectra peaking at around 1.6 Å, the neutron flux is larger for wavelengths between water and para-H_2 peaks, which could be of interest to some kind of experiments.

3.2 Moderators: layout

The target and moderator shall be integrated in a beryllium reflector to increase the neutron moderation efficiency. The target-moderator-reflector (TMR) system, hosted within a cylindrical stainless-steel vessel, will confine the neutron stream in a very small volume (see Figures 6a and 6b).

When a 31.5 MeV proton beam impinges on the beryllium plate a cloud of neutrons fills the target vessel and clash within the reflector. Figure 6c focuses on neutrons at the epithermal range and below (E < 400 meV) because they can be more efficiently moderated to get useful thermal and cold neutrons. Neutrons above this energy are too fast and could not be moderated in this small target vessel.

Figure 6d shows the neutron flux for several energy ranges. Results reveal that there is a wide angular range with a very predominant presence of fast neutrons with E > 400 meV around the normal of the target surface (135°±30° with respect to the proton beam). This angular range should be avoided in the selection of the final location of the moderators and NBEX of the instruments since these high energy neutrons can go through the beam ports and reach the sample and the detectors, conveying background issues. For this reason, the four instrument beamports must be distributed at angles outside this range.

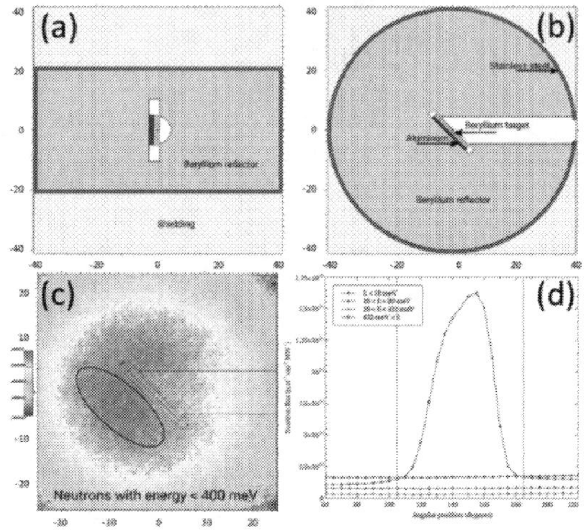

Fig, 6. (a) Side view and (b) top view of the MCNP model for ARGITU TMR; (c) and neutron cloud with energies lower than 400 meV; (d) Integrated neutron flux at several energy ranges depending on the angle with respect to the proton beam.

The selection of the instrument suite shall be made jointly with the local user community, and a final consensus is still to be reached. In any case, workhorse neutron scattering instruments, such as diffractometers, instruments to investigate large-scale structures (SANS and reflectometers) and spectrometers, share similar characteristics within a somewhat wide range, including the useful neutron wavelength range.[20] In any case, any potential selection should consider:

i. focusing on the neutron scattering techniques that could be more profitable for the scientific community.
ii. the technical limitations due to the moderate flux delivered by the source.

Based on the demands of the local scientific community, four instruments can be tentatively envisioned to be implemented in the target station of ARGITU: a thermal powder diffractometer, a single-crystal diffractometer (bispectral), a small angle scattering (SANS) instrument (cold), and a horizontal reflectometer (cold).

Figure 7 shows a proposed set-up of the four moderators and the four beamports for the four tentative neutron scientific instruments. Instrument moderator-beamports 1 and 3 (single crystal diffraction and neutron reflectometer) are above the equatorial plane (+Z) and beamports 2 and 4 (powder diffraction and SANS, respectively) are below the equatorial plane (-Z).

https://doi.org/10.1051/epjconf/202328602002

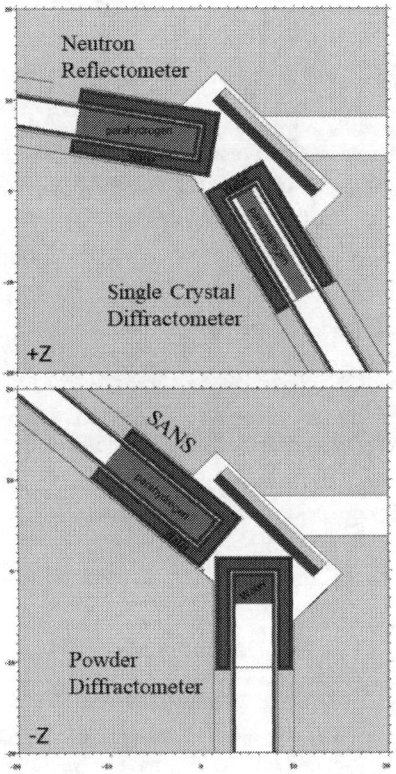

Fig. 7. MCNP layout for ARGITU concept of 4 neutron beam extraction lines (NBEX) designated for 4 different instruments (SANS, Reflectometer, Powder Diffractometer and Single Crystal Diffractometer)

Figure 8 depicts the neutron spectra of the four moderators at their downstream end surface (i.e., at the entrance of their respective NBEX). Results confirm a similar spectral shape and intensity values, with respect to the previous results for stand-alone moderators, except for the expected and intuitive slight reduction of the final neutron flux. Needless to say, the spectra provide a very stimulating result towards the development of the future neutron instrument suite of the ARGITU neutron source.

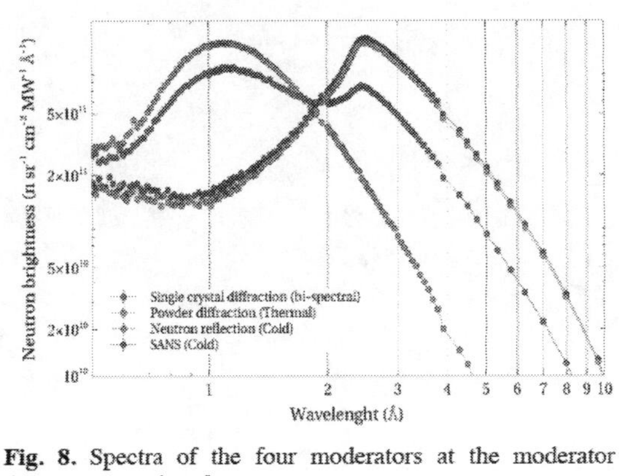

Fig. 8. Spectra of the four moderators at the moderator downstream end surface.

4 Conclusions

This work presents preliminary neutronics calculations for candidate moderators to be integrated in the target station of the ARGITU high-current accelerator-driven neutron source. After a first approach to the neutron yield of the Be target hit by the proton beam, next steps were oriented to find the best configuration, geometry, layout for 4 moderators. All moderators include a small wrapping layer of water (1 cm thick) that acts as a pre-moderator, providing a significant improvement in the efficiency of the neutron moderation, and thus their eventual neutron yield.

Results suggest that a long cylindrical (one-dimensional) para-H_2 moderator, with height of 13 cm is the ideal choice to deliver cold neutrons with $\lambda > 3$ Å, ideal for cold instruments like a SANS or a neutron reflectometer. Solid methane of 2-3 cm long is also a good candidate, that can be considered in, e.g., locations with particular spatial restrictions.

A 3 cm-long water moderator is considered as a suitable thermal moderator that can be integrated in, e.g., a thermal powder diffractometer.

Finally, a reduced-size cold moderator (like para-H_2 with smaller cylinder diameter of 3 cm diameter),[21,22,23] presents superior neutron yield at larger energy range for implementation in bispectral instruments (like, e.g., a single crystal diffractometer) compared to the hybrid thermal/cold (water/liquid methane) concept, that is only superior at a narrow wavelength range around $\lambda \approx$ 2 Å.

This study is ongoing and additional calculations are being carried out. The final design of the moderator will need to consider strong coupling between the moderator and the optics of the instrument that serves. However, as a first approach. ARGITU will enable the production of cold and thermal neutrons up to an adequate flux for the the performance of neutron scattering experiments, which will surely represent a significant stimulus for the local users community in the field of neutron science.

References

[1] T. Gutberlet, U. Rücker, E. Mauerhofer, P. Zakalek, T. Cronert, J. Voigt, J. Baggemann, J. Li, P. Doege, S. Böhm, M. Rimmler, O. Felden, R. Gebel, O. Meusel, H. Podlech, W. Barth & T. Brückel, Neutron News, **31**, 37 (2020).

[2] F. Ott, A. Menelle,C. Alba-Simionesco, EPJ Web Conf., **231** 01004 (2020).

[3] http://www.elena-neutron.eu

[4] M. Pérez, F. Sordo, I. Bustinduy, J. L. Muñoz, F. J. Villacorta, Neutron News, **31**, 19 (2020).

[5] I. Bustinduy, F.J. Bermejo, Phys. Procedia, **60**, 157 (2014)

[6] D.V. Baxter, J.M. Cameron, V.P. Derenchuk, C.M. Lavelle, M.B. Leuschner, M.A. Lone, H.O. Meyer, T. Rinckel, W.M. Snow, Nucl. Instrum. Meth. Phys. Res. B, **241**, 209 (2005)

[7] F. Maekawa, M. Harada, K. Oikawa, M. Teshigawara, T. Kai, S. Meigo, M. Ooi, S. Sakamoto, H. Takada, M. Futakawa, T. Kato, Y. Ikeda, N. Watanabe, T. Kamiyama, S. Torii, R. Kajimoto, Mi. Nakamura, Nucl. Instrum. Meth. Phys. Res. A, **620**, 159 (2010)

[8] T. Cronert, J. P. Dabruck, P. E. Doege, Y. Bessler, M. Klaus, M. Hofmann, P. Zakalek, U. Rucker, C. Lange, M. Butzek, W. Hansen, R. Nabbi, T. Bruckel, J. Phys.: Conf. Ser., **746** 012036 (2016)

[9] G. Muhrer, Nucl. Instrum. Meth. Phys. Res. A. **664**, 38 (2012)

[10] MCNP Version 6.2 Release Notes, Los Alamos National Laboratory report LA-UR-17- 29981 (2018)

[11] D.A. Brown, et al. 2018 Nucl. Data Sheets, **148**, 1 (2018)

[12] M. B. Chadwick et al., Nucl. Data Sheets, **107** 2931 (2006)

[13] Y. Wu, Fusion Sci. Technol., **74**, 321 (2018)

[14] Y. Wu, J. Song, H. Zheng, G. Sun, L. Hao, P. Long, L. Hu, Ann. Nucl. Energy, **82**, 161 (2015)

[15] F. Fernandez-Alonso, D. Price, Neutron Scattering - Fundamentals **49** (2013).

[16] W. B. Howard, S. M. Grimes, T. N. Massey, S. I. Al-Quraishi, D. K. Jacobs, C. E. Brient, J. C. Yanch, Nucl. Sci. Eng., **138**, 145 (2001)

[17] N. Ratcliffe, R. Barlow, A. Bungau, R. Cywinski, R. Edgecock, Conf. Proc. C, 1205201, 4154 (2012)

[18] I. Tilquin, P. Froment, M. Cogneau, Th. Delbar, J. Vervier, G. Ryckewaert, Nucl. Instrum. Meth. Phys. Res. A, **545**, 339 (2005)

[19] A.R. Páramo, F. Sordo, J.M. Perlado, A. Rivera, J. Nucl. Mater. **444**, 469 (2014).

[20] https://www.lens-initiative.org/wp-content/uploads/2021/02/LENS-Report-on-Low-Energy-Accelerator-driven-Neutron-Sources.pdf

[21] L. Zanini, K.H. Andersen, K. Batkov, E.B. Klinkby, F. Mezei, T. Schönfeldt, A. Takibayev, Nucl. Instrum. Meth. Phys. Res. A, **925** 33 (2019).

[22] M. Magán, F. Sordo, L. Zanini, S. Terrón, A. Ghiglino, F. Martínez, J.P. de Vicente, R. Vivanco, J.M. Perlado, F.J. Bermejo, F. Mezei, G. Muhrer, Nucl. Instrum. Meth. Phys. Res. A, **729**, 417 (2013).

[23] K. H. Andersen, M. Bertelsen, L. Zanini, E. B. Klinkby, T. Schonfeldt, P. M. Bentley, J. Saroun, J. Appl. Cryst., **51**, 264 (2018)

The High Brilliance neutron Source (HBS): A project for a next generation neutron research facility

Thomas Brückel[1*], *Thomas* Gutberlet[1], *Johannes* Baggemann[1], *Junyang* Chen[1], *Tania* Claudio-Weber[1], *Qi* Ding[1], *Monia* El-Barbari[1], *Jingjing* Li[1], *Klaus* Lieutenant[1], *Eric* Mauerhofer[1], *Ulrich* Rücker[1], *Norberto* Schmidt[1], *Alexander* Schwab[1], *Jörg* Voigt[1], *Paul* Zakalek[1], *Yannick* Bessler[2], *Romuald* Hanslik[2], *Richard* Achten[2], *Fynn* Löchte[2], *Mathias* Strothmann[2], *Olaf* Felden[3], *Ralf* Gebel[3], *Andreas* Lehrach[3], *Marius* Rimmler[3], *Holger* Podlech[4], *Oliver* Meusel[4], *Frédéric* Ott[5], *Alain* Menelle[5], and *Mariano Andrés* Paulin[5]

[1]Forschungszentrum Jülich GmbH, Jülich Centre for Neutron Science (JCNS-2 & JCNS-HBS), 52425 Jülich, Germany
[2]Forschungszentrum Jülich GmbH, Central Institute of Engineering, Electronics and Analytics (ZEA-1), 52425 Jülich, Germany
[3]Forschungszentrum Jülich GmbH, Nuclear Physics Institute (IKP-4), 52425 Jülich, Germany
[4]Goethe University, Institute of Applied Physics, 60438 Frankfurt am Main, Germany
[5]Centre d' Etudes de Saclay, Laboratoire Léon Brillouin CEA/CNRS, 91191 Gif sur Yvette Cedex, France

Abstract. The High Brilliance neutron Source (HBS) is a project for a next generation neutron research facility, based on new concepts and recent technological advancements. As elementary processes it uses neither fission nor spallation, but instead low energy nuclear reactions in a very compact Target-Moderator-Reflector (TMR) assembly. Our facility design results in very efficient production of neutron beams with high brightness. Key features of HBS are: (i) very competitive instrument performance, (ii) comparatively low construction and operation costs, (iii) resilience, (iv) sustainability, (v) flexibility, (vi) accessibility and (vii) scalability. Here we present the basic layout of the facility, elaborate on the mentioned key features and report on the commissioning of a small test setup.

1 Introduction

Large user facilities for research with neutrons, which offer instruments for neutron scattering, imaging, analytics, and nuclear physics or precision particle physics, are commonly based on fission in research reactors or spallation as the elementary processes to release neutrons from the atomic nuclei. In addition, there exist some smaller neutron sources, which use nuclear reactions after bombardment with electrons or rather low energy ions (mainly H^+ - protons, or D^+ at energies below 13 MeV). They are relatively small and are branded Compact Accelerator-driven Neutron Sources (CANS). Beams from such sources feature a neutron flux, which is orders of magnitude below the one of the larger user facilities. But CANS have an important local impact for specialized applications, training of users and method development. Japan hosts an entire network of CANS, organized in the Japan Collaboration on Accelerator-driven Neutron Sources (JCANS - http://jcans.net/) [1].

It is only in the last few years that an entirely new approach emerged to produce brilliant beams of cold, thermal, and epithermal neutrons. This approach is based on the same elementary process of low energy nuclear reactions but fully exploits the most recent technological developments in accelerator physics, target- and moderator technology, beam extraction, neutron optics and instrument design. Such facilities are labelled High Current Accelerator-driven Neutron Sources (HiCANS). While no HiCANS exists yet worldwide, several projects are being pursued in Europe aiming at the realization of such a facility. They are organized within the European Low Energy accelerator-based Neutron facilities Association (ELENA – http://www.elena-neutron.eu). HiCANS will produce pulsed neutron beams of high peak brilliance, comparable to the ones from spallation sources or research reactors. They have the potential to replace the former network of research reactors in Europe, as affirmed in the recent position paper of the League of advanced European Neutron Sources (LENS - https://lens-initiative.org/).

At first sight it might be surprising that the rather inefficient process of low energy nuclear reactions can produce neutron beams of such high brightness. As elaborated in [2], the main reasons are as follows:

* Corresponding author: t.brueckel@fz-juelich.de

 https://doi.org/10.1051/epjconf/202328602003

- A high current LINAC combined with a high-power target, resulting in a high reaction rate.
- An optimized proton pulse structure (pulse length and frequency) for the Target-Moderator-Reflector (TMR) assembly which feeds an adapted suite of instruments.
- An extremely efficient coupling of the thermal moderator to the compact target minimizing the loss of neutrons in the shielding.
- The use of highly efficient low dimensional "finger" moderators, which become part of the instrument, and deliver an optimized spectrum for the respective applications.
- A beam extraction optics starting very close to the moderator, so that a large phase space volume can be extracted and transported to the instrument when needed.

Due to these features, HiCANS have the potential to become the most efficient neutron facilities. Obvious savings with respect to spallation sources result from the low energy accelerator (MeV compared to GeV energy range) and reduced shielding. The most ambitious project for a HiCANS is being pursued at Forschungszentrum Jülich together with Helmholtz Zentrum Hereon and numerous partners from universities and research organizations. Here we report on the planned layout of the facility, labelled High Brilliance neutron Source (HBS - https://hbs.fz-juelich.de/) [3, 4], highlight the key features and report on the successful commissioning of a small test facility.

2 HBS facility layout

Fig. 1 shows the proposed layout of the HBS facility; table 1 lists some key parameters. These were chosen in a conservative approach as a compromise between state-of-the-art technological feasibility, operational practicability, cost, and maximization of neutron yield. For example, the proton accelerator is designed to operate in normal conducting technology and avoids expensive superconducting cavities. Although the neutron yield would still increase beyond a proton energy of 70 MeV, for reasons of operational economy, a superconducting section would be required after the normal-conducting cavities, such as is implemented at the ESS. A peak current of 100 mA for the selected accelerator type is challenging but feasible. It can provide the 100 kW thermal power per target station that our Ta target can still digest. Similar arguments apply to the choice of the other parameters in Table 1.

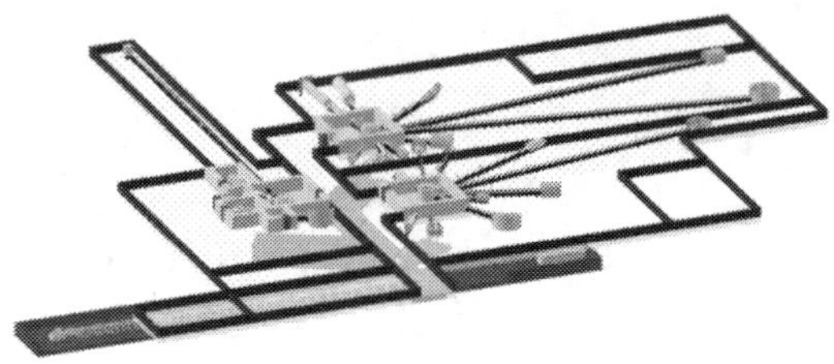

Fig. 1. Schematic layout of the HBS facility showing the LINAC in an underground tunnel (dark grey), the distribution of the proton beam to three target stations located in three experimental halls and a sketch of the proposed reference instrument suite.

The LINAC in an underground tunnel feeds the high-energy beam transport (HEBT) line with the underground multiplexer. The latter consists of a kicker-magnet and a three-sector septum magnet to separate an interlaced pulsed proton beam. The multiplexer supplies three target-moderator-reflector (TMR) units in a biological shielding with proton beams of different pulse structures. The beams impinge on the thin, flat, horizontal Ta target from below. The instruments are grouped around these TMR units in three experimental halls. The TMR unit in Hall 1 (front right in Fig. 1) operates at 96 Hz. It is supplied with proton pulses of 167 μs length, resulting in 250 μs thermal neutron pulse length. This pulse structure fits to instruments like diffractometers or direct geometry spectrometers. Hall 2 (back right in Fig. 1) hosts instruments that require low wavelength resolution e.g., small angle scattering, reflectometer, or spin echo spectrometer, but also diffractometer requiring pulse shaping. Its TMR unit operates at 24 Hz with a proton pulse length of 667 μs equal to the thermal neutron pulse length. A 96 Hz TMR station in Hall 3 (left in Fig. 1) is optimized to provide neutrons with energies above the thermal range. It hosts mainly imaging and analytics instruments. Each TMR station offers 12 beam ports. Each beamline hosts just one instrument. It can be equipped with an individual moderator, thus providing a beam with optimised pulse structure and neutron spectrum for the respective application. A complete reference suite of 25 instruments for diffraction, large-scale structures, quasi-, high-resolution and inelastic scattering, analytics, and imaging has been worked out by a large group of specialists (see acknowledgements).

Table 1. Key parameters of the HBS

Accelerator [5]	protons of energy 70 MeV, up to 100 mA peak current, beam duty factor of 6%, average proton beam power of 420 kW
High Energy Beam Transport (HEBT) [6]	multiplexer (kicker & septum magnet) distributing the beam to three TMR stations
Target [7]	Ta target with microchannel water cooling; 10x10 cm^2 active area; designed for 100 kW power
TMR stations [8]	TMR 2 : 24 Hz (667 µs p-pulse)) TMR 1 & 3 : 96 Hz (167 µs p-pulse) each TMR offers 12 beam ports
Moderators [9, 10]	thermal moderator: light water cold moderators: 1d «finger» moderators (solid mesitylene, solid methane, liquid para-hydrogen)
Instruments [11-13]	full suite of 25 instruments for scattering, imaging, and analytics

3 Key features of the HBS

Traditionally, neutron sources have been designed in such a way as to maximize the source strength. The number of free neutrons per second released by the source was a key performance indicator. For high flux sources, such as the HFR reactor of ILL or the FRM II reactor of MLZ, the source strength amounts to some $10^{18}\, n/s$. Such a high source strength leads to a large thermal flux of around $10^{15}\, n/(s \cdot cm^2)$ in a larger volume around the core, ideally suited for applications like doping of Si single crystals or the production of certain radioisotopes. However, only a minuscule fraction of the neutrons produced arrive at the sample for beamline instruments. These have typical flux values between 10^5 and $10^{10}\, n/(s \cdot cm^2)$, depending on the required beam tailoring. Thus, for beamline applications, reactor or spallation neutron sources are not very efficient, as they produce many neutrons which end up in the shielding. HiCANS, and especially the HBS, are based on fresh thinking and are following different design criteria. The HBS maximises efficiency and source brightness, the relevant quantity for beamline experiments.

Employing time-of-flight techniques with strong neutron pulses, a compact design of the TMR unit including shielding, and optimal moderation and beam extraction, the peak brilliance of HBS of around $10^{13}\, n/(cm^2 s Å sr)$ for para-H_2 at 0.3 nm wavelength, is comparable to some of the best neutron facilities, with clear exception of the flagship MW spallation sources. This performance is achieved with a source strength of about $10^{15}\, n/s$, three orders of magnitude smaller than in high flux reactors, which translates directly into cost savings for construction and operation. Due to the high source brightness, instrument performance at HBS compares well with some of the best existing instruments of their respective class. Table 2 presents some parameters for one selected instrument per instrument class.

Table 2. Some main parameters for a selection of instruments at HBS

Instrument type	Key features	Flux & resolution
Diffraction: macro-molecular diffractometer	• 96 Hz target station • p-H_2 moderator • Selene optics • beam spot at sample: $1x1\, mm^2$	$2 \cdot 10^7$ n/cm²s ±0.19 deg div.
Large scale structures: horizontal reflectometer	• 96 Hz target station • p-H_2 moderator • Selene optics • length 7m	10^7 n/cm²s 5.6% - 1.0% wavelength resolution
Spectroscopy: near backscattering spectrometer	• 24 Hz target station • solid methane mod. • primary flight path 85m	$7 \cdot 10^6$ n/cm²s 5 μeV energy resolution
Imaging: thermal neutron imaging	• 96 Hz target station • water moderator • high flux or high-resolution mode	$1 \cdot 10^7$ n/cm²s $L/D = 100$ in high flux mode

In addition to the efficiency, which translates into reduced costs, and the competitive performance of the instruments, the design of the HBS is characterized by several advantageous features:

- Resilience: HBS does not depend on the supply of regulated materials. It has a modular design. All components can be exchanged within hours or at most during a week maintenance time. To give some examples: The accelerator is driven by solid state amplifiers, which are redundant and can be easily exchanged in case of failure. For exchange of the target, a fully automated handling tool has been developed, which can pull the activated target out of the plug and insert a new target within a maintenance week. Same holds for the cold sources: a failure of a finger moderator affects only one instrument. During a maintenance week, the moderator insert can be exchanged from outside the TMR shielding.
- Sustainability: A concept for climate neutral construction is being worked out. Electricity from renewable energy sources is foreseen to drive the HBS accelerator. HBS does not use nuclear fuel and minimizes radioactive waste. The only heavily activated component is the target itself, about 5 kg of Ta. The remaining activity originates mainly from the decay of 182-Ta isotope with a half-life time of 114 days.
- Accessibility & flexibility: HBS will be licensed according to the German radiation protection ordinance (StrSchV) and not according to nuclear

law. This facilitates access for users from science and industry enormously. The relatively modest shielding allows for very flexible setups.
- Scalability: HiCANS are scalable, with respect to the proton energy (by adding additional Drift Tube Linac elements), the number of target stations (due to their rather low cost), and instruments.

4 JULIC neutron test platform

All critical components of HBS have been separately tested: (i) the accelerator is largely based on the MYRRHA (https://www.myrrha.be) design, (ii) cross sections for neutron release from different materials have been measured in absolute units [14], (iii) a prototype of the multiplexer has been realized and tested [6], (iv) the target has been tested up to the design value of $1\,kW/cm^2$ at an electron beam facility, (v) "finger" moderators for all moderator materials under consideration were built and tested. However, in order to verify the interplay of all components, to improve the design and perform method development, a test station has been realized on the premises of the Nuclear Physics Institute of Forschungszentrum Jülich. The shielding of the TMR unit features an automated opening mechanism, so that the thermal moderator and reflector can be accessed and exchanged to optimize the TMR arrangement. The setup will also be used to test the automated exchange of an activated target. First beam on target was achieved on December 12, 2022. A photo of the test setup is shown in Fig. 2.

Fig. 2. Photo of the JULIC neutron test platform after first beam on target. The detector test station is not visible as it is located behind the TMR unit.

A proton beam of 45 MeV energy and a current in the 100 nA range was provided by the JULIC cyclotron, which normally serves as injector for the COSY cooler synchrotron. The proton beamline incorporates several diagnostic tools and a prototype of the three-septum magnet, a key element of the multiplexer designed for HBS. In contrast to the HBS design, where the proton beam impinges the target from below, the proton beam enters the TMR unit of the test facility from the side. The TMR shielding shown in the photo is only slightly smaller than the one of HBS, which will also have an octagonal shape of 4 m in width and 3.2 m in height. At the test platform, neutrons were released from a prototype Ta target having the size and fishbone microchannel structure of the HBS target. The target is embedded in a Polyethylene (PE) thermal moderator and Pb reflector structure. Three beam port channels were equipped with instrumentation. A detector test station with a pinhole collimator received neutrons from the thermal moderator. This beam port was used to test the prototype of the SONDE detector for the SKADI instrument [15] of ESS in time resolved acquisition mode. Another beam port hosts the HERMES ToF-reflectometer. It is provided by the Laboratoire Léon-Brillouin and was previously situated at the now shut-down Orphée reactor. Reflectivity of a neutron guide section was measured during the commissioning of the test setup. The third beam port is equipped with a ToF- diffractometer. A solid methane CH_4 cold source feeds neutrons into a 6.5 m long ^{58}Ni neutron guide of cross section 30 x 50 mm^2. At the end of the guide a ToF detector and an angular dispersive diffractometer is located in a shielding box made of borated PE. Fig. 3 shows the neutron spectrum measured at different temperatures of the cold source.

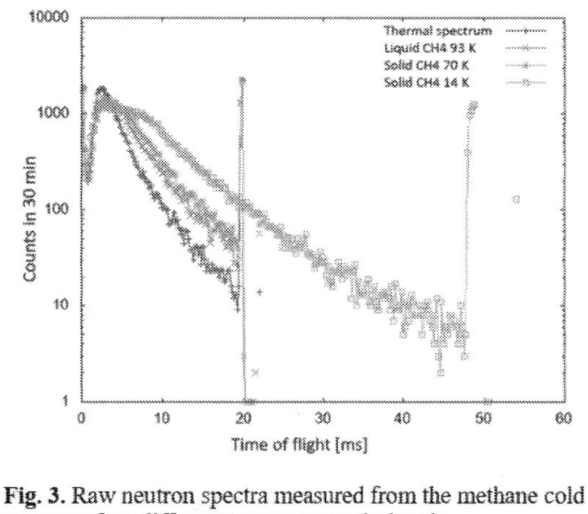

Fig. 3. Raw neutron spectra measured from the methane cold source at four different temperatures during the commissioning of the JULIC neutron test platform. The sharp peaks at the end of a TOF frame belong to the start of the next frame and result from the prompt pulse.

Clearly visible is the shift of the spectra towards longer ToF times, corresponding to longer wavelengths, with decreasing moderator temperature. Also visible is the double peak structure around thermal neutron energies. It comes from the fact that the wide neutron guide not only receives neutrons from the cold source, but also from the surrounding thermal moderator. This demonstrates that bispectral extraction is straight forward with such "finger" moderators. A clean cold spectrum can be obtained by placing a circular diaphragm in front of the neutron emitting surface of the cold moderator.

5 Summary, conclusions, and outlook

Projects for High Current Accelerator-driven Neutron Sources (HiCANS) are being pursued at several places within Europe. The LENS and ELENA Associations work hand in hand to make these facilities happen. HiCANS promise to become the next generation of neutron facilities, not as flagship sources, but as a replacement for the former network of reactor-based neutron facilities in Europe.

The HBS project in Germany is the most ambitious of the ongoing projects in terms of accelerator energy and power, number of TMR units and instrument suite. Simulations show that HBS will feature very competitive instrument performance, at comparatively low construction and operation costs. Resilience and flexibility through modular design, sustainability, accessibility, and scalability are clear plus points of the HBS concept. All critical components of the HBS have been tested. Recently the JULIC neutron test platform has been commissioned which serves to prove the interplay of all components, to test automated target exchange procedures, to experimentally determine instrument performance in absolute units and to serve as testbed for further method development.

Having developed a design for all components of the HBS and experimentally demonstrated the functionality of all critical components, we are now finalizing the Technical Design Report (TDR) of the HBS. It will be published in 5 volumes: (i) Accelerator, (ii) Target Stations and Moderators, (iii) Instrumentation, (iv) Infrastructure and Sustainability and (v) Summary. The TDR will demonstrate the technical feasibility of the HBS and be a blueprint for the construction of the facility. As with all major projects, implementation of the HBS requires a long lead time. Construction of the full HBS, as described in the TDR, will require 10 years from start of funding. However, the facility could also be implemented in stages, with the first stage having only one TMR unit in operation no sooner than four years after funding.

In summary, the facility developments within the ELENA Association provide an opportunity and starting point for the rejuvenation of the European neutron ecosystem based on a network of smaller local CANS, HiCANS as national facilities, and the ESS as the leading flagship facility for completely novel experiments.

Acknowledgements Funding by the Innovation Pool Project of the Helmholtz Society and the Federal Ministry of Education and Research is gratefully acknowledged. We are thankful to many colleagues from different universities and research centres who have contributed to the design of instruments: R. Bewley, J. Fenske, M. Feygenson, C. Franz, H. Frielinghaus, M. Ganeva, A. Glavic, S. Jaksch, N. Kardjilov, G. Kemmerling, H. Kleines, Z. Ma, S. Mattauch, S. Pasini, K. Schmalzl, T.E. Schrader, W. Schweika, M. Strobl, E. Vezhlev, N. Violini, R. Zorn.

References

1. Y. Kiyanagi, Eur. Phys. J. Plus **131**, 132 (2016)
2. T. Brückel, T. Gutberlet, S. Schmidt, C. Alba-Simionesco, F. Ott, A. Menelle, Neutron News **31**, 13 (2000)
3. T. Gutberlet et al., Neutron News, **31**, 37 (2020)
4. T. Brückel, T. Gutberlet (Eds.) *Conceptual Design Report Jülich High Brilliance Neutron Source (HBS),* Schriften des Forschungszentrum Jülich, General, Vol.8 (2020)
5. H. Podlech et al., *Conceptual Design of the Proton LINAC for the High Brilliance Neutron Source HBS*, Proc. IPAC'19, Melbourne, Australia, 910-913 (2019)
6. M. Rimmler, O. Felden, U. Rücker, H. Soltner, P. Zakalek, R. Gebel, T. Gutberlet, T. Brückel, J. Neutr. Res. **23**, 143 (2021)
7. P. Zakalek, P.-E. Doege, J. Baggemann, E. Mauerhofer, T. Brückel, EPJ Web of Conf. **231**, 03006 (2020)
8. P. Zakalek, J. Li, S. Böhm, U. Rücker, J. Voigt, E. Mauerhofer, T. Gutberlet, T. Brückel, J. Neutr. Res. **23**, 185 (2021)
9. T. Cronert, J.P. Dabruck, M. Klaus, C. Lange, P. Zakalek, P.-E. Doege, J. Baggemann, Y. Beßler, M. Butzek, U. Rücker, T. Gutberlet, R. Nabbi, T. Brückel, Physica B: Condensed Matter **551**, 377 (2018)
10. S. Eisenhut, M. Klaus, J. Baggemann, U. Rücker, Y. Beßler, A. Schwab, C. Haberstroh, T. Cronert, T. Gutberlet, T. Brückel, C. Lange, EPJ Web of Conf. **231**, 04001 (2020)
11. Z. Ma, K. Lieutenant, J. Voigt, T. Gutberlet, M. Feygenson, T. Brückel, Nucl. Instr. Meth. A, **1009**, 165479 (2021)

https://doi.org/10.1051/epjconf/202328602003

12. P. Zakalek et al, JPS Conf. Proc. **22**, 011025 (2018)
13. J. Voigt, S. Böhm, J.P. Dabruck, U. Rücker, T. Gutberlet, T. Brückel, Nucl. Instr. Meth. A, **884**, 59 (2018)
14. M. Rimmler, J. Baggemann et al., Nucl. Instr. Meth. A, **990**, 164989 (2021)
15. S. Jaksch, D. Martin-Rodriguez, A. Ostermann, J. Jestin, S.D. Pinto, W. Bouwman, J. Uher, R. Engels, H. Frielinghaus, Nucl. Instr. Meth. A, **762**, 22 (2014)

https://doi.org/10.1051/epjconf/202328602004

The High Brilliance Neutron Source Target Stations

Paul Zakalek[1,*], *Richard* Achten[2], *Johannes* Baggemann[1], *Yannick* Beßler[2], *Fabian* Beule[1], *Thomas* Brückel[1], *Junyang* Chen[1], *Qi* Ding[1], *Monia* El-Barbari[1], *Ralf* Engels[1], *Olaf* Felden[3], *Ralf* Gebel[3,5], *Kirill* Grigoryev[5], *Thomas* Gutberlet[1], *Romuald* Hanslik[2], *Vsevolod* Kamerdzhiev[3], *Peter* Kämmerling[1], *Harald* Kleines[1], *Jingjing* Li[1], *Klaus* Lieutenant[1], *Fynn* Löchte[2], *Eric* Mauerhofer[1], *Mariano Andrés* Paulin[4], *Ivan* Pechenizkiy[1], *Ulrich* Rücker[1], *Norberto* Schmidt[1], *Alexander* Schwab[1], *Alexander* Steffens[1], *Fréderic* Ott[4], *Yury* Valdau[5], *Egor* Vezhlev[1], and *Jörg* Voigt[1]

[1]Forschungszentrum Jülich GmbH, Jülich Centre for Neutron Science (JCNS), 52425 Jülich, Germany
[2]Forschungszentrum Jülich GmbH, Central Institute of Engineering, Electronics and Analytics (ZEA-1), 52425 Jülich, Germany
[3]Forschungszentrum Jülich GmbH, Nuclear Phsyics Institute (IKP-4), 52425 Jülich, Germany
[4]Centre d' Etudes de Saclay, Laboratoire Léon Brillouin CEA/CNRS, 91191 Gif sur Yvette Cedex, France
[5]GSI Helmholtzzentrum für Schwerionenforschung GmbH, Planckstraße 1, 64291 Darmstadt, Germany

Abstract. High Current Accelerator driven Neutron Sources (HiCANS) are a cost-efficient alternative for future large scale neutron facilities. They excel in transforming neutrons released by the primary nuclear reaction into a spectral range usable for applications. In particular, the cost of a target station represents only a minor fraction of the overall construction cost due to the lower energy of the primary neutrons, which requires less shielding. They can be designed to provide optimized pulse and spectral properties for applications in neutron scattering, analytics and imaging experiments. For the High Brilliance neutron Source (HBS) project at Forschungszentrum Jülich, we have developed a modular design that meets the radiation protection requirements while providing sufficient space in its core to adapt the target-moderator-reflector assembly to the different applications. In the following, we present the basic target station design which will be used at HBS for three different realizations with their own instrument suite. All relevant components have been designed, built and tested at the JULIC Neutron Platform which has produced neutrons since December 12th 2022. The simulated performance of a target station shows that the brightnesses of the moderators are in the range of modern research reactors and sub-MW power spallation sources.

1 Introduction

Research reactors like ILL in Grenoble, FRM II in Garching or HFIR in Oak Ridge release neutrons in a single core that is surrounded by moderators providing neutrons for experiments in multiple experimental areas. Increasing the experimental area for instruments surrounding the research reactor is realized by complex neutron transport systems, providing identical source properties for a large group of instruments. At short pulse spallation neutron sources like ISIS in Didcot and SNS in Oak Ridge, different moderators were developed to provide dedicated pulse properties for different applications. As individual beam ports feed only individual instruments, the number of instruments per target station is limited. The upgrade with a second target station done at ISIS [1] and planned at SNS [2] is a significant investment due to severe shielding requirements against the high energy particles produced during the spallation reaction. Compact Accelerator-driven Neutron Sources (CANS) like LENS in Bloomington or RANS in Wako, which are mainly used for education and methods developments, release neutrons via nuclear reactions at low proton beam energy in the MeV range requiring less shielding than spallation neutron sources thus having a low price tag. They however serve only a limited number of in-house users at a time but can therefore adapt the proton pulse properties for a specific application.

High-Current Accelerator-driven Neutron Sources (HiCANS) benefit also from the lower radiation protection requirements but aim at a high availability for a large number of users. Therefore the provision of multiple target stations, whose cost represents only a minor fraction of the overall investment, is considered for projects like the High Brilliance neutron Source (HBS) project or the ICONE project [3] from the very beginning. This requires a target station design which is modular and adaptable to the specific requirements. This allows a further optimization of the target station and tailoring of the neutron beams to its instruments. In the following, the basic target station design for the HBS project is presented which is planned to be built in three stages.

2 High brilliance neutron source project

The HBS project team is developing a HiCANS concept which should serve as a national user facility in Germany. The proposed layout is shown in Figure 1 with three target

*e-mail: p.zakalek@fz-juelich.de

 https://doi.org/10.1051/epjconf/202328602004

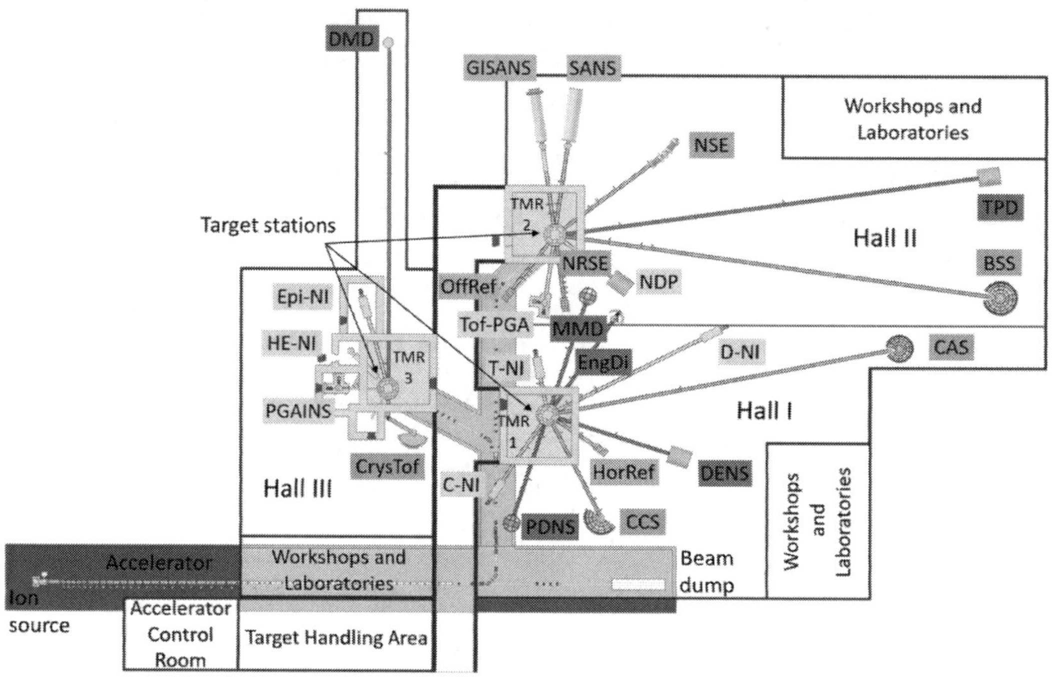

Figure 1. Proposed HBS facility with three target stations and 24 instruments.

Table 1. Main parameters for the HiCANS developed in the HBS project

Accelerator	70 MeV proton beam 100 mA peak current 20% RF-duty cycle
Target	Ta target with micro channel structure 100 kW average power deposiiton 10^{15} n/s source yield
TMR stations	3 individual TMR stations TMR 2: 24 Hz (667 μs pulse) TMR 1 + 3: 96 Hz (167 μs pulse)
Moderators	thermal moderator: light water cryogenic moderators: solid methane, solid mesithylene, liquid para-hydrogen
Instruments ´	TMR station optimized to instruments 6 - 12 instruments per TMR station scattering, imaging and analytics

stations and 24 instruments. The main parameters of this facility are summarized in Table 1.

The instruments are grouped regarding their requirements and built at optimized target stations. It was chosen to operate a target station (TMR-2) served by a 24 Hz proton beam with pulses of 667 μs length and two target stations (TMR-1 & TMR-3) served by a 96 Hz proton beam with pulses of 167 μs length. Instruments requiring a broad band of neutron energies are served by TMR-1 featuring a low frequency. The long neutron pulses of 667 μs makes it also the best choice either for low wavelength resolution applications, which can accommodate the full pulse, or for applications that require pulse shaping to perform very high resolution experiments. The other two target stations provide good resolution already using the full pulse of the target stations. The applications grouped around these target stations require typically a limited bandwidth. The difference between TMR-1 and TMR-3 is the focus on different applications and therefore energy ranges. TMR-1 serves mostly scattering instruments in the thermal and cold neutron energy range like diffractometers or direct geometry spectrometers and TMR-3 serves mostly analytical instruments in the epithermal and thermal energy range like imaging and PGNAA.

In order to minimize the development requirements for a facility with multiple target stations with different focuses, it was chosen to design a target station that can be used as a basis for different target station realizations.

3 Basic target station design

A target station at a HiCANS has to provide tailored neutron beams with the requested energy spectrum and time structure to its instruments. Furthermore, it needs to shield the surrounding target room and its equipment from radiation leakage, both during operation and after operation, due to the activated components to enable access to the the target room. As multiple target stations are planned for a HiCANS, the basic design needs to be easily modifiable and adjustable to the instrument suite and the proton pulse structure. Finally, the number of instruments supplied with neutrons by the target station needs to be maximized.

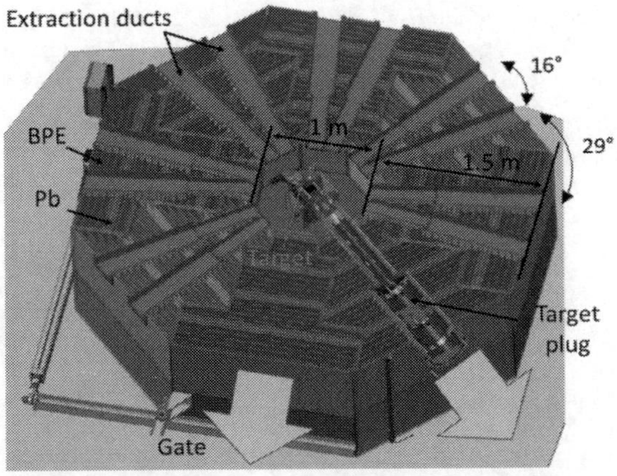

Figure 2. Basic design of the target station

In order to fulfill these requirements, the basic target station design has the following design features:

- Layered shielding of borated polyethylene and lead
- Vertical proton beam providing 360° for instrumentation and access to the target station
- Spacious inner core reserved for optimized moderator-reflector assembly to specific instrument suite requirements
- Large extraction ducts allowing an extraction of neutrons from different layers of the target-moderator-reflector assembly thus allowing fast, epithermal, thermal and cold neutron extraction
- Instrument specific moderators further tailoring neutron beams by optimized thermal or cryogenic moderator plugs
- Small neutron producing target resulting in efficient coupling of neutron production, moderation and extraction

The basic target station design for the HBS project is shown in Fig 2 which has an octagonal structure with a total width of 4 m and a height of 4.5 m. It features 4 double layers of borated polyethylene and lead with a total thickness of 1.5 m leaving an inner octagonal space with a width of 1 m for the moderator-reflector assembly. The octagonal structure is separated into 8 individual sections, one section dedicated for target access, one section that can be opened to allow access to the inner core and 6 sections usable for neutron extraction with two extraction ducts in each section.

3.1 Neutron producing target

The target is made of a single stationary structure consisting of tantalum as shown in Fig. 3. The total thickness is 21 mm, the width is 112 mm and the length is 120 mm with an irradiated area of 100 x 100 mm^2. In order to efficiently cool the 100 kW power injected by the proton beam in the small volume, the target is cooled with a micro channel

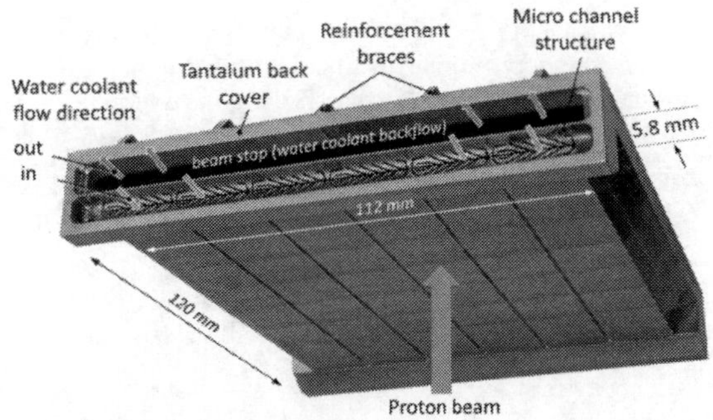

Figure 3. Tantalum target at HBS

cooling structure with a channel width of 0.35 mm inside the neutron producing layer allowing heat removal as high as 1 kW/cm^2. The neutron producing layer with the inserted cooling structure has a total thickness of 5.8 mm resulting in a neutron production within a volume of less then 100 ml thus maximizing the brightness and minimizing the activated amount of material. The exchange of the target is a standard operation that can be accomplished during regular maintenance periods. The time averaged neutron yield for a single target station is ~ 10^{15} n/s.

Due to the irradiation with protons and the high radiation level inside the target station, the target will undergo irradiation-induced damage. The estimated life-time of the tantalum target is 2.6 years [4] mainly due to the high proton induced displacement per atom (DPA). For the HBS project, the target change is planned annually, to have a high safety margin.

3.2 Target extraction

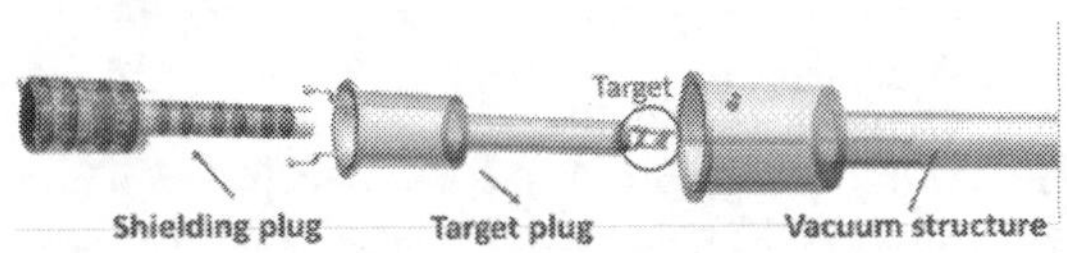

Figure 4. Target plugs

The exchange of the target is enabled by a plug-in-plug system as shown in Fig. 4. A vacuum structure made of aluminium is placed in one segment of the target station and shares the vacuum with the accelerator. This allows a windowless design preventing energy loss of the proton beam and minimizing radiation background. The target is placed at the front of the target plug made of aluminium with water pipes being fed through. The target plug closes the vacuum outside the target station which allows breaking the vacuum under safe radiation levels. A shielding plug is inserted into the target plug with the same layered structure of lead and borated polyethylene as the shielding of the target station.

The target plug together with the shielding plug can be extracted with an appropriate mechanism from the outside after the vacuum is broken. A mobile shielding is placed around the target during the extraction for transport to a storage or target handling area. The high activation of the Ta target decays with a half-life of Ta^{182} of 115 days, after which the target can be separated from the plug structure and decommissioned. This reduces the nuclear waste, that has to be decommissioned, to a mass of 3.5 kg.

3.3 Extraction ducts and moderator plug

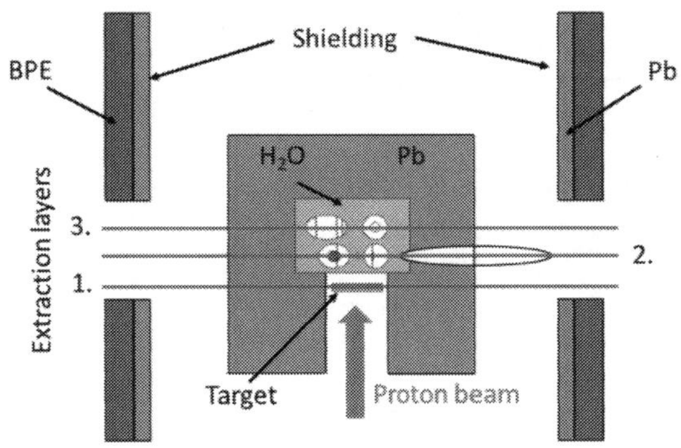

Figure 5. Target-moderator-reflector assembly with illustration of three extraction planes with a thermal light water moderator and a lead reflector.

The optimal extraction volumes for cold, thermal and epithermal neutron beams differ for each target station. Hence neutrons are extracted from different positions of the target-moderator-reflector assembly. This is realized by a large extraction duct through the shielding of the target station (Figure 2) with an inner width of 200 mm and an inner height of 400 mm, where the neutron guide can be placed to have an ideal view towards the extraction volume. The target is placed in the lower third of the extraction ducts height allowing to extract neutrons directly either from the target or from the moderator placed in front of the target as shown in Figure 5 with an example of three extraction layers.

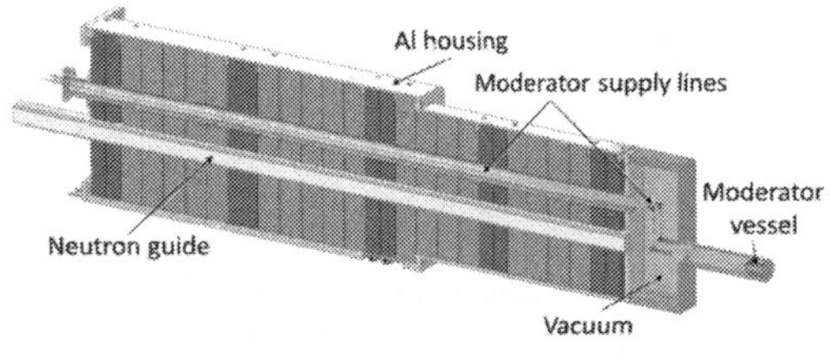

Figure 6. Moderator plug, which can be inserted into the extraction duct.

These extraction ducts are filled with instrument specific moderator plugs including the supply lines and neutron guides as shown in Figure 6. The guide has an undisturbed view to the moderator surface and the neutron spectra can be tailored to the instrument requirements as the moderator in the front can be changed. In such a design, the optical components like neutron guides can be placed as close as 400 mm from the moderator surface.

3.4 Moderator-reflector assembly

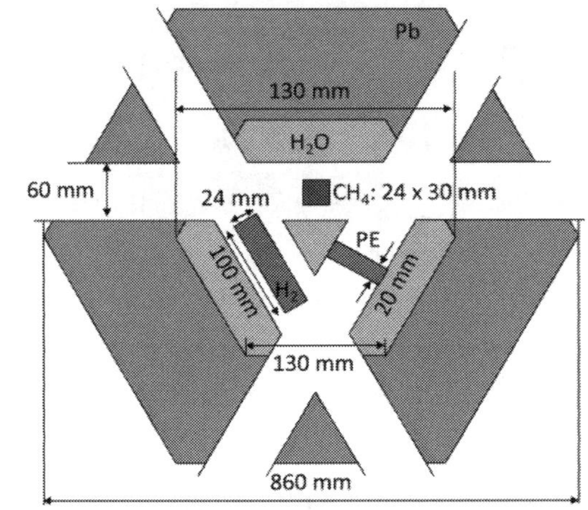

Figure 7. Example of moderator-reflector assembly with a cut through extraction plane 2 indicated in Figure 5 with a thermal polyethylene moderator and two cryogenic moderators, namely liquid hydrogen and methane.

In the core of the target station, different moderator-reflector assemblies can be tailored to the proton pulse structure and the instrument suite. In Figure 7, an example shows a cut through the second extraction plane of a moderator-reflector assembly. It uses a light water moderator to thermalize the primary neutrons coming from the target and to feed the moderators from the extraction ducts. A lead reflector around the thermal moderator feeds unmoderated neutrons back into the thermal moderator and acts also as the first shielding layer against fast neutrons and gammas. Three different moderators are used inside the extraction ducts, one polyethylene moderator to extract thermal neutrons and two cryogenic moderators (liquid hydrogen and methane) to extract cold neutrons.

The peak brightness for the different moderators calculated with MCNP6 is shown in Figure 8. It is in the range of 10^{13} $s^{-1}sr^{-1}Å^{-1}cm^{-2}$ which is comparable to modern research reactors and sub-MW power spallation sources. By choosing the moderator type, it is possible to shift the spectrum from thermal at 1.3 Å for polyethylene, cold at 3 Å for liquid hydrogen or a cold spectrum with a large number of thermal neutrons for methane due to its small volume.

3.5 Target test station

A mock-up of the target station for HBS was built at the JULIC cyclotron in Jülich as shown in Figure 9. It serves as a testbed for all critical components for the HBS like

https://doi.org/10.1051/epjconf/202328602004

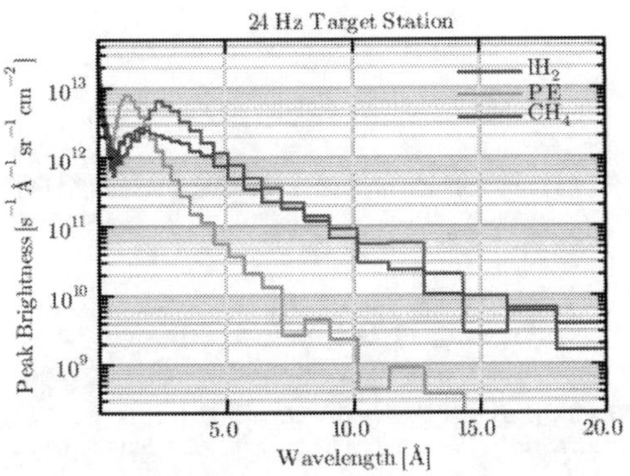

Figure 8. Simulated brightness for the moderator-reflector assembly for three different moderators.

target, cryogenic moderators, TMR assembly, target handling as well as the operation of a target station served by a low energy proton beam. The target station has reduced shielding dimensions compared to the HBS target station from 2 m to 1.5 m and the proton beam enters the target station horizontally instead of vertically.

On the 12th of December 2022, the first beam on target was realised with three experimental setups: a diffractometer setup at an extraction duct equipped with a cryogenic methane moderator, the HERMES reflectometer provided by CEA using a thermal extraction duct, which will be later upgraded with a liquid hydrogen moderator, and a detector test station looking at the thermal polyethylene moderator with a pin-hole camera with a width of 6 mm inserted into the extraction duct. The target test station together with the instruments is named JULIC Neutron Platform.

Figure 9. Target test station at the JULIC cyclotron.

Further optimization of the target test station with additional experiments are planned for 2023. It is planned to equip two extraction ducts with a thermal imaging station and an epithermal / fast imaging station. The planned experiments will allow to validate the simulation results obtained by Monte Carlo codes like MCNP, FLUKA, McStas or VITESS. Furthermore it will prove the feasibility of an accelerator-driven neutron source as the instrument performances can be scaled to the full fledged facility.

4 Summary, conclusion and outlook

The target station is the core of a HiCANS as it tailors the neutron beams to match requirements of specific instruments. The operation of multiple target stations at a HiCANS allows an unprecedended integral approach for the instrument design. At the same time, the modular design that is presented here allows the multiplexing of a whole target station including target, moderator-reflector assembly, shielding and moderator plugs at a cost of less than 10 M€. Dedicated target station setups can be derived from this basic design. It offers the flexibility by having a common outer shielding structure with an empty inner core. The inner core can be adjusted to the proton beam structure as well as the instrument suite optimizing the neutron beams to the instrument requirements. Large extraction ducts allow the extraction of the desired neutron beams from different positions of the moderator-reflector assembly and allow further tailoring of the neutron beams by inserting optimized moderators at the front of moderator plugs.

All components required for a target station have been designed, built and tested at the JULIC Neutron platform. The target test station allows further optimization of the critical components as well as to test the interplay between all of them. A much more detailed description will be published soon in a Technical Design Report (TDR). The performance of a HiCANS using such a target station was estimated for many instruments and will be published in a separate volume of the TDR. It shows that the flux at the sample position is comparable to modern research reactors or sub-MW power spallation sources with a very attractive price tag and a low background. Both the instruments and the facility allow an easy access thanks to the lower radiation levels.

The investigation of an accelerator-driven neutron source as well as the performed experiments indicate that a HiCANS is a valid option as a third corner stone of a healthy neutron landscape together with reactor neutron sources and spallation neutron sources ensuring neutron beam access for a large user community.

References

[1] A. Taylor, Physica B: Condensed Matter **276-278**, 36 (2000)

[2] R.K. Crawford, R. Dean, P. Ferguson, J. Galambos, F. Gallmeier, T. McManamy, M. Rennich, Journal of Physics: Conference Series **251**, 012054 (2010)

[3] Ott, Frédéric, Menelle, Alain, Alba-Simionesco, Christiane, EPJ Web Conf. **231**, 01004 (2020)

[4] N. Ophoven, E. Mauerhofer, J. Li, U. Rücker, P. Zakalek, J. Baggemann, T. Gutberlet, T. Brückel, C. Langer, Applied Physics A **127**, 576 (2021)

Thermal moderator-reflector design of the 24 Hz target station for the High Brilliance Neutron Source

Junyang Chen[1,*], *Ulrich* Rücker[1], *Jörg* Voigt[1], *Paul* Zakalek[1], *Egor* Vezhlev[2], *Jingjing* Li[1], *Thomas* Gutberlet[1], and *Thomas* Brückel[1]

[1]Jülich Centre for Neutron Science (JCNS-2/PGI-4), Forschungszentrum Jülich GmbH, 52428 Jülich, Germany
[2]Jülich Centre for Neutron Science (JCNS-4) at Heinz Maier-Leibnitz Zentrum (MLZ), Lichtenbergstr. 1, 85748 Garching, Germany

Abstract. The High Brilliance Neutron Source (HBS) project develops a high current accelerator driven neutron source (HiCANS), aiming to substitute existing neutron research reactors, that reach the end of life. This study focuses on the thermal neutron moderator design for a low frequency, long pulse target station of HBS. We investigate the neutronic characteristics of the D_2O+H_2O mixture moderator material using the Monte Carlo particle transport code PHITS. Our findings show that, compared to D_2O with lower neutron absorption, confinement of H_2O is the dominant factor in achieving a high neutron flux within the moderator and thus a high neutron brightness. We find that surrounding the central thermal water moderator with a D_2O+H_2O mixture can achieve a better balance between confinement and absorption. However, because of multiple angles of reflecting neutron, this design will also increase the divergence of the neutron beam.

1 Introduction

Neutrons are critical in understanding the structure and dynamics of matter and materials, leading to a high demand of neutron scattering and analytical experiments. The advent of high-current accelerator systems launched the development of high-current accelerator-driven neutron sources (HiCANS) utilizing low energy nuclear reactions. High current accelerator-driven neutron sources (HiCANS), are cost-effective to construct and to operate. At current there are mainly three HiCANS projects which are the High Brilliance Neutron Source (HBS) project [1] located in Jülich, the ICONE project developed by Université Paris Saclay [2] and the ARGITU project [3] located in Basque Country area (Spain). These developments can counteract the loss of existing fission-based neutron sources and a resulting decline in available neutron beam days as well as establishing HiCANS as a next generation neutron source.

Despite the lower primary neutron yield of the nuclear reactions compared to reactor or spallation neutron sources, HBS achieves a competitive neutron brightness by a compact moderator and reflector design, which makes a large fraction of the primary neutron spectrum available for thermal and cold neutron applications [1]. The proton power of HBS is 100 kW. The peak current is 100 mA. The HBS project features three target stations with different pulse frequency (24 Hz, 96 Hz and 96 Hz) and proton pulse length (667 μs, 167 μs, 167 μs) [1]. Also, the spectral characteristics are tailored for the instruments hosted at a specific target station. In this work, we will focus on the thermal moderator design for the 24 Hz target station, which provides a thermal neutron beam for many instruments.

Two crucial components of the neutron source design are the moderator, which moderates fast neutrons to the expected energy range, and the reflector, which reflects neutrons, which escape from the moderator, to increase the neutron intensity. The neutron flux, the neutron brightness and time structure of the neutron pulse are important metrics for evaluating the moderator-reflector design. The time distribution of the neutron pulse is determined by the proton pulse shape and the time the neutron spends inside the moderator and in the reflector. When high-energy neutrons collide with nuclei in the moderator, they lose energy until they reach thermal equilibrium. The time spent by neutrons moderating and diffusing inside the moderator can be broken down into two main components: the first is the thermalization time, which is the time taken for the neutron to decrease from the fast energy region to the thermal or cold energy region; the second is the residence time, which is the time the neutron can diffuse inside the material before it leaves the moderator volume or is absorbed [4]. Moderator materials with high elastic scattering cross section, high logarithmic energy reduction, and lower absorption cross section should be considered to achieve high neutron brightness. The nuclides H and D are considered as ideal candidates for moderating neutrons, and H_2O and D_2O are commonly used as moderator materials [4]. Due to its higher elastic cross section, H_2O moderates the neutrons faster and in a smaller volume yielding a higher density of the neutron cloud and a larger neutron flux in the center of the cloud, i.e. the volume, in which the neutrons travel before reaching thermal equilibrium. Compared to H_2O, D_2O has a smaller absorption cross section, making it easier for

*Corresponding author: ju.chen@fz-juelich.de

https://doi.org/10.1051/epjconf/202328602005

neutrons undergoing multiple collisions to survive longer in D_2O exploring a larger volume. The use of D_2O results in a longer tail for the neutron pulse. Alternatively, the use of H_2O provides a thermal neutron lifetime (time interval from neutron generation to absorption) below 200 μs, which is significantly shorter than the proton pulse length at the 24 Hz target station. That means the neutron lifetime can be relaxed by decreasing absorption. Therefore, using pure D_2O or pure H_2O as thermal moderator materials seems not an optimal solution. Instead, a mixture of D_2O and H_2O is expected to achieve higher neutron cloud density with suitable escaping time and lower absorption.
This work will first illustrate the research methods and related geometry, followed by an investigation into the neutronic characteristics for a mixture of materials as a thermal moderator. Finally, a new design of moderator and reflector that places H_2O and the D_2O+H_2O mixture in different geometries is explored.

2 Method

In this study, the Monte Carlo simulation code PHITS 3.27 is used for the neutron transport simulation. The simulations are based on the evaluated neutron cross section library ENDF/B-VII.0 and lwtr.20t and hwtr.20t for thermal scattering kernels [5]. A neutron source generated in advance created with a 70 MeV proton beam distributed over the surface of a cylindrical Ta target with 10 cm diameter and 0.5 mm thickness was used for the simulations, which can avoid tracing of charged particles. Behind the Ta target is a 0.5 cm water (H_2O) layer, which in reality serves for cooling the target. In this study, the ratio of the D_2O+H_2O mixture material is 70% D_2O and 30% H_2O realizing a mean lifetime of approximately 700 μs. Only a mixture 70% D_2O and 30% H_2O has been studied. The energy range for tally is from 1 meV to 500 meV.
For investigating the neutron flux inside H_2O and the D_2O+H_2O mixture material, the following geometry model is used as shown in Fig.1. The moderator has a cylindrical shape. The width and the thickness of the H_2O moderator and the D_2O+H_2O mixture moderator are optimized to maximize the neutron flux in the center of moderator. The width and thickness of the H_2O moderator is 18 cm × 14 cm. The width and thickness of the D_2O+H_2O mixture moderator is 22 cm × 22 cm. There are two [T-Track] tallies [6], one is put at the center of the moderator, called tally 1, the other is placed in the outside area at the same normal distance, but shifted 5 cm outwards within a plane parallel to the target plane, called tally 2, as shown in Fig.1. The maximum neutron flux along the depth direction was investigated initially, indicating an optimal depth of 2 cm for H_2O and 6 cm for the D_2O+H_2O mixture material. Thus, the normal distance from the target of tally 1 and tally 2 is set to 2 cm for H_2O and 6 cm for the D_2O+H_2O mixture material above the target. For investigating the effect of deuterium containing material surrounding a central water moderator, the geometries as given in Fig.2 have been simulated. In this design, a

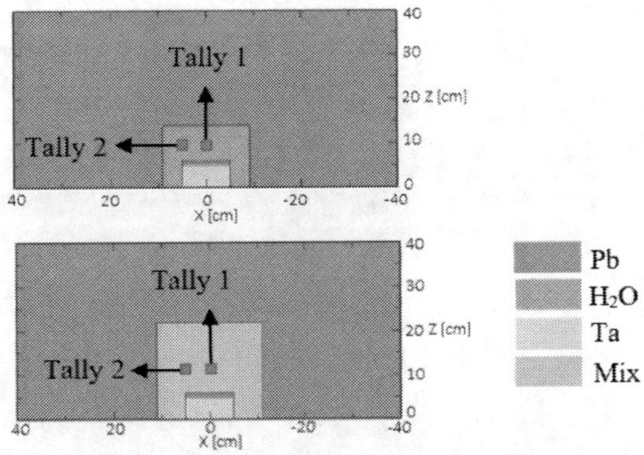

Fig. 1. Geometry model for the water moderator (upper) and mixture moderator (below).

single extraction channel (diameter 6 cm in cylindrical shape) is added into the center of the thermal moderator. The center of the extraction channel is 5 cm away from the target. An additional water cylinder is put in the center of the extraction channel, which acts as "scatterer" to increase the neutron emission along the channel.

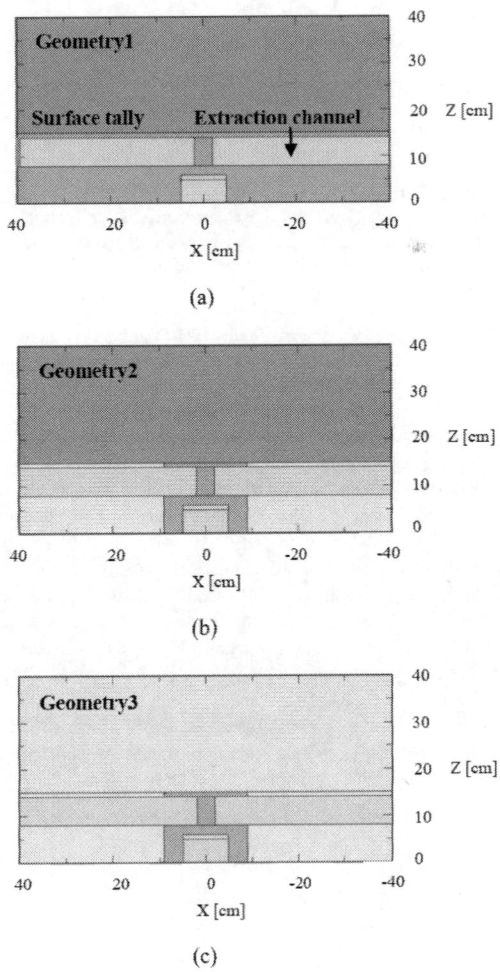

 https://doi.org/10.1051/epjconf/202328602005

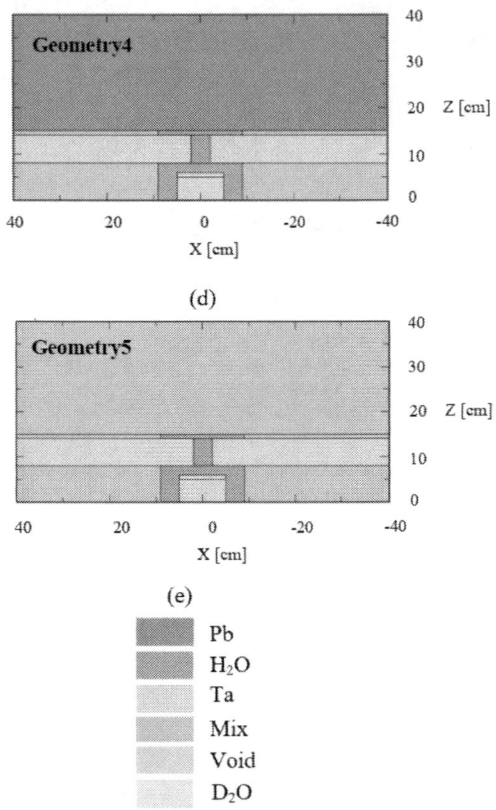

Fig. 2. Different geometry for water and D_2O+H_2O mixture moderator. (a) Geometry 1: water is used as moderator; (b) Geometry 2: D_2O is used around the centered water moderator; (c) Geometry 3: D_2O is used around the center water moderator and D_2O is used as reflector; (d) Geometry 4: D_2O+H_2O mixture material is used around the centered water moderator; (e) Geometry 5: D_2O+H_2O mixture material is used around the centered water moderator and D_2O+H_2O mixture material is used as reflector.

As shown in Fig.2(a) a water layer is surrounding the channel and Pb is used as a reflector. In Fig.2(b)-(e), different surrounding materials and reflector materials are used. In Fig.2(b)-(c), D_2O is used for the surrounding material, whereas in Fig.2(d)-(e) a D_2O+H_2O mixture material is used for it. As for the reflector, Pb is used just above the channel (Fig.2(b) and Fig.2(d)), while D_2O and D_2O+H_2O mixture material is used above and behind the channel (Fig.2(c) and Fig.2 (e)). The surface tally is put at the end of the channel (40 cm from centre). The tally records the neutrons emitted into 0°-90°, i.e. the full space in increasing x-direction, and into 0°-5°, which corresponds to a collimation that we consider a upper limit to be extracted by an instrument.

3 Results

The neutron flux versus time related to the geometry in Fig.1 is shown in Fig.3. For tally 1, the flux reached at saturation for H_2O is two times larger than in the mixture case. For tally 2, the saturated value for H_2O is 1.6 times larger than in the mixture case. This result indicates that with H_2O one can obtain a larger neutron flux even though the neutron pulse decays faster than in the D_2O+H_2O mix's case. As described above, H_2O confines the neutron cloud stronger. This effect cannot be compensated by the lower absorption during the pulse due the lower macroscopic absorption cross section of D_2O. In other words, the absorption in the central area is of minor importance. Thus, for the geometry shown in Fig.1, the confinement effect of H_2O plays a leading role for the neutron flux around the central area.

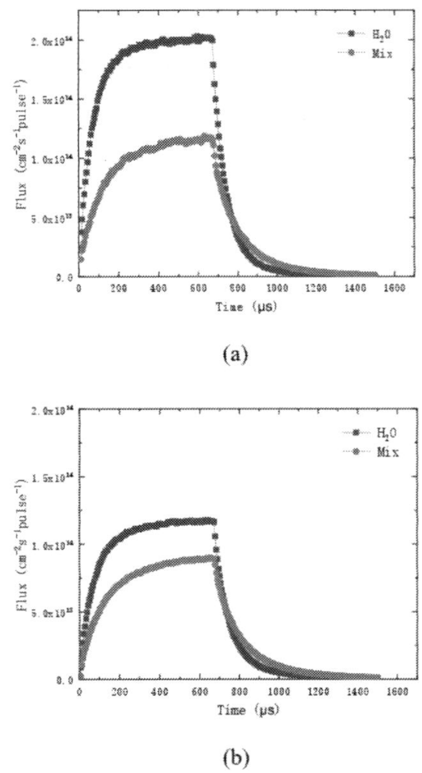

Fig. 3. Neutron flux versus time for geometry in Fig.1. (a) The result of tally 1. (b) The result of tally 2.

In order to take advantage of the lower neutron absorption of D_2O without diluting the neutron cloud around the central area, H_2O and D_2O+H_2O mixture material should be placed at different positions. For confining the neutron cloud, H_2O should be put at the center of the moderator to confine the neutron. To store more neutrons in the whole moderator and reflector structure, deuterium containing material is used, which decreases the absorption in this volume. The D_2O+H_2O mixture material surrounding the H_2O cannot influence the neutron cloud in the central area. Thus, the geometries as presented in Fig.2 are investigated.

The results for the different geometries introduced in Fig.2 are shown in Fig.4. Fig.4(a) presents the output of the surface tally into an emitted angle 0°-90° and Fig.4(b) is the result related to the tally emitted into an angle of 0°-5°, for which most of the events originate from the volume in the center of the extraction channel. Table 1 shows the related time integral value and decay time for these two tallies. The decay time is obtained by fitting an exponential decay function from the curve shown in Fig.4 [5]. The integration time range in Table 1 is from 0 ms to 1.5 ms.

As for the emitting angle 0°-90° shown in Table 1, the time integral value of geometry 2 is almost the same as geometry 1 (difference is smaller than 5 %). The maximum time integral value is for geometry 3, which increases by 18 % compared to geometry 1. It indicates

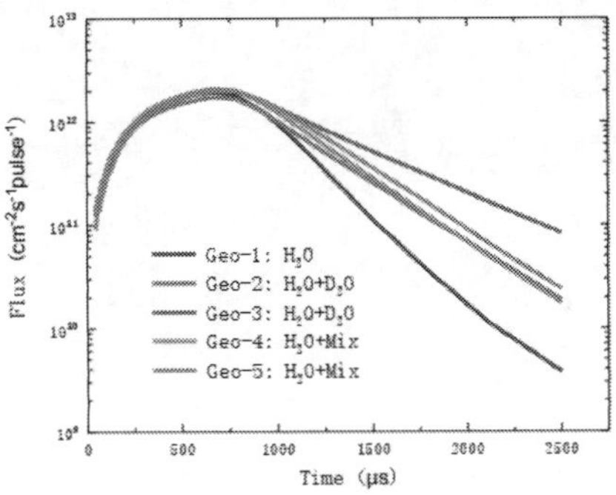

(a)

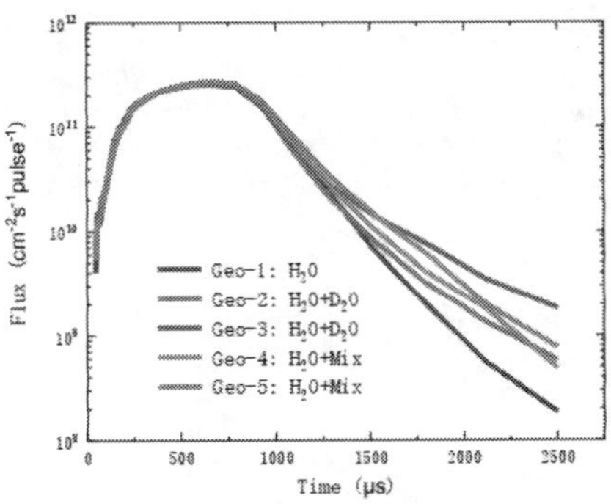

(b)

Fig. 4: Neutron flux versus time for an energy range of 1 meV to 500 meV for geometry in Fig.2. (a) Neutron flux versus time for emitting angle 0°-90°. (b) Neutron brightness versus time for emitting angle 0°-5°.

that the neutrons can travel for a long time in the D_2O volume to reach the surface tally. However, from the time distribution curve of geometry 3, the added neutrons contribute mostly to the tail of the neutron pulse. This is also demonstrated in Table.1. The decay time of geometry 3 in the emitting angle 0°-90° is 1.7 times larger than that of geometry 1. Furthermore, the time integral value of geometry 4 is 14 % higher than geometry 1. This means in the D_2O+H_2O mixture material the thermal neutrons can survive and be confined into the channel. The decay time of geometry 4 is just 1.3 times longer than that of geometry 1, which is smaller than the case using pure D_2O in geometry 3. That means the D_2O+H_2O mixture material along the extraction channel achieves an optimal balance between absorption and confinement. The time integral value of geometry 5 is also 14% higher than geometry 1, which is the same as in geometry 4. Thus, the neutron is not scattered back from the D_2O+H_2O mixture material. For the emitting angle 0°-5°, Table 1 indicates that the neutron flux varies less than 8 % between geometry 1 to 5. This difference is smaller than that when the geometry is not larger than 14 %. Fig.4(b) shows the emitting angle 0°-90°. The difference in the decay time for each

Table 1. Time integral neutron brightness and decay time for the geometries shown in Fig.2 for an energy range of 1 meV to 500 meV. The integration time was 1500 μs.

Emitting angle : 0°-90°		
	Time integral × 10^9 (cm^{-2} pulse^{-1})	**Decay time (μs)**
Geo-1: H_2O	1.35 ±0.03	318 ±20
Geo-2: H_2O+D_2O	1.31 ±0.02	400 ±23
Geo-3: H_2O+D_2O	1.64 ±0.03	536 ±26
Geo-4: H_2O+Mix	1.57 ±0.02	396 ±27
Geo-5: H_2O+Mix	1.57 ±0.02	456 ±31

Emitting angle : 0°-5°		
	Time integral × 10^9 (cm^{-2} pulse^{-1})	**Decay time (μs)**
Geo-1: H_2O	1.78 ±0.03	255 ±13
Geo-2: H_2O+D_2O	1.75 ±0.04	238 ±13
Geo-3: H_2O+D_2O	1.74 ±0.04	250 ±12
Geo-4: H_2O+Mix	1.86 ±0.04	257 ±14
Geo-5: H_2O+Mix	1.88 ±0.04	276 ±10

peak value among different geometries is similar to each other. These results indicate that the emission of a collimated beam is only weakly affected by the properties of the reflector for the HBS TMR geometry. The neutrons scattered back from the D_2O+H_2O mixture material do not contribute to the neutron pulse emitted into a narrow angle of 0°-5°.

4 Conclusion

In this study, we have investigated the neutronic characteristics of a thermal water moderator and a mixture material of D_2O and H_2O for a moderator-reflector design for a long pulse target at the HBS. It was found that confining the neutron cloud is more important than reducing the absorption even for a rather long neutron production time of nearly 700 μs.

Comparing the neutron flux emitted into a limited collimation cone, we find, that the reflector material affects mainly the tail of the neutron time distribution. That means that neutrons scattered back from the

https://doi.org/10.1051/epjconf/202328602005

reflector contribute hardly to the instantaneous neutron flux through the emission surface into the forward direction. Due to the small target volume, basically all neutrons are coupled into the light water moderator and only a tiny fraction of not-equilibrated neutrons escape towards the reflector at the HBS. Therefore, we observe an effect of the reflector mainly in the tail of the thermal neutron pulse, in particular if we consider the collimated emission at the extraction surface. This allows to optimize the reflector for the HBS based on different aspects, e.g., to shorten the pulse tail by appropriate thermal neutron absorbers, which decouple the reflector from the moderator to maximize the signal-to-noise ratio for the instruments.

References

1. T. Gutberlet, et al. Conceptual Design Report-Jülich High Brilliance Neutron Source (HBS), Forschungszentrum Jülich GmbH, Zentralbibliothek, Verlag, (2020).
2. Tran, Hoang Ngoc, et al. Neutrons production on the IPHI accelerator for the validation of the design of the compact neutron source SONATE. EPJ Web of Conferences. 231. EDP Sciences, 2020.
3. Pérez, M., et al. ARGITU compact accelerator neutron source: A unique infrastructure fostering R&D ecosystem in Euskadi. Neutron News 31.2-4 (2020): 19-25.
4. N. Watanabe. Neutronics of pulsed spallation neutron sources. Rep. Prog. Phys. 66.3, 339 (2003).
5. P. Zakalek, et al. Tailoring neutron beam properties by target-moderator-reflector optimisation. J. Neutron Res. 23.2-3, 185-200 (2021).
6. T. Sato, Y. Iwamoto, S. Hashimoto, et al. Features of Particle and Heavy Ion Transport code System (PHITS) version 3.02. J. Nucl. Sci. Technol. 55, 684-690 (2018).

Focusing High-Resolution Three-Axis Neutron Diffractometer for Investigations of Special Tasks of Powder Diffractometry

Pavol Mikula[1*], and *Vasyl* Ryukhtin[1] *Pavel* Strunz[1] and *Jan* Džugan[2]

[1]Nuclear Physics Institute CAS, 250 68 Řež, Czech Republic
[2]Comtes FHT, 334 41 Dobřany, Czech Republic

Abstract. Feasibility of focusing high-resolution three-axis diffractometer with the polycrystalline sample between the monochromator and the analyzer was tested for studies of finer effects of powder diffraction lines and for investigations of special tasks of powder diffractometry. The focusing 3-axis diffractometer set-up equipped with bent perfect crystal monochromator and analyzer offers the sensitivity in determination e.g. of strains in polycrystalline materials $\varepsilon=\Delta d/d$ close to 10^{-5}. Together with special tasks of strain/stress studies related namely, to plastic deformation, it permits to study a finer substructure of individual diffraction lines which can appear e.g. in case of polycrystalline alloys where more phases having very close values of the lattice spacing could exist. Resolution properties of the high-resolution diffractometer setting will be documented on several experimental results obtained on unconventional bulk samples.

1 Introduction

Neutron (as well as X-ray) powder diffraction is a wide spread method for determination of the structure and structural properties of polycrystalline materials. Conventional powder diffraction instrument is based on a two-axis diffractometer including a sophisticated monochromator, diffraction optics elements and a position sensitive detector (PSD) [1-4]. However, in some cases its $\Delta d/d$ resolution suffer limitations coming e.g. from a necessity of using samples of small dimensions, a requirement of a large monochromator take-off angle and a high spatial resolution of PSD, which can be solved acceptable at high-flux neutron sources. In the special cases of elastic and plastic deformation studies of bulk polycrystalline samples when the whole powder diffraction spectrum is not required and/or identification diffraction peaks related to phases having very close lattice spacings, a high-resolution three-axis alternative can be helpful [5,6]. The alternative consists of an unconventional three axis set-up employing optimally bent perfect crystal (BPC) monochromator and analyzer with a polycrystalline sample in between. The alternative provides a substantially higher resolution and though it is much more time consuming than the conventional powder diffractometer it can be used even at the medium power neutron source. The aim of this paper is to prove a feasibility of using the alternative diffractometer setting for studying special material research tasks e.g. determination and volume content of different material phases having very close values of lattice spacing, or microstructure analysis of plastically deformed samples on the basis of diffraction profile analysis etc. [7-9].

2 Basic description of the 3-axis diffractometer set-up

As schematically drawn in Fig. 1, the testing neutron optics diffractometer installed at the research reactor LVR-15 in Řež and operating at the constant neutron wavelength of 0.162 nm was used for the feasibility studies. The diffractometer is equipped with the BPC Si(111) monochromator of the dimensions of 200 x 40 x 4 mm^3 (length x height x thickness) and with the fixed radius of curvature R_M of about 12 m. Ge(311) BPC analyzer was in the form of a rather thin slab of the

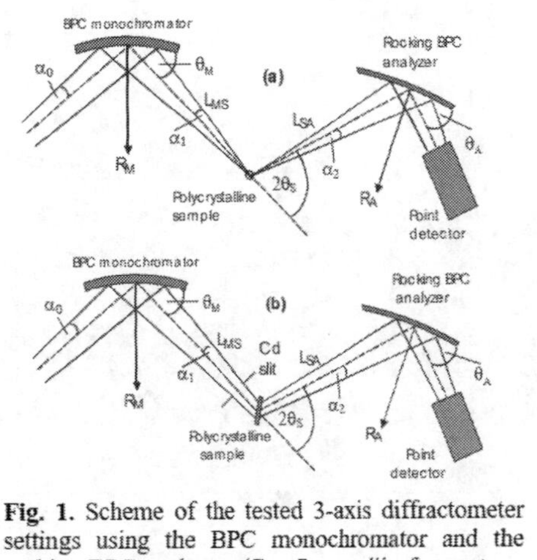

Fig. 1. Scheme of the tested 3-axis diffractometer settings using the BPC monochromator and the rocking BPC analyzer (R_M, R_A - radii of curvature, θ_M, θ_A - Bragg angles) for vertical – (a) and horizontal – (b) positions of a polycrystalline sample.

* Corresponding author: mikula@ujf.cas.cz

https://doi.org/10.1051/epjconf/202328603001

dimensions of 200 x 40 x 1.3 mm^3 which permitted us to bend it by a four point bending device in a large range of curvatures. The monochromator - sample distance, L_{MS}, being 1.7 m and the sample - analyzer distance, L_{SA}, being 0.5 m were fixed. The diffractometer operates in the open beam mode without any Soller collimators.

3 Resolution of the 3-axis set-up for bulk samples

During the extensive experimental studies several well annealed α-Fe standard polycrystalline samples of the cylindrical form and 4 bent BPC analyzers (Ge(111), Ge(311), Si(220) and Si(400)) were used for investigation of basic resolution characteristics for α-Fe(110) reflection and the detailed resolution properties can be found in ref. 6 and 10. Following the drawing shown in Fig. 1a the samples of several diameters of from 2 mm to 8 mm were situated in the vertical position. It is clear that the best resolution of the setting can be obtained when minimizing dispersion of the whole system. When solving the problem in momentum space, it means that the orientation of the Δk momentum elements related to the monochromator and analyzer should be matched to that of the sample, and this can be achieved by proper optimum radii of curvatures of the bent perfect crystal monochromator and analyzer [5]. Then, it results in a high-resolution (a minimum *FWHM* of the analyzer rocking curve) and a maximum detector signal, simultaneously. Fig. 2 shows the best basic resolution characteristics as dependent on the curvature of the Ge(311) BPC analyzer. Inspection of Fig. 2 reveals that as a result of a simultaneous impact of the focusing in real and momentum space the best resolution of the setting as well as a maximum intensity was found for the same radius of curvature of the analyzer of R_A=9 m. Furthermore, it has been found, that much larger widths (up to 20 mm) of the irradiated gauge volumes can be investigated by the three-axis setting when contrary to the conventional diffractometry, the resolution is much less affected (see Fig. 3). The width of the samples was simulated by situation of the samples in the horizontal position (see Fig. 1b) in combination with different widths of the Cd slits before the sample. In this case the contribution of the vertical dimension to the resolution was neglected. It can be seen from Fig. 3 that even for 20 mm width of the sample, the resolution represented by *FWHM* of the rocking curve is still good.

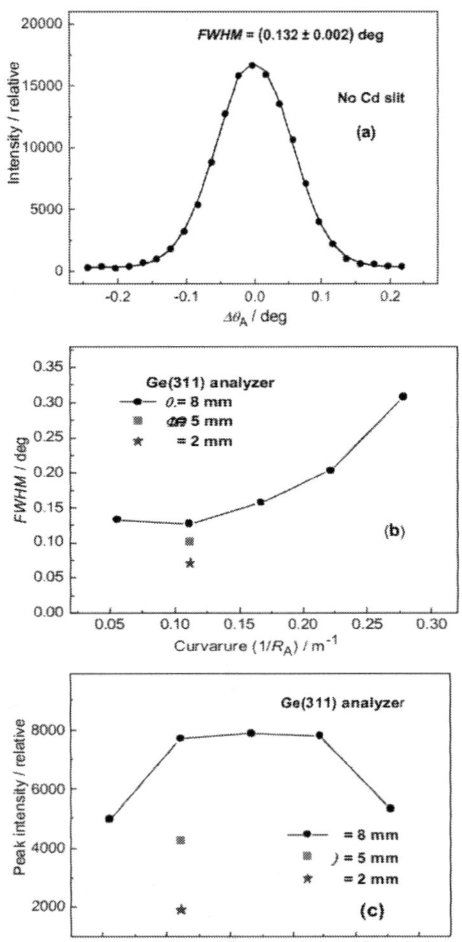

Fig. 2. Analyzer rocking curve for 8 mm α-Fe(110) standard sample and the optimum radius of curvature of the analyzer of 9 m – (a), its *FWHM* dependence – (b) and the peak intensity – (c) on the analyzer curvature. The inserted points correspond to the other diameters of the sample.

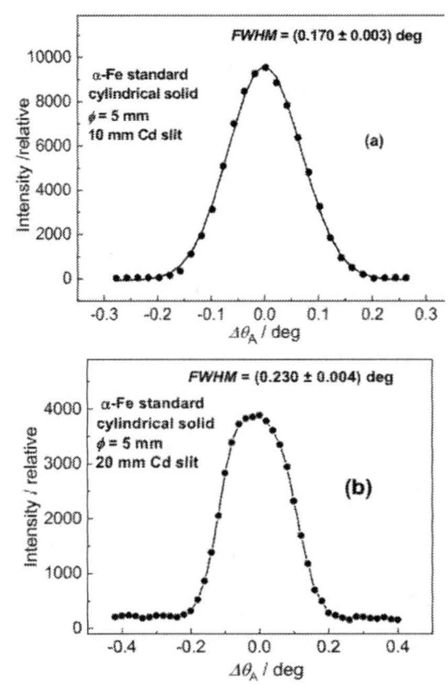

Fig. 3. Analyzer rocking curve related to 5 mm α-Fe(110) standard sample situated at the horizontal position for the optimum radius of curvature of the analyzer of 9 m and for two widths of Cd slits: (a) – 10 mm and (b) – 20 mm.

4 Few possible applications

First, we used several samples of the 3D printed material of gamma titanium aluminides (Ti48Al2Cr2Nb) loaded on the bulk Ti substrate. A powder fraction 45 – 150 μm was used. Manufacture of specimens was carried out by using the DED (Directed Energy Deposition) printing device InssTek MX-LAB. Laser power, speed of the printing and the dwell time were 130 W, 849 mm/min and 3.5 sec, respectively. Parameters related to the samples were: Both S1 (a preheat of the platform on 500 °C) and S2 (without a preheat) used cladding head SDM 800 and the 500 μm hatch distance for the 800 μm spot size and the 250 μm height of the layer. The following analyzer rocking curves (see Fig. 4) show how the differences in preparation of the samples reflect their *FWHM*. As to this sample it should be noticed that TiAl powder sample has not appeared to be a good scatterer for neutron diffraction.

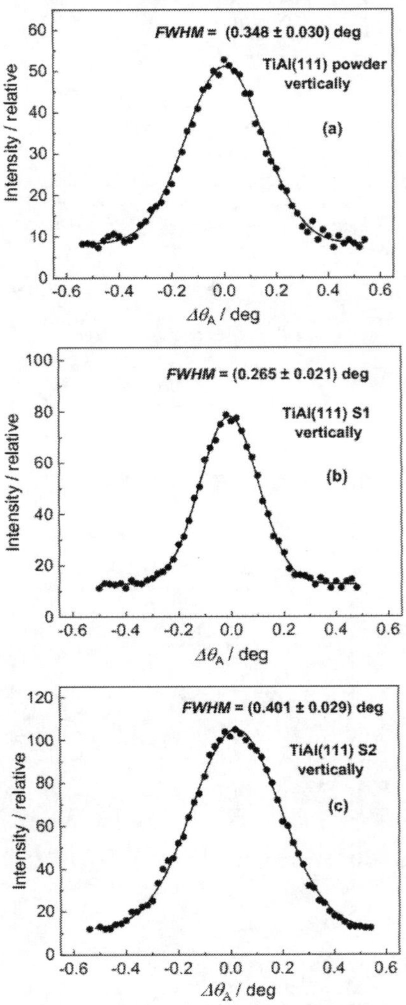

Fig. 4. Analyzer rocking curves related to the individual TiAl samples the $\phi = 9$ mm diameter and 20 mm height: (a) – powder sample, (b) – sample S1 and (c) – sample S2.

Further experimental applications of this setting are demonstrated by using unconventional samples rarely used in conventional powder diffractometry. First α-Fe(110) powder in 5 mm V container situated in a vertical and horizontal position was used (see Fig. 5). It can be seen from Fig. 5 that even for a sample 20 wide (simulated by the Cd slit) the $FWHM_A$ of the analyzer

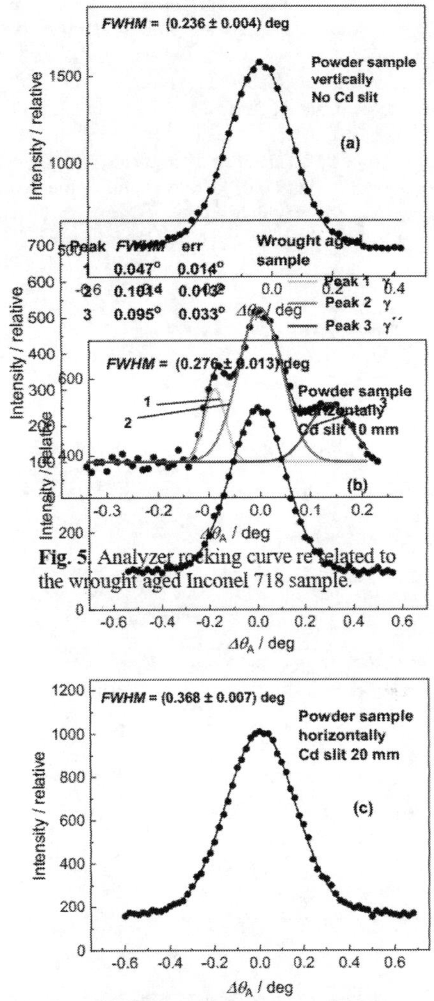

Fig. 5. Analyzer rocking curves related to the α-Fe(110) powder in a Vanadium container of the 5 mm diameter situated on the sample position : (a) - vertically and (b) and (c) – horizontally.

rocking curve could be acceptable in some medium-resolution powder diffraction experiments.

For next samples, we used the steel cylindrical solid of the diameter of 18 mm (see Fig. 6), the steel tube with the external diameter of 16 mm and the thickness of the wall of 1.6 mm (see Fig. 7) set both in the vertical and horizontal position and α-Fe plate of the thickness of 5.2 mm in the vertical position. Detailed inspection of the Fig. 6a reveals that the rocking curve is left-right asymmetrical (has no Gaussian form) due to the absorption of neutrons overcoming different path lengths in the material when looking along the axes of the incident and diffracted monochromatic beams. Similarly, in case of Fig. 7a the diffraction profile of the left part of the tube is slightly separated from the right one related to the irradiated right part when looking along the axis of the diffracted monochromatic beam. Therefore, the diffracted neutrons are impinging on spatially different parts of the bent analyzer. On the other hand, for samples horizontally situated in the setting the rocking curves are smooth and of course, their *FWHMs* depend of the width of the irradiated samples simulated by the width of the Cd slit. Furthermore, it has been observed that, as expected, in all cases with the large width of the irradiated sample (e.g. the case of the Cd slit of 20 mm) the rocking curves do not become Gaussian. However, for some types of experiments (e.g. when studying changes of the lattice spacing) it should not be a principal problem.

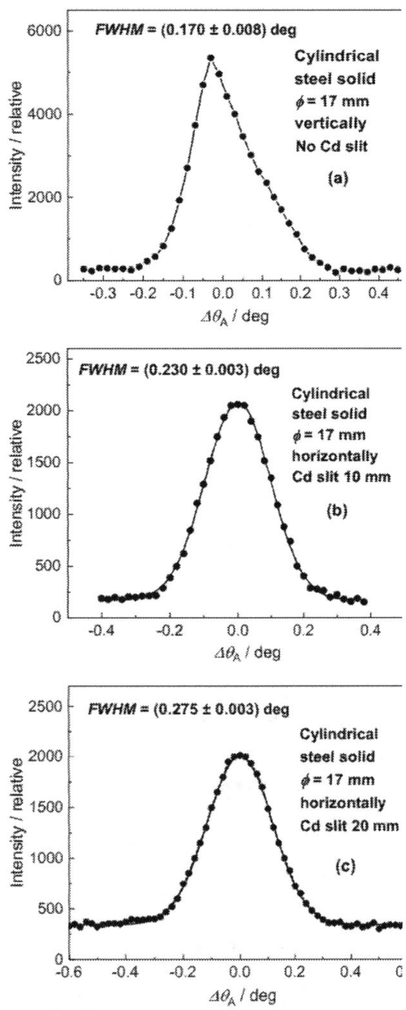

Fig. 6. Analyzer rocking curves related to the cylindrical steel solid of the 17 mm diameter situated: (a) – vertically and (b) and (c) – horizontally.

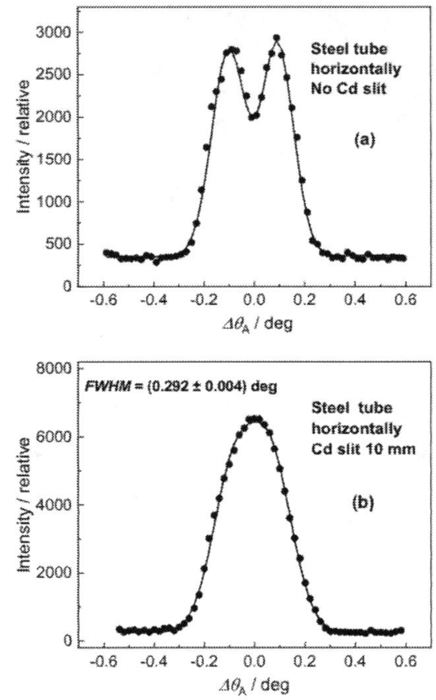

Fig. 7. Analyzer rocking curves related to the $\phi =$ 16 mm diameter steel tube (thickness of the wall of 1.6 mm) situated: (a) – vertically and (b) - horizontally.

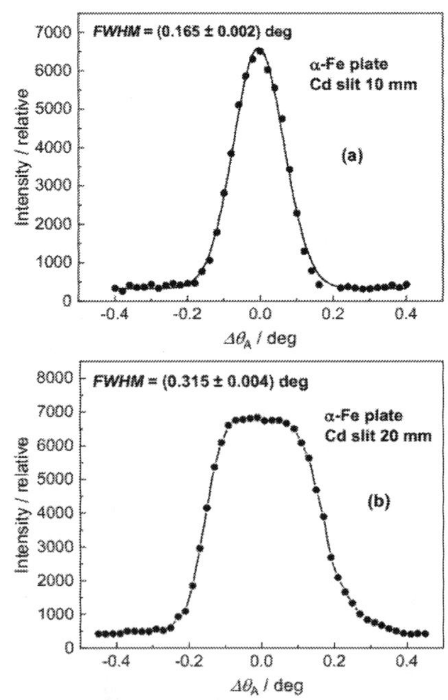

Fig. 8. Analyzer rocking curves related to the α-Fe(110) plate of the 5.2 mm thickness situated vertically for Cd slits of the width: (a) - 10 mm and (b) – 20 mm.

5 Conclusion

Resolution properties of the three-axis setting employing BPC Si(111) monochromator and BPC Ge(311) analyser with the α-Fe(110) polycrystalline standard samples between them were tested on the NPI neutron optics diffractometer. Some basic resolution results are shown in Figs. 2 and 3. It is necessary to point out that the optimisation of the setting was limited by the fixed curvature of the installed Si(111) monochromator and its take-off angle as well as the mutual distances between the monochromator, the sample and the analyzer. Then, in some special cases of unconventional bulk samples (up to about 20 mm width) the feasibility of using such setting for investigations requiring higher resolution than conventional instruments is documented. The resolution of the setting for the Ge(311) analyzer shows a minimum in the vicinity of $1/R_A = 0.11\ m^{-1}$.

The resolution represented by *FWHM* of the analyzer rocking curve depends on many parameters [5] and for a design of the fully optimized three-axis setting which incorporate all free parameters, Monte Carlo simulation is recommended [11].

As expected, thanks to rather low scattering angle, the setting with α-Fe(110) sample appears quite luminous with a high angular resolution *FWHM* because it exploits much more (at least partly) focusing in real and momentum space, simultaneously. In this case the lattice spacing of α-Fe(110) is much closer to the one of the Ge(311) analyzer. The curvature $1/R_A$ related to the best achieved resolution is situated within the range of the available analyzer curvatures. Therefore, some increase of the luminosity of the setting can be achieved by using a thicker crystal analyzer, e.g. up to 4 mm.

The following best values of *FWHM* obtained for different diameters of the standard sample were: 0.13 deg - $\phi = 8$ mm, 0.10 deg - $\phi = 5$ mm and 0.07 deg - $\phi = 2$ mm for Ge(311) analyzer.

Feasibility of the three-axis setting for some high-resolution studies on the bulk samples of rather large width simulated by different slit widths (see Fig. 3) was documented. I was shown that even for 20 mm slit width, the mildly relaxed resolution is still sufficient for e.g. possible application of the analysis of the position and shifts of the diffraction peak profiles [7-9]. In comparison with the conventional two-axis set-up, this alternative instrument could be successfully applied, namely, in the strain/stress measurements in bulk samples exposed to a thermo-mechanical external load, e.g. in tension/compression rig, in aging machine etc. It means in the cases when the whole powder diffraction spectrum is not required and the position of the sample is fixed. Thus, it can provide average strain/stress values over a large irradiated volume. Otherwise, for comparison of different samples attention should be paid to keeping the central point of the irradiated gauge volume in the same position with a high accuracy to avoid a shift error to $\Delta\theta_A$. Naturally, the high-resolution setting can also be attractive for studies of a fine substructure of e.g. alloys which could hardly be seen by using the conventional powder diffractometer [12]. Of course, that the pre-sented setting can also be used for macro-strain scanning, but due to the analyzer step-by-step analysis, the measurement would be rather time consuming.

In conclusion, we hope that the presented neutron diffraction setting can offer an additional support and complement the information obtained using other methodologies.

Measurements were carried out at the CANAM instrument of NPI CAS Řež installed the CICRR infrastructure, which is financially supported by the Ministry of Education and Culture - project LM2023041. The authors acknowledge support from ESS participation of the Czech Republic – OP (CZ.02.1.01/0.0/0.0/16_013/0001794) and from the project ESS Scandinavia-CZ II (LM2018111), respectively. Furthermore, they acknowledge support from the Czech Academy of Sciences in the frame of the program "Strategie AV21, No. 23". The authors thank B. Michalcová from NPI CAS for significant help with measurements and data elaborations.

References

1. I. C. Noyan, J. B. Cohen, Residual stress: Measurement by diffraction and interpretation, 1st edn; 1987, New York, Springer-Verlag. ISBN: 9780387963785.
2. M. T. Hutchings, A.D. Krawitz (Eds.), "Measurement of residual and applied stress using neutron diffraction", NATO ASI Series, Applied Sciences **26**, Kluwer Academic Publisher, 1992. DOI: 10.1007/978-94-011-2797-4.
3. T. Pirling, G. Bruno, P. J. Withers, Materials Sci. Eng. **A 437**(1), 139-144 (2006). DOI: 10.1016/j.msea.2006.04.083.
4. R. C. Wimpory, P. Mikula, J. Šaroun, T. Poeste, J. Li, M. Hoffmann, R. Schneider, Neutron News, **19**, 16-19 (2008). DOI:10.1080/10448630701831995.
5. M. Vrána, P. Lukáš, P. Mikula, J. Kulda, Nucl. Instrum. Methods in Phys. Research, **A 338**, 125-131 (1994). DOI:10.1016/0168-9002(94)90172-4.
6. P. Mikula, J. Šaroun, V. Ryukhtin, P. Strunz, Materials - MDPI, **13**, 5449-5456 (2020). DOI: 10.3390/ma13235449.
7. R. Delhez, T. H. De Keijser, E. J. Mittemeijer, Fresenius' Zeitschrift für analytische Chemie, **312**(1) 1-16 (1982). DOI:10.1007/BF00482725
8. K. Macek, P. Lukáš, J. Janovec, P. Mikula, P. Strunz, M. Vrána, M. Zaffagnini, Mat. Sci. Eng. **A 208**, 131-138 (1996). DOI:10.1016/0921-5093(95)10047-4.
9. V. Davydov, P. Lukáš, P. Strunz, R. Kužel, Mat. Sci. Forum, **571-572**, 181-188 (2008). DOI:10.1088/0953-8984/21/9/095407.
10. P. Mikula, J. Šaroun, V. Ryukhtin, J. Stammers, Powder Diffraction, **35**, Issue 3, 185-189 (2020). DOI:10.1017/S0885715620000329.
11. J. Šaroun, J. Kulda, Physica B, **234-236**, 1102-104 (1997). DOI:10.1016/S0921-4526(97)00037-9.
12. P. Mikula, V. Ryukhtin, Procedia Structural Integrity, **43**, 119-123 (2023). DOI:10.1016/j.prostr.2022.12.245.

Modelling of The New Engineering Diffractometer eMAP at ISIS Target Station 2

Tung Lik Lee[1*], *Saurabh* Kabra[1], *Joe* Kelleher[1], *Winfried* Kockelmann[1] and *Robert* Bewley[1]

[1]ISIS Neutron and Muon Source, Science and Technology Facilities Council, Rutherford Appleton Laboratory, Didcot, OX11 0QX, UK

Abstract. eMAP is a new proposed high flux medium-resolution time-of-flight engineering diffractometer that will be built at ISIS Target Station 2. Monte Carlo simulations were performed to optimise the neutron guide to maximise brilliance transfer. eMAP features great flexibility in tuning for flux or resolution. The diffractometer achieves a figure of merit that surpasses the ENGIN-X diffractometer at ISIS by more than a factor 10. The new instrument will bring additional capacity and advanced stress measurement capabilities to the engineering diffraction user programme at the ISIS Neutron and Muon Source.

1 Introduction

The ISIS Neutron and Muon Source is now seeking its next phase of new instruments and significant upgrades to existing instruments: a portfolio of projects that is called the Endeavour Programme [1]. Endeavour will increase both capacity and capability of the facility to address 21st century challenges and enable research in areas such as advanced materials and manufacturing, clean energy technologies, and biosciences and healthcare. Starting in the 2023/24 financial year, the Endeavour Programme will construct four new instruments and five significant upgrades to existing instruments, over a 10-year period.

ENGIN-X [2] is currently the only instrument at ISIS dedicated to neutron diffraction strain scanning of engineering components. Strain scanning at more modest resolution will also be available on IMAT [3] once the current project to install $2\theta=\pm90°$ detectors and collimators is completed, though beamtime will be shared with the imaging programme on the instrument. A majority of the experiments conducted on ENGIN-X have industrial collaboration. However, ENGIN-X, which serves as the primary instrument, is over-subscribed, particularly for industrial access. To address the growing demand from industry, a new engineering instrument is necessary to provide additional capacity and new capabilities for stress measurement in the interior of full-size engineering components and characterisation of dynamic material behaviours at extreme conditions, allowing industrial manufacturers and materials engineers to design components that are more advanced and machines with enhanced performance, durability, and capabilities.

Here we report on the modelling of the new proposed time-of-flight engineering diffractometer, eMAP, which will be built at ISIS Target Station 2 (TS2). TS2 is a pulsed source operating at 10 Hz. The name of the instrument refers to strain mapping, with the letter 'e' representing the Greek letter 'ε' (strain). eMAP is designed with a performance complementary to the diffraction capabilities offered by ENGIN-X and IMAT. The high diffraction resolution of ENGIN-X is likely to be difficult to match. Therefore, eMAP is optimised as a medium resolution instrument offering higher flux and figure of merit than ENGIN-X. eMAP offers tuneable functionality to measure residual stress in large, heavy, thick, and intricately shaped components, offer improved 2D/3D strain mapping, and enable the study of short-timescale phenomena (e.g. in-situ welding) under extreme sample conditions relevant to industrial processes. The performance of eMAP will result in improved data acquisition times and greater depth penetration capability compared to ENGIN-X, representing a significant advancement in our ability to study real engineering components at ISIS.

2 Modelling of the instrument

2.1 Instrument layout

Fig. 1 shows the basic layout of eMAP. The instrument will be viewing the 120×120 mm² 40 K decoupled poisoned methane moderator at TS2. The primary flight path (L_1) is 40 m. eMAP features an elliptic guide to help maximise incident vertical divergence. Due to the low horizontal divergence required for strain scanning, the sides of the guide are straight. The guide has a cross-section of 40×60 mm² (width × height) at the entrance of the guide at L_1=1.7 m from the moderator face. The

* Corresponding author: Tung-lik.lee@stfc.ac.uk

height of the guide expands to a maximum of 149 mm at mid-length of the instrument. The cross-section at the end of the guide is 40×56.5 mm². The guide ends at 1.5 m (L_1=38.5 m) from the sample position to ensure ample sample area to accommodate various sample environment and specimen sizes. The coating at the top and bottom of the guide has non-uniform distribution that features m=3 at mid-length and increasing m-value of up to m=5 at both ends of the guide to maximise brilliance transfer efficiently. The sides of the guide have m=1.5 coating due to the low horizontal divergence requirement for strain scanning. The guide is designed with a removable section upstream as a contingency measure so that a T0 chopper could be retrofitted, if necessary. Two single disk choppers operating at the source frequency of 10 Hz are positioned at L_1=8 m and 10 m to define the wavelength band, prevent frame-overlap of neutrons between successive neutron pulses and neutron leaking through the disk choppers at long wavelengths. The chopper positions have been optimised using path-time and phase space simulations to ensure no frame overlap and no neutron leaks at long wavelengths of up to 200 Å. The wavelength bandwidth of the instrument is ~9 Å.

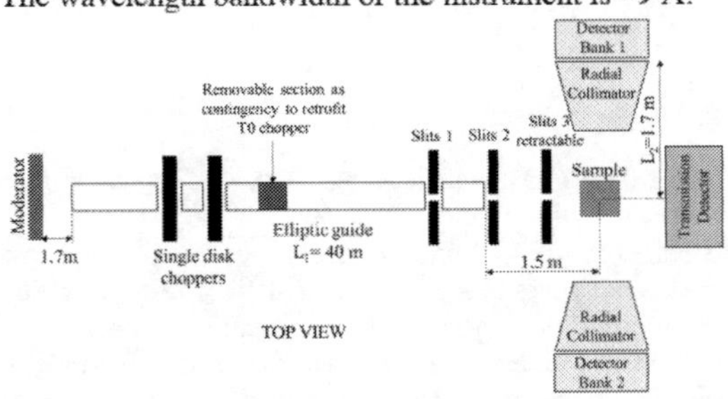

Fig. 1. Schematic of eMAP showing the main components of the instrument.

It is envisaged that the samples studied on eMAP will have different levels of sample-induced broadening which can lead to varying counting time requirements. Therefore, this calls for an instrument with a variable divergence which can be adjusted to match the peak broadening introduced by the sample since typically the incident flux can be increased if one is willing to increase the divergence. A total of three sets of slits will be positioned downstream the guide to control divergence. In order to minimise the number of gaps required in the guide system, only the first set of slits is positioned in the guide. The second set will be installed immediately after the guide exit. The final set of slits will be retractable between the second set of slits and sample, and defines the beam size at sample position.

Wavelength shifting fibre coupled ZnS:Ag/^{6}LiF detectors will be installed on eMAP. Two diffraction detector banks at 2θ=±90° will enable the measurement of two orthogonal strain components simultaneously to minimise data acquisition times. The detectors will be positioned at 1.7 m from the sample position and will be mounted with removable radial collimators to define the size of the gauge volume (GV) along the beam. A total of five sets of radial collimators providing 0.5, 1, 2, 3, and 4 mm gauge width options will be made available. The instrument will provide a large sample area to accommodate unique sample environments and large engineering components. For example, the available space between the radial collimator faces for a 2 mm gauge collimator is estimated to be about 1 m similar to the specifications at IMAT [4]. As a comparison, the same collimator gauge size on ENGIN-X provides a gap of about 0.6 m between the collimator faces [2].

To match the incident beam divergence and meet the requirements for texture measurement, a detector pixel size of 3×100 mm² (width × height) was chosen. With 100 mm high pixels at 2θ=90°, eMAP will achieve an angular resolution of about 3.4° which helps with recording of sharp textures. A large detector coverage is required to ensure a high count rate and facilitate texture measurement. However, the error introduced into strain measurements through averaging over a range of detector angles increases with the angular span. At eMAP, the detectors will have angular coverage of ±14° horizontal and ±21° vertical similar to that of ENGIN-X which is well optimised for strain measurements. Previous studies have shown that detector arrays of up to ~±20° (horizontally and vertically) give errors that are small compared to the uncertainty in strain from a typical engineering instrument [5]. It is worth noting that elastic strains in engineering materials are typically of the order of 10^{-3} to 10^{-4}, requiring strain measurement uncertainties (strain resolution) of the order of 10^{-4} to 10^{-5}. Such accuracy in Bragg peak shift determination can be achieved by good Bragg peak definition and thus instruments of different diffraction resolutions and count rates achieve the same strain resolution by counting for different times. For example, to achieve the same strain resolution of 50 microstrain, ENGIN [6] had to count >10 times longer than ENGIN-X. Therefore, the overriding requirement of an engineering diffractometer is to obtain peak positions with a given uncertainty for a minimum counting time period. Thus, a figure of merit (FOM) is defined as:

$$FOM = I / \omega^2 \quad (1)$$

where I is the Bragg peak intensity and ω is the peak width to describe the performance of an engineering instrument [7].

Additionally, a transmission detector will be made available to enable for Bragg edge mapping.

2.2 Instrument location at TS2

Fig. 2 shows the three potential locations for constructing eMAP at TS2. Position 1 features an L_1 of 64 m, which is the longest possible length for an instrument facing this viewport. However, building a new external structure to accommodate the instrument at this position would incur an additional cost of about £5 million. Positions 2 and 3 have shorter L_1 of 40 m and are located within the R80 building. After careful consideration, position 3 that views the W2 ('West' 2) beam port on TS2 was selected as the location to build eMAP due to its advantage of providing height space to

 https://doi.org/10.1051/epjconf/202328603002

include an overhead crane, while requiring fewer building and floor alterations compared to position 2.

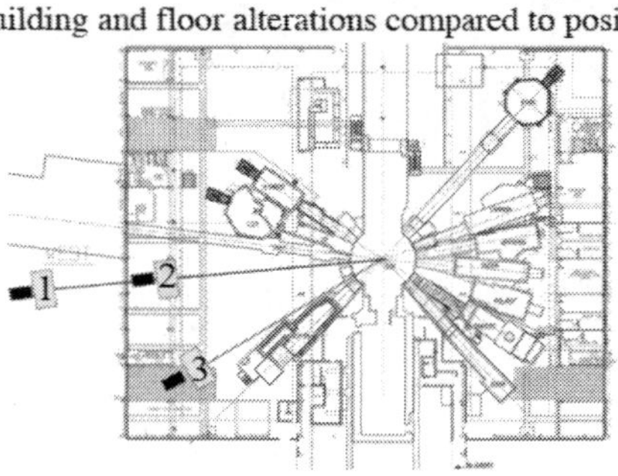

Fig. 2. Potential locations of eMAP at ISIS Target Station 2.

2.3 Modelling approach

The Monte Carlo modelling package McStas 2.5 [8] was used to model and optimise the instrument parameters. The decoupled poisoned methane moderator viewed by eMAP was modelled using the 'Commodus_I' [9] component and the methane moderator file (W4_Base_2022_wDecFL_wPOISeMapSide_off6mm _hExc12mm.mcstas) produced by the ISIS Neutronics Group. In the simulations, 2×10^{10} neutrons were generated in the wavelength bandwidth of 0.1-9.4 Å. Experimentally determined neutron guide coating reflectivity profiles provided by SwissNeutronics AG were used. The choppers were disabled in the simulations but the gaps in the guide at the chopper positions were retained. A beam size of 4×10 mm^2 (width $\times$ height) was defined at the sample position using the final set of slits set at 200 mm from the sample position. The TOF_spher detector component that describes a spherical detector arrangement with upper and lower limits in 2θ and ϕ, and which creates time-focused 1D spectra detected at $2\theta=\pm90°$ was used in the model. The TOF_spher detector component is modified from the TOF_logspher component to include the contributions from the detector pixel size. The detector radial collimators were excluded from the simulations to reduce computation times. A $4\times10\times4$ mm^3 (width $\times$ height $\times$ depth) CeO_2 sample was used to emulate the dimensions of the GV defined at the sample position. The flux and Bragg peak profiles of existing ISIS instruments, namely WISH [10] (views the same methane moderator) and ENGIN-X, were simulated with the aforementioned assumptions. Comparison with experimentally measured profiles (Fig. 3) shows good agreement was found which gives confidence that the predictions from the simulations of eMAP presented herein are representative of the actual performance.

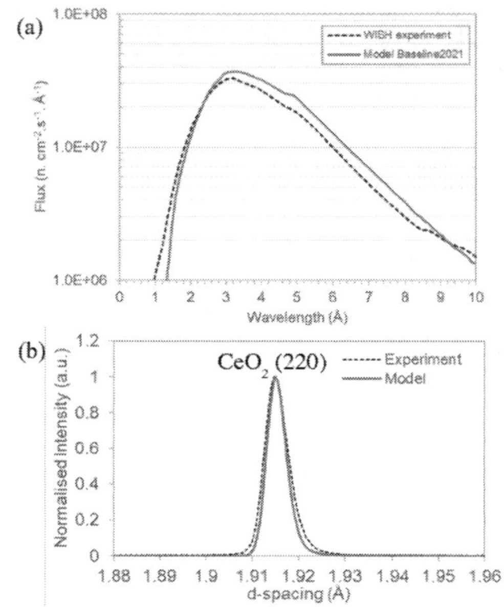

Fig. 3. (a) Neutron flux distribution of WISH at sample position, and (b) Bragg peak from ENGIN-X.

2.3.1 Guide geometry

An elliptic guide was selected for eMAP to maximise flux. The elliptic guide design was adjusted to feature a constant width along its length to due to the low horizontal divergence requirement for strain scanning. Simulations were performed to compare its performance with other guide geometries such as the straight guide of rectangular cross-section and ballistic guide (divergent at the start, straight in middle then converging at end). Fig. 4 shows the elliptic guide generally gives higher brilliance transfer (BT) at λ>0.8 Å and achieves >90% BT at λ=2-3 Å where most dominant Bragg peaks of engineering materials are located. The BT simulations were performed using a divergence of $0.12°\times0.6°$ half width at half maximum (HWHM) at sample position. The small 0.12° horizontal divergence was selected to achieve good diffraction resolution and a homogeneous beam profile. It was found that a uniform beam can be defined at the sample position by controlling the vertical divergence up to ~0.4°. Nevertheless, the vertical divergence was increased to 0.6° in the BT simulations to account for the need of higher flux with certain experiments.

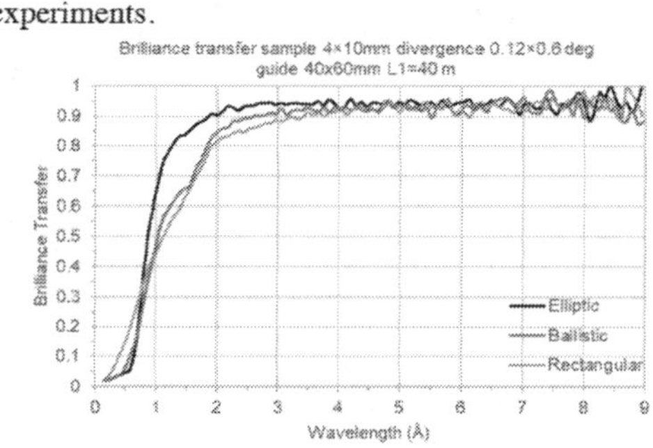

Fig. 4. Brilliance transfer of the straight m=3 guide in various geometries. 40×60 mm^2 guide cross-section defined at the entrance of the ballistic and elliptic guides. Ballistic guide is elliptic in the first and final 10 m of L_1.

The conventional design for engineering diffractometers often includes a curved section in the guide to eliminate line of sight. This is to reduce the exposure of the sample to significant levels of neutron and gamma radiation that could potentially compromise further processing or characterisation of engineering components or valuable specimens after neutron measurements. However, a straight guide combined with a T0 chopper could be considered as an alternative. Fig. 5 shows the BT from straight and curved elliptic guides (m=3 coating). The straight guide was designed with a 300 mm gap in the upstream section of the guide to fit a T0 chopper. The straight guide design proved to be more effective compared to a curved guide, particularly at λ<2 Å where the curved guide demonstrated poorer BT. As a result, a straight elliptic guide was chosen for eMAP. It is worth noting that experimental measurements showed that prompt pulse neutron and gamma fluxes are low on an existing instrument at TS2 that has similar straight elliptic guide with the same L_1 and views the methane moderator. Thus, fast neutron and gamma contributions to signal-to-noise at eMAP are expected to be low and the T0 chopper could be omitted from the straight guide. Nevertheless, a removable section of the guide is included in the guide design as a contingency measure so that the T0 chopper could be retrofitted, if necessary.

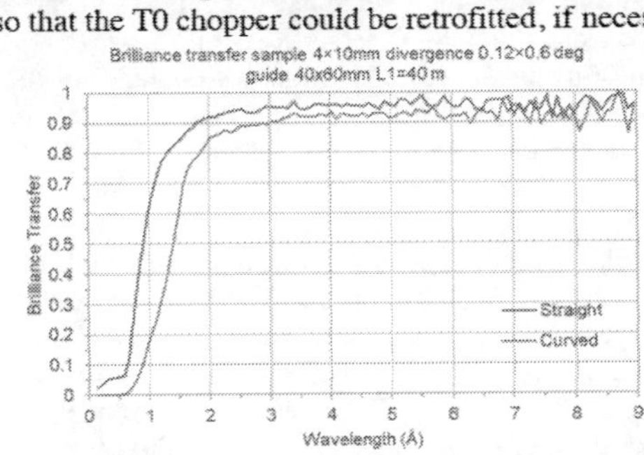

Fig. 5. Brilliance transfer of a straight versus curved m=3 elliptic guides.

2.3.2 Elliptic guide cross-section dimensions

Fig. 6 shows the BT of a straight elliptic guide with m=3 coating simulated with various guide cross-section dimensions. The guide height refers to the dimension at the entrance of the elliptic guide. In the downstream sections, the guide height varies according to the dimension defined at the entrance. A 40 mm wide guide achieves the maximum BT of ~95% at λ>2 Å. The 60 mm height at the guide entrance gives the maximum BT at λ>2 Å and high BT at low λ region. Heights larger than 60 mm generally gives similar or slightly lower BT at λ>1 Å. Therefore, the 40×60 mm^2 guide cross-section at the entrance of the elliptic guide was chosen for eMAP.

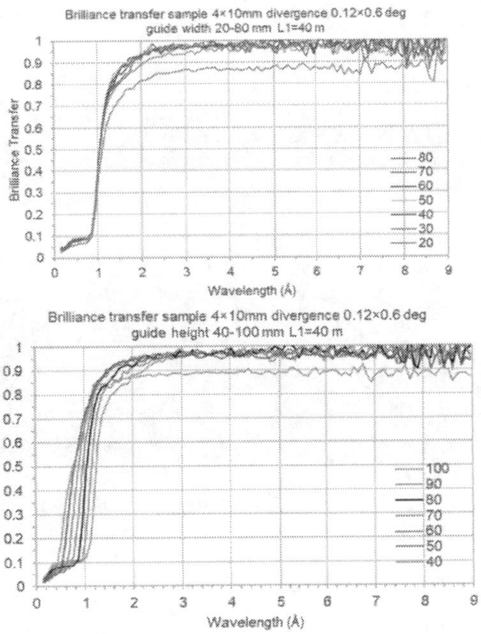

Fig. 6. Brilliance transfer of the straight m=3 elliptic guide simulated with various guide heights and widths. The guide height refers to the height defined at the entrance of the elliptic guide. In the downstream sections, the guide height varies according to the dimension defined at the entrance.

2.3.3 Guide coating

Fig. 7a shows the BT of the straight elliptic guide specified with various uniform guide coating up to m=5. Coatings higher than m=5 was not considered due to high costs. Based on the BT, a minimum of m=3 is recommended as it gives significant gains compared to m=2 at λ<3 Å and reasonable BT (~40%) at around 0.8 Å. This λ range (0.8-3 Å) is particularly important for engineering materials, for e.g. Fe. Although BT at short wavelengths generally improves with higher m-values, it is often more cost effective to apply non-uniform guide coating distribution with higher m-values at both ends of the elliptic guide. Fig. 7b and 7c show the various coating distributions considered for the elliptic guide. Fig. 7d shows that the non-uniform coating distribution is effective at increasing BT at λ<3 Å. Model A (up to m=5 at the ends and m=2 in the central region) showed higher BT of a few percent at around λ=2 Å when compared to a uniform m=2 guide coating. Nevertheless, the BT of Model A at λ=~0.8 Å is poor compared to a uniform m=3 coating. Therefore, the non-uniform coating distribution that features m=3 in the central region (Model B) was selected for eMAP. Further optimisation of the coating distribution will be performed to further improve BT.

A low m-value coating of m=1.5 was chosen for the sides of the guide due to the small horizontal divergence. This arrangement gives a slight decrease of ~3% in BT that is mostly limited to λ=1-2 Å region compared to m=3 coating but delivers significant cost savings.

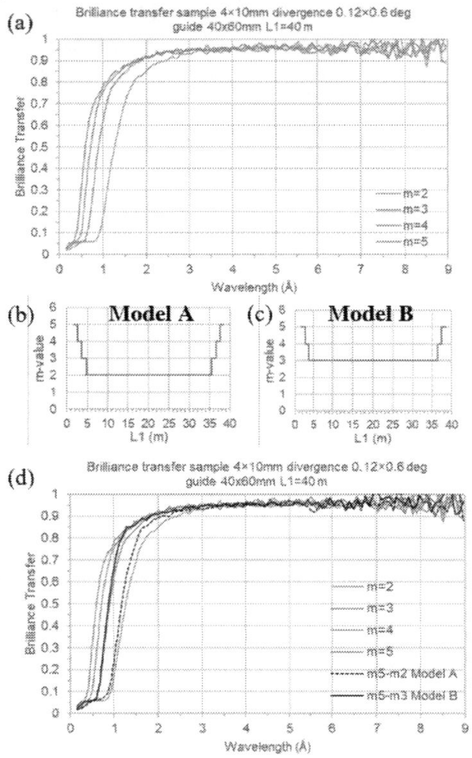

Fig. 7. (a) Brilliance transfer of the straight elliptic guide using various guide coating. (b) & (c) Non-uniform elliptic guide coating distribution profiles. (d) Brilliance transfer of the elliptic guide using the coating distribution profiles in comparison with uniform coatings.

3 Performance

Fig. 8a shows the resolution function of eMAP in comparison with ENGIN-X. The resolution data were derived from fitting selected CeO_2 Bragg peaks produced in the simulations. The diffraction resolution of eMAP, operating at 0.12°×0.4° HWHM divergence, is about 0.3% while ENGIN-X has a resolution of ~0.25%. Data points at λ<1 Å are not shown due to relatively low Bragg peak intensities from the simulations. Fig. 8b shows the FOM of eMAP surpasses ENGIN-X by more than a factor 10 at higher λ while achieving about a factor 4 at λ<2.5 Å. The FOM of ENGIN-X running at 25 Hz was used in this comparison because this mode is optimal for macroscopic strain scanning in most engineering materials [2].

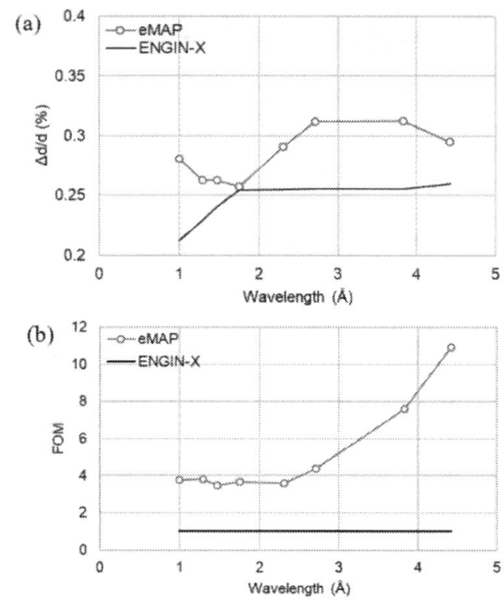

Fig. 8. Comparison in performance between eMAP and ENGIN-X. (a) Resolution function of both instruments. (b) Gain in figure of merit on eMAP compared to ENGIN-X (operating at 25 Hz).

At $\Delta d/d$=0.31% (at 2.7 Å), the integrated flux in the full λ bandwidth on eMAP is about 1.2×10^6 $n.cm^{-2}.s^{-1}$. A uniform 4×10 mm^2 beam at sample position can be defined using the small divergence in this mode (see Fig. 9). The flux profile of eMAP shows higher flux than ENGIN-X throughout the λ range. This is reflected in the simulated CeO_2 diffraction patterns showing the Bragg peak intensities are up to a factor 5 higher on eMAP. The vertical divergence can be increased by tuning the vertical gaps of the slits to double the flux to about 2.5×10^6 $n.cm^{-2}.s^{-1}$ at $\Delta d/d$=0.39% (at 2.7 Å). Bragg peak intensities on eMAP in this mode are almost a factor 10 higher than ENGIN-X. Though the beam becomes less uniform in the vertical direction due to increased divergence (see Fig. 10), the improved time resolution in this 'high flux' mode is crucial for studying short-timescale phenomena such as those occurring in in-situ loading measurements where beam uniformity is not a critical requirement.

https://doi.org/10.1051/epjconf/202328603002

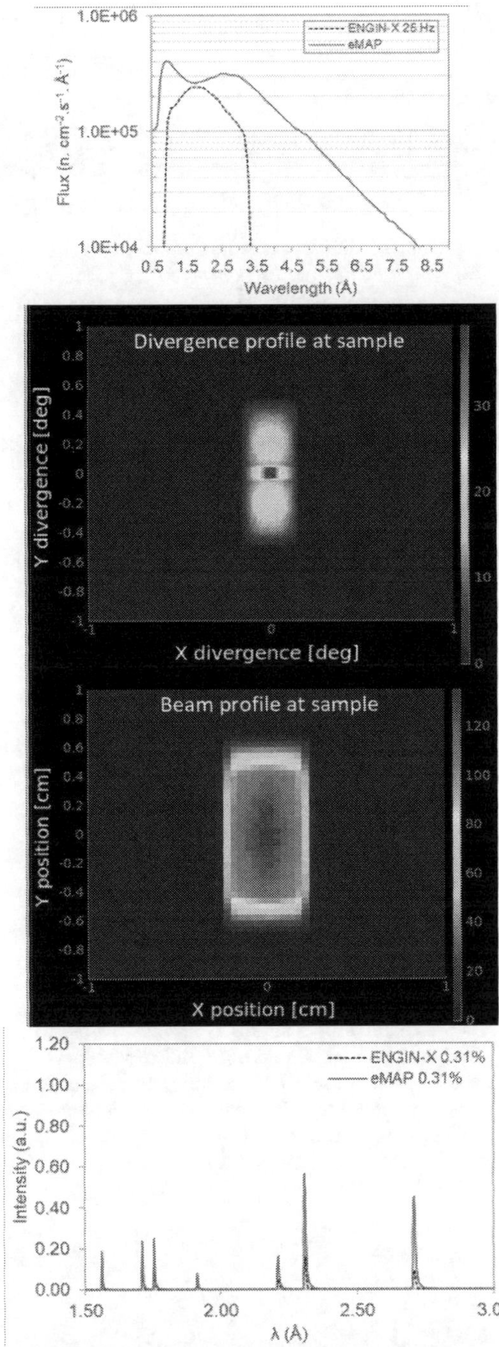

Fig. 9. The simulated flux, divergence, beam profile at sample position, and CeO_2 diffraction pattern on eMAP at $\Delta d/d$=0.31% (at 2.7 Å). Simulated flux and diffraction pattern on ENGIN-X (operating at 25 Hz) at the same resolution are shown for comparison.

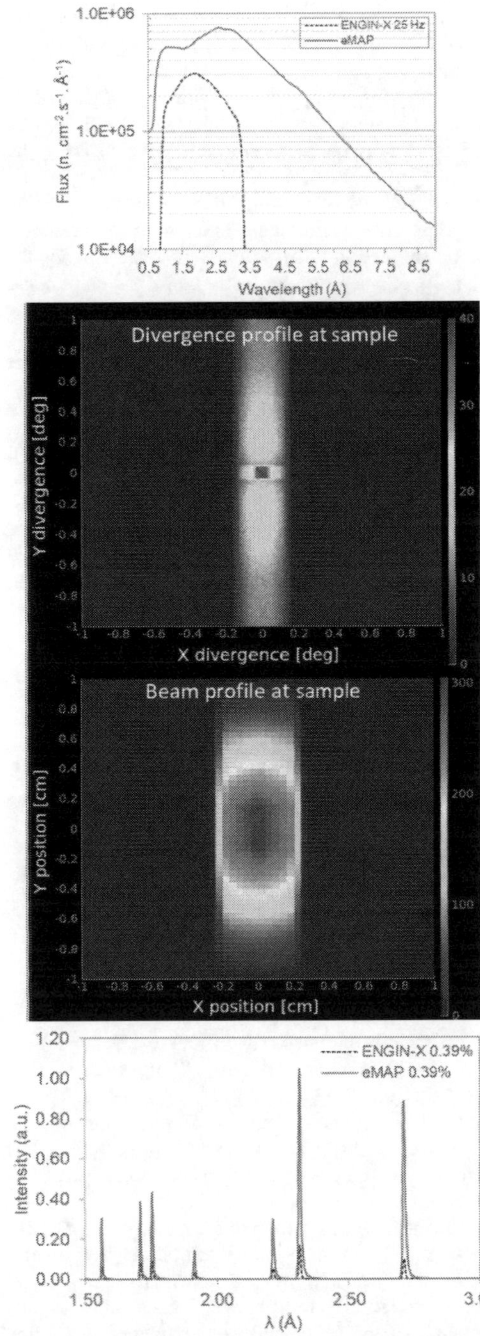

Fig. 10. The simulated flux, divergence, beam profile at sample position, and CeO_2 diffraction pattern on eMAP at $\Delta d/d$=0.39% (at 2.7 Å). Simulated flux and diffraction pattern on ENGIN-X (operating at 25 Hz) at the same resolution are shown for comparison.

Fig. 11 shows a comparison of the performance of eMAP with several world-leading time-of-flight engineering diffractometers [11-14]. Flux data from other instruments at a diffraction resolution comparable to eMAP are shown. It should be noted that data of other instruments may correspond to different wavelength intervals because of the different lengths and source frequencies. At ISIS, ENGIN-X will serve as the main instrument for high resolution ($\Delta d/d=0.25\%$) engineering materials studies while IMAT will provide high flux at more modest resolution ($\Delta d/d=0.7\%$). The performance gap between them will be addressed by eMAP which will be the high flux medium resolution ($\Delta d/d=0.3$-0.4%) instrument. The user programme at ISIS for engineering diffraction will benefit from the availability of a selection of instruments that encompass a larger range of resolution and flux to increase the range of industrial problems that can be studied. Nevertheless, ISIS instruments work with a lower source brilliance and lower time-averaged neutron flux than many of the diffractometers at other sources.

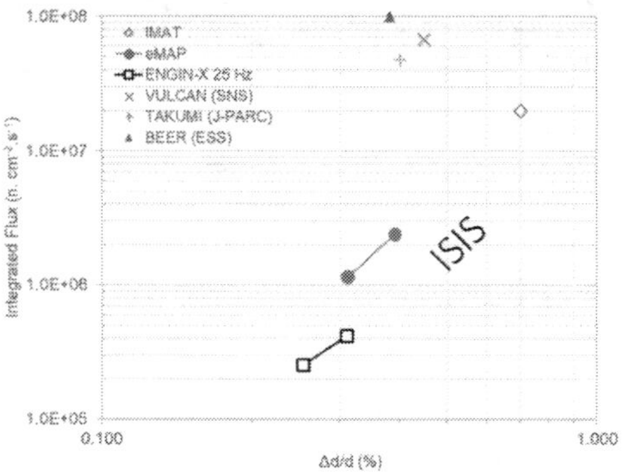

Fig. 11. Integrated flux at sample position as a function of diffraction resolution for eMAP. Data available for BEER (ESS) [11], VULCAN (SNS) [12], TAKUMI (J-PARC) [13], IMAT [14] and ENGIN-X (25 Hz) are shown for comparison.

4 Summary

eMAP is the new proposed high flux medium resolution engineering diffractometer that will be built at ISIS Target Station 2. Monte Carlo simulations were performed to optimise the straight elliptic neutron guide to achieve a brilliance transfer of >90%. The diffractometer achieves a figure of merit that surpasses the ENGIN-X diffractometer at ISIS by more than a factor 10. eMAP has a diffraction resolution of about 0.3% (at 2.7 Å) and integrated flux in the full wavelength bandwidth of ~9 Å is 1.2×10^6 n.cm^{-2}.s^{-1}. It features great flexibility in increasing vertical divergence to double the flux at a diffraction resolution of about 0.4%. The new instrument will bring additional capacity and advanced stress measurement capabilities to the engineering diffraction user programme at the ISIS neutron and muon source.

This work was supported by the Science and Technology Facilities Council (UK). TLL would like to thank the ISIS Neutron and Muon Instrument Development Group for the valuable discussions and suggestions on the instrument design. The authors are grateful to the ISIS Neutronics Group who provided the moderator files used in the simulations.

References

1. https://www.isis.stfc.ac.uk/Pages/Endeavour.aspx (Last accessed on 5 February 2023)
2. J.R. Santisteban, M.R. Daymond, J.A. James, L. Edwards, J. Appl. Crystallogr. **39(6)**, 812-825 (2006)
3. W. Kockelmann, et al., J. Imaging **4(3)**, 47 (2018)
4. W. Kockelmann, et al., *Status of the neutron imaging and diffraction instrument IMAT*, in Proceedings of the 10th World Conference on Neutron Radiography, Phys. Procedia **69**, 5-10 October 2014, Grindelwald, Switzerland (2015)
5. M.R. Daymond, Physica B **301(3-4)**, 221-226 (2001)
6. M.W. Johnson, L. Edwards, P.J. Withers, Physica B **234**, 1141-1143 (1997)
7. M.W. Johnson, M.R. Daymond, J. Appl. Crystallogr. **35(1)**, 49-57 (2002)
8. P.K. Willendrup, K. Lefmann, J. Neutron Res. **22(1)**, 1-16 (2020)
9. https://isisneutronmuon.github.io/mcstas/How-to-Use-the-ISIS-moderator-component (Last accessed on 5 February 2023)
10. L.C. Chapon, et al., Neutron News **22(2)**, 22-25 (2011)
11. J. Fenske, et al., *BEER – The Beamline for European Materials Engineering Research at the ESS*, in Proceedings of the 6th European Conference on Neutron Scattering, J. Phys.: Conf. Ser. **746(1)**, 30 August – 4 September 2015, Zaragoza, Spain (2016)
12. X-L. Wang, et al., Physica B **385**, 673-675 (2006)
13. S. Harjo, Nihon Kessho Gakkaishi **50(1)**, 40-45 (2008)
14. W. Kockelmann, et al., *IMAT – a new imaging and diffraction instrument at ISIS,* in Proceedings of the 7th International Topical Meeting on Neutron Radiography, Phys. Procedia **43**, 16 – 24 June 2012, Kingston, Canada (2013)

The HERMES reflectometer at the JULIC Neutron Platform

Mariano Andrés Paulin[1*], *Ivan* Pechenizkiy[2], *Paul* Zakalek[2], *Klaus* Lieutenant[2], *Peter* Kämmerling[2], *Alexander* Steffens[2], *Harald* Kleines[2], *Ulrich* Rücker[2], *Thomas* Gutberlet[2], *Sébastien* Gautrot[1], *Alain* Menelle[1], and *Frédéric* Ott[1, 3]

[1]Laboratoire Léon Brillouin, UMR12 CEA-CNRS, 91191 Gif sur Yvette Cedex, France
[2]Jülich Centre for Neutron Science, FZJ, 52428 Jülich, Germany
[3]Université Paris Saclay, 91191 Gif sur Yvette Cedex, France

Abstract. HERMES is a time-of-flight reflectometer that operated at the Orphée reactor until 2019. In 2022, HERMES was installed at the JULIC (Jülich Light Ion Cyclotron) Neutron Platform as part of a collaboration between the Laboratoire Léon Brillouin and the Jülich Centre for Neutron Science. The main goal of the current setup is to probe the viability of neutron instrumentation at a High Current Compact Accelerator-driven Neutron Source (HiCANS). As the flux at the JULIC neutron platform is several orders of magnitude lower than the original Orphée flux or the expected flux for a HiCANS, our current objective is to perform reflectivity experiments with supermirrors as a proof of concept. Nevertheless, Monte-Carlo simulations showed that the HERMES instrument's performance at a HiCANS such as HBS or ICONE could match that of reflectometry instruments operating at research reactors or spallation sources. An experiment with a supermirror carried out in December 2022 allowed us to preliminary prove the feasibility of this kind of experiments at an accelerator-driven neutron source.

1 Introduction

High current Compact Accelerator-driven Neutron Sources (HiCANS) have risen in recent years as a possible answer to the closure of various research reactors in Europe and the drop in beam time availability for the neutron user's community. This interest was motivated by the previous experience in compact accelerator-driven neutron sources [1, 2]. Several projects in numerous laboratories around Europe [3-6] are currently at different stages of their development and are merged within the European Low Energy accelerator-based Neutron facilities Association (ELENA) [7].

The Laboratoire Léon Brillouin is currently evaluating the use of HiCANS to provide the French neutron scattering community with a suite of world-class instruments. For that purpose, the performances of potential instruments at this novel type of source must be evaluated.

To achieve this goal and through a collaboration with the Jülich Centre for Neutron Science (JCNS), the HERMES reflectometer (see Fig. 1) has been installed at the JULIC neutron platform (JNP) at the JULIC accelerator [8] of the Institute for Nuclear Physics (IKP) Forschungszentrum Jülich.

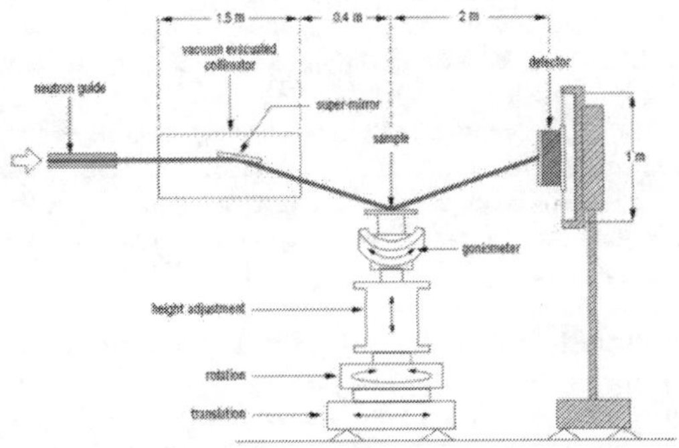

Fig. 1. Scheme of the HERMES reflectometer.

HERMES [9-11] is a time-of-flight horizontal reflectometer designed for soft-matter studies that operated at the Orphée reactor [12] until 2019.

The JULIC Neutron Platform (JNP) shown in Figure 2 is conceived as a technological test platform for the High Brilliance Neutron Source project (HBS) [13].

* Corresponding author: mariano-andres.paulin@cea.fr

Fig. 2. The JULIC Neutron Platform at the FZJ.

Taking advantage of the JULIC proton cyclotron accelerator at Forschungszentrum Jülich, the Nuclear Physics Institute (IKP) has built a dedicated proton beamline to inject protons into the target-moderator-reflector (TMR) assembly built by JCNS.

Neutrons are generated by the impact of 45 MeV protons on a tantalum target and moderated by a polyethylene block surrounded by a lead reflector. The TMR assembly is embedded in a multilayer shielding with eight extraction channels designed to fit several instruments.

Furthermore, the TMR assembly can accommodate dedicated cold moderators for each channel. The source is able to deliver a cold spectrum with pulses in the 100 µs – 2 s range and is very well suited to evaluate the feasibility of reflectivity experiments at a HiCANS.

Reflectometry measurements are especially suited to evaluate the performances of a HiCANS source: (i) The operation in Time–of–Flight (ToF) mode is very efficient for reflectivity measurement; (ii) Some samples are very reflecting so that it is possible to perform demonstration measurements even with a very low incident flux; (iii) As the instrument is in principle built to measure reflectivity over 6 orders of magnitude in intensity it is very easy to assess the impact of the background noise on the measurements which is a key information which will determine the ultimate performances of instruments around HiCANS.

2 Monte-Carlo simulations

2.1 Neutron Source

Monte-Carlo ray tracing simulations were carried out to evaluate the instrument performance using McStas [14] software. In Table 1, the main parameters for the neutron source are shown for two scenarios. The first one is the projected flux for the High Brilliance Neutron Source (HBS) [4, 15]. The second one is the best-case scenario expected for the JULIC Neutron Platform which is currently only operating at a very low power.

Table 1. Neutron source parameters for HBS and the JULIC neutron platform (JNP) at the maximum projected current.

	HBS	JNP
ν [Hz]	24	50
Duty Cycle	1.6 %	4 %
E_{proton} [MeV]	70	45
$I_{proton, peak}$ [mA]	89.3	0.01
$\phi_{0, cold}$ (60 K) [n/cm².sr.pulse]	7.4 10^{10}	3.0 10^{6}
$\phi_{0, thml}$ (305 K) [n/cm².sr.pulse]	2.3 10^{10}	9.5 10^{5}
$\phi_{0, um}$ (UM unmoderated) [n/cm²sr.pulse]	1.6 10^{9}	6.5 10^{4}

The bi-spectral source is modeled with three components (Maxwellian) corresponding to the cold-moderated neutrons (60 K), the thermal-moderated neutrons (305 K), and the under-moderated ones (UM). This last component represents the neutrons that are not properly moderated by the thermal premoderator and the cold moderator. The flux corresponding to each component (cold, thermal, and under-moderated) is detailed in Table 1 and the simulated spectrum is shown in Figure 3.

Both for HBS and JNP, the flux is calculated considering a para-hydrogen cold moderator inserted in the extraction plug dedicated to the instrument.

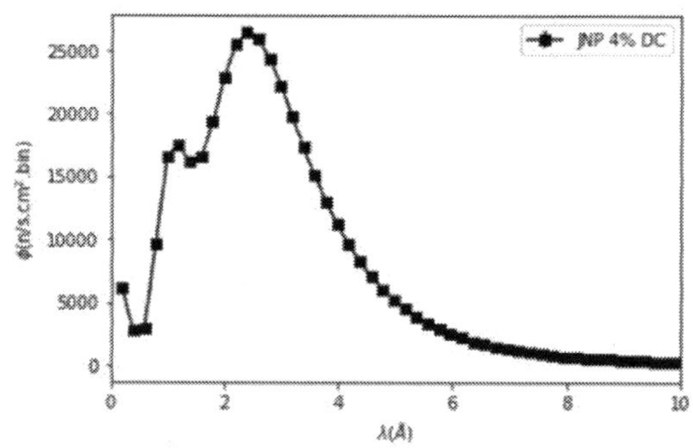

Fig. 3. Simulated source spectra at the JNP at 10 µA and 4 % duty-cycle (bin size: 0.2 Å).

2.2 Guide system

The design of the TMR allowed us to install a dedicated 2.5 m neutron extraction guide for HERMES. Being a horizontal reflectometer, only a minimum beam vertical divergence is useful, and therefore m = 1 coating for the top and bottom mirrors was chosen. Several geometries were analyzed for the lateral walls of the guide including semi-elliptical, trumpet, and straight ones (see Fig. 4). For the lateral walls, a higher m value would increase the total flux on the sample and thus m = 2 was considered for that purpose. Choosing a higher m-value (m > 2) would increase the horizontal divergence of the beam and thus the flux at the guide's exit. Nevertheless,

as HERMES was originally built with m = 2 guides along the instrument, the increase in flux wouldn't necessarily benefit the instrument's performance.
The case of a straight guide with m = 1.2 on all sides was also considered as it matches the current extraction guide configuration at JNP.

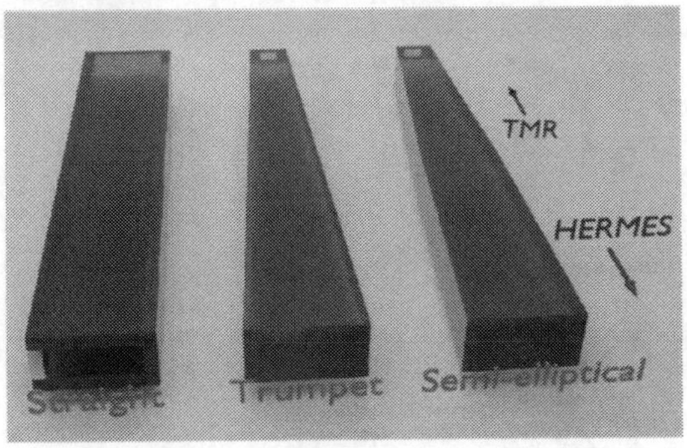

Fig. 4. Scheme of the different guide geometries evaluated to increase HERMES performance.

As it is shown in Table 2, except for the m = 1.2 guide, no major differences are observed for the different guide configurations. Due to the complications and costs associated with complex guide geometries, the choice was to stick to a straight guide.

Table 2. Estimated flux at the sample position for different guide configurations at JNP at 10 μA and 4 % duty-cycle.

Géométrie	ϕ_{sample} [n/s.cm²]
Straight m = 1.2	2.16 10³(1)
Straight	3.00 10³(1)
Trumpet	3.27 10³(1)
Semi-elliptical	3.40 10³(1)

2.3 Instrument

A 20 cm gap was left between the end of the guide and the collimator to allow the insertion of a frame-overlap chopper. The HERMES collimator consists of four variable slits with a maximum collimation distance (d_{14}) of 1.5 m. The slits have a 3 cm horizontal opening and, for these simulations, a 2 mm vertical opening was selected corresponding to a vertical divergence of 0.15°.

Two horizontal m = 3.5 supermirrors sit between slits 2, 3 and 4 and can be used to deflect the beam for air-liquid reflectivity experiments. Also, vertical m = 2 supermirrors transport the beam along the collimator to take advantage of the beam's horizontal divergence. A schematic top view of the instrument is shown in Figure 5 including its principal dimensions.

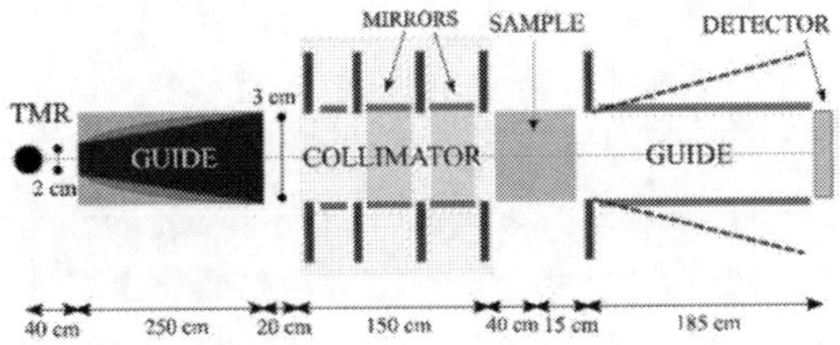

Fig. 5. Top view scheme of HERMES reflectometer at JNP with different guide configurations: straight (grey), trumpet (black) and semi-elliptical (red).

The sample sits at a distance of 40 cm from the last slit. An "anti-background" slit with an adjustable vertical opening is positioned at 15 cm from the center of the sample. Finally, a single He^3 tube detector (P = 10 bar) shielded with B_4C is located at 2 m from the sample position.

The beam reflected by the sample has a rather high horizontal divergence (see blue dashed line in Fig. 5) and thus a large detector would be needed to count all the reflected neutrons. A neutron guide with m = 2 supermirrors on the sides and B_4C on top and bottom is placed between the "anti-background" slit and the detector. This guide allows to harness all the horizontal divergence of the beam with a rather small detector area thus avoiding the background signal due to spurious neutrons arriving at the detector.

2.4 Reflectivity for a 20 nm Ni layer

For these Monte-Carlo simulations, we chose a 10 x 10 cm² sample consisting of a Ni layer (20 nm) on Si. Due to ToF resolution limitations with long pulses, a 200 μs pulse length was chosen for these simulations. This implies a duty-cycle of 1% for JNP and 0.48% for HBS.

In Table 3, the estimated flux at the source and at the sample position is presented for the two scenarios.

Table 3. Estimated flux at the source and the sample position for the JNP at 1% duty-cycle and HBS at 0.48% duty-cycle.

Source	ϕ_{source} [n/s.cm²]	ϕ_{sample} [n/s.cm²]
JNP	1.01 10⁵ (1)	7.60 10²(1)
HBS	1.42 10⁹(1)	1.07 10⁷(1)

If higher fluxes are needed, it can be easily increased by at least a factor 4 by just increasing the duty cycle.

In Figure 6, we show the simulated reflectivity curve for the JNP configuration for a θ = 0.8° incident angle.

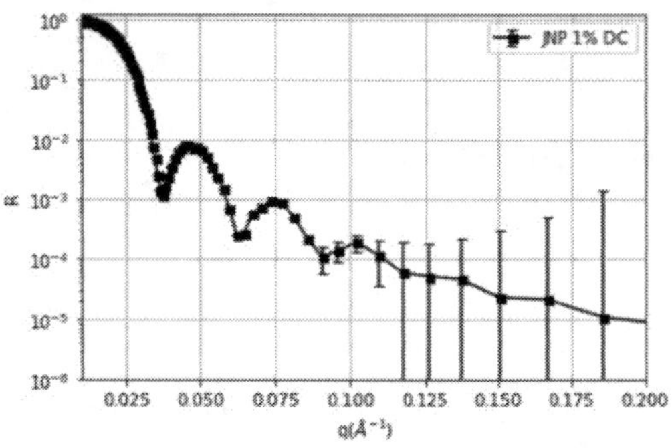

Fig. 6. Simulated reflectivity curve for a 20 nm Ni layer on Si on HERMES at the JULIC neutron platform.

Error bars were estimated assuming a 1 h collecting time. The Kiessig fringes are noticeable below q = 0.1 Å^{-1}, and reflectivities down to 10^{-4} should be measurable provided that the background noise is kept low enough. We can also extrapolate this performance to the HBS flux and, in that case, reflectivities below 10^{-5} should be easily achieved.

2.5 Frame-overlap

Count rate values in McStas are expressed in n/s, but actually, only one single neutron pulse is simulated by the code. If we want to consider the effect of frame-overlap (FO) in the case of the absence of a FO chopper, we need to add manually the frames. In this case, shown in Figure 7, the FO effect is clear for a 50 Hz frequency (20 ms frames).

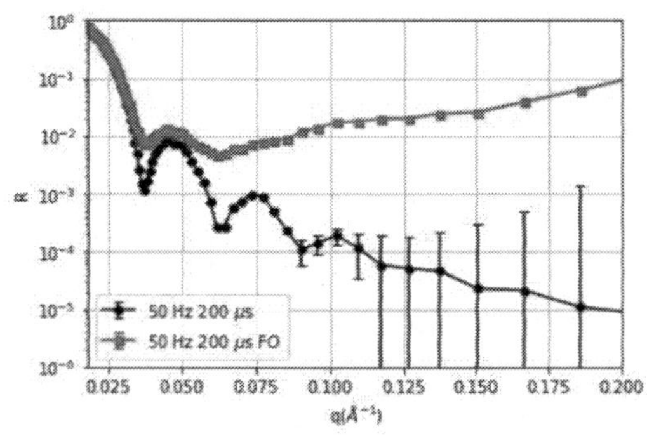

Fig. 7. Simulated reflectivity curve for a 20 nm Ni layer on Si with and without frame-overlap effect.

The effect of the FO entails that only one fringe is clearly observed, thus indicating that the use of a FO chopper or similar device is needed if we want to be able to resolve this kind of structure operating at those frequencies. Another simple solution would be to reduce the operation frequency, at the cost of reduced efficiency.

2.6 Pulse length and resolution

As it was mentioned before, an easy way to increase the flux at the sample position is to increase the proton pulse length. The tradeoff is that the instrument resolution is reduced thus limiting the samples that can be characterized.

In order to evaluate this effect, we simulated different pulse lengths ranging from 200 μs to 800 μs. The wavelength resolution for those pulse lengths is shown in Figure 8.

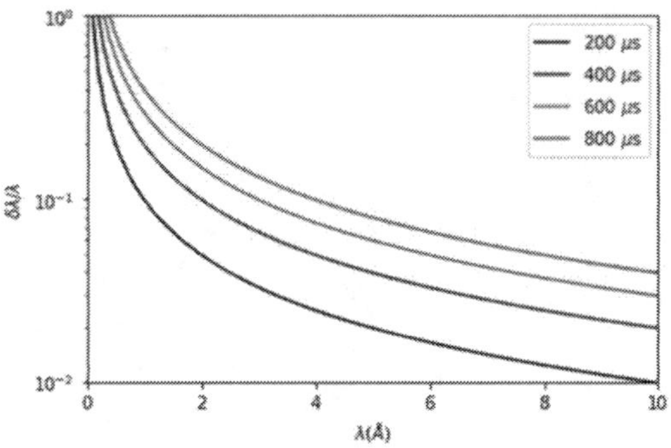

Fig. 8. Wavelength resolution for different pulse lengths.

The simulated reflectivity curves for the extreme cases are shown in Figure 9. For the reference sample (20 nm Ni on Si) and, assuming that there is no background noise or rugosity on the sample that could worsen the reflectivity curve, we are still able to distinguish the Kiessig fringes up to 0.1 Å^{-1} even for long pulses and gain a factor 4 in total flux.

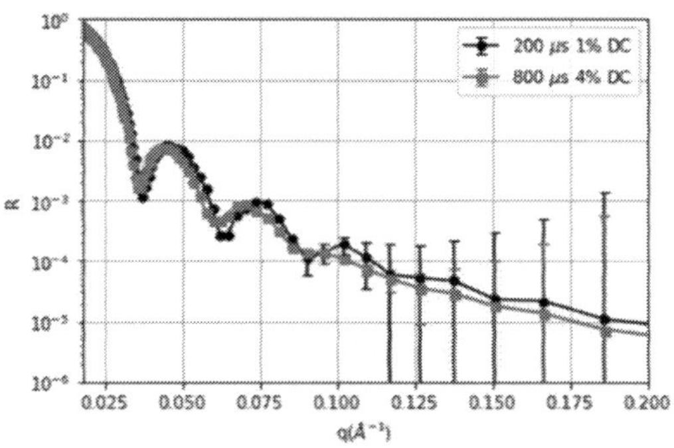

Fig. 9. Simulated reflectivity curve for a 20 nm Ni layer on Si for different proton pulse lengths.

3 First experiments at the JULIC neutron platform

HERMES was installed at the extraction channel 3 of the TMR in December 2022. Due to the lack of a cold moderator at that time, experiments were performed with thermal neutrons. The parameters for the JULIC proton pulse were $I_p \approx 250$ nA, $\tau = 400$ μs, $\nu = 125$ Hz corresponding to a power of 0.6 W on the target. The current neutron guide is an m = 1.2 straight guide with a 2.5 cm x 3 cm section and 2.5 m in length. A 40 x 250 mm^2 m = 4 supermirror provided by SDH was measured for 2.5 hours. The incident angle θ = 0.6° was chosen to match the peak flux with the expected critical edge. The direct and the reflected beam are shown in Figure 10, the high count rate at low ToF is due to the fast neutrons from the prompt pulse being moderated by materials close to the detector.

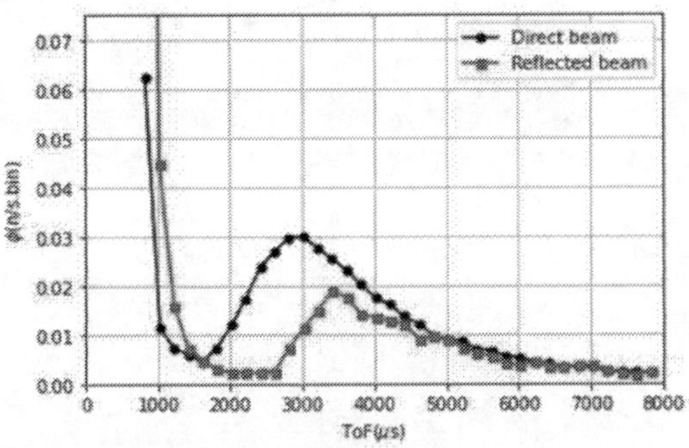

Fig. 10. Direct beam and reflected beam by an m = 4 supermirror measured at HERMES (bin size: 200 μs).

In Figure 11, the measured reflectivity curve along with Monte-Carlo simulations are shown. McStas simulations allowed us to distinguish the frame-overlap component from the contribution of the prompt neutron pulse.

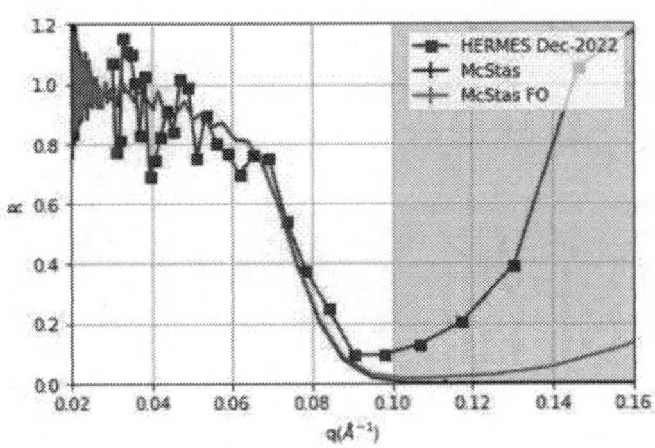

Fig. 11. Measured reflectivity curve for an m = 4 supermirror and McStas simulations.

It can be inferred that, although there is a contribution from the frame-overlap effect, the apparent rise in the reflectivity for q > 0.1 $Å^{-1}$ is mainly due to the fast neutrons being moderated close to the detector creating an intense peak for short times. Dealing with background neutrons at HiCANS is one of the most challenging tasks and needs to be accounted for to reach reflectivities below 10^{-5}. For these experiments, the detector's shielding was merely a layer of B_4C that proved to be effective when the instrument was installed at Orphée. This might not be enough for this type of facility and additional shielding is currently being evaluated.

4 Conclusion and prospects

The very first experiments performed with HERMES at the JULIC neutron platform are encouraging. An operation at a power of 0.6 W allowed measuring reflectivities over one order of magnitude in reflectivity. A rough scaling to a power of 100 kW (HBS design) suggests that it should be possible to readily measure reflectivities below 10^{-5}. While these performances would not yet match existing state-of-the-art instruments, the current measurements have been performed in non-optimal conditions.

Several upgrades can be implemented to HERMES to improve its performance and are currently undergoing. The first major upgrade will be the installation of the para-hydrogen cold source, which should provide an increase of at least one order of magnitude in flux and would allow us to use higher reflection angles to avoid the direct beam.

The HERMES shielding was originally optimized for its position at the Orphée reactor in the guide hall, far away from the reactor's core. An improved shielding is currently being designed and built in order to improve the signal-to-noise ratio at JNP to reduce the influence of the prompt neutron pulse.

Although the background component is currently dominating over the frame-overlap effect, the incorporation of a frame-overlap chopper or a frame-overlap mirror to cut higher wavelengths is being evaluated.

Finally, the JULIC accelerator and the TMR are technically capable of significantly increasing the proton current, which would directly increase the neutron flux.

Based on the Monte-Carlo simulations we are confident that an optimized instrument at the JULIC neutron platform should be able to deliver sufficient performances to characterize large surface samples such as neutron mirrors, something that is currently demanded by the industry.

Acknowledgments This work is part of the collaboration within ELENA and LENS on the development of HiCANS. It has been funded by the "CANS Inflexion" program at the CEA and the "IPHI-Neutron" SESAME project of the Île de France region.

References

1. D. V. Baxter, Neutron News **31**, 44-47 (2020)
2. Y. Otake, Neutron News **31**, 32-36 (2020)
3. F. Ott, M. Plazanet, ICONE, une nouvelle source de diffusion neutronique française (2023)
4. T. Gutberlet, U. Rücker, P. Zakalek, T. Cronert, J. Voigt, J. Baggemann, P.E. Doege, E. Mauerhofer, S. Böhm, J.P. Dabruck, R. Nabbi, M. Butzek, M. Klaus, C. Lange, T. Brückel, Phys. B: Condens. Matter **570**, 345-348 (2019)
5. T. Gutberlet, U. Rücker, E. Mauerhofer, P.Zakalek, T. Cronert, J. Voigt , J. Baggemann, J. Li, P. Doege, S. Böhm, M. Rimmler, O. Felden, R. Gebel, O. Meusel, H. Podlech , W. Barth, T. Brückel, Neutron News **31**, 37-43 (2020)
6. M. Pérez, F. Sordo, I. Bustinduy, J.L. Muñoz, F. J. Villacorta, Neutron News 31, 19-25 (2020)
7. http://www.elena-neutron.eu

8. O. Felden, N. Demary, N.-O. Fröhlich, R. Gebel, M. Rimmler, Y. Valdau, 22nd Int. Conf. on Cyclotrons and their Applications (2019)
9. F. Cousin, F. Ott, F. Gibert, A. Menelle, Eur. Phys. J. Plus **126**, 109 (2011)
10. G. Battaglin, A. Menelle, M. Montecchi, E. Nichelatti, P. Polato, Glass Technol. **43**, 203-208 (2002)
11. D. Lairez, A. Chennevière, F. Ott, J. Appl. Crystallogr. **53**, 464-476 (2020)
12. B. Farnoux, D. Cribier, Phys. B+C **120**, 31-36 (1983)
13. T. Brückel, T. Gutberlet, Conceptual Design Report Jülich High Brilliance Neutron Source (HBS) **8** (2020)
14. P. Willendrup, K. Lefmann, J. of Neutron Res. **22**, 1-16 (2020)
15. T. Brückel, T. Gutberlet, K. Lieutenant, J. Voigt, Technical Design Report HBS **3** (2023)

https://doi.org/10.1051/epjconf/202328603004

Polarisation Development at the European Spallation Source

Wai Tung Lee[1*], *Joel* Hagman[1], *Damian* Martin Rodriguez[1], *Annika* Stellhorn[1,2], *Alex* Backs[1,2], *Thomas* Arnold[1,3], *Elizabeth* Blackburn[2], *Pascale* Deen[1], *Celine* Durniak[1], *Mikhail* Feygenson[1,4,5], *Alexander* T. Holmes[1], *Judith* Houston[1], *Sebastian* Jaksch[1,4], *Oliver* Kirstein[1], *Dan* Mannix[1], *Martin* Månsson[6], *Manuel* Morgano[1], *Gøran* Nilsen[3], *Daria* Noferini[1], *Tommy* Nylander[2], *Dmytro* Orlov[2], *Valentina* Santoro[1], *Søren* Schmidt[1], *Michael* Schulz[7], *Werner* Schweika[1,4], *Markus* Strobl[8], *Aureliano* Tartaglione[1,7] , *Rasmus* Toft-Petersen[1,9], *Félix J.* Villacorta[1,10], *Peter* Willendrup[1,9], *Maximillian* Wolff[5], and *Robin* Woracek[1].

[1]European Spallation Source ERIC, 224 84 Lund, Sweden
[2]Lunds Universitet, 221 00 Lund, Sweden
[3]ISIS Neutron and Muon Source, STFC, Didcot, OX11 0QX, UK
[4]Forschungszentrum Jülich, D-52425 Jülich, Germany
[5]Uppsala Universitet, 751 05 Uppsala, Sweden
[6]KTH Royal Institute of Technology, SE-100 44 Stockholm, Sweden
[7]Technische Universität München, 85747 Garching, Germany
[8]Paul Scherrer Institut, 5232 Villigen PSI, Switzerland
[9]Technical University of Denmark, 2800 Kongens Lyngby, Denmark
[10]ESS Bilbao, 48160 Derio, Spain

Abstract. To meet the ever-increasing user demand, eleven of the fifteen European Spallation Source (ESS) instruments under construction aim to offer polarised neutrons for user experiments. They include an imaging instrument, a SANS instruments, two reflectometers, three diffractometers, and four spectrometers. In conjunction with in-kind contributions and instrumentation grants, the ESS Polarisation Project will support the incorporation of polarisation analysis on eight of the eleven instruments. The project aims to deliver polarised neutrons for first-science experiments as instruments enter operation. Different polariser and polarisation analyser techniques will be available to accommodate the specifics of experiments on a given instrument. Polarised ^{3}He neutron spin filter using either Metastable Optical Pumping (MEOP) or Spin-Exchange Optical Pumping (SEOP) techniques will provide shared-use equipment among many instruments, with SEOP's main application being *in situ* beam-polarisation. Several instruments will also use polarising-supermirror devices. To provide wide-bandwidth spin-flipping capability to the time-of-flight instruments, Adiabatic Fast Passage (AFP) neutron spin flippers, also known as gradient-field radio-frequency spin flippers will be the main method of choice. Devices based on the same AFP principle will also be used to flip ^{3}He nuclear spins. We are constructing our first ^{3}He polariser setup, including field coils to produce highly uniform magnetic field. Monte Carlo simulations are being done for the supermirror polarisers. To ensure science-focused development, we are working with university partners in doing scientific experiments with polarised neutrons. These are some of the activities developing polarisation analysis for ESS instruments in our project.

1 Introduction

Polarised neutrons are used extensively to study magnetism: in diffraction to study atomic-scale magnetic densities [1] and structures, Small Angle Neutron Scattering (SANS) to investigate magnetic nano-scale structures [2], reflectometry to probe magnetic multilayers and interfaces [3], spectroscopy to study magnetic dynamics [4], imaging to observe buried magnetic domains, mesoscopic structures and magnetic field distributions [5], and magnetic contrast in reflectometry is instrumental for solving the inverse-problem to obtain depth profiles directly with minimal use of modelling [6, 7]. Polarised neutrons have also proven to be an important complementary tool to deuteration, leading to accurate separation of coherent from spin-incoherent scattering, thereby separating coherent motion from single-particle motion [8]. Understanding these motions is of outmost importance for the field of sustainable energy materials, e.g., H-storage, rechargeable batteries, photovoltaic materials. In addition, the investigation of dynamical processes greatly benefits from the capabilities of high-resolution Neutron Spin Echo spectroscopy [9]. In fundamental physics, the search for the neutron electric dipole moment, the study of symmetry violation, and neutron life-time measurement are some of the topics that use polarised neutrons.

Neutron facilities have a long history of supporting their users' research by providing polarised neutrons. There has been a doubling of this capacity in the past two decades. This expansion continues: in existing

* Corresponding author: waitung.lee@ess.eu

facilities, 25 new and 13 existing unpolarised instruments plan to include polarised neutron capabilities. Together with 12 ESS instruments, there will be 50 more instruments around the world equipped with polarised neutron capabilities in the next 10 to 15 years. This increase is driven by increasing user demand, which in turn can be accounted for by the broader availability of polarised neutron technology, the two reinforcing each other in a virtuous cycle.

Few instruments had polarised neutrons capabilities in the early 1990s. This was due both to the relatively low neutron flux available for intensity-limited polarised neutron applications and to the limitations of the polarised neutron technology to accommodate experimental and instrumentational requirements. Since then, advances in polarising supermirror technology [10–12] – higher m-values, stacked supermirrors, and more complex geometries have expanded the energy range, angular divergence, and beam cross-section while reducing the length of the devices. At the same time, polarised ^{3}He neutron spin filter techniques have evolved far beyond the polarisation threshold and production volume required for scattering applications. Metastable Optical Pumping (MEOP) has achieved high throughput at high polarisation [13]. *In situ* Spin Exchange Optical Pumping (SEOP) was developed to continuously polarise the ^{3}He gas in the spin filter on the neutron beam. It keeps the neutron polarisation and transmission of the spin filter a constant during an experiment [14, 15]. Implementation of a nuclear magnetic resonance technique called "Adiabatic Fast Passage" (AFP) to change the ^{3}He polarisation with high efficiency [14, 15] allows the selection of the transmitted neutron spin state over almost any neutron energies. These continuous technological advancements have transformed polarised neutrons from a scarce to a widely available resource.

The increase in polarised neutron usage can also be quantified to some extent by searching a range of keywords relevant to polarised neutron in Google Scholar. The search results are only indicative and not an exact counting of publications that reported studies using polarised neutrons. The results nonetheless show a trend in the increasing use of polarised neutrons in research. Between the year 1999 and 2019, the polarised neutron scattering keywords hit-count almost tripled to 889 while "neutron scattering" hit-count almost doubled to 12800. Polarised neutron usage has outpaced the increase in using neutron scattering. While polarised neutron publications amount to 7% of neutron scattering publications, a significant number of them are published in high-impact journals. This is not a coincidence; the unique ability of polarised neutrons is often used to address the most difficult scientific questions that cannot be resolved using other techniques, resulting in a higher impact of polarised neutron publications.

2 Instruments with polarised neutrons

The ESS embraces a road map that will see twelve of the fifteen instruments that are currently under construction [16], have polarised neutron beam capabilities [17].

Polarisation components for the MAGiC diffractometer, the ESTIA reflectometer and the T-REX spectrometer are fully included in their in-kind instrument construction projects. They all involve using polarising supermirrors, and T-REX will also use polarised ^{3}He devices. The details of these instruments are given in another article [16] and not covered in this article.

Table 1. ESS instruments with polarised neutron option.

Class	Name	Application
Diffractometer	DREAM	Bispectral diffraction + nm-SANS
	HEIMDAL	Hybrid diffraction + SANS + Imaging
	MAGiC	Magnetism single-crystal diffractometer
Imaging	ODIN	Imaging
Reflectometer	ESTIA	Polarised neutron focusing reflectometer
	FREIA	Liquids reflectometer
SANS	SKADI	SANS (magnetism + soft-matter)
Spectrometer	BIFROST	Extreme environment spectrometer
	CSPEC	Cold chopper spectrometer
	MIRACLES	Backscattering spectrometer
	T-REX	Bispectral chopper spectrometer

This article focuses on the ESS Polarisation Project, which will provide polarisation components to the other eight instruments in conjunction with in-kind contributions to the SKADI polariser, a grant-supported DREAM polariser by the DREAM team, a grant supporting ODIN, SKADI and LoKI polarised instrumentation, and a grant for BIFROST and CSPEC analysers.

The characteristics of each instrument, the characteristics of the available technologies for polariser, analyser, and spin-flipper, and the requirements of the spin-transport and sample magnetic field, determine the layout and configuration of the polarisation setup. A detailed evaluation of the equipment choices however, is not the focus of this article. Here we focus on specifying the polarisation instrumentation for the nine instruments in the project and providing an update on some of the developments.

It is important that the needs for scientific experiments should guide the instrumentation development. In this project, we have been engaging researchers in universities to carry out scientific studies using polarised neutrons in other neutron facilities. In particular, our collaborations have reported both polarised SANS and polarised imaging studies in this conference [18, 19]. In carrying out these experiments, the experimental needs have informed the designs of the polarisation setup for our instruments. In return, our instrumentation knowhow and capabilities have assisted in both analysing the measurement data and developing the experimental setup.

2.1 Diffractometers

The DREAM [20] and HEIMDAL [16] neutron powder diffractometers have complementary science cases, with design features suitable for adopting similar polarization equipment. The wavelength band of both instruments covers 0.5 Å to 4 Å and 4 Å to 10 Å, respectively, with the shorter wavelengths for diffraction and pair-distribution function measurements, and the longer wavelengths for SANS measurements, and in the case of HEIMDAL for imaging. DREAM features a low-angle scattering detector for covering limited q-range and suitable for nanoscale materials studies (nm-SANS), while the HEIMDAL SANS detector provides a conventional SANS q-coverage. The ^{3}He polariser for cold neutrons at DREAM is provided by a grant (see Acknowledgement). The thermal neutron polariser cell for DREAM and HEIMDAL will be provided by the ESS Polarisation Project. AFP is implemented in the ^{3}He polariser to select the neutron spin state. This method flips the ^{3}He polarisation and switches which neutron spin state is filtered. Polarising supermirror device for cold neutrons are currently being evaluated for both instruments.

Diffraction instruments have large detector coverage. SANS and imaging applications need to avoid background due to potential SANS scattering from supermirrors. Polarised ^{3}He spin filters are therefore best suited for the analyser for DREAM and HEIMDAL. The project will provide the analyser setup to both instruments. The wide-angle component will cover the large scattering angles in diffraction and a shorter cold neutron analyser will support SANS and imaging.

A coil set producing a 3D field with mT field strength will provide the uniform holding field for the ^{3}He analyser and allow the user to apply the field in any direction at the sample position. For diffraction, the technology to flip the analyser polarisation in a wide-angle analyser is currently not available, making it challenging to separate magnetic and nuclear scattering. The 3D field will allow the use of the "XYZ method" [21–23] to extract the magnetic scattering signals. Using finite-element field calculations, we have designed a coil set providing a large unobstructed angular range in both the horizontal and vertical directions.

The ^{3}He analyser for SANS and imaging has a narrower profile and can be placed further from the sample position. The AFP technique to flip the ^{3}He polarisation for such geometry is readily available.

2.2 Imaging instrument

At ODIN [16], a supermirror polariser will polarise the incoming beam where it enters the instrument cave. The polariser operates in a wavelength band from 2.5 Å to 11 Å. Despite the beam divergence at this location, Monte Carlo simulations using McStas [24, 25] show that a multi-channel v-cavity can deliver a beam with polarisation > 95% (see Fig. 1) and transmission > 40% (not shown, see the following referenced article) for an area of 10 cm × 10 cm at 5 m from the pinhole [26]. More calculations are underway to optimise the polariser to the beam geometry.

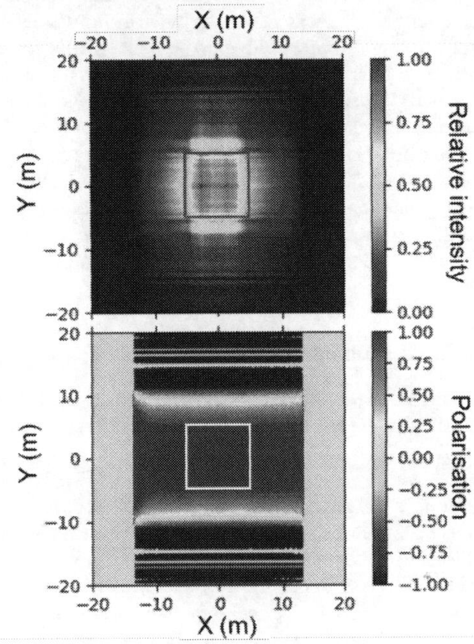

Fig. 1. ODIN polariser Monte Carlo simulation results showing (a) relative beam intensity normalise to the highest intensity value and (b) polarisation at 5 m from the pinhole. The beam is diverging at this location. A square is superposed to the figures to identify the centre 10 cm × 10 cm area. The beam has high intensity and the polarisation is > 95% in that area. Reference [26] gives the details of this simulation and also includes simulations for BIFROST and MIRACLES.

Other than the "remnant polariser" [27], a setup using a polarising supermirror device such as the ODIN polariser described above requires a neutron spin flipper to enable measurements with both spin+ and spin− states consecutively. Based on the same principle, an AFP neutron spin flipper [28] has a wide neutron wavelength working range and places no material in the neutron beam. It is hence a suitable choice for the ESS instruments and in particular for imaging.

To obtain the highest possible spatial resolution in neutron imaging, the sample is typically placed as close to the detector as possible. In polarised neutron imaging, the analyser has to be placed behind the sample and in order to minimise the sample-to-detector distance, a customised compact supermirror analyser that matches the beam characteristics will be made available.

To allow for further extension of the wavelength range, the guide fields will be designed to be highly uniform to accommodate a ^{3}He analyser. In addition, polarised imaging applications often use custom designed, application specific, spin manipulation equipment that requires precise control of the magnetic field. Using another uniform field compensation coil to enclose the spin manipulation components and applying a field in the opposite direction, the highly uniform guide field can be cancelled and thus preventing the guide field from interfering with the spin manipulation devices.

Polarised ^{3}He is susceptible to magnetic interference. This will restrict how close a magnetised sample in a sample field can be placed to the ^{3}He analyser and therefore to the detector. This in turn limits the spatial resolution of the measurements. Wolter optics [5, 29] is currently being developed at a number of facilities. It has many advantages, including the ability to focus a neutron beam at the detector. This will allow greater distances between the sample, the analyser and the detector without significantly degrading spatial resolution. Thereby allowing the full utilisation of the ^{3}He analyser.

2.3 Reflectometer

The main use for polarised neutrons on the FREIA reflectometer [16] is the magnetic reference layer technique [6, 7]. The wavelength band usually spans from 2 Å to 11.3 Å. The unique design of FREIA uses an elliptical guide to deliver a wide (around 3.5°) divergence to the sample, from which up to three collimated beams are selected in order to vary the incident angle without the need to move the sample. Since the polarised ^{3}He technique is considerably less sensitive to beam divergence, it is the preferred method for this instrument. An *in situ* SEOP polariser with AFP flipping of the ^{3}He polarisation will be used to keep the polarisation constant during an experiment. We will also investigate an analyser setup that can operate within the geometry of the scattered beam.

The magnetic reference layer method requires a sample field to manipulate the magnetisation of the reference layer. The location and magnetic shielding of the polariser, and the sample magnet design will prevent the magnetic fringing field from interfering with the polariser. There are also various steel components in some of the slits in the current collimation system design, which are close to the polarised beam. The effect on the spin-transport guide field has been investigated using finite-element magnetic field calculation which shows that strengthening the guide field at several locations will mitigate the interference to the polarised beam.

2.4 SANS instruments

The wavelength range for this project are similar for both SKADI [30] and LoKI [16], from 3 Å to 10 Å. SKADI will be the main polarised SANS instrument for users and comes with its own supermirror polariser and incident beam neutron spin flipper. While LoKI does not focus on polarisation in its application, the earlier commissioning of LoKI however allows us to test the polarisation equipment for SKADI in advance.

Both SANS instruments will use ^{3}He analysers to avoid background due to potential SANS scattering from supermirrors. There are two options provided: a uniform 3D field coil set that allows a large scattering angle for SANS, and a magnetostatic cavity that can work with a Tesla-range sample field will be provided for the analyser. The analyser will feature *in situ* SEOP to maintain a static analysing efficiency during an experiment. The selection of neutron spin state will be facilitated by AFP flipping of ^{3}He polarisation.

2.5 Spectrometers

The polarisation setup for three spectrometers, BIFROST [16], CSPEC [31], and MIRACLES [16] is supported by this project, with an additional grant partially supporting the polarisation setup on BIFROST and CSPEC. The shared polarised ^{3}He equipment can be used on T-REX if needed.

BIFROST will use polarising supermirrors for both its polariser and wide-angle analyser, which match the beam characteristic and geometry of the instrument. Both devices will be designed to cover an energy band from 2.3 meV to 13 meV. The polariser, a multi-channel v-cavity device will be located at the end of the straight section of an elliptical neutron guide at 19 m from the sample position. At this location, the neutron beam will have less than 0.2° divergence. Monte Carlo simulations using McStas show that a multi-channel v-cavity polariser can produce a beam with polarisation > 95% and transmission > 40% [26].

The polarisation analyser of BIFROST will accept ±14° vertical angle. It is therefore viable to use a supermirror analyser and benefit from the high transmission and analysing efficiency. The analyser is partially supported by a grant which supports the construction of a similar device at the ISIS Neutron and Muon Source in UK. A preliminary magnetic field calculation using finite element calculation software COMSOL Multiphysics [32] was performed to evaluate the uniformity and field strength required to magnetise the supermirrors.

CSPEC plans to use a supermirror polariser at a focusing part of the neutron guide. The design will focus on 3 Å to 10 Å neutrons. Given the large detector coverage, a ^{3}He wide-angle analyser will be used together with the 3D uniform field coil mentioned above. The grant mentioned above also supports the development of the wide-angle analyser.

The polarisation setup at MIRACLES is illustrated in Fig. 2. The polariser is located before the focusing section of its elliptical neutron guide at 19 m from the sample position. Likewise, we have used McStas to simulate the polariser and found that a v-cavity polariser will polarise neutrons at wavelengths longer than 5 Å with polarisation > 95% and transmission > 40% [26].

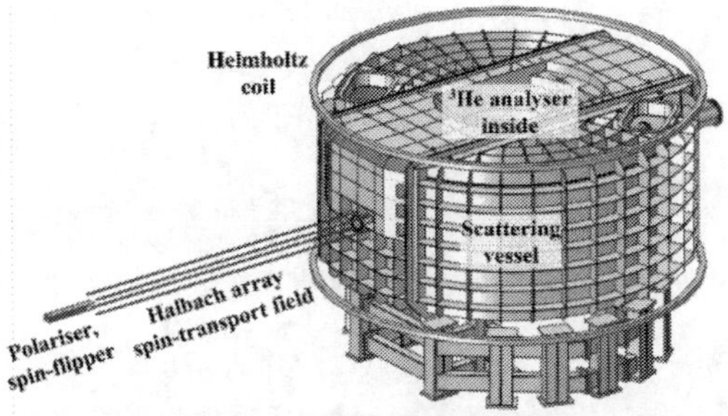

Fig. 2. Layout of MIRACLES polarisation setup. The analyser cell is located inside the sample chamber.

A ^{3}He wide-angle analyser will be located in the sample chamber and configured as a sample environment insert. The setup simplifies the setting up experiments without the need to break the vacuum in the scattering vessel. Helmholtz coils outside the scattering vessel will provide the magnetic field environment needed for the ^{3}He analyser and for spin-transport. Rather than placing the coils inside the scattering vessel, the placement avoids the issue of cooling large DC coils in the vacuum environment and potential interference with neutron flight paths by the coils' support structure.

To ensure neutron spin-transport from the polariser on MIRACLES, and similarly on BIFROST to their respective sample position 19 m away, Halbach array consisting of four lines of permanent magnets will provide the guide field along the neutron flight path.

2.6 Test beam line

The Test Beam Line is one of the instruments in the approved suit of ESS instruments. Besides several other tasks, it will provide a test bed for polarisation equipment. We are investigating the beam equipment that can be tested within the constraints of the instrument (the average length is ~3 m in front of the beam stop).

3 Shared polarised ^{3}He equipment

Polarised ^{3}He equipment can be shared among different instruments without sacrificing the performance. Shared devices are used in our design where it is applicable and beneficial.

3.1 Neutron spin filter

Polarised ^{3}He neutron spin filters are separated into two classes associated with the ^{3}He cell geometry: cylindrical cells and wide-angle cells.

Cylindrical cells are typically 5 cm to 15 cm in diameter and 5 cm to 15 cm in length. The neutron beam enters and exits the cell through the circular faces. The applications are polarisers, and analysers for SANS and imaging. The technology for using this device with *in situ* SEOP gas-polarising is well-developed and readily available [14, 15]. A majority of applications of cylindrical cells at the ESS will be *in situ* SEOP setup.

Without continuously polarising the gas, ^{3}He polarisation will decay over time, resulting in low neutron transmission and polarisation. Consider a polariser with an initial ^{3}He polarisation of 80%, at a selected wavelength where the neutron polarisation would be 90%, the unpolarised incident beam transmission would be 37%. If a similar analyser is used, the combined transmission would be 24% for the high-transmission spin state. If one of them has a decay time constant of 150 hours while the other is an *in situ* setup, the transmission would be 19% after 24 hours; a drop of 20% from the initial transmission. If both spin filters are *ex situ* setups, the transmission drops to 15% in 24 hours, i.e., a loss of 40% in transmission. To allow for no less than 19% transmission, both cells will need to be changed every 12 hours. Experiences show that an experiment would be interrupted between 15 minutes to a couple of hours in order to change to a newly polarised cell or batch of ^{3}He gas. The use of *in situ* SEOP setup, where it is applicable, would reduce such disruption to experiments.

The other class of cells are the wide-angle analyser cells to accommodate a large range of scattering angles in diffractometers and spectrometers. The cells consist of concentric arcs of glass walls with flat top and bottom. Scattered neutrons pass the curved walls and the volume of polarised ^{3}He gas inside. A typical cell has an inner diameter of 6 cm to 7 cm and an outer diameter of about 20 cm. The arc covers an angle between 90° to 160°. The cell height is 6 cm to 12 cm.

These cells typically work with *ex situ* gas polariser using a Metastable Optical Pumping (MEOP) filling station. A MEOP station is too large to be located on an instrument but has several other advantages. There are two modes of MEOP operation. In the cell-exchange mode, a spin filter cell is filled in the MEOP station, then taken to the instrument and exchanged with the cell in the instrument. In the gas-exchange – or what is known as "local filling" mode [33] – the gas in the spin filter cell in the instrument is exchanged with the newly polarised gas in a transport cell. While the cell-exchange method may be simpler when there is easy access to the location of the cell, it is difficult to access the cell at most ESS instruments that will use the wide-angle ^{3}He analyser. The more complex local-filling setup would be the method of choice. Despite the complexity in the setup, the local-filling method allows a shorter interruption to the experiment and have lower gas polarisation loss during the exchange.

Wide-angle SEOP cells have been developed in several facilities [34]. The polarising condition of 160 °C to 200 °C and the need to use laser along the field axis make it impractical for *in situ* application in close proximity to a sample or sample environment equipment. The cell seals in ^{3}He gas and SEOP requires at least 1 day to polarise a cell. This does not compare favourably to the MEOP method where the optimal gas pressure is chosen according to experimental needs and the significantly shorter 1 to 2 hours of gas-filling time. In addition, it is challenging to work with high-stress aluminosilicate glass to make SEOP cells. For these reasons, a majority of the existing wide-angle cells are MEOP based.

The ESS polarised ^{3}He neutron spin-filter line-up thus includes both *in situ* SEOP setup for a majority of applications using cylindrical cell and MEOP setup for wide-angle analysis cells. There are some exceptions that use cylindrical cells but cannot use *in situ* SEOP. They will use the MEOP method. As a first step, a SEOP setup is being built to provide the platform for development and testing. And discussion is underway with the Institute Laue Langevin for a collaboration to construct a MEOP station.

3.2 Magnetic field device

^3He polarisation is highly susceptible to magnetic field gradients. The environment of the spin filter must be well controlled to minimise magnetic interference. In the case where an external field gradient cannot be eliminated, magnetic shielding made of high permeability materials such as μ-metal should be used to shield the ^3He cell.

First and foremost, the magnetic field device that provides the holding field to the ^3He cell must be highly uniform. At the cell, the angular gradient of the applied magnetic field would ideally be less than 2×10^{-4} rad/cm [35]. There are two main class of magnetic field devices for polarised ^3He applications: field coils and magnetostatic cavity.

Magnetostatic cavities are used with cylindrical cells, either as polariser or as analyser at a distance downstream from the sample. The μ-metal that is part of its constituents provides magnetic shielding to the ^3He cell, allowing the device to be used with Tesla-range sample magnets. The two most common cavity types are "Magic Box" [36] and solenoid [37]. In a magic box, field coils wrapped around μ-metal plates provide the magnetic flux. The flux is channelled by the μ-metal plates to produce a uniform field over the entire volume of the cell. The other type uses a compensated solenoid to produce the uniform field, where a μ-metal outer shell both improves the uniformity and serves as flux return.

3D uniform field coils are used with the sample environment at its centre. Uniform field coils have long been investigated and many configurations can be found in literature. Some of the existing coils used in ^3He spin filter applications are the "Pastis" coils [38, 39] and there are also hybrid approaches that use both coils and μ-metal plates to produce a 3D uniform field with large horizontal angular span [40, 41].

In our on-going development of uniform field coils, our design of a rectangular coil set has, thus far, reached a uniform-field volume of 22 dm^3 (Fig. 3a). This is larger than coils published in literature for rectangular coil sets with similar outer dimensions [42]. Two 15 cm diameter and 10 cm long ^3He cells, i.e., a polariser and an analyser, can be placed in the uniform field region at 10 cm to 12 cm apart, with space to spare.

Likewise, our coil design of circular coil sets for wide-angle scattering use has reached a uniaxial uniform volume 24 cm in diameter and 12 cm tall, with an unobstructed vertical angle at ±28° (Fig. 3b). In contrast, a similar coil called "tetracoil" [43], which is known for its large uniform volume, has a considerably smaller unobstructed vertical angle of ±17°. We know of no other published coil with similar performance.

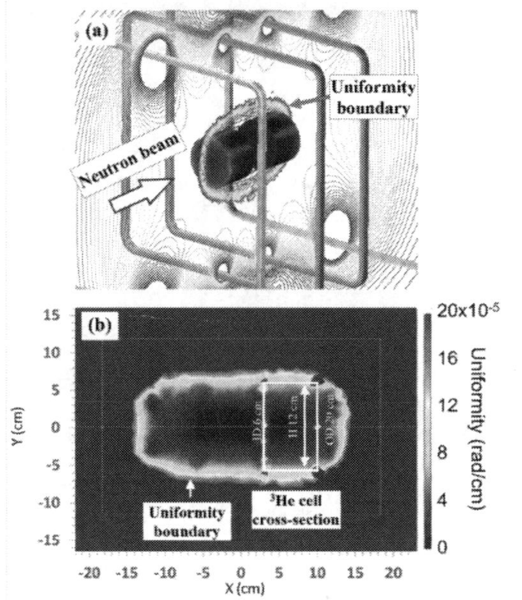

Fig. 3. Uniformity volume in our designs of (a) rectangular coil set, showing ^3He cells in their operation position, (b) circular coil set, showing the cross-section of a wide-angle cell. Within the red boundary, the uniformity is better than 2×10^{-4} rad/cm.

We have incorporated additional design to provide a 3D field. There is additional loss of vertical opening yet it still yields unprecedented performance. As this is an on-going work, the details of these coils are out of the scope of this overview article.

4 Polarised data processing

A major bottleneck in using polarised neutrons is data reduction. Several measurements of different incident and scattered spin states need to be reduced together, as oppose to unpolarised neutron measurements where each measurement is reduced independently. Without an appropriate tool to reduce the polarised data during an experiment, users cannot easily separate the effects due to finite neutron polarisation from sample magnetic scattering. The situation is even more complicated if polarised ^3He with changing polarisation is used.

The mathematics to reduce polarised data is readily available [44]. It applies across different scattering methods and can be automated. The key equation is

$$\bar{I} = \widehat{T^A}\ \widehat{F_2}\ \widehat{F_1}\ \widehat{T^P}\ \bar{S} \tag{1}$$

with $\bar{I}$ being the measured intensities,

$$\bar{I} = \begin{pmatrix} I_{++} \\ I_{+-} \\ I_{-+} \\ I_{--} \end{pmatrix} \tag{2}$$

and $\bar{S}$ being the intensities after polarisation data reduction,

$$\bar{S} = \begin{pmatrix} S_{++} \\ S_{+-} \\ S_{-+} \\ S_{--} \end{pmatrix} \tag{3}$$

where the first and second subscripts of + and – denote the neutron spin state before and after the sample, respectively. The matrices connecting $\bar{I}$ and $\bar{S}$ are:

Polariser transmission matrix,

$$\widehat{T^P} = \begin{pmatrix} T_+^P & 0 & T_-^P & 0 \\ 0 & T_+^P & 0 & T_-^P \\ T_-^P & 0 & T_+^P & 0 \\ 0 & T_-^P & 0 & T_+^P \end{pmatrix}; \tag{4}$$

Incident neutron spin-flipper transmission matrix, assuming spin+ neutrons are transmitted with the flipper off,

$$\widehat{F_1} = \begin{pmatrix} 1 & 0 & 0 & 0 \\ 0 & 1 & 0 & 0 \\ 1-F_1 & 0 & F_1 & 0 \\ 0 & 1-F_1 & 0 & F_1 \end{pmatrix}; \tag{5}$$

Scattered neutron spin-flipper transmission matrix, assuming spin+ neutrons are transmitted with the flipper off,

$$\widehat{F_2} = \begin{pmatrix} 1 & 0 & 0 & 0 \\ 1-F_2 & F_2 & 0 & 0 \\ 0 & 0 & 1 & 0 \\ 0 & 0 & 1-F_2 & F_2 \end{pmatrix}; \tag{6}$$

Analyser transmission matrix,

$$\widehat{T^A} = \begin{pmatrix} T_+^A & T_-^A & 0 & 0 \\ T_-^A & T_+^A & 0 & 0 \\ 0 & 0 & T_+^A & T_-^A \\ 0 & 0 & T_-^A & T_+^A \end{pmatrix}. \tag{7}$$

F_1 and F_2 in Equations 5 and 6 are the neutron spin-flipper efficiency before and after the sample, respectively. Their values are measured experimentally, with 0 being non-functioning and 1 being perfectly efficient. A properly designed flipper often has a high efficiency of more than 0.99.

For polarising supermirror devices, the transmission of spin+ and spin– states are given by

$$T_\pm = \frac{1 \pm E}{2} \tag{8}$$

where E is the polariser or analyser efficiency. $E = 1$ corresponds to perfect efficiency and $E = 0$ means the device is not spin-filtering neutrons. The efficiency is determined experimentally. Here we omit the designation of polariser P and analyser A for simplicity.

For polarised ^{3}He, the spin+ and spin– transmissions are given by

$$T_\pm = T_g \exp\left(-O \pm O P_{He}\right). \tag{9}$$

T_g is the neutron transmission of the cell body. Its value is measured experimentally. P_{He} is the ^{3}He polarisation and O is the ^{3}He opacity, given by

$$O = n\, \sigma_0\, \lambda\, l, \tag{10}$$

where n is the number density of ^{3}He gas in the cell. σ_0 *is* the absorption cross section of 1 Å neutron by ^{3}He, $\sigma_0 = 2966 \times 10^{-24}$ cm^2. λ is the neutron wavelength. l is the path length through the cell. Helium has the least deviation from the ideal gas law among all gases, with less than 0.1% deviation in density in standard conditions. We consider n to be related to the cell pressure p by the ideal gas law, given by $p = nk_BT$, with Boltzmann constant k_B and temperature T. A practical adaptation of Equation 10 is the follow equation:

$$O = 0.0733\, p\, l\, \lambda \tag{11}$$

where p is the cell pressure at 20 °C in bars, neutron wavelength λ is in Å and the path length l is in cm.

The path length l is known when the cell body is made. ^{3}He pressure p is measured when the cell is filled. Knowing the neutron wavelength λ in the measurement, the opacity O can be calculated from Equation 11.

During the experiment, ^{3}He polarisation P_{He} can be obtained by measuring unpolarised neutron transmission T_N and using the following equation to analyse the result:

$$T_N = T_g \exp(-O) \cosh(O P_{He}). \tag{12}$$

Subsequently one can calculate the spin filter's transmission matrices in Equations 4 and 7 and use Equation 1 to obtain the reduced data $\bar{S}$.

This formalism has been successfully applied to analyse polarised diffraction data measured using polarised ^{3}He neutron spin filters [45]. Working together with the data scientists at ESS' Data Management & Software Centre (DMSC), we intend to develop and incorporate the data reduction algorithm to the instrument data processing software so that users can reduce polarised data during the experiment.

In a number of publications, including the seminal article cited above, Equation 1 was incorrectly given as $\bar{I} = \widehat{F_1}\,\widehat{F_2}\,\widehat{T^P}\,\widehat{T^A}\,\bar{S}$. While the two spin-filter matrices commute, $\widehat{T^P}\,\widehat{T^A} = \widehat{T^A}\,\widehat{T^P}$, and the two spin-flipper matrices also commute, $\widehat{F_1}\,\widehat{F_2} = \widehat{F_2}\,\widehat{F_1}$, the polariser or analyser matrix and either spin flipper matrix do not, i.e. $\widehat{F_1}\,\widehat{T^P} \neq \widehat{T^P}\widehat{F_1}$ and so on. This error was not discovered earlier perhaps due to the fact that neutron spin flippers often have high efficiency. $\widehat{F_1}$ and $\widehat{F_2}$ are close to being a unit matrix so the error was small. The equation however gives unphysical results in some settings so it is important to use the correct formula.

5 Summary

In this article, we provide an overview of the polarised neutron instrumentation for nine ESS instruments supported by the Polarisation Project. We specified the equipment selected for each instrument and provided further information on the polarised ^{3}He equipment. Finally, we discussed polarised data reduction, especially the methodology when using polarised ^{3}He neutron spin filters.

Acknowledgements

The authors acknowledge the support from the following grants and in-kind project:

1. Support for DREAM polariser: Röntgen Ångström Cluster: nPDFSAS (grant no 2019-06117_VR). https://portal.research.lu.se/en/projects/npdfsas-simultaneous-polarized-sans-and-npdf-methods-to-study-nov
2. Support for ODIN, SKADI/LoKI polarisation: W.- T. Lee, E. Blackburn, T. Nylander, D. Orlov, J. Houston, S. Jaksch, M. Manual, C. Pappas, S. Schmidt, M. Schulz, M. Strobl, A. Tartaglione, R. Woracek, Tillväxtverket European Regional Development Fund project SREss3 (2020-2023).
3. Support for BIFROST, CSPEC polarisation: E. Blackburn, M. Månsson, W.-T. Lee, G. Nilsen, P. Deen, R. Toft-Petersen, M. Pascal, Swedish Research Council (VR) "Cooperation with ISIS: instrumentation and methods for ESS" Grant # 2021-06157.
4. Swedish Research Council (VR): In-Kind Contribution to the ESS, "Neutron Polarisation Capability Development for User Experiment Applications on ESS Instruments", 2023-2025.

References

1. B. Gillon, P. Becker, *Magnetization Densities in Material Science,* in Modern Charge-Density Analysis, C. Gatti, P. Macchi, *ed.*, Dordrecht, Springer, 277 (2011).
2. A. Wiedenmann, Physica B **356**, 246 (2005).
3. C. Majkrzak, Physica B **173**, 75 (1991).
4. L. Regnault, *Inelastic Neutron Polarization Analysis*, in Neutron Scattering from Magnetic Materials, Tapan Chatterji, *ed.*, Elsevier Science, 363 (2006).
5. M. Strobl, H. Heimonen, S. Schmidt, M. Sales, N Kardjilov, A. Hilger, I. Manke, T. Shinohara, J. Valsecchi, J. Phys. D **52**, 123001 (2019).
6. T. Nylander, R. Campbell, P. Vandoolaeghe, M. Cárdenas, P. Linse, A. Rennie, Biointerphases **3**, FB64–FB82 (2008).
7. S. Holt, A. Le Brun, C. Majkrzak, D. McGillivray, F. Heinrich, M. Lösche and J. Lakey, Soft Matter **5**, 2576 (2009).
8. A. Arbe, G. J. Nilsen, J. R. Stewart, F. Alvarez, V. Sakai and J. Colmenero, Phys. Rev. Res. **2**, 022015 (2019).
9. F. Mezei, Z. Physik, **255**, 146 (1972).
10. P. Böni, Physica B **234-236**, 1038 (1997).
11. T. Krist, C. Lartigue, F. Mezei, Physica B **180-181**, 1005 (1992).
12. T. Bigault, G. Delphin, A. Vittoz, V. Gaignon, P. Courtois, J. Phys.: Conf. Ser. **528**, 012017 (2014) and references therein.
13. K. Andersen, R. Chung, V. Guillard, H. Humblot, D. Jullien, E. Lelièvre-Berna, A. Petoukhov, F Tasset, Physica B **356**, 103 (2005).
14. G. Jones, J. Baker, W.-C. Chen, B. Collett, J. Cowan, M. Dias, T. Gentile, C. Hoffmann, T. Koetzle, W.-T. Lee, K. Littrell, M. Miller, A. Schultz, W.M. Snow, X. Tong, H. Yan, A. Yue, Physica B **356**, 86 (2005).
15. W.-T. Lee, X. Tong, J. Pierce, M. Fleenor, A. Ismaili1, J. Robertson, W.-C. Chen, T. Gentile, A. Hailemariam, R. Goyette, A. Parizzi, V. Lauter, F. Klose, H. Kaiser, C. Lavelle, D. Baxter, G Jones, J. Wexler, L. McCollum, J. Phys.: Conf. Ser. **251**, 012086 (2010).
16. K. Andersen, *et. al.*, Nucl. Instrum. Methods A **957**, 164302 (2020).
17. W.-T. Lee, A. Arbe, T. Arnold, E. Blackburn, S. Carretta, P. Deen, E. Lelièvre-Berna, S. Mattausch, G. Nilsen, C. Pappas, S. Schmidt, X. Tong, Report on ESS Polarisation Workshop, ESS-3549713 (2020).
18. A. Backs, S. Sebold, M. Busi, W.-T. Lee, M. Strobl, D. Orlov, *this proceeding* (2023).
19. A. Stellhorn, W.-T. Lee, E. Kentzinger, W.C. Chen, J. Gaudet, K. Krycka, E. Blackburn, *this proceeding* (2023).
20. W. Schweika, N. Violini, K. Lieutenant, C. Zendler, D. Nekrassov, A. Houben, P. Jacobs, P. Henry, J. Phys.: Conf. Ser. **746**, 012013 (2016).
21. W. Schweika, J. Phys.: Conf. Ser. **211**, 012026 (2010).
22. G. Ehlers, J. Stewart, A. Wildes; P. Deen; K. Andersen, Rev. Sci. Instrum. **84**, 093901 (2013).
23. O. Schärpf, H. Capellmann, Phys. Status Solidi A **135**, 359 (1993).
24. P. Willendrup, and K. Lefmann, J. Neutron Res. 22, 1-16 (2020).
25. P. Willendrup, and K. Lefmann, J. Neutron Res. 23, 7-27 (2021).
26. D. M. Rodríguez, P. Willendrup, W.-T. Lee, A. Backs, F. J. Villacorta, R. Toft-Petersen, M. Morgano, *this proceeding* (2023).
27. P. Böni, D. Clemens, M. Kumar, C. Pappas, Physica B **267-268**, 320 (1999).
28. V. Luschikov, Yu. Taran, Nucl. Instrum. Methods Phys. Res. **228**, 159 (1984).

29. P. Jorba, M. Schulz, D. Hussey, M. Abir, M. Seifert, V. Tsurkan, A. Loidl, C. Pfleiderer, B. Khaykovich, J. Magn. Magn. Mater. **475**, 176 (2019).
30. S. Jaksch, A. Chennevière, S. Désert, *et. al.* Appl. Sci., **11**(8), 3620 (2021).
31. P. Deen, S. Longeville; W. Lohstroh, F. Moreira, G. Fabrèges, L. Loaiza, D. Noferini, Rev. Sci. Instrum. **92**, 105104 (2021).
32. COMSOL Multiphysics® v. 6.1. www.comsol.com. COMSOL AB, Stockholm, Sweden.
33. K. Andersen, R. Cubitt, H. Humblot, D. Jullien, A. Petoukhov, F. Tasset, C. Schanzer, V. Shah, A. Wildes, Physica B **385–386**, 1134 (2006).
34. Q. Yea, T. Gentile, J. Anderson, C. Broholm, W.C. Chen, Z. DeLand, R. Erwinb, C. B. Fue, J. Fuller, A. Kirchhoff, J. Rodriguez-Rivera, V. Thampy, T. Walker, S. Watson, Physics Procedia **42**, 206 (2013).
35. L. Schearer, G. Walters, Phys. Rev. **139**, 1398 (1965).
36. A. Petoukhov, V. Guillard, K. Andersen, E. Bourgeat-Lami, R. Chung, H. Humblot, D. Jullien, E. Lelievre-Berna, T. Soldner, F Tasset, M. Thomas, Nucl. Instrum. Methods Phys. Res. A **560**, 480 (2006).
37. W.C. Chen, Md. Hassan, R. Erwin, S. Watson, T. Gentile, G. Jones, Rev. Sci. Instrum. **91**, 105102 (2020).
38. D. Jullien, A. Petoukhov, M. Enderle, N. Thiery, P. Mouveau, U. Hansen, P. Chevalier, P. Courtois, Nucl. Instrum. Methods Phys. Res. A **1010**, 165558 (2021).
39. J. Stewart, K. Andersen, E. Babcock, C. Frost, A Hiess, D. Jullien, J. Stride, J.-F. Barthélémy, F. Marchal, A. Murani, H. Mutka, H. Schober, Physica B **385–386**, 1142 (2006).
40. E Babcock, Z Salhi, P Pistel, G Simeoni, A Ioffe, J. Phys.: Conf. Ser. **528**, 012018 (2014).
41. K. Andersen, D. Jullien, A. Petoukhov, P. Mouveau, F. Bordenave, F. Thomas, E. Babcock, Physica B **404**, 2652 (2009).
42. R. Merritt, C. Purcell, G. Stroink, Rev. Sci. Instrum. **54**, 879 (1983).
43. G. Gottardi, P. Mesirca, C. Agostini, D. Remondini, F. Bersani, Bioelectromagnetics **24**, 125 (2003).
44. A. Wildes, Rev. Sci. Instrum. **70**, 4241 (1999).
45. J. Auckett, W.-T. Lee, K. Rule, A. Bosak, C. Ling, Inorganic Chemistry **58**, 12317 (2019).

Silver Jubilee for the OSIRIS spectrometer: Achievements and Outlook

Franz Demmel[1,*], *Adrien* Perrichon[1], *David* McPhail[1], *Paula* Luna Dapica[1], *Nick* Webb[1], *Andy* Cook[1], *Erik* Schooneveld[1], *Johnny* Boxall[1], *Nigel* Rhodes[1], *Cyril* Lockett[2], *Colin* Dabinett[2], *Joel* Hodder[2], *Daniel* Nye[1], *Sanghamitra* Mukhopadhyay[1], *Ian* Silverwood[1], *Mona* Sarter[1], *Victoria* Garcia Sakai[1], and *Felix* Fernandez-Alonso[3,4,5]

[1]ISIS Facility, STFC, Didcot, OX11 0QX, United Kingdom
[2]Technology Department, STFC, Didcot, OX11 0QX, United Kingdom
[3]Materials Physics Center, CSIC-UPV/EHU, Paseo Manuel de Lardizabal 5, 20018 Donostia - San Sebastian, Spain
[4]Donostia International Physics Center (DIPC), Paseo Manuel de Lardizabal 4, 20018 Donostia - San Sebastian, Spain
[5]IKERBASQUE, Basque Foundation for Science, Plaza Euskadi 5, 48009 Bilbao, Spain

Abstract. In December 1997 the Osiris beamline at the ISIS facility, UK, recorded its first neutron spectrum. The instrument enjoyed a first stint as a cold neutron diffractometer before the spectroscopic capabilities were fully commissioned. Osiris soon became a workhorse quasielastic spectrometer as well as a highly successful low-energy spectrometer. The status of the instrument is recognized by the user community with high-impact publications ranging from energy materials over life science to quantum matter. To enhance the existing capabilities a silicon analyzer is under construction. The primary spectrometer will be upgraded with a new supermirror guide providing a factor 10 to 14 increased flux in combination with a new hydrogen moderator. Beyond these developments further improvements of the energy resolution with a combination of a fast pulse shaping chopper and using a direct backscattering geometry are being investigated.

1 Introduction

The Osiris instrument at the ISIS facility recorded its first spectrum in December 1997. At that time it went into operation as a cold neutron diffractometer and a few years later the spectrometer capabilities were installed and commissioned [1–7]. This indirect spectrometer is based on the concept of analyzing the incoming energy through the time-of-flight (tof) and the fixed final energy through Bragg scattering from a near-backscattering crystal analyzer. Further improvements were made with the installation of a movable Beryllium filter to remove the higher order reflections of the pyrolytic graphite analyzer [8].

The strength of this spectrometer concept lies in the combination of a relative high energy resolution due to the near-backscattering geometry (FWHM = 25 μeV for the PG002 reflection) over a wide range of momentum transfers with a large dynamic range (up to several meV in energy transfer) due to the tof-technique. These characteristics allowed successful experiments in the field of high resolution low-energy spectroscopy. The spin dynamics of strongly correlated electrons materials was unraveled, which often required single crystal experiments with high magnetic fields in combination with dilution fridge temperatures [9–12]. Materials for energy research had a prominent role in the research programme of Osiris from the beginning [13]. The high sensitivity of the spectrometer, which reaches a factor 10000 in signal to background [6], benefitted QENS-studies on ion mobility. Novel electrode materials for batteries [14, 15] have been investigated, the mobility of protons and hydride ions was studied [16, 17] and details about materials for future photovoltaic applications could be revealed [18]. More recently the quasielastic response to pressure of barocaloric materials was resolved [19]. QENS was also applied studying the diffusion of small molecules in framework materials for catalytic applications and attracted a growing user community [20, 21]. Within softmatter and bioscience the relaxations and mobility of large molecules and the water environment were investigated [22–24]. This short overview of the science programme demonstrates that Osiris contributes to a wide and diverse range of challenging problems. To stay at the forefront of scientific discoveries upgrade projects are under way or are proposed.

Here we present an overview of upgrade activities to further expand the capabilities of the Osiris beamline. At present, the energy resolution will be improved with the installation of a silicon analyzer. Within the Endeavour project of the ISIS facility the primary spectrometer will be upgraded with a huge gain in intensity and we are looking further into the future where plans emerge to improve the energy resolution towards μeV.

2 The silicon analyzer project

Opposite to the pyrolytic graphite analyzer side inside the vacuum tank there is an empty space, which will be filled with a silicon analyzer. To enhance the performance of Osiris with respect to energy resolution and allow novel

*e-mail: franz.demmel@stfc.ac.uk

challenging studies such as slow diffusive motions of ions in battery materials, an upgrade is under way to implement a Si111 analyzer unit with an energy resolution of $\Delta E \approx 11\ \mu eV$ [25]. On pulsed sources silicon as analyzer material is applied to increase the energy resolution towards μeV [26–28]. In combination with position sensitive detectors the third direction in momentum space can be resolved, which is not possible with the detectors at the existing pyrolytic graphite analyzer. Figure 1 shows a schematic of the parts which will be installed into the vacuum tank. The silicon analyzer crystals are arranged

Figure 1. A schematic engineering drawing of the new Si analyzer bench with the two detector arrays and the radial collimator is shown inside the vacuum tank. On the right side of the picture the existing graphite analyzer bench is drawn.

on a radius of 880 mm from the sample axis position and will have a total height of about 700 mm. Hence the total covered solid angle of the secondary spectrometer will be around 2 sr. The silicon analyzer is split into an upper and a lower bench. The neutrons will therefore be reflected into two radial detector arrays, which consist of position sensitive half-inch diameter ^{3}He-detector tubes with a sensitive length of 160 mm (the tubes are colored in red in figure 1). Each analyser bench consists of 4 frames. Two frames are mounted onto a single tower mechanics (grey structure on the left in figure 1). The Bragg angle (about 83°) of the analyzer frames can be aligned through a stepping motor. The preamplifier electronics will be mounted near to the helium tubes and will be housed in an air box (yellow box in figure 1) . Between the analyzer and detectors a coarse radial collimator will be positioned (aqua color in figure 1).

There are several advantages in using silicon compared to pyrolytic graphite. Since Si111 has no second order reflection, there is no need for the installation of a cooled beryllium filter. In addition, due to the lack of thermal diffuse scattering from the analyzer crystals itself no cooling of the whole analyzer bench is needed. However, as-grown silicon wafers are perfect crystals and as such would select a too small wavelength range in comparison with the incoming wavelength band. The reflected wavelength band is given by the Darwin width which corresponds to $\frac{\Delta\lambda}{\lambda} \approx 1.2 \cdot 10^{-5}$ for the Si111 reflection [29].

To overcome the mismatch in resolution between the primary and secondary spectrometer, elastic deformation of the silicon crystals is applied, which increases the reflected intensity substantially. The deformation is achieved through gluing the wafers onto the spherical frame [30]. The elastic deformation results in a gradient in lattice spacing, Δd, which depends on the curvature radius of the analyzers, R_A, and the thickness of the wafers, D, according to Eq. (1) [30], with $\mu_{eff} \approx 0.4$ being an average Poisson ratio and the small contribution from the Darwin width is omitted.

$$\frac{\Delta d}{d} = \mu_{eff}\frac{D}{R_A} \quad (1)$$

With a wafer thickness of D=0.8 mm and a radius of R_A=880 mm we might obtain $\frac{\Delta d}{d} = 3.6 \cdot 10^{-4}$, which provides a large increase in reflected intensity.

To reduce background from the aluminium frame the backside of the silicon wafers will be covered by an absorbing layer. Several tests with neutrons have been performed to select the best combination of absorbing material and glue. It became clear that spraying provides a more homogenous coating than adding the paint with a brush. Also, tests with homemade water based resins could be excluded.

Finally, a Gd_2O_3-paint emerged as the best option. Several wafers have been glued onto aluminium disks with a radius of R=880 mm to confirm the viability of the whole process. Figure 2 shows the results of some tests with Si-wafers. One pure Si-wafer, not glued onto a disk, was also measured for comparison reasons. The wafers cut in hexagonal shape had a diameter of about 100 mm. The beam cross section of 20×40 mm^2 only covers the middle part of the wafers. The measurements were made on Osiris using the transmission monitor for transmission measurements and the diffraction detectors for measurements of the reflected background. A few of the 960 available pixels from the diffraction detector were used which are not affected by Bragg reflections and hence monitor the incoherent background from hydrogen scattering. The detector is about 1 m away from the wafers in a backscattering geometry, hence similar to the later application except that no collimator is installed. The reflected signal is used as a measure for the background caused by the glue. A wavelength range around 6.3-8 Å was used. Panel (a) shows the transmitted signal over time-of-flight of the neutrons for a pure Si-wafer (circle) and a Si-wafer (circle) glued on the aluminium disk without an absorbing layer. The transmitted signal is nearly identical and might allow the neutrons to be scattered inelastically in the thick aluminium frame and come back towards the detectors. This would deteriorate the instrument background at least for inelastic spectroscopy. All the other wafers have been covered by absorbing paint, where the *thin layer* wafer (triangle down) shows that a coverage of about 0.8g of absorbing paint per wafer reduces the amount of transmitted neutrons by a factor 30. However, the three *thick layer* wafers (stars) provide a transmission reduction of about a factor 1000. Practically no neutron scattered back from the aluminium frame has a chance to reach the detectors. Typically 2g of absorbing paint has been applied onto these three wafers. In panel (b) of figure 2 the measured re-

https://doi.org/10.1051/epjconf/202328603005

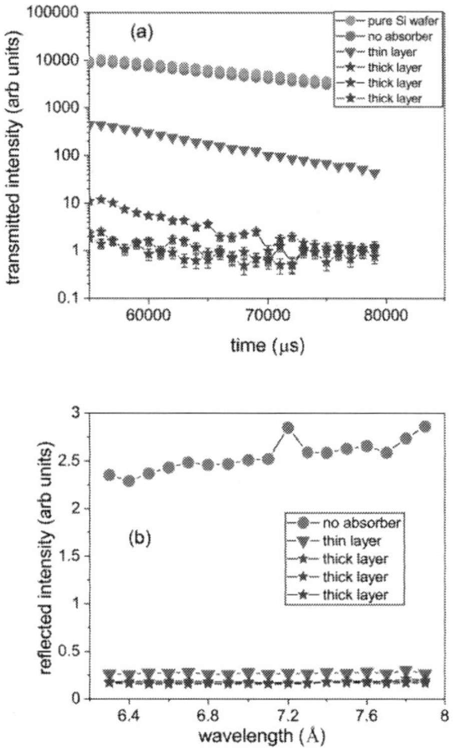

Figure 2. Panel (a) shows transmission measurements on a logarithmic scale against the time-of flight for Si-wafers. In panel (b) the reflected signal from the glued wafers is plotted against the wavelength, which is a measure for incoherent background from the silicon analyzer.

flected background from the wafers is depicted against the wavelength. These results clearly demonstrate that the absorbing layer reduces the overall background by an order of magnitude compared to the glued wafer without an absorbing layer. Note that these measurements were made without a collimator between wafer and detector and hence will be a relative measure for the background with the silicon analyzer. There are practically no differences in reflected background neutrons within the three wafers with thick absorbing layers, which confirm the reliability of the whole process. As a conclusion from all the tests we have chosen an absorbing layer of $\approx$ 120 μm thickness using a commercially available absorbing paint with a 47% Gd_2O_3 content, which is sprayed onto the wafers before they are glued onto the aluminium frame. A square test frame with 300 mm side length has successfully been produced and we are now in the process of building the whole analyzer frame. Installation and commissioning of the Si-analyzer bench is foreseen for 2024.

3 The upgrade of the primary spectrometer

The current Osiris guide is a straight m = 2 supermirror guide with a m = 3.6 linear focusing section at the end, where m stands for m-times the critical angle of total reflection of a natural-Ni coated guide. The non-negligible number of reflections of long wavelength neutrons in a small cross section guide will reduce the transmission of neutrons decisively. This is the cause for the quite modest gain factor in flux of about 1.6 at $\lambda \approx$ 6 Å for Osiris in comparison with the natural Ni guide on the Iris beamline [5].

New guide geometries have less reflections and can transport cold neutrons very effectively over long distances [31, 32]. The performance of these novel guide geometries has been assessed several times in the past (see for example [33–35]) and the elliptic geometry emerged as the most favorable one. Therefore, a new guide with an elliptic defocusing and focusing geometry has been proposed to replace the present Osiris guide [36–38]. The position of the secondary spectrometer will not be moved and to avoid direct line of sight a curved section with a curvature radius of R=1208 m sits between the elliptic sections. Furthermore, the elliptic geometry allows to focus neutrons on a smaller sample size, a necessity when it comes to investigate small sample quantities. Extensive MonteCarlo simulations have been performed to optimize the geometry and coating [36, 38]. The resulting gain factors, defined as the intensity ratio between new and present guide at sample position are shown in figure 3. The simulations predict intensity gain factors between 5 and 7. In addition, a new hydrogen moderator is in process of being installed at target station 1, which is predicted to provide an intensity gain factor of about 2 [39]. All together, one can expect intensity gain factors between 10 and 14 at the sample position, which will enable new experiments, that are nowadays not possible. This intensity increase will keep Osiris competitive with the most modern spectrometers at more powerful sources, albeit not achieving the high energy resolutions offered on these spectrometers [38]. The

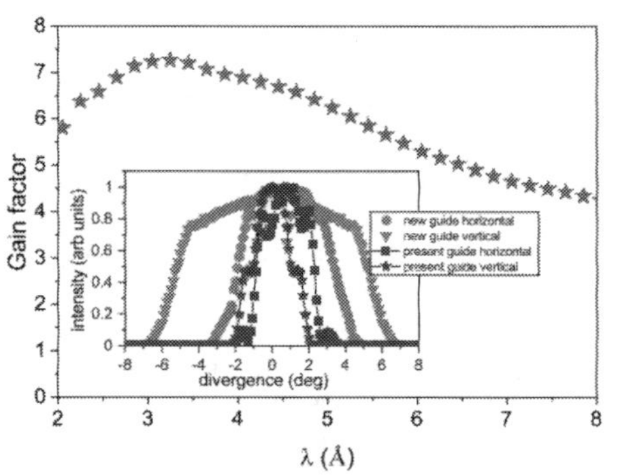

Figure 3. The intensity gain over wavelength is plotted for the new guide. The insert shows the increase in divergence with the new geometry in comparison with the existing guide.

increased flux comes with the prize of an increase in divergence, which is often not detrimental for most of the QENS experiments. The insert in figure 3 shows the vertical and horizontal divergences of the present guide and the new guide. The divergence might increase by up to factor 3 with a more smooth distribution compared to current guide with a linear focusing section. A slit system with 3 movable slits will be integrated into the guide offering a reduction of the divergence by up to a factor 3, which might be necessary when low energy dispersive excitations should be resolved. Of course, the divergence reduction is accompanied by an intensity reduction, because the divergence reduction can be directly related to a reduction in gain. The new guide will have an increased height with a maximum value of 200 mm, which requires to change the bandwidth chopper system. The increased opening and closing times of these slow rotating choppers would reduce the dynamic range of the spectrometer. For this reason a double disc chopper system will be installed with a slightly larger diameter per disc.

4 Towards μeV resolution

To achieve an even higher energy resolution on Osiris one has to consider the main factors to the energy resolution for a tof-backscattering instrument [6, 40]. For the incoming energy it is the pulse width from the moderator, which sets a limit for the energy resolution. To reduce this term one can choose a poisoned moderator and increase the length of the instrument [26] or use a fast pulse shaping chopper [27, 28], which then also provides some flexibility in trading resolution against intensity. With these steps in shortening the pulse width an energy resolution of a few μeV was achieved in a near-backscattering geometry. The main contribution to the final energy uncertainty stems from the Bragg reflection in combination with a divergence term: $\cot(\theta_B)\Delta\theta$ [6]. The divergence term $\Delta\theta$ is dominated by the sample height for a silicon analyzer. The energy resolution term with $\cot(\theta_B)$ will become smaller with increasing Bragg angle towards 90°, but there are technical limitations to be observed. In a near-backscattering geometry the maximum Bragg angle θ_B is dictated by the sample to analyzer distance and the need to install the detectors near the sample position without shadowing the solid angle towards the analyzer bench. Increasing the distance will allow larger Bragg angles, but will limit the solid angle and hence the measured intensity. Geometries with a Bragg angle of 87° and a distance of 2500 mm have been realized and represent the limit [26, 27].

Backscattering spectrometers at reactor sources achieve a higher energy resolution of less than 1 μeV by using a direct backscattering geometry [29, 30, 41]. Consequently, to achieve a lower energy resolution one way forward is to abandon the near-backscattering geometry and apply the direct backscattering geometry at a tof-backscattering spectrometer. For direct backscattering it has been shown that the geometric contribution to the energy resolution is given by [42]:

$$\frac{\Delta E}{E} = 2\{\frac{\Delta\tau}{\tau} + \frac{1}{8}\Delta\theta^2\} \quad (2)$$

Here $\frac{\Delta\tau}{\tau}$ represents the contribution from the deformation of the lattice constants due to the bending of the wafers, see Eq. 1. The second term is due to the divergence $\Delta\theta$ which is mainly caused by the size of the sample. It is worth mentioning that in the derivation of this equation different contributions to the energy resolution are added directly.

Now we are in a position to calculate the expected energy resolution for the Osiris instrument, using a fast pulse shaping chopper and direct backscattering from the silicon analyser. Assuming a pulse width of $\Delta t = 20\,\mu s$ and a sample height of 30 mm in a distance of 880 mm we calculate the energy resolution for the near-backscattering geometry ($\theta_B = 83^o$) and for the direct backscattering geometry. The results are shown in figure 4. For the near-backscattering geometry we obtain values of $\approx 12\,\mu$eV, which are similar to the values predicted previously for this geometry, except that this value is now achieved for a large sample of 30 mm height. The reduction in energy resolution is due to the pulse shaping chopper although this pulse width shortening will cancel any eventual gain in intensity due to the larger sample. Hence, for this range in energy resolutions the installation of a pulse shaping chopper appears not to be particularly beneficial. Previously we demonstrated that the contribution from the secondary spectrometer for the silicon analyzer near-backscattering geometry of Osiris with a 10 mm large sample has a lower limit of around 6 μeV [25] and is still short of a μeV resolution.

However, when a direct backscattering geometry can be achieved in combination with a fast pulse shaping chopper an energy resolution of $\approx 2\,\mu$eV can be obtained (see figure 4). For this calculation a sample height of 30 mm and a width of 20 mm are assumed.

This upgrade of the Osiris instrument does not need a complete rebuild of the instrument and could be implemented in an evolutionary way. The new 50Hz counter rotating double disc bandwidth choppers could be replaced by similar fast rotating choppers. The installation of the silicon analyzers includes motors for changing the Bragg angle and the mechanics is designed in such a way that the change from near-backscattering to direct backscattering could be done during an experiment without direct intervention inside the vacuum tank. For the direct backscattering geometry a further detector array positioned opposite to the existing detector arrays of the silicon analyzer needs to be installed, which should be movable out of the scattering plane not to block the graphite side when used. Now MonteCarlo simulations are envisaged to assess the performance of this novel setup in detail.

5 Conclusions

Over the past 25 years of operation Osiris was living through a continuous evolution in capabilities. From a cold neutron diffractometer Osiris morphed into a world

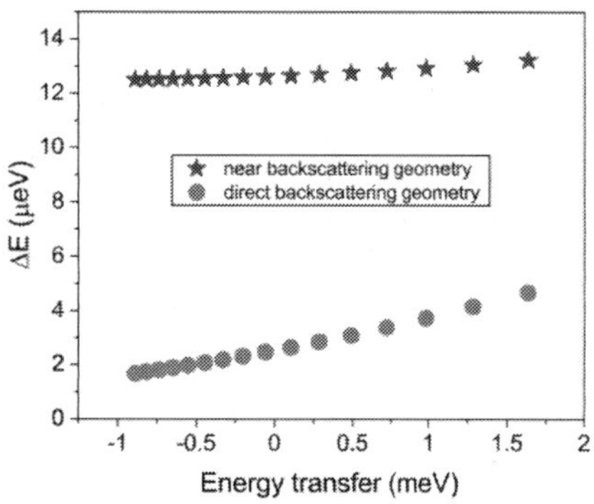

Figure 4. The calculated energy resolution of the Osiris spectrometer is shown for the case using a near-backscattering geometry of the Si analyzer and the resolution when direct backscattering will be employed for a sample with a height of 30 mm. A short pulse of $\Delta t = 20\ \mu s$ is applied through a fast pulse shaping chopper.

leading low-energy spectrometer. This enabled a successful science programme around quantum materials and energy research with high impact publications. There is no standstill and currently the capabilities will be further enhanced through the installation of the silicon analyzer. This upgrade will allow access to longer relaxation times and the survey of full four-dimensional $S(Q,\omega)$ maps of single crystals. The next step will be a decisive increase in intensity with the upgrade of the primary spectrometer. This will assure the competitiveness of Osiris with similar spectrometers at more powerful sources. Further into the future there exists the possibility to extend the resolution capabilities towards μeV. All these improvements will allow Osiris to stay at the forefront of low-energy spectroscopy in the coming decades.

References

[1] D. Martin y Marero, S. Campbell, C.J. Carlile, *J Phys Soc Jpn* **65** 245 (1996)

[2] D. Martin y Marero, D. Engberg, *Physica B* **267-268** 134 (1999)

[3] D. Martin y Marero, *Appl Phys A* **74** Materials Science S289 (2002)

[4] K.H. Andersen, D. Martin y Marero, D. Barlow, *Appl Phys A* **74** Materials Science S237 (2002)

[5] M.T.F. Telling and K.H. Andersen, *Phys Chem Chem Phys* **7** 1255 (2005); M.T.F. Telling, S.I. Campbell, D. Engberg, D. Martin y Marero, K.H. Andersen, *Phys Chem Chem Phys* **18** 8243 (2016)

[6] F. Demmel and K. Pokhilchuk, *Nucl Instr Meth A* **767** 426 (2014)

[7] F. Demmel D. McPhail, C. French, D. Maxwell, S. Harrison, J. Boxall, N. Rhodes, S. Mukhopadhyay, I. Silverwood, V. Garcia Sakai, F. Fernandez-Alonso, *J Phys Conf Ser* **1021** 012027 (2018)

[8] F. Demmel, D. McPhail, J. Crawford, D. Maxwell, K. Pokhilchuk, V. Garcia-Sakai, S. Mukhopadyay, M.T.F. Telling, F.J. Bermejo, N.T. Skipper, F. Fernandez-Alonso, *EPJ Web of Conferences* **83** 03003 (2015)

[9] R. Coldea, D.A. Tennant, E.M. Wheeler, E. Wawrzynska, D. Prabhakaran, M. Telling, K. Habicht, P. Smeibidl, K. Kiefer, *Science* **327** 177 (2010).

[10] C. Stock, C. Broholm, Y. Zhao, F. Demmel, H.J. Kang, K.C. Rule, C. Petrovic, *Phys Rev Lett* **109** 167207 (2012)

[11] C. Stock, E.E. Rodriguez, N. Lee, M.A. Green, F. Demmel, R.A. Ewings, P. Fouquet, M. Laver, Ch. Niedermayer, Y. Su, K. Nemkovski, J.A. Rodriguez-Rivera, S.-W. Cheong, *Phys Rev Lett* **117** 017201 (2016)

[12] W.J. Gannon, I.A. Zaliznyak, L.S. Wu, A.E. Feiguin, A.M. Tsvelik, F. Demmel, Y. Qiu, J.R.D. Copley, M.S. Kim, M.C. Aronson, *Nat Comm* **10** 1123 (2019)

[13] G.S. MacGlashan, Y.G. Andreev, P.G. Bruce, *Nature* **398** 792 (1999)

[14] T.J. Willis, D.G. Porter, D. J. Voneshen, S. Uthayakumar, F. Demmel, M.J. Gutmann, M. Roger, K. Refson, J. P. Goff, *Scientific Reports* **8** 3210 (2018)

[15] M.K. Gupta,S. K. Mishra, R. Mittal, B. Singh, P. Goel, S. Mukhopadhyay, R. Shukla, S.N. Achary, A. K. Tyagi, S.L. Chaplot, *Phys Rev Mat* **4** 045802 (2020)

[16] R. Laven, U. Häussermann, A. Perrichon, M. S. Andersson, M. Sannemo Targama, F. Demmel, M. Karlsson, *Chem Mater* **33** 2967 (2021)

[17] G.J. Irvine, F. Demmel, H.Y. Playford, G. Carins, M.O. Jones, J.T.S. Irvine, *Chem Mater* **34** 9934 (2022)

[18] A.M.A. Leguy, J. Moore Frost, A.P. McMahon, V. Garcia Sakai, W. Kockelmann, C. Law, X. Li, F. Foglia, A. Walsh, B.C. O Regan, J. Nelson, J.T. Cabral, P.R.F. Barnes, *Nat Comm* **6** 7124 (2015)

[19] B. E. Meijer, G. Cai, F. Demmel, H.C. Walker, A.E. Phillips, *Phys Rev B* **106** 064302 (2022)

[20] A.J. O Malley, V. Garcia Sakai, N. Dimitratos, W. Jones, C.R.A. Catlow, S.F. Parker, *Phil. Trans. R. Soc. A* **378** 20200063 (2020); A.J. OMalley, S.F. Parker, C.R.A. Catlow, *Chem Comm* **53** 12164 (2017)

[21] S.K. Matam, C.R.A. Catlow, I.P. Silverwood, A.J. OMalley, *Topics in Catalysis* **64** 699 (2021)

[22] V. Arrighi, J. Tanchawanich, M.T.F. Telling, *Macromolecules* **46** 216 (2013)

[23] M. P. M. Marques, A.L.M. Batista de Carvalho, V. Garcia Sakai, L. Hatter, L.A.E. Batista de Carvalho, *Phys Chem Chem Phys* **19** 2702 (2017)

[24] F. Foglia, A.J. Clancy, J. Berry-Gair, K. Lisowska, M.C. Wilding, T.M. Suter, T.S. Miller, K. Smith, F. Demmel, M. Appel, V. Garcia-Sakai, A. Sella, C. A. Howard, M. Tyagi, F. Cora, P.F. McMillan, *Sci Adv* **6** eabb6011 (2020)

[25] A. Perrichon, F. Fernandez-Alonso, M. Wolff, M. Karlsson, F. Demmel, *Nucl Instr Meth A* **947** 162740 (2019)

[26] E. Mamontov and K.W. Herwig, Rev Sci Instrum **82** 085109 (2011)

[27] N. Takahashi, K. Shibata, T.J. Sato, Y. Kawakita, I. Tsukushi, N. Metoki, K. Nakajima, M. Arai *Nucl Instr Meth A* **600** 91 (2009); K. Shibata, N. Takahashi, Y. Kawakita, M. Matsuura, T. Yamada, T. Tominaga, W. Kambara, M. Kobayashi, Y. Inamura, T. Nakatani, K. Nakajima, M. Arai, *Jap Phys Soc Conf Proc* **8** 036022 (2015)

[28] N. Tsapatsaris, R.E. Lechner, M. Markó, H.N. Bordallo, *Rev Sci Instrum* **87** 085118 (2016); N. Tsapatsaris, P.K. Willendrup, R.E. Lechner, H.N. Bordallo, *EPJ Web of Conferences* **83** 03015 (2015)

[29] B. Frick, E. Mamontov, L. van Eijck, T. Seydel, *Zeit Phys Chem* **224** 33 (2010)

[30] A. Meyer, R.M. Dimeo, P.M. Gehring, D.A. Neumann, *Rev Sci Instrum* **74** 2759 (2003)

[31] C. Schanzer, P. Böni, U. Filges, T. Hils, *Nucl Instr Meth A* **529** 63 (2004)

[32] K.H. Kleno, K. Lieutenant, K.H. Andersen, K. Lefmann *Nucl Instr Meth A* **696** 75 (2012)

[33] H. Jacobsen, K. Lieutenant, C. Zendler, K. Lefmann *Nucl Instr Meth A* **717** 69 (2013)

[34] C. Zendler, D. Nekrassov, K. Lieutenant, *Nucl Instr Meth A* **746** 39 (2014)

[35] D.M. Rodriguez, D.D. DiJulio, P.M. Bentley, *Nucl Instr Meth A* **808** 101 (2016)

[36] A. Perrichon, F. Fernandez-Alonso, M. Wolff, M. Karlsson, F. Demmel, *Journal of Surface Investigation: X-ray, Synchrotron and Neutron Techniques* **14** S169 (2020)

[37] A. Perrichon, and F. Demmel, *Technical Report TR RAL* 002 (2020)

[38] A. Perrichon and F. Demmel, *Nucl Instr Meth A* **1039** 167014 (2022)

[39] G. Skoro, S. Lilley, R. Bewley, *Physica B* **551** 381 (2018)

[40] F. Demmel and K. H. Andersen, *Meas Sci and Tech* **19** 034021 (2008)

[41] J. Wuttke, A. Budwig, M. Drochner, H. Kämmerling, F.-J. Kayser, H. Kleines, V. Ossovyi, L.C. Pardo, M. Prager, D. Richter, G.J. Schneider, H. Schneider, S. Staringer, *Rev Sci Instrum* **83** 075109 (2012)

[42] M. Birr, A. Heidemann, B. Alefeld, *Nucl Instr Meth A* **95** 435 (1971)

Upgrade of JCNS SANS instrument KWS-2 for improved performance and beam-time efficiency

Aurel Radulescu[1*], *Jia-Jhen* Kang[1], *Marie-Sousai* Appavou[1], and *Aristeidis* Papagiannopoulos[2]

[1]Jülich Centre for Neutron Science (JCNS) at Heinz Maier-Leibnitz Zentrum (MLZ), Forscungszentrum Jülich, 85748 Garching, Germany

[2]Theoretical and Physical Chemistry Institute, National Hellenic Research Foundation, 11635 Athens, Greece

Abstract. KWS-2 is a classical small angle neutron diffractometer where, following recent upgrades, the pinhole mode with different neutron-wavelengths and detector distances can be combined with focusing mode with MgF_2 lenses to reach a wide Q-range between $2x10^{-4}$ and 1.0 Å^{-1}. Further upgrades of the detection system and sample positioning system are currently in progress. A wide-angle detection prototype is currently being tested and optimized aiming at measurement conditions over an extended Q-range up to 2.0 Å^{-1}, which will be beneficial for semi-crystalline materials and small biological morphologies. The high neutron flux provided by the FRMII reactor and transported by the optimized neutron guidance system requires an adjustment of the measurement routine for a more rational use of the beam time. A new versatile in-beam sample positioning system, which includes a multi-position carousel with temperature control, robotic elements, and a storage pool for sample cuvettes, has just been installed at the instrument's sample position. This allows the instrument to be continuously supplied with samples and provides the ability to combine experiments on similar samples or topics into a common long experimental session. Here we report in detail on the expanded Q-range option of the KWS-2 from the perspective of current performance and planned upgrades, as well as the new sample positioning system with robotic elements.

1 Introduction

The understanding of soft- and bio-materials, such as semi-crystalline polymers in melt and solution, block copolymer assemblies, polymer gels and membranes, or proteins, polysaccharides and lipids complexes, requires the global knowledge of their microstructural features on a wide length scale between several Å and µm sizes. Such a wide length scale can be explored at the KWS-2 small-angle neutron (SANS) diffractometer of Jülich Centre for Neutron Science (JCNS) at Heinz Maier-Leibnitz Zentrum (MLZ) [1-5].

KWS-2 instrument is designed for high intensity studies of the mesoscopic structures with their local conformation order and long-range assemblies and correlations and their changes due to kinetic processes under various stimuli, such as temperature, hydration level, deformation, rapid mixing, etc., in the fields of soft condensed matter, chemistry, and biology. The instrument was recently optimized for structural investigations over a wide momentum transfer Q range, between $2x10^{-4}$ Å^{-1} and 1.0 Å^{-1}, by combining classical pinhole and focusing (with neutron lenses) methods, while simultaneously providing high-neutron intensities with an adjustable resolution – $\Delta\lambda/\lambda$ between 2% and 20%, using a versatile velocity selector in combination with a primary double-disc chopper with variable slit opening and TOF data acquisition [3]. A secondary single-disc compact chopper is available at the sample position and allows for the separation of the inelastic and elastic scattering with consequence in reducing the incoherent background from hydrogenated samples [5].

To further improve instrument performance and beam time efficiency, new upgrades to the detection system and sample environment are currently under development. A wide-angle detection option is currently in the installation phase and will allow measurements over an extended Q-range of up to 2.0 Å^{-1}, which will be beneficial for the characterization of semi-crystalline materials and small biological morphologies. In addition, the high neutron flux demanded optimization of the measurement procedure to make the best use of the beam time. A new versatile sample positioning system in the beam, which includes a thermostated multi-position carousel, robotic elements, and a storage pool of sample cuvettes has just been installed at the instrument's sample position. This allows a continuous supply of samples to the instrument and the pooling of experiments from different user teams when similar experimental conditions are required. In this paper, we focus on the extended Q-range measurement approach on the instrument and report on combined simulations by McStas and proof-of-principle measurements to provide a comprehensive understanding of the instrument's capabilities in terms of characterizing structures and morphologies over a large length scale, including the current status and expected performance due to ongoing upgrades.

* Corresponding author: a.radulescu@fz-juelich.de

2 The Q-range of KWS-2

2.1 Methodical aspects

The information in the elastic scattering experiments with neutrons is contained in the neutron intensity measured as a function of the momentum transfer Q:

$$Q = \frac{4\pi}{\lambda}\sin\theta$$

where λ is the neutron wavelength and 2θ is the scattering angle. Aiming for mesoscopic scale, SANS is optimized for observation at small angles and thus large scattering objects. A figure of merit of such kind of instrument is the Q-range that can be covered, thus the scale of lengths that can be characterised in real space. Typically, in pinhole SANS instruments at a reactor source for routine experiments, a specific wavelength corresponding to the maximum flux on the sample is selected from the broad flux distribution transported by the neutron guide system. This defines the typical Q-range that the instrument can cover by performing a series of measurements in which the detector is placed at different predefined positions after the sample so that a wide angular range 2θ can be scanned. Shorter and longer wavelengths can be used to complement the measurements, depending on the method used for obtaining monochromatic beams, and available flux at the instrument, pushing Q_{max} to higher values (for shorter λ) or Q_{min} to lower values (for longer λ). For KWS-2, $\lambda = 5$ Å is routinely used, allowing a Q range between 0.002 and 0.55 Å^{-1} to be covered in routine experiments, since the detection length L_D can be varied between 1.0 m and 20.0 m downstream of the sample. Since the velocity selector (Astrium GmbH) can be operated at different velocities and inclination angles against the beam direction, neutrons with $\lambda = 2.8$ Å and 10.0 Å can be used complementarily to extend the Q range in pinhole mode from 0.001 to 1.0 Å^{-1}. Tilting of the selector with respect to the beam axis enables the shift of the λ_{min} from 4.5 Å in the standard configuration to 2.8 Å. However, tilting of the velocity selector is a two-ways approach: the provision of shorter wavelengths compared to the standard configuration is accompanied by the worsening of the $\Delta\lambda/\lambda$ wavelength spread and the limitation of the largest wavelength available. Thus, the velocity selector in operation at KWS-2 delivers a monochromatic λ in the range 4.5 - 20 Å with a $\Delta\lambda/\lambda$ = 10% when it is operated in the standard configuration, with its axis parallel to the beam axis, and with rotation speeds within the range 8000 – 28200 rpm, while when used in a tilted configuration with the tilting angle $\xi = -10°$ against the beam axis, the λ range is shifted to 2.8 – 7 Å with $\Delta\lambda/\lambda$ between 14% and 18.5%, for the same range of rotation speed. Larger values than $\lambda = 7$ Å in the tilted configuration are prohibited due to vibrations from resonance frequencies of the selector.

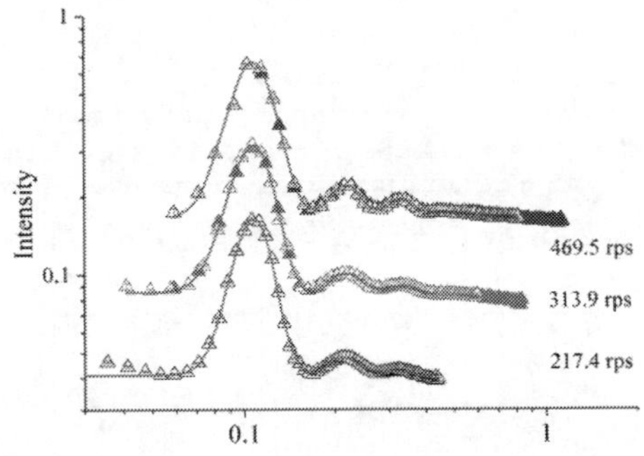

Fig. 1. Experimental SANS data (symbols) for the reference AgBeh sample measured on KWS-2 at an L_D of 1.0 m (full symbols) and 1.60 m (open symbols), with the velocity selector positioned at inclination angle of −10° with respect to the beam axis and operated at different rotation speeds (expressed in rotation per second units). The fit of the data (red lines) delivered the λ and the $\Delta\lambda/\lambda$ values reported in Table 1.

Fig. 1 presents the scattering patterns from AgBeh measured in the high Q regime at KWS-2 at different L_D and using the velocity selector at different rotation speeds, in inclined configuration. AgBeh is a wavelength calibrant used in small angle scattering experiments with X-rays and neutrons. The Bragg spacing of AgBeh is known to be $d_{AgBeh} = 58.38$ Å [6, 7]. AgBeh can also be used as reference material for determining the instrumental resolution parameters, as shown in [8], when Bragg peaks with several orders are observed at large angles (thus at $Q \geq 0.1$ Å^{-1}). In this range the uncertainty in Q is mainly governed by the wavelength spread $\Delta\lambda/\lambda$. The peak analysis of the scattering patterns shown in Fig. 1 was done by a simultaneous fit of the three observed peaks in all scattering patters measured with the same λ according to the equation

$$I_{AgBeh}^{model}(Q) = \sum_{i=1}^{N} \frac{A_i}{\sqrt{2\pi}\,\Delta Q(Q)} exp\left(-\left[\frac{Q-Q_i^{peak}}{\sqrt{2}\,\Delta Q(Q)}\right]^2\right) + bckd$$

with A_i – the area of the peak, N – the number of observed peaks, which was always 3, and *bckd* a constant background describing the flat behavior of the scattering patterns at high Q. The $(\Delta Q(Q))^2$ was considered as a sum of the wavelength and the geometrical terms of the Q resolution [3], namely $(\Delta Q_\lambda)^2$ – the wavelength spread term that depends on $\Delta Q_\lambda = Q\,\Delta\lambda/\lambda$, and $(\Delta Q_{geo})^2$ – the geometrical term that depends on the experimental conditions such as collimation and detection lengths, apertures sizes, etc. [3]. Because ΔQ_{geo} was known, it was considered fixed parameter in the analysis, while $\Delta\lambda/\lambda$ was considered variable parameter and was thus delivered by the fitting procedure, as reported in Table 1. The effect of the wavelength cut-off shown by the neutron guide system at KWS-2 [2] is clearly observable in the variation of the

$\Delta\lambda/\lambda$, which increases in increasing the λ between $\Delta\lambda/\lambda$ of 14% and 18.5%.

Table 1. KWS-2 velocity selector parameters tilted under the angle $\xi = -10°$ against the beam axis: the rotation speed, the wavelength and the wavelength spread.

Speed [rps]	λ [Å]	$\Delta\lambda/\lambda$ [%]
469.5	2.8	14.0
313.9	4.0	16.5
217.4	5.0	18.5

For lower Q values, pinhole operation can be supplemented by focusing mode: a package of MgF_2 lenses can focus a small entrance aperture (typically 4 x 4 mm^2) placed 18.5 m in front of the lens system onto a two-dimensional high-resolution scintillation detector (HRD) with 1.0 mm position resolution placed 18.5 m behind the lens system, so that neutrons scattered from a sample placed immediately behind the lenses can be detected at a much smaller angle than in pinhole operation. The small entrance aperture and the HRD are placed in fixed positions in the two focal points of the lens system, thus at 20 m and 17.0 m in front and after the sample, respectively. The focusing operation mode at KWS-2 is schematically depicted in Fig. 2 together with the conventional pinhole mode.

At KWS-2 26 MgF_2 lenses with a diameter $\Phi = 5$ cm and a parabolic shape, to avoid geometrical aberrations [9], are operational. The lenses can be grouped in different packages, to match the focusing conditions for different wavelengths λ in the configuration of entrance aperture – sample position – HRD which is mentioned above and schematically shown in Fig. 2. Thus, for λ = 17.5 Å, 4 lenses, for λ = 11 Å, 10 lenses, and for λ = 7 Å, 26 lenses are used, respectively. The lenses are kept at 77 K, to increase the transmission due to decrease in the scattering on phonons in the lens material [9]. The measured transmission of 26 lenses at room temperature was about T = 32%, when the full lens area was used, thus a circular sample aperture of 5 cm diameter was placed after the lens system. The transmission increased to T = 65% when lenses were used at 77 K, while in the same condition, but with only 1 cm x 1 cm beam size after the lens system, more realistic for routine applications, the transmission was T = 92%. For an entrance aperture of 4 mm x 4 mm at 20 m in front of the sample position, a neutron flux of 10^4 n s^{-1} cm^{-2} for λ = 7 Å and 600 n s^{-1} cm^{-2} for λ = 17.5 Å was estimated on the sample, respectively, by measuring the transmitted beam on the HRD detector and taking into account the detector efficiency of about 95%, which was considered constant for this wavelength range. If one multiplies the neutron flux estimated for λ = 7 Å when lenses are in beam with the ratio between the entrance aperture size (4 mm x 4 mm) and the cross-section of the guide feeding the instrument (50 mm x 50 mm) and normalizes the obtained value to the lens transmission, one can find a nominal neutron flux of 1.7 x 10^6 n s^{-1} cm^{-2}, which is pretty much the measured neutron flux on the sample position for λ = 7 Å at a collimation length of 20 m [2]. The same calculation holds for λ = 17.5 Å. This shows the good neutron transmission of the lens system.

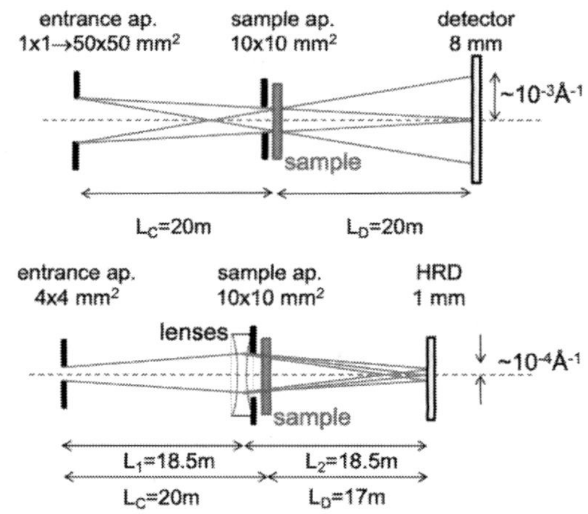

Fig. 2. Schematic presentation of the pinhole (top) and focusing (bottom) operation modes at KWS-2. The main technical elements and dimensions are indicated.

Lenses are focusing ideally only the central wavelength in the broad $\Delta\lambda/\lambda$ that is delivered by the velocity selector. A large wavelength spread yields on the detector a "peanut like" transmitted beam, due to defocusing of shorter and longer wavelengths and gravity effects. To tackle this effect, the quality of the transmitted beam by the lens system was checked on a first attempt for a longer neutron wavelength, λ = 17.5 Å, and a broad wavelength distribution $\Delta\lambda/\lambda$ = 20% first on the main detector at KWS-2, which was placed at L_D = 17.0 m, in the image focal point of the lens system, then on the HRD, at the same distance after the sample. Fig. 3 shows the transmitted beam with λ = 17.5 Å and $\Delta\lambda/\lambda$ = 20% through the lens system, so with 4 lenses in beam, as collected on the main detector (resolution 8 mm). The "peanut like" shape of the direct beam can be easily recognized on the left side panel presenting a sector of the detector. When the beam is analysed in TOF mode by using the resolution chopper [3], the broad distribution $\Delta\lambda/\lambda$ = 20% delivered by the selector is split into narrower slices and several beams with slightly different λ, each characterized by the narrow distribution $\Delta\lambda/\lambda$ = 5%, are recorded on the detector at different TOF. As shown in the other panels of Fig. 3 the gravity effect is clearly recognized due to the vertical distribution of these narrow beams that are characterized by an isotropic geometrical shape: the longer the λ, the lower the beam positioning on the detector.

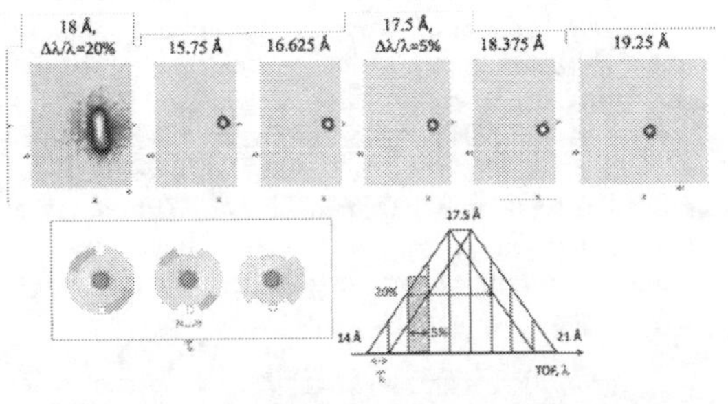

Fig. 3. The profile of the direct beam through the 4-lens system for $\lambda = 17.5$ Å and $\Delta\lambda/\lambda = 20\%$ measured in standard (upper left panel) and TOF (other upper panels) data acquisition modes, when different λ with $\Delta\lambda/\lambda = 5\%$ were sorted out from the initial broad wavelength distribution (lower right scheme) by setting the proper opening time τ of the chopper slit (lower left scheme).

Results of McStas simulations of the direct beam transported through the full guide system of KWS-2 and the system of lenses and detected at $L_D = 17.5$ m after the sample position on a two-dimensional detector (pixel size = 0.5 mm) are reported in Fig. 4 as detector images and wavelength distributions at the detection positions in the graphical format delivered by the simulation package. The simulations were carried out with a 20 m collimation length with a small entrance aperture (4 mm x 4 mm) and a 15 mm x 15 mm sample aperture for different wavelengths and wavelength distributions. The McStas component lens.comp (spherical lens, diameter $\Phi = 2.5$ cm) was used in the simulation. Fig. 4a,g report the direct beam image when lenses are out of beam for $\lambda = 17.5$ Å and $\lambda = 7$ Å, always with $\Delta\lambda/\lambda = 20\%$, while Fig. 4d,i show the wavelength distribution computed at the detection position. The effect of lenses on the beam profile is observed in Fig. 4b,c for $\lambda = 17.5$ Å (4 lenses in beam) with $\Delta\lambda/\lambda = 20\%$ and 5%, respectively, with the corresponding wavelength distribution shown in Fig. 4e,f. Fig. 4h,j report the beam profile and wavelength distribution computed at the detection position for $\lambda = 7$ Å (26 lenses in beam) with $\Delta\lambda/\lambda = 20\%$. In parallel, the measured direct beam for $\lambda = 17.5$ Å with $\Delta\lambda/\lambda = 20\%$ and 5% (in TOF mode), are presented in Fig. 4k,l. Focusing effect induced by lenses is clearly observed when compared the results to those obtained for the lenses free instrument. Also, the gravitational and polychromatic effects are obvious in case of using the lenses for focusing the long wavelength $\lambda = 17.5$ Å with a broad wavelength distribution $\Delta\lambda/\lambda = 20\%$: while the focusing is achieved on the horizontal beam spot dimension, the vertical spot dimension is still large, comparable to the lenses free vertical size of the beam. When the wavelength distribution is narrowed down to $\Delta\lambda/\lambda = 5\%$, beam focusing is obtained in both directions. The effect is observed in both simulated and measured results, and is striking in the case of the shorter wavelength $\lambda = 7$ Å even for the broader wavelength distribution $\Delta\lambda/\lambda = 20\%$. We can add that the experimental background observed in Fig. 4k,l outside of the direct beam spots is due to the γ-sensitivity (10^{-4}) of the scintillation-type HRD, which however does not affect the quality of the measured data, when proper corrections for the detector sensitivity, "dark current" and empty sample container are applied.

The good agreement between the simulation of an ideal instrumental setup and the measurement carried out in similar conditions proves the normal behavior of the lens system at KWS-2.

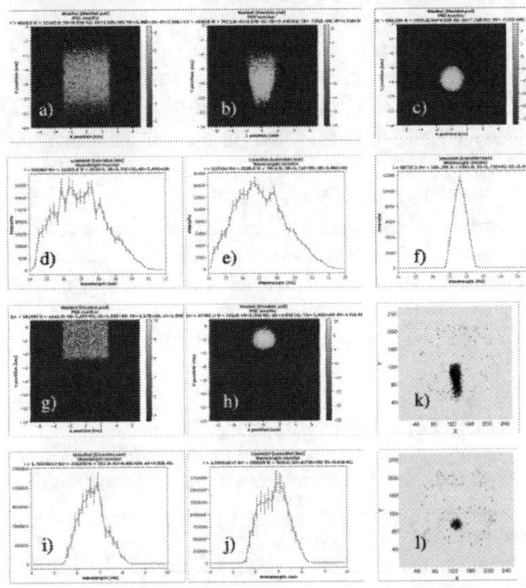

Fig. 4. Results of McStas simulations (taking into account the gravity) and neutron measurements of the direct beam profile on the detector placed at 17.5 m for 4-lens system in beam. Panels a) and g) show the direct beam on the detector when lenses are out of beam for $\lambda = 18$ Å and $\lambda = 7$ Å, respectively, for a $\Delta\lambda/\lambda = 20\%$ in both cases, while panels e) and g) present the wavelength distribution in both situations. Panels b) and c) show the direct beam on the detector when lenses are in beam, for a $\Delta\lambda/\lambda = 20\%$ and $\Delta\lambda/\lambda = 5\%$, respectively, while panels e) and f) present the corresponding wavelength distribution. Panels h) and j) present the same data for $\lambda = 7$ Å with a $\Delta\lambda/\lambda = 20\%$. Panels k) and l) show measured direct beam profile on the HRD for $\lambda = 18$ Å, with $\Delta\lambda/\lambda = 20\%$ and $\Delta\lambda/\lambda = 5\%$ (using TOF mode).

2.2 Current performance

With its high flux [4], comparable to that of the world's leading pinhole SANS instruments, and its extended *Q*-range between 2 x 10^{-4} Å^{-1} and 1.0 Å^{-1}, which can be covered by combining pinhole and focusing modes, KWS-2 is a high-throughput instrument that typically measures a large number of samples during short experiments. As an example of typical experimental results, Fig. 5 shows scattering patterns recorded at room temperature over the entire *Q* range currently available at the instrument from samples representative of those characterized on the KWS-2: a

https://doi.org/10.1051/epjconf/202328603006

sample consisting of a uni-lamellar vesicle (ULV) forming mixture of lipid molecules in D_2O buffer solution (Fig. 5a), showing adhesion (docking) effects when myelin basic protein (MBP) is added to the solution (Fig. 5b), resulting in vesicle assemblies and multilamellar domains, as nicely shown by the cryo-TEM image in Fig. 5b. The detailed sample composition and preparation protocols as well as the experimental procedure and data analysis are reported elsewhere [10]. Here, only the experimental details related to the methodological approach and beam time efficiency are discussed to support the currently ongoing instrument developments.

SANS measurements on the sample with mixed lipids in D_2O buffer solution (Fig. 5a) were performed in pinhole mode using $\lambda = 5.0$ Å with $L_D = 2$ m and 8 m, and $\lambda = 10.0$ Å with $L_D = 20$ m. For all experimental conditions, $\Delta\lambda/\lambda$ was 10% (standard selector configuration). The lipid molecules lead to the formation of ULVs as shown by cryo-TEM observation (inset in Fig. 5a). The scattering pattern in Fig. 5a is well described by the vesicle form factor (indicated by the red line) and shows all the typical scattering features of a vesicular morphology: the total size of vesicles, indicated by the Guinier plateau observed at low Q, and the thickness of the vesicle bilayer, indicated by the deviation of the intensity from the Q^{-2} power law characteristic of the two-dimensional bilayer morphology. It is also worth noting that SANS measurements performed with a higher lipid concentration in solution did not show vesicle aggregation and the scattering features of individual ULVs were still clearly observed [10]. The data acquisition time was 10 min for $L_D = 2$ m and 8 m, and 15 min for $L_D = 20$ m, respectively. According to our experience and the opinion of our users, the acquisition time for such samples is comparable to that at other high-flux pinhole SANS diffractometers such as the D22 at ILL-Grenoble. When MBP is added to the ULV solution, much larger morphologies are formed, as no Guinier regime of vesicle size is detectable anymore in the scattering pattern measured in standard pinhole mode. The SANS pattern was completed with measurement at very low Q, using the focusing mode with $\lambda = 7.0$ Å and 26 MgF_2 lenses in the beam, while data were collected on the HRD instrument for data acquisition in standard mode (not TOF mode) over a measurement time of 15 minutes. In the extended scattering pattern, a Guinier regime was observed at about an order of magnitude lower Q compared to the ULV solution case. In addition, two broad maxima, indicated by the vertical arrows in Fig. 5b, are observed at intermediate and high Q values. The Guinier regime observed at very low Q in the case of the MBP-doped vesicle solution indicates the formation of vesicle aggregates characterized by a radius of gyration about ten times larger than that of the ULVs themselves, as also shown by the cryo-TEM images, an effect attributed to the addition of MBP to the solution after ULV formation. Relying on the cryo-TEM observations and the model interpretation of the scattering data (red line in Fig. 5b) based on a partitioning between single and docked ULVs (blue line) as described in [11], combined with a Beaucage form factor [12] of the large-scale vesicle aggregates, we report here only the nature of the two broad maxima clearly seen in Fig. 5b: the higher Q maximum is due to vesicle adhesion (docking) mediated by the MBP, which promotes a local multi-lamellarity effect and the formation of larger vesicle aggregates, while the intermediate Q maximum is a detail of the vesicle form factor accentuated by the additional scattering due to vesicle aggregates at low Q, which overlaps the form factor of the ULVs.

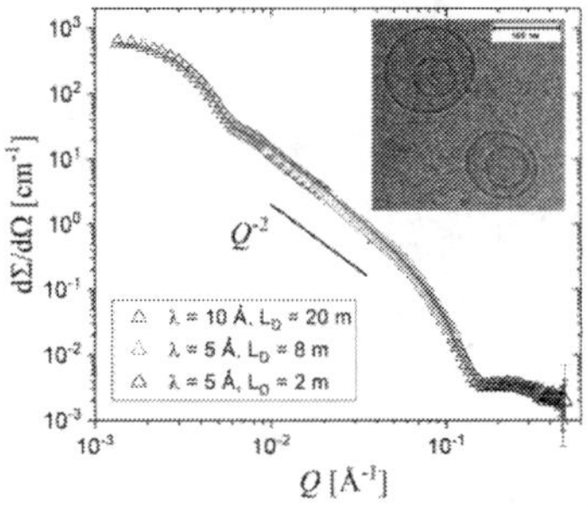

a)

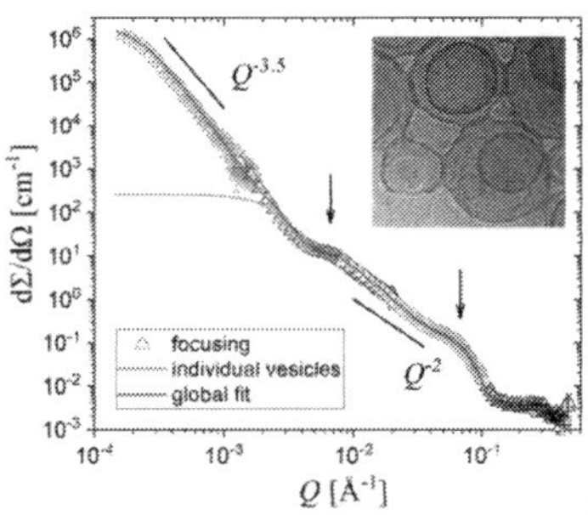

b)

Fig. 5. SANS data of lipid (a) and lipid + MBP (b) solutions recorded at KWS-2 in pinhole mode (a) and in combined pinhole and focusing modes (b). The red curves show the model fit of the experimental data according to the vesicle form factor (a) and the unified model of the docked vesicles, as the product of vesicle form factor and docking structure factor, and the large-scale aggregates (b). The light blue curve in panel (b) shows the scattering contribution of the docked vesicles. The insets show cryo-TEM images of the same samples, where the scale bar valid for both images is 100 nm.

Thus, experimental studies as described in Fig. 5a,b can be performed in a very short time at the KWS-2, and it typically takes up to one hour per sample to cover the entire Q range between 2.0 x 10^{-4} and 1.0 Å^{-1} using the combination of pinhole and focusing modes. The total estimated time for a complete measurement of a sample also includes the time lost by reaching different instrument configurations. The most time consuming is the movement of the main detector between different detection positions L_D and the time needed to move the HRD in the beam, while the lenses can be moved in or out of the beam in a few seconds, as well as all other elements such as collimation apertures and guide segments or the beam-stop, which has to be repositioned according to the λ and L_D conditions. It is also worth mentioning that the selection of a new λ via the velocity selector is done simultaneously with the main detector movement, so no additional time is lost by this adjustment. The versatile NICOS [13] instrument control software allows optimization of complex experimental protocols, including pinhole and focusing settings, reducing time-consuming adjustments during the investigation of a full set of samples. As with any other SANS diffractometer, whether it is a pinhole or TOF instrument, most of the lost time at KWS-2 is due to changing samples or the experimental team at the instrument. Therefore, one of the upgrades currently underway aims to change the concept of sample delivery to the instrument to use the beam-time more efficiently, and will be discussed in more detail in the next section.

Finally, it should be mentioned that an extended *Q*-range structural study at lower *Q*-values towards 10^{-4} $Å^{-1}$ is currently used routinely only at a very limited number of SANS pinhole instruments besides KWS-2, namely SANS-J-II at the JRR-3 reactor, Tokai, Japan [14], with neutron lenses and HRD, and QUOKKA at ANSTO, Australia [15], with neutron lenses but without HRD. To our knowledge, no TOF-SANS instrument operating at research reactors or spallation sources can cover such a wide *Q* range down to these low *Q* values. Although some TOF-SANS instruments at spallation sources can cover a very wide *Q* range in a single measurement without requiring a configuration change, such as the BL-15 TAIKAN at J-PARC, which uses a broad range of wavelengths along with a large number of detectors in fixed positions so that a wide angular range in scattering can be covered [16], the Q_{min} achieved at such instruments is nevertheless much larger than that achieved at KWS-2 with the combined pinhole focusing modes. Most TOF-SANS instruments at spallation or steady-state (reactor) neutron sources, such as SANS-2D at ISIS, UK [17], EQ-SANS at SNS, USA [18], D33 at ILL-Grenoble, France [19] or BILBY at ANSTO, Australia [20], use mobile main detectors. Therefore, a combination of different experimental setups must be used to cover the specified *Q* range, which translates into a long acquisition time per sample comparable to the time required at KWS-2. However, none of these instruments can achieve a Q_{min} value of less than 1.0 x 10^{-3} $Å^{-1}$. Pinhole SANS instruments at reactors using a complementary focusing method are far superior to these instruments in terms of extended *Q* range toward lower *Q* values. On the other hand, the advantage of TOF-SANS instruments, which use a broad wavelength band, is the ability to cover an extended *Q* range with higher *Q* values than pinhole instruments. Some of these instruments, such as BL-15 TAIKAN at J-PARC, can reach *Q* values up to several $Å^{-1}$. As we have just reported [21], the ability to cover a *Q* range up to 2.0 $Å^{-1}$ offers unique advantages in the characterization of semi-crystalline polymer systems under application-relevant conditions using the neutron contrast variation method. Several pinhole or TOF instruments are already equipped with wide-angle detectors to cover higher *Q* values than the usual *Q* range. Therefore, another ongoing upgrade at KWS-2, presented in the next section, is the installation of a wide-angle detector system to increase the instrument performance up to Q_{max} = 2.0 $Å^{-1}$.

2.3 Developments in progress

A fully automated sample exchange and positioning system was developed for KWS-2 to allow continuous supply of samples to the instrument while minimizing sources of error and scientist involvement in sample positioning, and to provide the ability to pool experiments from different user teams when similar experimental conditions are required.

Fig. 6. The fully automated new sample positioning system at KWS-2, including the multi-position sample carousel, the robotic arm and the cuvette storage pool.

The new system, which includes a multi-position carousel with sample temperature control, robotic elements and a storage pool for sample cuvettes, was installed in the sample area (Fig. 6). The new compact setup will remain in operation at the sample area as a standard option for all routine experiments, i.e. about 80% of the beam time at KWS-2. For the remaining experiments requiring more specialized sample environments such as rheometers, stop-flow devices, pressure or humidity chambers, the ease of removal and precise repositioning by a zero-point clamping system will allow a smooth change of options. A 48-position COLOSSUS sample carousel (Quantum Northwest Inc., Liberty Lake, WA) with temperature controllers was installed on the sample table (Fig. 6). The instrument is divided into six sections, each with 8 positions. Each section is equipped with two Peltier elements and a temperature controller so that the sections can be controlled independently. This allows up to six temperatures between 5 °C and 110 °C to be set simultaneously on different samples, with a precision of 0.01 °C, an accuracy of 0.5 °C and a temperature reproducibility of 0.05 °C in the temperature interval from 5 °C to 80 °C. The geometry of each sample chamber equipped with quartz windows has been designed to allow an angular range for dispersion greater than ± 50 °C. Mechanical control and temperature regulation are provided by the KWS-2 control software

NICOS [13]. The instrument is designed to hold Banjo quartz cuvettes in special Al carriers, which are moved by a robotic arm between COLOSSUS and a nearby cuvette storage pool. A nearly continuous supply of samples to the instrument is ensured by regular refreshing of the cuvette storage pool by the experimentalist when the ongoing experimental routine allows this without loss of time, e.g. during the movement of the main detector between different positions L_D, which takes several minutes. Loading and unloading of the COLOSSUS with sample cuvettes from the cuvette storage is performed by a compact six-axis Meca500 robotic arm (Mecademic Inc, Canada) equipped with an electric parallel gripper MEGP 25LS (Fig. 6). A proof-of-principle of the new sample changing concept at KWS-2 can be seen in [22]. With the COLOSSUS, the robotic arm and the cuvette storage pool installed at the sample position, a nearly fully automated instrument supply of samples is possible at the KWS-2. At least 96 samples can be measured without interruption, which, taking into account the foreseen temperature variations at the sample, means that almost a week of beam-time can be spent at the instrument without the need for the instrument or guest scientists at the sample position. Renewal of the cuvette storage pool is easily possible at any time, as mentioned above, which further increases the degree of freedom. In our experience, an estimated 3-4 experiments of the same type can be pooled, minimizing the lost time and improving the beam-time planning at the instrument.

To further improve the performance of the instrument, a wide-angle detection option has just been tested and optimized to enable measurements over an extended Q-range of up to 2.0 $Å^{-1}$. This will be beneficial for the characterization of semi-crystalline materials and small biological morphologies. In 2015, a new high count rate detection system was commissioned at KWS-2 [4]. This system consists of an array of 144 position-sensitive ^{3}He detection tubes (PSDs) and fast readout electronics, providing high count rates up to several MHz, high count stability and, based on the large detection area (almost 1.0 m^2, with a diamond shape for optimal filling of the vacuum tank cross-section), measurements up to Q_{max} = 1.0 $Å^{-1}$ at λ = 2.8 Å, with a resolution $\Delta\lambda/\lambda$ = 14.5%. However, as reported in [21], an efficient combined SANS/WANS approach on the same instrument with the ability to use contrast variation to obtain comprehensive information on hydrocarbon systems under conditions relevant to the application would require a Q_{max} of at least 2.0 $Å^{-1}$ at KWS-2. Therefore, the installation of a WANS option at KWS-2 is currently in progress. Given the great versatility of the instrument, which is equipped with two types of chopper systems, the resolution double-disk chopper [3] and the background single-disk chopper [5], WANS and SANS measurements at KWS-2 will be aimed at in various combinations of standard and TOF modes.

Considering the instrument concept and the available conditions in the FRM II reactor guide hall, two large-area detector wings are currently being designed to be installed in inclined fixed positions behind the sample (Fig. 7), so that the scattered neutrons can be detected in a wide angular range up to $2\theta_{max}$= 50°. The two detector wings are placed above and below the beam axis at a distance of 1.25 m downstream of the sample position, which provides the possibility to record the scattered neutrons in the high-Q region of the instrument in TOF mode, if required, and to discard the inelastic component from the measured data to improve the signal-to-noise ratio at the instrument, as described in [5]. Each wing consists of an array of 80 ^{3}He PSDs. The PSDs have a thickness of 6 mm and an active length of 60 cm and are supplied by Canon Electronics Tubes and Devices Inc. (CETD), Japan, along with fast readout electronics that can operate in vacuum (better than 0.002 mbar, as recommended by the manufacturer) or in air or He atmosphere. The latter configuration would not require a cooling system for the readout electronics, as demonstrated by the use of the same detection system for the WANS detection banks of the BL-15 TAIKAN, J-PARC [16], and would help avoid technical complications associated with operation in vacuum. A prototype of such a detection module (an 8-pack of PSDs) was tested at MLZ in Garching using a stationary ^{252}Cf neutron source and a borated mask with narrow slits. The test resulted in the expected position resolution of about 4-5 mm in the middle and at the end of the PSD tubes. A schematic representation of the proposed WANS option operating complementary and simultaneously with the SANS option at KWS-2 is shown in Fig. 7, where the sample location (1), the fixed WANS panels (2), the movable SANS detector (3), and the He-boxes (4) can be easily identified. In the final concept of the WANS option, the WANS detectors are operated in the air with He-boxes installed in front of them so that the air path between the sample and the PSDs is minimized. By using He gas at a pressure of 1 bar inside the boxes, very thin Al windows can be used at the entrance to the sample and at the exit to the PSD array, keeping the background at a very low level. A detailed discussion of the background sources in SANS experiments, including the air path at the sample position and the windows, and how their contribution can be minimized can be found in [23]. This measurement geometry also allows optimization of the pyramid-shaped nose of the main detector vacuum tank so that a smaller and thinner Si entrance window than the actual one can be installed to maintain the large pressure difference between the ambient pressure at the sample position and the 10^{-2} mbar vacuum in the tank. In this experimental configuration, a large vacuum shutter is no longer needed, which will simplify the technical implementation of the WANS option and the operation of the instrument during the routine user program. It is planned to install mechanical external shields that can be moved in and out of the window front and be interlocked with the condition of the access lead shielding door, so that damage to the windows can be avoided during activities at the sample position. In this configuration, the closest position of the main SANS detector to the sample position is L_D = 1.25 m, which allows acquisition of SANS data with low parallax distortions for a maximum of $2\theta_{SANS}$ = 24° [19]. The Q

range can be completed to higher Q values with data measured with the WANS detectors.

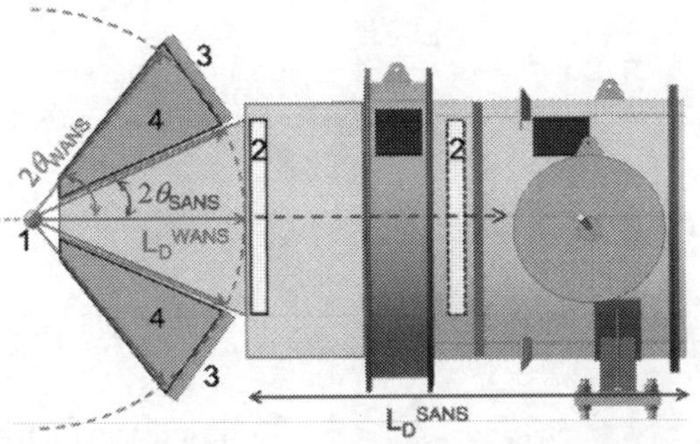

Fig. 7. Schematic representation of the concept of the planned WANS option at KWS-2: the wide-angle detector panels (3) are arranged in a fixed inclined position above and below the beam axis at L_D^{WANS} = 1.25 m downstream of the sample (1), so that the scattered neutrons can be detected up to an angle $2\theta_{WANS}$ = 50°. The mobile SANS detector (2) that can be moved between L_D = 1.25 m and L_D = 20 m is marked in yellow. The He boxes (4) in front of the WANS detection arrays are marked in green.

McStas simulations were performed to evaluate the concept of combining SANS and WANS at KWS-2. The crystalline scattering pattern of AgBeh in the high Q region, up to Q_{max} = 2.0 Å^{-1}, is characterized by a large number of narrow peaks that make wide-angle scattering test analysis with neutrons difficult, even with a high-resolution approach at a conventional SANS diffractometer. Instead, the fullerene C60 powder in this Q region is characterized by isolated and very sharp crystalline reflections, making such a sample ideal for feasibility tests and simulations of the experimental SANS-WANS approach at KWS-2. We chose to follow the reflection from the crystalline planes (111) that appears in XRD patterns at 2θ = 10.78° [24, 25]. In Q, this crystalline reflection appears isolated at Q_{C60} = 0.76 Å^{-1}. In a SANS experiment, this reflection from a fullerene C60 powder sample is observed as a ring at a scattering angle of 2θ = 19.8° when neutrons with a wavelength of λ = 2.8 Å are used, while in the WANS regime it is observed at 2θ = 35.7° for neutrons with a wavelength of λ = 5.0 Å.

The results of the McStas simulations for the setup schematically shown in Fig. 7 with the detectors at a distance L_D = 1.25 m from the sample, in an oblique position with respect to the beam axis, and using the Powder1.comp component of McStas for the fullerene C60 sample are shown in Fig. 8. Different wavelength distributions were used in the simulations, always with λ = 5 Å. The ring scattering of the powder sample due to the crystalline 111 reflection is well reproduced. As can be seen, the quality of the results obtained at high resolution ($\Delta\lambda/\lambda$ = 5%) suggests that useful structural data on semi-crystalline systems with small crystalline units yielding crystalline reflections above Q = 1.0 Å^{-1} can be obtained with this setup using the TOF data acquisition and the resolution chopper of the instrument.

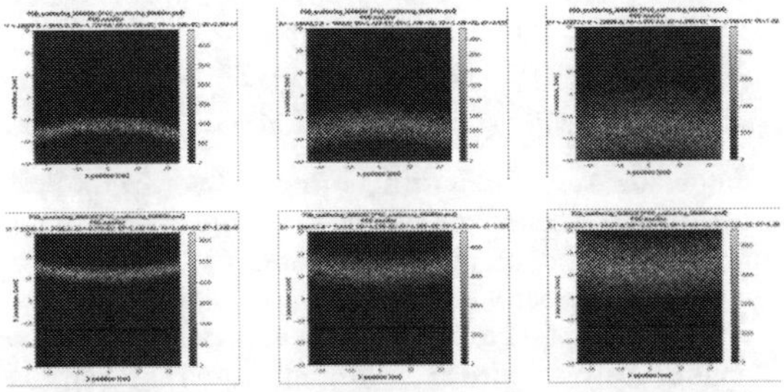

Fig. 8. McStas simulation results of scattering of fullerene C60 samples on two detection panels positioned as indicated in Fig. 7, using neutrons with a wavelength λ = 5 Å with a resolution of $\Delta\lambda/\lambda$ = 5% (left panels), 10% (middle panels) and 20% (right panels).

Combined with simultaneous SANS data acquisition and the contrast variation technique, this will contribute to a more thorough characterization of such materials. Treatment of the incoherent background was not considered in the McStas simulation because one of the planned ways to work with the WANS detectors at KWS-2 is to use the background chopper and TOF data collection [5] to sort out the neutrons inelastically scattered from the hydrogen-containing samples and thus reduce the incoherent background. Neutrons inelastically scattered from hydrocarbon materials such as polymer or protein solutions contribute significantly to the recorded intensity in a SANS experiment. This contribution is due to incoherent and multiple scattering from hydrogen or deuterium nuclei and contributes as a flat background to the measured data. The proof of principle has been provided elsewhere [5], using the typical high-Q setup at KWS-2, i.e., short input wavelengths, λ = 5 - 7 Å, and short detection distances, L_D = 1.1 - 1.6 m. Positioning the future WANS detectors at KWS-2 at a distance of L_D = 1.25 m from the sample will allow the proper TOF conditions to be established for these wavelengths to provide insight into the inelastic scattering and resulting background, which is useful when data collected at high Q do not show a flat profile.

3 Conclusions

The KWS-2 is a classical 40-m pinhole small-angle neutron diffractometer that can cover a wide Q range by varying the sample-to-detector distance between 1 m and 20 m and the wavelength λ between 2.8 and 20 Å. With recent upgrades following the move to the FRMII reactor, the pinhole mode has been combined with the focusing mode using MgF_2 lenses, and a wide Q range between 2 x 10^{-4} and 1.0 Å^{-1} has become available on the instrument, enabling structural studies of complex morphologies with multiple size levels over a broad length scale from nm to micron. Further upgrades to the detection system and sample environment are currently underway. A wide-angle detection option is currently being tested and optimized and will allow measurements over an extended Q-range up to 2.0 $Å^{-1}$, which will be beneficial for the complete characterization of semi-crystalline materials and small biological morphologies. The high neutron flux, which is comparable to that at the world-leading SANS instruments, was crucial for optimizing the measurement procedure to make the best use of the beam time. A new versatile system for positioning samples in the beam, consisting of a temperature-controlled carousel with multiple positions, was recently put into operation. It allows samples to be continuously supplied to the instrument and experiments on similar samples or effects to be pooled in a common long experimental session using robotic elements and a storage of sample cuvettes that can be easily renewed by the experimentalist.

The participation of Dr. Benjamin Krugmann and Dr. Andreas Stadler (both FZJ) in the experiments on lipid solutions with MBP is gratefully acknowledged. We thank Simon Staringer, Thomas Kohnke, Georg Brandl, Vladimir Ossovyi, Frank Suxdorf, Harald Kleines, Jacqueline Lippertz, Andreas Thelen, Herbert Feilbach, Jos Dahmen, Bingöl Kendal, Dr. Alexander Weber, Dr. Günter Kemmerling, and Dr. Ralf Engels (all FZJ) for their active participation in upgrading the sample positioning system and WANS at KWS-2. Dr. Alexander Weber (FZJ) is thanked for the photo in Fig. 6. We thank Dr. Stefan Mattauch and Prof. Stephan Förster (both FZJ) for their continuous support in the upgrade of KWS-2.

References

1. A. Radulescu, V. Pipich, H. Frielinghaus, M-S. Appavou, J. Phys.: Conf. Ser. **351**, 012026 (2012)
2. A. Radulescu, V. Pipich, A. Ioffe, Nucl. Instrum. Methods Phys. Res. Sect. A **689**, 1-6 (2012)
3. A. Radulescu, N. K. Szekely, S. Polachowski, M. Leyendecker, M. Amann, J. Buitenhuis, M. Drochner, R. Engels, R. Hanslik, G. Kemmerling, P. Lindner, A. Papagiannopoulos, V. Pipich, L. Willner, H. Frielinghaus, D. Richter, J. Appl. Cryst. **48**, 1849–1859 (2015)
4. J. E. Houston, G. Brandl, M. Drochner, G. Kemmerling, R. Engels, A. Papagiannopoulos, M. Sarter, A. Stadler, A. Radulescu, J. Appl. Cryst. **51**, 323-336 (2018)
5. L. Balacescu, G. Brandl, A. Radulescu, J. Appl. Cryst. **54**, 1217–1224 (2021)
6. Gilles, R., Keiderling, U., Strunz, P., Wiedenmann, A., Fuess, H., Mater. Sci. Forum, **321–324**, 264–269 (2000)
7. Okabe, S., Karino, T., Nagao, M., Watanabe, S., Shibayama, M., Nucl. Instrum. Methods Phys. Res. Sect. A **572**, 853–858 (2007)
8. Grillo, I. (2008). In Soft Matter Characterization, edited by R. Borsali & R. Pecora. Berlin, Heidelberg: Springer.
9. H. Frielinghaus, V. Pipich, A. Radulescu, M. Heiderich, R. Hanslik, K. Dahlhoff, H. Iwase, S. Koizumi, D. Schwahn, J. Appl. Cryst. **42**, 681-690 (2009)
10. B. Krugmann, A. Radulescu, M.-S Appavou, A. Koutsioubas, L. R. Stingaciu, M. Dulle, S. Förster, A. M. Stadler, Scientific Reports, **10**, 16691 (2020)
11. K. Komorowski, A. Salditt, Y. Xu, H. Yavuz, M. Brennich, R. Jahn, T. Salditt, Biophysical Journal **114**, 1908–1920 (2018)
12. B. Hammouda, J. Appl. Cryst. **43**, 1474-1478 (2010)
13. G. Brandl, et al., NICOS – The Instrument Control Solution at the MLZ, in Proceedings of the 10th International Workshop on Personal Computers and Particle Accelerator Controls, Karlsruhe, Germany, 14-17 October 2014.
14. S. Koizumi et al., J. Appl. Cryst. **40**, s474-s479 (2007)
15. K. Wood et al., J. Appl. Cryst. **51**, 293-314 (2018)
16. S. Takata et al., JPS Conf. Proc. **8**, 036020 (2014)
17. R. K. Heenan et al., Proc. ICANS – XVII, 780-785 (2005).
18. W. T. Heller et al., J. Appl. Cryst. **51**, 242-248 (2018)
19. C. D. Dewhurst et al., J. Appl. Cryst. **49**, 1-14 (2016)
20. Sokolova A. et al., J. Appl. Cryst. **52**, 1-12 (2019)
21. Schiavone et al., J. Appl. Cryst. **56**, 947-960 (2023)
22. https://mlz-garching.de/media/roboter_v1.5.5_short.mp4
23. J. G. Barker and D. F. R. Mildner, J. Appl. Cryst. **48**, 1055-1071 (2015)
24. M. Sathish and K. Miyazawa, Cryst. Eng. Comm. **12**, 4146-4151 (2010)
25. A. Sanz, H. C. Wong, A. J. Nedoma, J. F. Douglas, J. T. Cabral, Polymer **68**, 47-56 (2015)

A design study of a 1-m^2 Multi-Wire-Proportional-Chamber Position-Sensitive-Neutron-Detector (MWPC-PSND)

Gregor Nowak[1*], *Irina* Stefanescu[2], *Andreas* Beldowski[1], *Jochen* Fenske[1], *Richard* Hall-Wilton[3] and *Martin* Müller[1]

[1]Helmholtz-Zentrum Hereon, WPE & TKA, 21502 Geesthacht, Germany
[2]European Spallation Source ERIC (ESS), Detector Group, 22100 Lund, Sweden
[3]Fondazione Bruno Kessler, Sensors & Devices Centre, 38123 Trento, Italy

Abstract. Neutron detection by MWPC based on $^{10}B_4C$ coatings can surpass in performance the one utilising ^3He-gas in terms of position resolution and count rate capability at similar detection efficiency [1-3]. The use of m^2-size foils coated by $^{10}B_4C$ demands the development of a mechanical concept to avoid deformations of the neutron sensitive surface due to their own weight and the electrostatic forces resulting from high voltage for operation. To preserve a flat shape of the converter, a pressure gradient is applied between its both sides [4]. The design shall allow a stacking of detection modules consisting of sequenced elements [converter/x-y-multi-wire planes/converter] in few mm proximity. A multiple stacking of the detector modules is needed to accumulate a sufficient high detection efficiency for diffraction with thermal neutrons over the entire detector [2]. The MWPC is planned to have fifteen detection modules corresponding to an efficiency equivalent of 30 converters coated with 1.2 μm of $^{10}B_4C$. Efficiency simulations of a converter coating thickness-profile from 0.6 μm up to 1.2 μm prognoses a conversion rate up to 50 % for thermal neutrons and the presented design parameters of the 1-m^2 MWPC allow a position resolution of 2 mm.

1 Introduction

Multi-Wire-Proportional-Chambers (MWPC) are used for a two-dimensional position-sensitive detection of radiation fields [5]. They consists of corresponding x- and y-sensitive detection wire planes immersed in stopping-gas, which releases charged particles when hit by ionizing radiation. The released primary electrons from ionised gas are accelerated towards the closest detection wire of a wire plane, which is set under high voltage (HV). The radial gradient of the electric field surrounding a detection wire is the main source of kinetic energy gain for the primary ionisation electrons. The acceleration and charge multiplication process is confined to a distance of a couple of hundreds μm around the detection wire [6]. The charge multiplication starts when the kinetic energy transferred by the electric field gradient to the primary ionisation electrons is high enough to further ionise the gas atoms and give rise to a charge avalanche. The process of generating electrons close to the detection wire in a sufficient number to trigger electric pulses in modern amplification electronics is called gas amplification. In the following sections, the neutron-sensitivity and neutron-event position-encoding techniques of a MWPC will be discussed. This introduces the special type of a MWPC, which offers a 2D-position sensitive neutron detection.

1.1 Techniques for neutron-sensitivity of MWPC

Neutron-sensitive MWPCs for a 2D-position registration are very useful for application in various neutron scattering techniques as conducted in neutron scattering research centers like ILL (France) [7], SNS (USA) [8], MLZ (Germany) [9] or J-PARC (Japan) [10] and in many more places worldwide and soon at the European Spallation Source (ESS) in Lund (Sweden) [11]. Their advantage over other neutron detector technologies are the low gamma sensitivity and a fast response compared to scintillator detectors as well as a spatially more homogenous response with a higher spatial resolution compared to tube detectors on the costs of sophisticated manufacturing techniques. The sensitivity of a MWPC to (thermal) neutrons is achieved by the introduction of an additional material in the gas volume, which converts the uncharged neutrons into charged primary particles by an appropriate conversion reaction [12]. The neutron sensitivity of a MWPC goes on top on the sensitivity for ionizing radiation. The design of a MWPC-Position Sensitive Neutron Detector (PSND) therefore has to trade-off strongly the sensitivity for neutrons over the sensitivity for other ionizing radiation fields as emitted by many parasitic sources including high-energy cosmic particles [13] in the field of use. Typical isotopes for this application are the ones with a large cross-section for neutron induced

* Corresponding author: Gregor.Nowak@hereon.de

fission, possessing simultaneously a low-Z number and being in a low density aggregate state, thus offering a low absorption probability for gamma radiation [6]. Therefore, they are the choice for neutron converting materials [14]. Reactions of the type $^{x}Y(n,\alpha)^{(x-3)}$ are in the focus for neutron conversion processes. Conventionally, the ^{3}He-gas or a ^{10}B-containing material are used for the conversion of neutrons into charged particles [15-16]. In addition, fissionable materials like $^{233-238}U$, ^{157}Gd, or ^{6}Li-isotope, etc. are in use [17].

Neutron detectors based on ^{3}He- or $^{10}BF_3$-gas have been extensively used for detection of thermal neutrons. However, further deployment of such detectors is currently constrained by the dramatic increase of ^{3}He-price (so called "^{3}He-crisis" [18]). Moreover, in case of $^{10}BF_3$-gas, the toxicity of larger gas volumes at places of use gets an issue with improved modern safety standards [19]. One way to construct large area PSND under these circumstances is to switch from gas to solid-state converter containing ^{10}B. A drawback of the switch to solid-state converter is the accompanied strongly limited (to micrometres) ion-escape-length (=ion travel range) from the conversion point in the coating prior the primary ions enter the stopping-gas-volume for the detection process. The short ion escape length in the converter material limits the probability for the ions created far away from the converter/stopping-gas-interface to enter the adjacent detection gas volume to trigger the detection process. The maximum efficiency of this detection process with ion generation in solid converter and their transfer to stopping gas scales with the ion population reaching out the stopping gas volume. Nevertheless, ion-range simulations and the good controllable deposition properties of $^{10}B_4C$ coatings on the lightweight material aluminium [1, 20-22] shifted the development of PSND in the last decade towards $^{10}B_4C$-coating-based detection techniques [22-24]. The good sputtering properties supports the manufacturing of large area converters. In ^{10}B-based neutron converters, the following Neutron- induced fission-reaction takes place [25] (with kinetic energy in the round brackets):

$$n + {}^{10}B \rightarrow \begin{cases} {}^{7}Li\,(840\,keV) + \alpha(1470\,keV) + \gamma(480\,keV), & 94\ \%\ (A) \\ {}^{7}Li(1015\,keV) + \alpha(1770\,keV), & 6\%\ \ (B) \end{cases} \tag{1}$$

The charged-particles emerging from the $(n,^{10}B)$-reaction ionize the stopping-gas thereby generating the primary electrons, which set off the charge avalanche near the wires and makes possible the detection process [5].

1.2 Size of active area and position resolution

The size of the active area of a PSND at a constant sample-detector distance defines the accessible range of wave-vector transfers covered in one detector configuration for a given scattering/diffraction experiment at a chosen pulse and wavelength band [26]. Therefore, the application of large area (m^2-size) detectors at <u>T</u>ime-<u>o</u>f-<u>F</u>light (ToF)-diffractometers contributes substantially to a shortening of the acquisition time in a measurement [27]. The current state-of-the-art engineering neutron-diffraction beam lines like BEER [28], TAKUMI [29], IMAT [30], SMARTS [31], ENGINE-X [32] or VULCAN [33] offer a 1-2 m sample to detector distance to accommodate the typical size, complex engineering in-situ sample environment. These instruments offer a resolution $\Delta Q/Q$ down to 0.1% for thermal neutrons. The approximation for the resolution of the wave-vector transfer for a ToF-diffractometer is expressed by Eq.(2) (see [26], page 142 or [34], page 107). This relation can be used to estimate the upper limit for the spatial resolution of the detector for a set of source and diffractometer parameters (operational wavelength band, pulse length) in order to achieve the required $\Delta Q/Q$:

$$\left(\frac{\Delta Q}{Q}\right)^2 = \left(\frac{\Delta L}{L}\right)^2 + \left(\frac{\Delta t}{ToF}\right)^2 + (\Delta\Theta \cot(\Theta))^2 \tag{2}$$

with Δt = chopper pulse length, ToF = Time of Flight from chopper to detector, L= Length of flight path, ΔL = uncertainty in flight path from the chopper to the detector, $\Delta\Theta$ = recorded Bragg-peak width at the detector position, folded with detector resolution, incident beam divergence and scattered beam size and Θ = Bragg angle of the reflection. For long instruments (>100 m), the first term of Eq.(2) contributes with about 0.01 % to $\Delta Q/Q$. The main variable contributions to $\Delta Q/Q$ in Eq.(2) result from the variations of the pulse length Δt and to the same order of magnitude from the Bragg-peak width $\Delta\Theta$ in the scattering plane, which is typically dominated by the fixed horizontal beam divergence of about 5 - 25 mrad. At a distance of 2 m between sample and the detector, a position resolution of a couple of mm in the horizontal scattering plane contributes only with a few mrad to the Bragg-peak width $\Delta\Theta$, but it allows for a sufficient spatial sampling of the Bragg-peak shape in the scattering plane. This motivates the envisaged couple of mm large position resolution of a modern ToF-diffraction detector.

1.3 Techniques for 2D-position encoding of MWPC-PSND

For the two-dimensional neutron-event position determination in a MWPC-PSND, two methods of position encoding are conventionally used.

The most common position encoding technique is the resistive-wire method. There, the accumulated charge is deposited on the detection wire closest to the location of the neutron capture by the converter material. The deposited electron charge is measured after its travel to both ends of the wire as a voltage pulse-height [35]. The voltage pulse-heigths measured at each end of the detection wire are proportional to the ratios of the electric charges that pass through the corresponding fraction of the total length measured from the detection point and their specific Ohmic-resistance. The relative position of the neutron event along the wire is given by the ratio of the partial and the total (sum) pulse-heights.

This method requires that the number of readout channels and Analog-to-Digital Converters (ADCs) equals the number of wire-ends in the detector (e.g., twice the number of single wires).

A second type of position encoding in a MWPC-PSND is the delay-line method [6]. In this method, one end of every single detection wire is connected to a unit cell of a delay-line ladder circuit consisting of a resistive, capacitive and inductive element. The number of the unit cells in the delay-line ladder circuits corresponds to the number of wires in each of the both x- and y-wire planes. At both ends of this ladder circuit (= delay line) a pre-amplifier is connected. Its function is to increase the microvolt pulse passed and damped through a certain number of unit-cells to a millivolt level as needed for the further signal processing. The resistive, capacitive and inductive elements within the delay-line unit cell are chosen so that the propagation speed of a typically shaped event pulse is slowed down to about 1 ns per unit cell transient. Therefore, the traveling time of an event-pulse towards the ends of the delay-line fed-in by a detection wire at a certain position of the delay-line circuit increases by the integer number m of transient times (here: 1 ns) resulting from the passage through m delay-units [6]. The precise measurement of the time difference between the delayed pulses at the ends of the delay-line allows determining the number m of passed delaying unit cells of the delay-line circuit and therefore determining the number m of passed wires with their specific geometric distance from each other. For a single detection module, the x- and y-position components of a certain neutron event can be determined within the same time-correlation radius corresponding to the time-delay of a whole x- or y-delay-line length. The registered arrival-time differences for the x- and y-components become converted by the number of passed delay-time steps (m_x and m_y) and by the corresponding geometric wire-pitch in a wire-plane (here: 2 mm) to (x,y)-event positions in mm within one detection module. The position determination of a neutron event using x- and y-delay-lines for one detection plane is reduced practically to a time difference measurement with a Time-to-Digital-Converter (TDC) unit instead of pulse-height determination using many ADCs, as in the case of the resistive wire scheme.

The advantage of the delay-line position-encoding over resistive-wire scheme becomes evident for detector systems consisting of a large amount of detection wires. In these systems, the delay-line techniques requires a much smaller number of amplifiers compared to resistive-wire method since one delay-line serves as pulse collector for many hundreds of detection wires and needs only one amplifier at every one of its two ends. In contrast to this, the resistive-wire method needs one amplifier at each end of every one of the many hundreds of detection wires. This leads to a much lower number of read-out channels and connecting feedthroughs between the detection volume and the outside space of the detector for the delay-line method compared to the resistive-wire method. With modern electronics and Printed Circuit Board (PCB) layout techniques, delay-lines circuits with their resistive, capacitive and inductive character can be realized on PCB without using individual resistors, capacitors and inductances. A corresponding, geometric guidance of the conductor paths on the PCB-board with terminals for the detection wires replaces all three electric elements.

1.4 Count-rate capability of a MWPC

The count-rate capability of a single detection module is derived from the detection performance of the TDC for a single pulse in the x- and y-delay-line read-out. For typical conversion process times (from neutron hit to electron avalanche end on the detection wire) of the order of 80 - 100 ns and a total length of the delay-line of e.g., 500 ns, the count-rate capability of a single-detection module can be estimated to 150 – 200 kcps [36-38]. Therefore, the full detector assembly, which should consist of fifteen independent detection modules giving rise to a global count rate capability of 15 times 150 – 200 kcps can be estimated to max. 3 Mcps.

The constructional advantage of delay-line read-out holds as long as the event-rate on a single detection-wire-plane stays below the above estimated value. For higher count-rate situations, the single-resistive-wire read-out coupled to fast but costly ADC is preferable.

2 Development of the detector design

The conception and description of the detector design discussed in this study follows the basic, methodical engineering order as applied for the design development of complex machines: from interior to the exterior [39-40]. Therefore, the simulations of the ion traces from the converter to the detection wires are the starting point in chapter 2.1 for the interior detector design. These simulations define the converter thickness and the converter-to-wire distance. In chapter 2.2, the next higher hierarchic level of the detector assembly is considered; here the detection performance is simulated resulting from the geometric design of the detection module and the internal organisation of detection modules within the detector. Finally, the mechanical design of the detection modules and for the detector housing (the exterior) is discussed in chapter 2.3. The mechanical design of the exterior shall ensure a stable operation of the detector in its environment.

2.1 Simulations of the ion travel ranges involved in the detection process

The ion-stopping process in solids and gases is governed by the strong and repulsive Coulomb-interaction between the positive ions and the positive nucleus of the stopping medium. This give rise to a stopping-power curve with a Bragg-like shape including a sharp cut-off energy. This leads to a strong correlation between the ion-energy and the ion-travel range in the stopping medium [41]. To ensure an efficient and localized electron collection by the detection wire system, the ranges and shapes of the traces of the released ions in and out of the conversion coating need to be evaluated. The ion ranges in the stopping gas are important to fix an appropriate detection-wire to converter distance. The divergence of the ion tracks allow us to estimate the lower limit of the position resolution given by the gas amplification. The kinetic energies of the ions escaping from the point of conversion in the converter, which is given by the conservation of momentum and energy for the nuclear reaction Eq.(1) and the stopping power in solid $^{10}B_4C$ are reported in [42-43]. For the branches (A) and (B) of the conversion reactions in Eq.(1), the maximum ranges of α and ^7Li ions in solid ^{10}B coating are [43]:

Branch (A) of Eq.(1):

- Ion range for α-particle: 3.2 µm and ^7Li-ion: 1.5 µm

Branch (B) of Eq.(1):

- Ion range for α-particle: 3.9 µm and ^7Li-ion: 1.7 µm

In order to allow the ions produced in both reaction branches described above to escape from the conversion layer with enough kinetic energy and ionise the adjacent detection gas, the converter thickness is chosen below the shortest ion-range in the coating. As demonstrated in [1,6,14,44], the appropriate $^{10}B_4C$ converter thickness for thermal neutron detectors with a gas-amplification processes at 1 bar pressure in adjacent to the stopping gas volume is 1.2 µm $^{10}B_4C$. This $^{10}B_4C$ converter coating thickness is a reasonable trade-off between neutron capture probability in the converter and a high enough ion escape efficiency out of the converter combined with a high enough kinetic energy of the ions emitted from a single $^{10}B_4C$ layer.

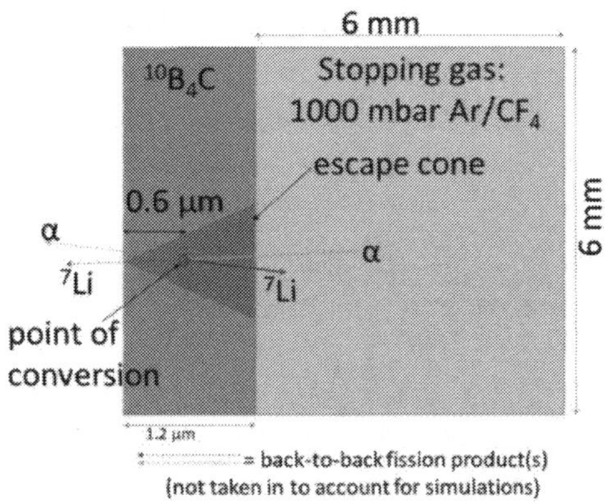

Fig. 1. Simulation set-up with point of conversion, 1.2 µm thick $^{10}B_4C$ coating, 6 mm x 6 mm stopping gas volume and the anticipated escape cone of the ions.

The simulations of ion-tracks consider the $^{10}B(n,\alpha)^7Li$-conversion in Eq.(1) to take place at 0.6 µm depth from the surface of a 1.2 µm thick $^{10}B_4C$ layer. The ions enter from left orthogonally (Fig. 1) in to a 6 mm x 6 mm extended stopping-gas volume. This configuration should cover the most in all likelihood conversion situation with the ion-travel range in the stopping-gas. Indeed the conversion can take place at any depth of the 1.2 µm thick $^{10}B_4C$, but statistically the probability for the conversion reaction increases with higher penetration depth of the neutron. At the same time for the neutron detection process also the ion-escape-process and its descending probability with the larger depth of the conversion point from the surface has to be taken in to account. The ion ranges from branch (A) and (B) represent the upper limit for travel paths of the corresponding ions in the solid converter. Because of this, these maximum travel ranges and the converter thickness span in the directional space an escape cone (with maximum probability for the ion transfer in to the stopping gas) oriented with its axis parallel to the converter normal direction, with an opening angle resulting from 2 times the cosine between the converter thickness (here: 1.2 µm) and the maximum travel range (here for the α-particle from branch (B): 3.9 µm). That is an angle of 72° from the normal direction of the converter plane or a cone described by a solid angle of 144° as shown in Fig. 1.

Typically, the ions from the solid converter within this solid angle/cone will enter the stopping-gas volume. But only those with the shortest travel range in the solid (elongation of the orthogonal direction to the converter plane) have the highest kinetic energy left for the travel in the stopping gas and generate consequently the longest ion tracks there. These ions are taken into account in the TRIM-simulations.

Due to the large surface of the envisaged detector housing and the m^2-size neutron entrance window, the limitations by breaking stress of constructional materials need to be considered. The structural materials taken into account allows for a maximal possible pressure gradient between the inside space of the detector housing and a variable ambient pressure environment to about $\Delta \pm 100$ mbar. In summary, the simulations of ion ranges in the detector volume filled with the stopping-gas are performed under ambient pressure condition, at 1000 mbar Ar/CF_4 (80%/20%) using the TRIM-software [45-46] an under the assumptions that:

- the conversion process takes place 0.6 µm from the surface of the solid, 1.2 µm thick $^{10}B_4C$-converter
- the ions travel after generation through the solid conversion coating and enter orthogonally oriented into the 6 mm x 6 mm extended stopping gas volume
- the ions travel and get stopped in the stopping gas volume adjacent to the $^{10}B_4C$ surface consists of Ar/CF_4 (80%/20%) at 1000 mbar

The corresponding ion tracks and their ranges in the stopping gas volume, called "Target depth" for a volume box of 6 mm x 6 mm are presented in Fig.2 and Fig.3. The visualization of the ion tracks in Fig.2 demonstrates clearly that the high kinetic energy of the α-particles gives rise to a less divergent ion beam in the stopping process compared to the ^7Li-ions normalized to an identic ion range. A strong straggling effect of the ion trajectories appears in the last tenths of the travel range. In Fig.3 is demonstrated, that the straggling parameter decreases with both, the decrease of the ion weight and increase of kinetic ion energy. Fig.2 and 3 represent the primary charges only.

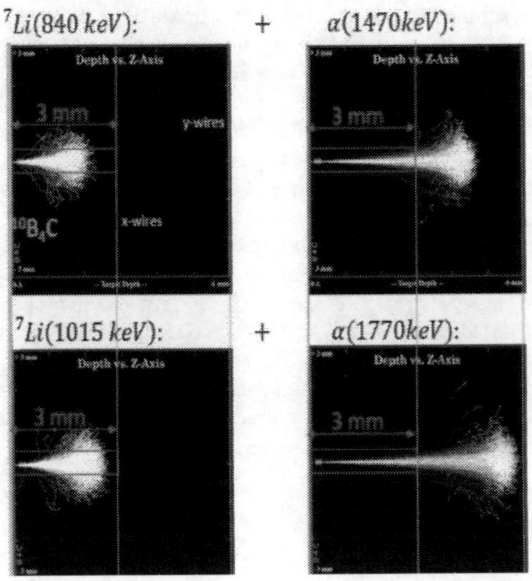

Fig. 2. Simulated tracks for α and ^{7}Li-ions by TRIM-software for (A) and (B) branch of the $^{10}B(n,\alpha)^7Li$-reaction in Eq.(1) taken place at 0.6 µm depth from the surface of a 1.2 µm thick $^{10}B_4C$ layer. The ions enter from left orthogonally in to a 6 mm x 6 mm extended Ar/CF_4 (80%/20%) at 1000 mbar gas volume (black background). The green bar on the right hand side in all plots marks the $^{10}B_4C$ coating and the yellow bar marks the x-detection wires at 3 mm and y-detection wires at 6 mm away from the ion emitting $^{10}B_4C$ coating. Top row for branch (A) and bottom row for (B) of reaction Eq.(1).

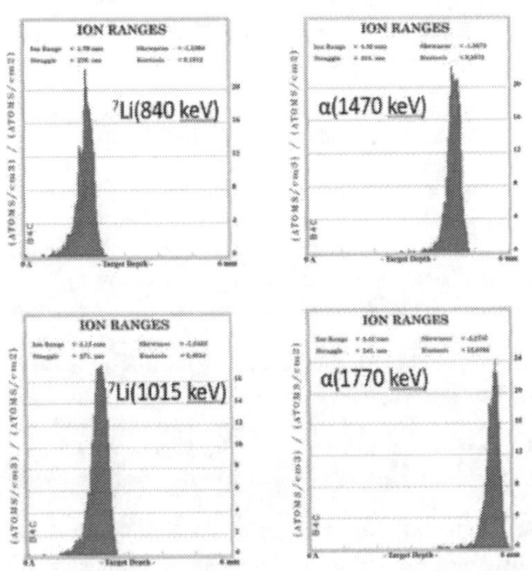

Fig. 3. Ion ranges histograms simulated by SRIM software for (A) and (B) branch of the $^{10}B(n,\alpha)^7Li$-reaction in Eq.(1) taken place at 0.6 µm depth from the surface of a 1.2 µm thick $^{10}B_4C$ layer. The ions enter from left orthogonally in to a 6 mm extended Ar/CF_4 (80%/20%) at 1000 mbar gas volume. The x-axis of each of these histograms represent the normal distance from the $^{10}B_4C$ converter surface (x = 0 mm) in the stopping gas volume within the range of 0 – 6 mm. Top row for branch (A) and bottom row for (B) of reaction Eq.(1).

The gas ionisation and amplification processes are not included in the TRIM-calculations in Fig. 2 and 3. However, the primary ion tracks can be regarded as origins of the gas amplification and charge avalanches if they are located close enough to the detection wire. Therefore, the radial distribution of these charges close to the detection wires influences the number of wires collecting the signal resulting from the detection process. This is quantified by the radial straggle parameter (in µm) [47] mentioned in Fig.3. Therefore, the simulated radial straggling parameter within the range of 210 – 270 µm can be regarded as the lowest value for a position resolution limit resulting from the gas ionization process when projected on a detection wire-plane. At the end of the ion-tracks, the charge cloud is enlarged to a diameter of about 1.8 mm which is less than the detection wire pitch. The range of the straggling parameter and the shape of the ion tracks ensures that the envisaged 2 mm position resolution of the detector, which corresponds to the wire-pitch in a x- or y-wire plane, is not impaired by the gas ionization process. For the stopping gas considered in these calculations, Ar/CF_4 (80%/20%) at 1000 mbar pressure, the ions are confined in a range of 1.8 - 5.5 mm distance in forward direction from the $^{10}B_4C$ converter surface.
It is worth mentioning that the position resolution that could be achieved with MWPC with $^{10}B_4C$-coated cathodes was intensively studied in the GEANT4 simulations of a detector designed for neutron diffraction studies [48]. In that study, the detector unit consisted of two thin identical wire counters sharing a segmented cathode and enclosed in a common housing filled with Ar/CO_2 (80%/20%) gas. The wire planes were positioned at a distance of 3.5 mm from the Boron converter and the distance between the wires was 17 mm. The full detector had cylindrical geometry and consisted of several detector units placed vertically around the sample and titled such that the angle between the Boron converter and the incoming neutron beam was 10°, as shown in Fig. 3 of [48]. Due to the tilting of the segment, the projection of the coated side on the entrance window of the detector becomes proportional to the product between the wire pitch and tan(10°). This gives rise to a horizontal spatial resolution of ~2.1 mm (FWHM). GEANT4 predicted a value of 2.96(6) mm (FWHM) for the spatial resolution of the proposed design, which was investigated by scanning across the wires with a collimated beam and recording the events collected by each wire, see Fig. 7 in Section 3.2 of [48]. That result is a good indication that spatial resolutions below 3 mm can be achieved with MWPC with ^{10}B-coated cathodes. We believe that this value can be further refined by replacing the CO_2 quenching gas with one with a higher stopping power, such as CF_4.
For an optimal trade-off between position resolution and charge accumulation on the detection wire, distances of 3 mm and 6 mm for the x- and y-wire-planes with respect to the converter surface seem to be the best compromise. The relative geometric configuration (stacking sequence) of the x- and y-detection wire-plane confined between the two, face-to-face oriented converters, within one detection module is as following:

[*n*-converter/3mm-gap/
/x-wire-plane/**3mm-gap**/y-wire-plane/
/3mm-gap/(*n*+1)-converter] (3)

where *n* and (*n*+1) numerate the converter with in the full detector assembly in the detection module *n* (with *n* = 1 for the first module hit by neutrons). The stacking sequence is mirror-symmetric with respect to the central **3mm-gap** in Eq.(3), which is found between the x- and y-wire planes. This mirror symmetric configuration with respect to the central **3mm-gap** allows for an equal charge distribution on both x- and y-detection wire planes from both face-to-face oriented $^{10}B_4C$ coatings.

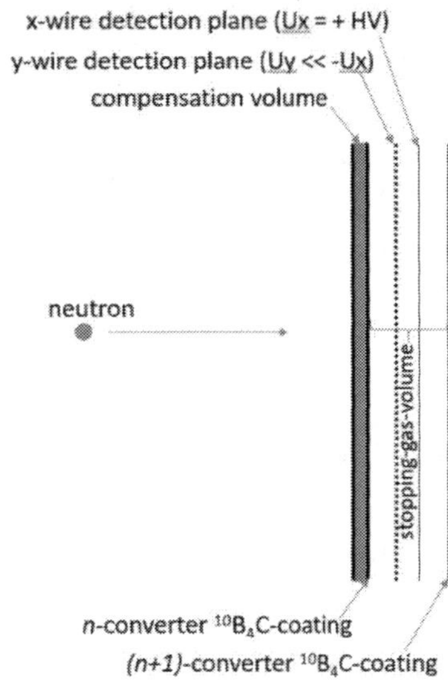

Fig. 4. Cross-section cut through one detection plane with stopping-gas-volume and two adjacent compensation volumes (blue shaded space) with $^{10}B_4C$-converter coatings and high voltage supply for x- and y-wires planes. The sequencing and spacing is as described in formula (3) above.

The stacking sequence Eq.(3) is a building block of a modular-designed detector, which consist of *n* of such detection modules. These detection modules are immersed in a common stopping-gas volume. The resulting converter-face-to-face-distance of 9 mm (see Eq.(3)) defines the ToF-timing precision within one detection module and therefore for the entire detector assembly, since the origin of an conversion event is untraceable for the detection wire system to the particular *n* or (*n*+1) converter. The back-to-back distance of adjacent detection modules is 10 mm (not included in the stacking sequence Eq.(3) above). This 10 mm-space contains the compensation volumes, which will be introduced in the next section.

The gas-gaps between the wire-planes and the converter-surfaces have to be kept constant during the detector operation. As already mentioned above, during the operation of the detector, one of the two wire-planes of a detection module is set under HV in order to accelerate the primary electrons and increase their kinetic energy to a value, which is high enough to set-off secondary gas ionisation and amplification. The electric field between the converter plane, held at ground potential, and the wire-plane under high-voltage, gives rise to an attractive force between these two components. During the detector operation this can lead to a non-uniform converter-to-wire distance (gas-gap), which results in an inhomogeneous detection efficiency distribution over the entire detection plane. To counteract the approach of the detection wires to the converter surface, a patented compensation volume design, had been developed by the Hereon research centre [4]. The compensation volume is allocated, as presented in Fig. 4, behind each converter plane and it is completely separated from the detection gas volume. The gas pressure in these compensation volumes can be adjusted individually and independently from the gas pressure in the detection volume in front of the converter. The application of a lower pressure in the compensation volume compared to the detection volume pulls the converter plate away from the detection wire planes and counteracts the attractive force resulting from the electric field between these two detection module components. The size and direction of the pressure gradient between the detection and compensation volumes separated by the converter plate can be set according to a feedback signal resulting from the converter deflection measured by means of laser spot shifts on the backside of the converter plate. This deflection control system for the converter plate is installed in every one of the fifteen detection modules. It allows the individual adjustment of every module. For a detailed description of the construction and operation of this system, the reader is asked to consult the reference [4].

2.2 Efficiency simulations of the detection modules using GEANT4

The ions-track simulations in the previous section for the $^{10}B_4C$-coating and the Ar/CF_4 (80%/20%) stopping-gas were performed with the help of the TRIM/SRIM software package to determine the distance between the converter surface and the detection wire-planes that is necessary in order to achieve the desired electron yield from the detection event on the detection wires.

In this chapter, the simulations on the next higher hierarchic level (at the level of the whole detector) are described. They were conducted in order to assess the neutron-detection performance of the complete detector assembly consisting of the *n* detection modules. These simulations assumed neutron incidence perpendicular on the converter plane and are performed by using the GEANT4 [49-52] software package. The main goal of these simulations was to find a reasonable balance between the number *n* of detection modules and the refined thickness of the $^{10}B_4C$ converter coating in the subsequent detection modules traversed by the neutron beam at the desired detection efficiency for a specific neutron spectrum foreseen for the MWPC-PSND.

The efficiency of the proposed design was investigated by building the theoretical model of the detector and tracking the slowing down of the ^{7}Li- and α-reaction products through the converter and counting gas until they came to total rest.

The energy deposited in the gas through ionization was used to generate the combined (both ions) pulse-height spectrum. The efficiencies represented graphically in Fig. 6 of the manuscript were calculated by taking the ratio between the total number of counts in the pulse-height spectra recorded for each configuration and the total number of incident neutrons impinging upon the detector. The results represent a measure for the absorption, conversion, and escape efficiency only, i.e., $\varepsilon_{abs} * \varepsilon_{conversion} * \varepsilon_{escape}$, not the total detection efficiency, ε_{det}, which also depends on the ability to collect the charge generated in the gas volume, $\varepsilon_{det} = \varepsilon_{abs} * \varepsilon_{conversion} * \varepsilon_{escape} * \varepsilon_{collection}$. Previous experience with operating a similar MWPC design indicated that the efficiency for collecting the charge generated in the gas is greater than 90 % [53]. Moreover, in the real detector, an energy threshold must be also applied to the individual pulses in order to eliminate those generated by the γ-background. Such a threshold was not considered in the present calculations. However, typical values for the n-γ discrimination threshold in the ^{10}B-based gas detectors lie between 100-150 keV, which leads to a reduction of the detection efficiency by less than 10%, as shown in Fig. 4 of [54].

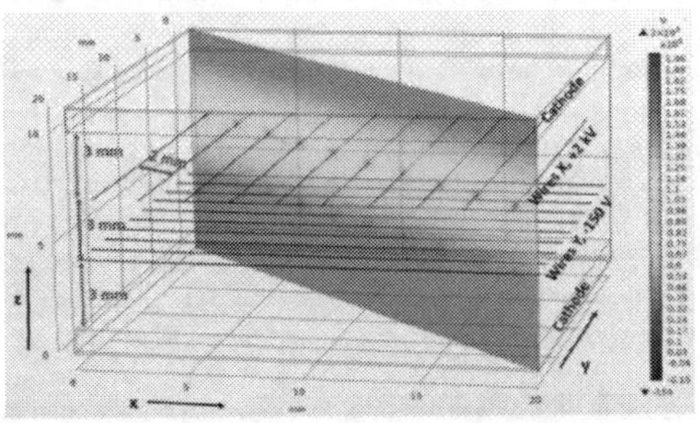

Fig. 5. COMSOL-simulations of the electric potential (V) (red=positive, blue=negative) for the detector design proposed in the present work. The detector design parameters are marked in the figure.

The electric field configuration for one x/y-wires-plane detector module under HV (U_x= +2000 V / U_y= -150 V) is presented in Fig.5. The electric field configuration inside one detection module had been calculated using the COMSOL-simulation-package and serves to visualise the E-field homogeneity and demonstrates the input configuration for the GEANT4 simulations.

Fig.6(A) represents the simulated efficiency contributions from each subsequent module consisting of a stack of 15. Each module contains two 1.2 µm-thick $^{10}B_4C$ converter layers. The calculations were performed with monochromatic neutrons at wavelengths 1, 2.5, 4 and 5 Å. The plots indicate that the efficiency contribution drops from one to the next module even though the modules contain a neutron absorbing coating of the same thickness. This behaviour can be observed for each of the four values of the neutron-energy considered in the calculations. The efficiency drop has its origin in the continuous extraction of neutrons from the transmitted neutron beam by every subsequent

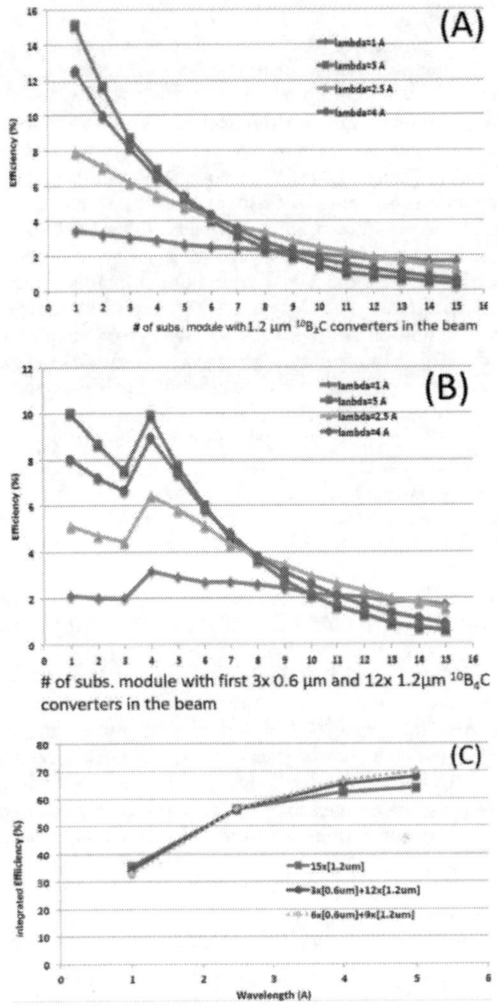

Fig. 6. (A) Efficiency contributions from single, subsequent detection modules consisting of two converters coated by 1.2 µm $^{10}B_4C$ for monochromatic neutrons at 1 Å (blue), 2.5 Å (green), 4 Å (violet) and 5 Å (red). The front $^{10}B_4C$-double layer is marked with the number "1". (B) Efficiency contributions of subsequent modules consisting of two different coating thicknesses of the $^{10}B_4C$ converters for monochromatic neutrons at 1 Å (blue), 2.5 Å (green), 4 Å (violet) and 5 Å (red). The first three modules have 0.6 µm and the further twelve modules have 1.2 µm thick $^{10}B_4C$-coatings. (C) Efficiency spectrum integrated over 15 modules with 1.2 µm thick $^{10}B_4C$ (red), over 3 modules coated by 0.6 µm and 12 modules coated by 1.2 µm thick $^{10}B_4C$ (red) and finally over 6 modules coated by 0.6 µm and 9 modules coated by 1.2 µm thick $^{10}B_4C$ (green; dashed).

module. This observation is in-line with the well-known Beer-Lambert-law for neutron absorption in matter. In Fig. 6(B) the individual efficiencies were simulated for the same neutron energies as in Fig. 6(A), but the thickness of the converter in the first three modules was reduced by a factor of two (i.e., 0.6 µm of $^{10}B_4C$) while the thickness of all the other layers was left unchanged (1.2 µm of $^{10}B_4C$). Consequently, the efficiency contribution of the first three modules dropped down compared to Fig.6 (A). However, the 4th module and the following ones have a higher efficiency contribution when compared to the 4th module in Fig. 6(A): compare 10 % at 5 Å in Fig. 6(B) to 7 % at 5 Å in Fig. 6(A). Thus, the variation of the converter thickness leads to a change in the efficiency distribution over the entire detector stack, and consequently to a change in the count-rate per detection module. This effect can help improve both the local and global count-rate capability of the detector, as the number of detected neutrons is more evenly distributed over the first 4-5 modules and not anymore concentrated in the first two (see Fig. 6(B)). The spread of the count-rate over the depth of the detector becomes more important when operating at cold neutron energy. Cold neutrons have a high probability to be detected within a few µm of $^{10}B_4C$, leading to much larger count-rates in the first two detection modules compared to the subsequent ones. Furthermore, using thinner converter coatings in the detection modules that are first exposed to the neutron beam leads to a slight improvement of the overall detection efficiency (~8 %) for cold neutrons, as seen in Fig. 6(C).

In the thermal regime, however, a detector stack consisting of 15 detection modules, all comprising $^{10}B_4C$ converter layers with 1.2 µm thickness, has an overall detection efficiency that is slightly higher than the stacks built with converter $^{10}B_4C$-coatings of variable thickness. The explanation for this is that for higher energy neutrons, the ion-escape length dominates over the neutron-absorption length. For the low-energy neutrons, the conversion point is statistically closer to the converter surface and the ion-escape probability increases compared to the case of higher neutron energies. Numerical details of these considerations can be found in [2].

The results of these simulations indicate that the overall efficiency of the detector stack is influenced by reducing the thickness of the $^{10}B_4C$ converter coating applied on the first few modules located close to the entrance window of the detector. The economic advantage of doing this is that it reduces the sputtering (coating) time and the amount of converter material used in the coating process, with positive effects on the total costs and manufacturing time of the detector. However, this simulation case also makes it clear that the converter coating thickness in the stacked detection modules needs to be adapted to the neutron energy-range and application for which the detector is intended to be used. The simulated efficiency values are summarized in Tab.1.

Tab.1. Summary of plot Fig. 6(C).

Neut. wave-length [Å]	**15x [1.2 µm] $^{10}B_4C$ integr. efficiency [in %]**	**3x [0.6 µm] + 12x [1.2 µm] $^{10}B_4C$ integr. efficiency [in %]**	**6x [0.6 µm] + 9x [1.2 µm] $^{10}B_4C$ integr. efficiency [in %]**
1	36.6	34.8	32.8
2.5	56.8	56.9	56.9
4	62.5	65.7	66.5
5	63.6	67.7	69.9

2.3 Mechanical design and strain simulations of the detector modules and detector housing

The hardware exterior of the detector is a gas-tight housing. The detector housing represents the interface between the detector and the ambient space. It consists of three sub-components 1. neutron entrance window, 2. top-cover and 3. the back-plate. The functions of the first two components are a loss-free transmission of neutrons to the detection modules in the housing. The second function is the protection of the well-defined stopping-gas volume from changes of ambient conditions and parameters like temperature and atmospheric pressure. The back-plate has three functions. The first function is a mechanical, rigid support for the detection modules to ensure minimal displacement of wires and converters during transport and operation. The second function of the back-plate is to offer a mechanical interface and a distribution base for loads during vertical mounting of the detector in a support structure for operation at a beamline or for transportation purpose. Last but not least, the third function of the back-plate, is the provision of a mechanical platform for the electronic and media feedthroughs to the inside components like the wire-planes with their signal- and HV-lines and the detection modules with their gas- and USB-data-lines. The shape of these three mechanical sub-components, their material grade and thicknesses had been designed for ± 0.1 bar pressure difference between the ambient pressure (corresponding to the average air pressure in Lund/SWE of 1000 mbar) and the pressure of the stopping-gas inside the detector. The weight limitations for the whole detector assembly and the large surface area of the detector housing as well as high mechanical load resulting from the envisaged pressure difference of max. 0.1 bar between in- and outside of the detector, makes a (Computed Aided Design) CAD-supported design simulation essential. Therefore, the mechanical loads in the housing and particularly in the converter support-frames of the detection modules were simulated by means of the (Finite Element Method) FEM-techniques using the concept of the von-Mises-tension [55]. These simulations delivered the von-Mises tensions and the local deformations of the components at the operational conditions as presented in Tab.2

https://doi.org/10.1051/epjconf/202328603007

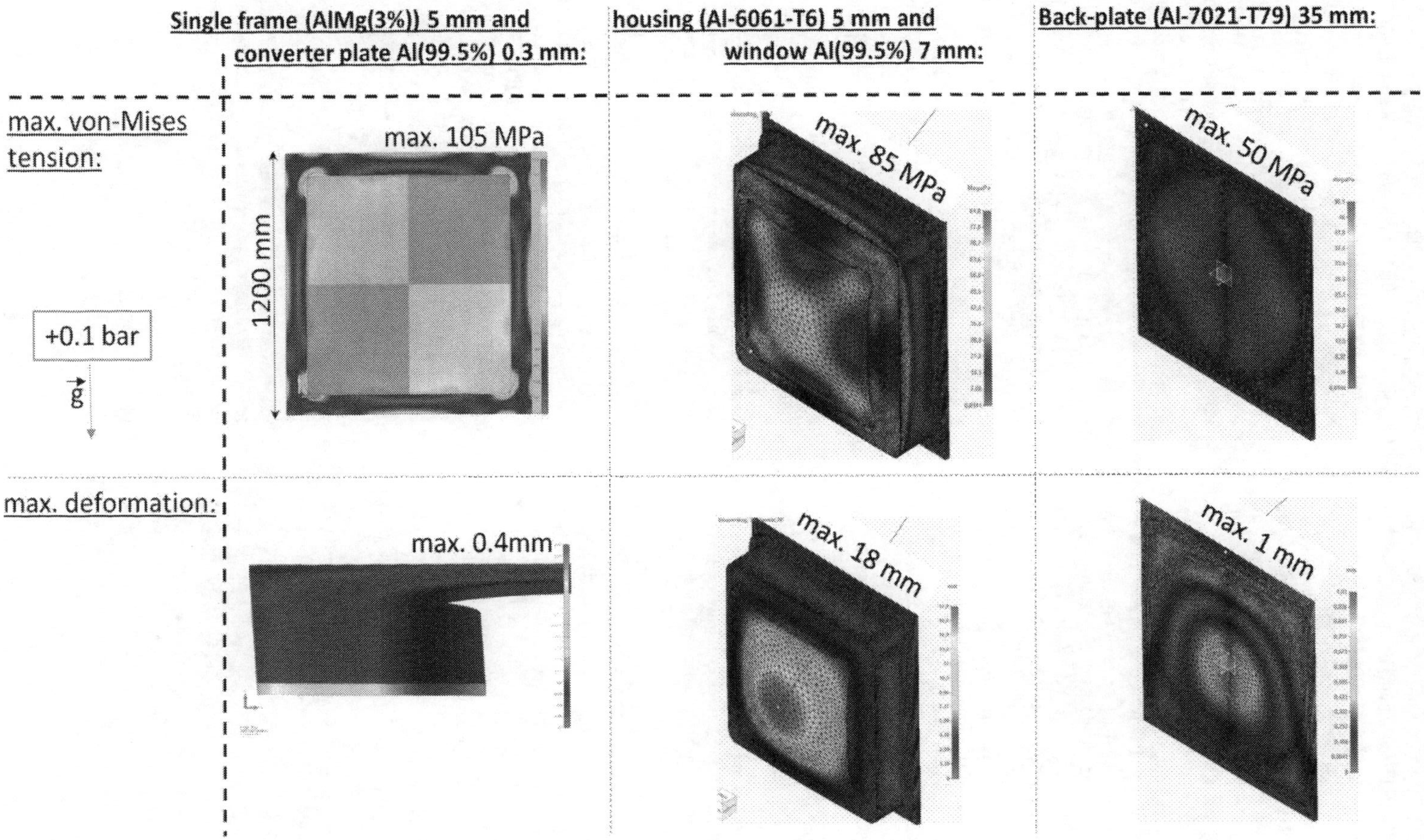

Tab. 2. FEM-simulations of the von-Mises-tension (top row) and deformation (bottom row) for: the single frame-converter element where the converter plate is clamped in the converter support frame (left column), the top housing with neutron entrance window (middle column) and back-plate (right column). The simulations were performed in gravity field and an overpressure of +0.1 bar in the detector housing as indicated on the left side in Tab.2.

for the corresponding envisaged material grades [56]. The FEM-simulations had been performed using the "Solid Edge ST(110.00.12.004 x64)FEMapp (11.3.1)"-software and the NX-Nastran (11.0) solver. For the operational parameters, it was assumed that the detector is erected vertically so that its entrance window surface normal is oriented orthogonally to the gravity (as indicated in Tab.2) and the ambient pressure is 0.1 bar higher than the stopping-gas pressure inside the detector. As described in reference [4], the $1m^2$-large Al-converter plates coated by $^{10}B_4C$ were clamped on support frames, which ensure a flat mounting of the converter surface in front of the wire-planes inside the detection module. The result of the FEM-optimization processes for supporting the 0.3 mm thick converter (Al 99.5 %), is a 5 mm thick AlMg(3%) frame with a shape as presented in the column labelled "Single frame-converter plate" in Tab.2. The frame has inside rounded corners to redistribute and reduces the von-Mises-tensions to 105 MPa in this work piece. This shape should ideally offer the required stiffness for the envisaged converter flatness. The connected maximum deformation of the single frame and converter-plate assembly was calculated to about 0.4 mm at the outer edges of the converter and frame. This is expected to be in the acceptable range for operation.

For the top part of the housing, a 5 mm thick Al-6061-T6 base material and for the neutron entrance window a 7 mm thick Al(99.5%) plate turned out to be appropriate (see column "housing and window" in Tab.2). This choice of material thicknesses and grades lead to a maximal deformation in the centre of the window of 18 mm at an acceptable, von-Mises tension level of maximum 85 MPa. For the "Back-plate" the FEM-simulations yields a 35 mm thick Al-7021-T79 as a useful thickness and material grade (see column "Back-plate" in Tab.2). In this configuration, a deformation of 1 mm in the central region of the Back-plate at a maximal von-Mises tension of 50 MPa is expected. This tension is far below the considered material grade limits. These evaluations were performed for an overpressure of +0.1 bar in the housing compared to the outside pressure and in the gravity field oriented as above mentioned.

3 Conclusions

This parametric study demonstrate the technical feasibility of a $1m^2$-size neutron detector for a perpendicular neutron incidence using stacked detection modules consisting of multi-wire detection planes confined between $^{10}B_4C$-coated converters. The conditions of operations (max. pressure difference and gravity field) had been considered in the design. The neutron-conversion and detection processes specific to the $^{10}B_4C$ converter coating and the characteristics of the Ar/CF_4-stopping-gas indicate that a converter-to-wire distance of 3 mm offers a sufficient high charge yield from both $^{10}B_4C$-converters on the detection wires of the x- and y-wire-plane in a single detection module. The ToF-timing precision of one detection module and therefore for the entire stacked detector assembly is consequently restricted to the neutron time-of-flight through the converter face-to-face distance of 9 mm. The position resolution of the detection wire-planes given by their 2 mm wire-pitch is not impaired by the pressure-range of the stopping-gas, which determine the maximal possible pressure load on the detector housing, as demonstrated in Sections 2.1 and 2.3. The thickness of the $^{10}B_4C$-converter in the stacked detection modules assembly was optimised for a specific use-case neutron spectrum with the help of GEANT4 simulations. The results of this calculation provides the refined values for the $^{10}B_4C$ coating thicknesses in the sequenced converters. The $^{10}B_4C$ thickness refinement leads to improved local and global count-rates loads at reduced manufacturing costs while still preserving the high overall detection efficiency of the entire detector assembly. Similar optimisations are also typically done for the gas-converter-based thermal neutron detectors such as ^{3}He-tubes, by carefully inspecting the dependence of the gas-pressure vs. neutron efficiency. As seen in Section 2.2, the integrated neutron efficiency that can be achieved with a detector assembly like the one described in the present work is similar to what is currently possible to achieve with the pressurized ^{3}He-detectors. The manufacturing of the required converter shapes and sizes as well as the wire-planes is feasible by using conventional production techniques. The required stabilities of involved materials for the operational case of the detector are not in conflict with the commercially available lightweight material grades as shown in chapter 2.3. Hereon, formerly HZG and GKSS, successfully applied the MWPC-technology with 2D-delay-line read-out to position-sensitive Neutron detectors based on ^{3}He in the past to the REFSANS and SANS-1 instruments at MLZ [57] as well as at the ARES instrument of the Geesthacht Neutron Facility (GeNF) at FRG-1 [58]. These results motivated the researches to make the next certainly challenging but interesting step and to combine the delay-line technology with the solid state $^{10}B_4C$-conveters. A successful realization of these kind of detectors would merge a 2D-sensive MWPC-neutron detector with a relative small number of read-out channels and with an affordable ^{3}He-free neutron converter technology. This detector design would then contribute to overcome the "^{3}He-crisis" mentioned in Chapter 1.1. The applications of these kind of detector technology is aiming for diffraction instruments with moderate position resolution requirements ($\geq$ 2mm) at ESS and of cause at other neutron research facilities.

The acknowledgements of the authors goes to the ESS ERIC for their ESS in-kind program offering the technically challenging research topic as well as to the German Federal Ministry of Education and Research (BMBF) for the financial support.

References

1. G. Nowak et al., J. Appl. Phys. **117**, 034901 (2015)
2. F. Piscitelli, et al., J. Instr. **13**, P05009 (2018)

3. European Patent Application 2 997 174 (14.07.2014)
4. European Patent Application 3 662 306 (10.03.2021)
5. G. Charpak, et al., Nucl. Instrum. Meth. **62**, 262 (1968)
6. F. Sauli, "Principles of Operation of Multiwire Proportional and Drift Chambers", Lecutres given in the Academic Training Programme of CERN, Geneva (1977)
7. https://www.ill.eu/de/neutrons-for-society/neutron-technology/detectors
8. Su-Ann Chong et al., Nucl. Instrum. Meth. A, **1039**, 167052 (2022)
9. J.-F. Moulin et al., Journal of Large-Scale Research Facilities, **1**, A9 (2015)
10. K. Toh et. al., Journal of Physics: Conference Series, **528**, 012045 (2014)
11. K.H. Andersen et al., Nucl. Instrum. Meth. A, **957**, 163402 (2020)
12. M. O. Klein, Ph.D.-Thesis, Ruprecht Karls Universität, Heidelberg, (2000)
13. G. Mauri et al., EPJ Techniques and Instrumentation, **6**:3 (2019)
14. R. Arnaldi, et al., Nucl. Instrum. Methods Phys. Res., Sect. B **213**, 284–288 (2004)
15. J. H. Coon, Phys. Rev. **75**, 1355 (1949)
16. J. A. Bistline Jr., The Rev. Sci. Instr. **19**, 12 (1948)
17. G. F. Knoll, "Radiation Detection and Measurement", 4th ed. John Wiley & Sons, p. 534 (2010)
18. D. A. Shea and D. Morgan, Report No. R41419, (2010)
19. See http://eval.esss.lu.se/cgi-bin/public/DocDB/ShowDocument?docid=274, "European Spallation Source—Technical Design Report", ESS-2013-001 (2013)
20. C. Höglund et al., J. Appl. Phys. **111**, 104908 (2012)
21. C.-C. Lai et al, Surf. Coat. Tech. **433**, 128160 (2022)
22. M. Anastasopoulos et al., J. Inst. **12**, P04030 (2017)
23. J. C. Lacy et al., IEEE Nucl. Sci. Conf. Rec., 3971 (2010)
24. J. C. Lacy et al., IEEE NSS MIC Conference Seattle, WA, USA, (2014)
25. H. J. Taylor, Proc. Phys. Soc. **47**, 873 (1935)
26. B.T.M. Willis and C. J. Carlile, „Experimental Neutron Scattering", Oxford University Press (2009)
27. I. Stefanescu et al., J. Inst. **12**, P01019 (2017)
28. J. Fenske et al., Journal of Physics: Conference Series, **746**, 012009 (2016)
29. Moriai, S. Torii, H. Suzuki, S. Harjo, Y. Morii, M. Arai, Y. Tomota, T. Suzuki, Y. Akiniwa, H. Kimura, K. Akita: Physica B, **385-386**, 1043-1045 (2006)
30. W. Kockelmann, S.Y. Zhang, J.F. Kelleher, J.B. Nightingale, G. Burca, J.A. James, Physics Procedia, **43**, 100-110 (2013)
31. M.A.M. Bourke, D.C. Dunand, and E. Ustundag, Appl. Phys. A: Mat. Sci. Proc. **74**, 1707-1709 (2002)
32. J. R. Santisteban, M. R. Daymond, J. A. James and L. Edwards, J. Appl. Cryst., **39**, 812-825 (2006)
33. X.-L. Wang, T.M. Holden, G.Q. Rennich, A.D. Stoica, P.K. Liaw, H. Choo, C.R. Hubbard, Phys. B: Cond. Mat., **385-386**, 673-675, (2006)
34. M. T. Hutchings, P. J. Withers, T. M. Holden and T. Lorentzen, "Introduction to the characterization of residual stress by Neutron diffraction", Taylor & Francis Group, LLC, (2005)
35. J. Lindsay et al., CERN 74-12 (1974)
36. A. Epstein and C. Boulin, IEEE Trans. Nucl. Sci. **43**, 159 (1996)
37. J. B. Khazaie and C. Boulin, IEEE Trans. Nucl. Sci. **43**, 197 (1996)
38. J. B. Khazaie and C. Boulin, IEEE Trans. Nucl. Sci. **43**, 193 (1996)
39. G. Pahl und W. Beitz, „Konstruktionslehre" - Grundlagen, 7. Auflage. Springer Verlag, (2007)
40. R. G. Budymas and J. K. Nisbett, "Shigley's Mechanical Engineering Design", TBS; 8th edition (2006)
41. W. H. Bragg, Phil. Mag. **10**, (57): 318 (1905)
42. J. P. Biersack and L. Haggmark, Nucl. Instrum. Meth. **174**, 257 (1980)
43. J. F. Ziegler, J. P. Biersack, and U. Littmark, "The Stopping and Range of Ions in Solids", Pergamon, New York, Vol. 1. (1985)
44. F. Piscitelli et al., J. Inst. 8, P04020 (2013)
45. J. F. Ziegler, M. D. Ziegler and J. P. Biersack, Nucl. Instrum. Meth. Phys. Res. Sect. B, **268**, 1818 (2010)
46. http://www.srim.org/;
47. TRIM/SRIM, Textbook on http://www.srim.org/
48. Stefanescu et al., J. Inst. **14**, P10020 (2019)
49. https://geant4.web.cern.ch/node/1
50. S. Agostinelli et al., Nucl. Instrum. Meth. A, 506, **3**, 250–303 (2003)
51. J. Allison et al., IEEE Trans. Nucl. Sci. 53, **1**, 270-278 (2006)
52. J. Allison et al., Nucl. Instrum. Meth. A, **835**, 186-225 (2016)
53. J. Orban et al., Nucl. Instrum. Meth. A, **632**, 124 (2011)
54. I. Stefanescu et al., J. Inst. **8**, P12003 (2013)

55. R. v. Mises, Z. Angew. Math. Mech. **8**, 161 (1928)
56. F. U. Mathiak,“ Die Methode der finiten Elemente (FEM)“, Neubrandenburg (2010)
57. R. Kampmann, M. Marmotti, M. Haese-Seiller, V. Kudryashov, Nucl. Instrum. Meth. A, **529**, 342-347 (2004)
58. M. Marmotti, M. Haese-Seiller, R. Kampmann, Physica B, **276-278**, 210-211 (2000)

Conceptual design of supermirror polarizers at the European Spallation Source

Damián Martín Rodríguez[1*], *Peter* Willendrup[1, 2], *Wai-Tung* Lee[1], *Alex* Backs[1, 3], *Félix* Jiménez Villacorta[4], *Rasmus* Toft-Petersen[1, 2], and *Manuel* Morgano[1]

[1]European Spallation Source ERIC, Partikelgatan 2, 22484 Lund, Sweden
[2]Technical University of Denmark, Anker Engelunds Vej 101, 2800 Kongens Lyngby, Denmark
[3]Lund University, Division of Materials Engineering, Ole Römers väg 1, 22363 Lund, Sweden
[4]ESS-Bilbao, Parque Laida Bidea, Edificio 207-B, 48160 Derio, Spain

Abstract. Polarized neutrons will be made available on many European Spallation Source (ESS) instruments. There are a number of technologies available for polarizers and polarization analyzers which will be used at the ESS. The selection of the technology for an instrument is based on the performance and the constraints of the instrument. We will focus on the design of polarizing supermirror devices using Monte Carlo ray tracing simulation as an integral part of instrument design process. A McStas module has been developed to simulate a multichannel V-cavity polarizer, seeking the appropriate parameters to be incorporated into the respective instrument. The performance of such polarizers is studied for three instruments at ESS (MIRACLES (backscattering spectrometer), BIFROST (indirect geometry spectrometer) and ODIN (imaging)) with different requirements and constraints, where the suitability of this kind of devices can be assessed. For the first two instruments, where there is no strong constraint on the placement of the polarizer, the optimal configurations show excellent performance over the whole required wavelength ranges. However, in ODIN, due to more strict constraints in the placement of the polarizer, the performance is more dependent on the wavelength in the required wavelength range and other options may need to be considered.

1 Introduction

Polarized neutrons have been used in many applications for research using neutron scattering techniques [1] and have enabled a wider science case in many facilities. They also have enabled new experimental techniques like those using Larmor precession encoding [2].

There are different technologies available to polarize the neutron beam like polarizing supermirrors [3], polarized ^{3}He neutron spin filters [4] and Heussler monochromators [5] among others. The choice will depend on the requirements and constraints of the instrument (To be discussed in Section 2).

In the specific case of the European Spallation Source (ESS), an instrument suite has already been defined and it is being constructed [6]. In many instruments, the scope has to be reduced and incorporate the use of polarization for a later time. This imposes certain constraints in the choice of technology and needs careful design to still have a competitive performance.

In this work, we will present the design of the polarizer for three instruments: MIRACLES (backscattering spectrometer) [7], BIFROST (indirect geometry spectrometer) [8] and ODIN (imaging) [9].

MIRACLES is a backscattering spectrometer that is located with other long instruments at ESS, where the sample position is around 160 m away from the moderator. Therefore, it has a long guide [10] where a polarizer can be placed where the beam is less divergent, ensuring a high polarization at the required operating wavelength range. The inclusion of polarization analysis in backscattering instruments is becoming more important and would enhance the current capabilities of the instruments [11].

The case of BIFROST is very similar as the case of MIRACLES in terms of constraints. Located in the same area, and at a similar distance from the moderator, it also has certain flexibility in the placement of the polarizer, and also can be included where the beam is less divergent. The inclusion of a polarization analysis setup with a convenient analyzer technology, like the one proposed by P. Böni [12], would be beneficial for increasing the capabilities of the instrument.

In contrast to the other cases, the constraints in the ODIN beamline are stricter as it is a shorter instrument with a complex chopper cascade, which gives less flexibility to place the polarizer where the beam is less divergent. In addition to that, unlike the other two cases, the sample size is bigger and imposes an additional requirement of a polarized beam over a large region. The inclusion of polarized imaging will enhance the capabilities of the instrument further [13].

For the purpose of designing the polarizers, Monte Carlo ray-tracing codes [14, 15] are an essential tool,

* Corresponding author: damian.rodriguez@ess.eu

providing important insight to facilitate the best design choice to achieve the performance requirements.

2 Technical considerations, simulation method and design principles

Within the project to provide polarization at ESS [16] there are different technologies considered for polarization. There is going to be shared equipment for ^{3}He polarization, but some instruments can explore the possibilities for supermirror polarization. ^{3}He polarization has an important advantage that it can adapt to many beam characteristics (with more or less divergence) as the cell containing the ^{3}He gas can be manufactured in different geometries. On the other hand, supermirror polarizers need specific beam characteristics, like low beam divergence, in order to have a good polarization performance. However, supermirror polarizers have the advantage of being passive elements that can produce polarized neutrons after the polarizer is placed accurately in the beam. In the case of ^{3}He polarization, it needs some time to be set up before the experiment as the gas needs to be polarized by optical pumping and polarization can change as a function of time. In the case of MIRACLES and BIFROST, as it will be discussed below (figs. 4 and 8), the choice of supermirror polarizers is obvious as it can be placed where the neutron beam has low divergence. In the case of ODIN, with the limitation of placing the polarization in the cave, this is not that obvious and it is worth exploring this possibility to use the advantages of supermirrors while minimizing its inconveniences. The requirements coming from the instruments are summarized in Table 1.

Table 1. Requirements for the polarizers

* Field of view at 5 m from the pinhole position

Instrument	Operating $\Delta\lambda$ (Å)	Minimum polarization	Sample size (cm²)
MIRACLES	5.0 – 7.0	0.95	3.0 x 3.0
BIFROST	2.0 – 6.0	0.95	3.0 x 3.0
ODIN	2.5 – 7.5	0.95	10.0 x 10.0 *

All the three instruments have requirements of a neutron beam with a high degree of polarization for the wavelength range where they have to operate (see third column in table 1). To work with the pre-determined beam line geometry, a transmission geometry of the polarizer is chosen as it does not change the characteristics of the beam going through the polarizer. The use of V-cavity geometry [17] has the advantage of providing beam uniformity, and in its multichannel version [18] it makes it more compact.

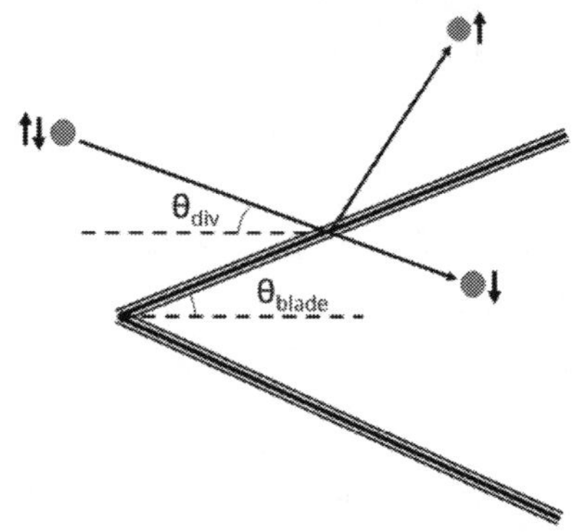

Fig. 1. Sketch of the neutron going through a supermirror polarizing cavity. For a polarizing supermirror, the incoming neutron beam is split into a reflected spin-up (polarized parallel to the applied saturated magnetic field) and a transmitted spin-down beam (polarized anti-parallel to the applied saturated magnetic field) spin state.

Fig. 1 shows the schematics on how the polarizing cavity works. The neutrons with the correct spin state will go through the mirror with the divergence distribution unchanged while the ones with the wrong spin state will be reflected with a higher divergence. In the case of the reflected beam we have two cases. In the case where the polarizer is placed within the guide system, the reflected beam's angle is higher and will be absorbed by the guide system, especially in its focusing section, until certain wavelength, where the critical angle is high enough to allow more divergent beams to be transported. Of course, multiple reflections within the polarizer also means that the surviving beam will be even more divergent, and therefore, its transport through the guide system will take place at even higher wavelengths. Furthermore, as the reflectivity for spin down (i. e., the wrong spin state) has $m < 1$, there will be a cut-off wavelength where total reflection of the beam in the cavity will take place [19]. We have checked that in the case of MIRACLES and BIFROST (which have long focusing sections of more than 10 m long) for the wavelengths they have to operate (table 1, third column) only the beam with the correct spin state reaches the sample position. In the case of ODIN, where the polarizer will be placed in the cave (see below, fig. 12), both contributions will be separated when arriving to the sample, and therefore, there is enough space to shield the beam with wrong spin state.

McStas code has been extensively used for the design of many instruments in different facilities around the world and it has been the preferred choice of software for the teams building the studied instruments. The simulation files of the instrument guide systems have been provided by the instrument teams and is the basis to add the polarizers in the instrument files.

In these simulations we used the McStas 2.7.1 version and two components had to be modified in order to perform this study. First, the polarizing cavity component available for McStas (*Pol_guide_vmirror*) was modified in order to allow multiple cavity channels. A new parameter has been added to specify the number of channels of the polarizer. This modification has been included in the following version (McStas 2.7.2). The second modification was the development of the component to obtain polarization maps (*PSD_pol_monitor*), which was not available in the McStas and has been developed by one of the co-authors (Alex Backs). This component has not been included yet, but there are plans to do it in the near future.

First of all, the V-cavities are constructed with Si blades which are coated from both sides. This is done to improve the polarization and to compensate the internal stresses of the wafers. Therefore, as an initial approximation we will use the reflectivity curve of a double mirror in the simulation, i. e., the reflectivity of the neutron going through two single reflective surfaces with the same layer sequence. This may not be accurate, but as a first approximation it will fulfil the aim of the study to find trends of different parameters and choose the optimal ones for the polarizers.

It is important to consider that the *Pol_guide_vmirror* component has an important limitation. Polarizing mirrors and cavities are sputtered over a Si wafer with a thickness of around 300 µm. The Si wafer will have some absorption that can be quantified but it is not taken into account in the component. It is possible to calculate the contribution of the absorption using different resources [20-22] and Fig. 2 shows the dependence of the penetration depth of the Si (the thickness needed to attenuate the incoming intensity by a factor 1/e) as a function of the neutron wavelength.

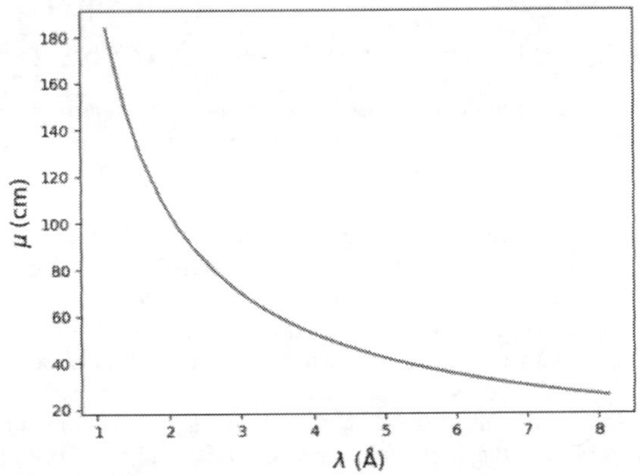

Fig. 2. Penetration depth (μ) of the Si as a function of the incoming neutron wavelength.

In this way, we can roughly estimate the absorption of the Si wafer.

The transmission T will be calculated as the ratio between the intensity of the beam at the sample position and area (defined in table 1, fourth column) with the polarizer and without the polarizer. Assuming that the incoming neutron beam has no divergence, the corrected transmission can be calculated as:

$$T_{corr} = T\ exp(-t\ /(\mu \sin \theta_{blade})) \quad (1)$$

where t is the substrate thickness and θ_{blade} is the angle of the polarizing cavity blade. The assumption made to derive the equation is more accurate in the case of a low divergence beam at the polarizer as the differences in thickness are low. This is not happening for higher divergences, but it will give an initial approximation on the effect of absorption.

A feature not taken into account in the simulation are the separators between channels in the V-cavities, as it appears, for example, in Ref. 18. If the separators are absorbing, there is going to be an additional decrease on transmission that is not taken into account in the simulations. These will depend on the number of channels and length of the polarizer.

Another important factor that is not taken into account in the simulations are the depolarization effects in supermirrors [23]. These come from the off-specular scattering in the multilayers and can be reduced by applying a magnetic field. The required strength of the magnetic field will depend on different factors like the angle and the m index of the coating. According to the results in Ref. 22, a magnetic field of around 30 mT is enough to reduce these depolarization effects to an acceptable level, but it will increase for high m coatings.

In addition to that, guide fields have to be designed to reduce depolarization effects during the path from the polarizer to the sample, which requires careful design using finite elements calculations. These are out of the scope of this work and the simulation considers that there is no depolarization in the path to the sample.

The choice of the type of polarizing mirror used in the device can be important. Fe/Si supermirrors [3] are widely used as a good compromise between performance and cost. In addition to that, they have no activation issues like the mirrors containing Co.

The inclination of the blades (θ_{blade}) will be calculated using a criterion of high polarization over the operating wavelength range. The formula to estimate the needed inclination of the blades in degrees is as follows:

$$|\theta_{blade}| = m \times \lambda_c \times 0.1^{\circ} - |\theta_{div,max}| \quad (2)$$

Where m is the critical angle of the coating expressed as a function of the critical angle of Ni, λ_c is the critical wavelength where the reflectivity will be high (which will be slightly lower than the lowest limit of the operating wavelength range of the instrument) and $\theta_{div,max}$ is the maximum divergence of the beam.

The design considerations are simpler than in the study made by Dewhurst [24], where the maximum divergence of the beam has a linear dependence with wavelength. This study takes the assumption of extended source geometry and straight guide geometry, which is not the case at ESS. The moderator at ESS uses a complex geometry with thermal and cold zones with a reduced dimensionality in the vertical direction in order to increase the brightness of the source [25]. As in the design of MIRACLES and BIFROST guides [8, 10], the low dimensionality of the moderator in the vertical direction enables extracting the beam with a defocusing elliptical guide section, converting it into a lower divergence beam with no wavelength dependence. In the case of the of the horizontal divergence, a defocusing guide can be placed after the curved section used to filter fast and epithermal neutrons. Therefore, it is possible to obtain a value of the maximum divergence which is independent of the wavelength, making the optimization easier.

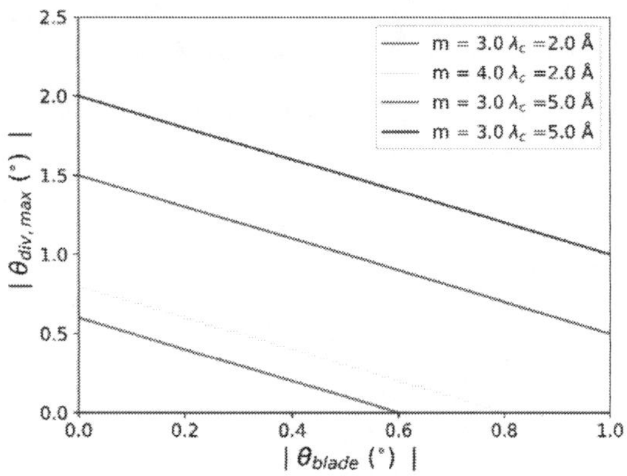

Fig. 3. Evolution of θ_{blade} as a function of $\theta_{div,max}$ as described in equation (2) for different examples.

The trends of equation (2) are shown in fig. 3. For given values of m and λ_c, the values of θ_{blade} depend linearly with $\theta_{div,max}$. In terms of manufacturing, a low value of θ_{blade} would imply a larger surface to coat and more channels, which would increase costs and would be more difficult to manufacture. However, increasing the m coating also can increase the costs. Therefore, a compromise has to be reached.

3 Results and discussion

3.1 Polarizer for instrument MIRACLES

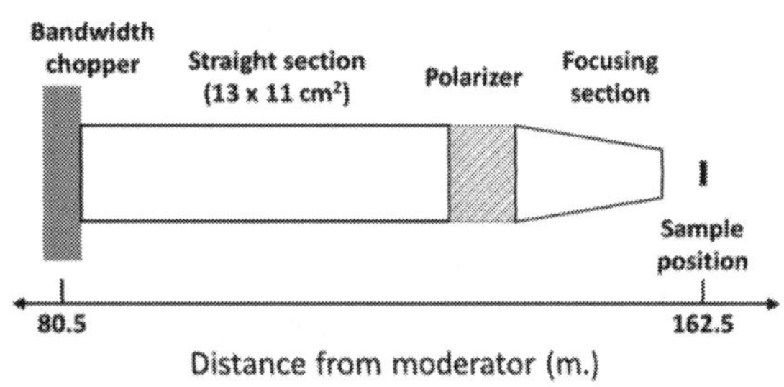

Fig. 4. Sketch of the location of the polarizer of MIRACLES instrument

Fig. 4 shows the location of the polarizer within the guide system of MIRACLES. After the bandwidth chopper there is long straight section with a large cross section (13 × 11 cm^2) before it starts the focusing section. The straight section also has a coating of m = 1.5 in all the sides. The focusing section will be an elliptic guide section that will focus the beam to the sample. The beam with wrong spin state will be absorbed by the focusing part within the operating wavelength range.

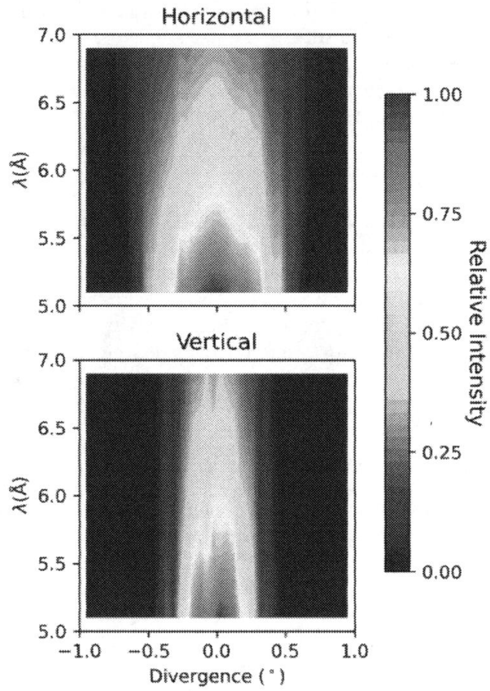

Fig. 5. Divergence distribution as a function of the wavelength of the neutron beam at MIRACLES before the polarizer.

Fig. 5 shows the divergence distribution as a function of wavelength in both horizontal and vertical direction before the polarizer. It shows that in the case of the vertical direction the maximum divergence is smaller and is wavelength independent. In the case of the distribution in the horizontal direction, the maximum divergence is larger and has a slight dependence with wavelength. If we take into account the trends in fig. 3, where it is important to seek a large θ_{blade}, a small maximum divergence would be more beneficial, apart from a wavelength independent divergence distribution. For this reason, the mirrors of the cavity will be vertically arranged (i. e., where the plane of the mirrors contains the horizontal direction axis) and therefore, in the calculation of equation (2) we will take $\theta_{div,max} = 0.3°$. In addition to that, we will use λ_c = 4.5 Å, in order to obtain the required polarization at the operating wavelength range.

Table 2. Parameters of the simulated polarizers for MIRACLES

m	θ_{blade}(°)
2.4	0.78
2.6	0.87
2.8	0.96
3.0	1.05

Table 2 shows the parameters of the simulated polarizers. In order to observe the angular dependence of the performance, we will also perform simulations for each polarizer with values of $\theta_{blade} \pm 0.02°$ for a given value of m. This will be represented in the following figures as a shade. The polarizers will be simulated as single channel as the optimal length will be later determined by the engineering requirements and constraints. The total length of the device will depend on the available space and a compromise has to be reached in order to have feasible alignment requirements. Of course, beam losses due to channel separators have to be considered in further studies. The polarizer will have the V-cavity pointing upstream.

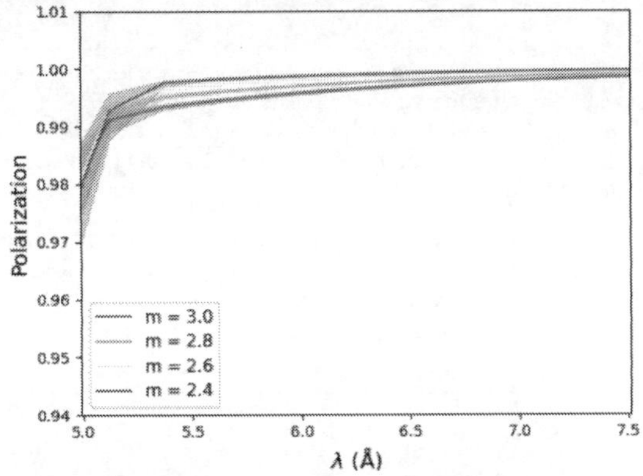

Fig. 6. Polarization at the sample position as a function of wavelength for the simulated polarizers in MIRACLES described in table 2. The shady areas show the variation of the polarization modifying the nominal θ_{blade} shown in table 2 in a range of ± 0.02°.

Fig. 6 shows the polarization of the simulated polarizers described in table 2 as a function of wavelength. The shady areas show variations in the polarization performance if θ_{blade} is modified in the range of ± 0.02° from the nominal value. In this case, all the studied polarizers produce a neutron beam with the required polarization. The curves look very similar and variations in θ_{blade} still fulfil the requirements.

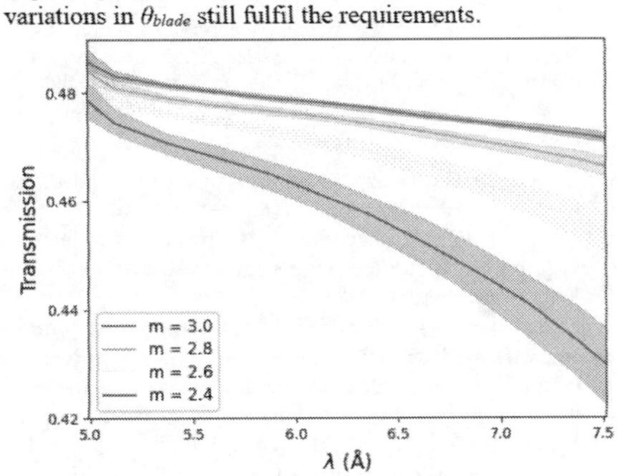

Fig. 7. Corrected transmission at the sample position as a function of wavelength for the simulated polarizers in MIRACLES described in table 2. The shady areas show the variation of the transmission modifying the nominal θ_{blade} shown in table 2 in a range of ± 0.02°.

Fig. 7 shows the corrected transmission (i. e., the transmission taking into account the Si absorption) as a function of the wavelength for the studied polarizers. As in fig. 7, the shady areas correspond to the variations in transmission performance if θ_{blade} is modified in the range of ± 0.02° from the nominal value found in table 2. For this case, the transmission curves have a strong variation. While the polarizer with m = 3 has a high transmission for the whole operating wavelength range, in the case of m = 2.4, it decreases continuously. The angular variation in transmission also changes as a function of m. The polarizer with m = 2.4 has a strong variation with θ_{blade}, but that is not the case with the polarizer with m = 3.

The results in MIRACLES show that in all cases, a high degree of polarization is achieved. However, the main driving difference in the performances of the polarizers is in its transmission. In order to achieve a competitive performance for the instrument the polarizer with m =3 would be the best option among the ones analyzed as it would also have a competitive transmission beyond the operating wavelength range (between 5 – 7 Å) in order to analyze the tail of the quasielastic peak (from 7-8 Å). In addition to that, the variation with angle is the lowest, as it has a negligible shade (see fig. 7), making it also more robust against the effects of surface waviness of the blades. A higher m would extend the range with high transmission, but it would not widen the science case of the instrument.

3.2 Polarizer for instrument BIFROST

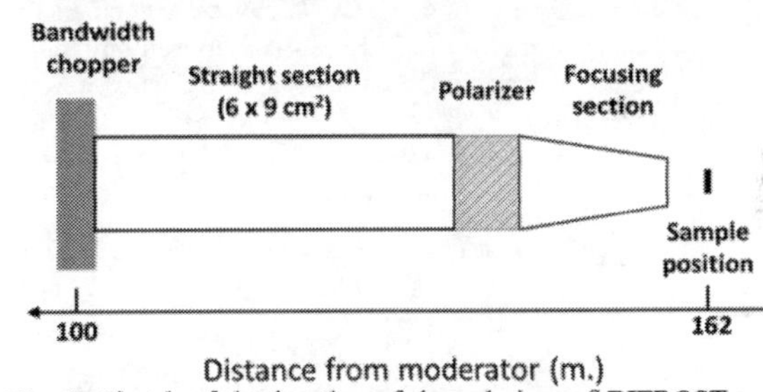

Fig. 8. Sketch of the location of the polarizer of BIFROST instrument

As it is shown in fig. 8, the case of BIFROST is also very similar to MIRACLES, where there is a long guide section after the last bandwidth chopper. In this case, the straight guide has a cross section of 6 x 9 cm^2. After that, a focusing section follows before the neutron beam reaches the sample position. The guide section enclosing the polarizer has a coating of m = 2.

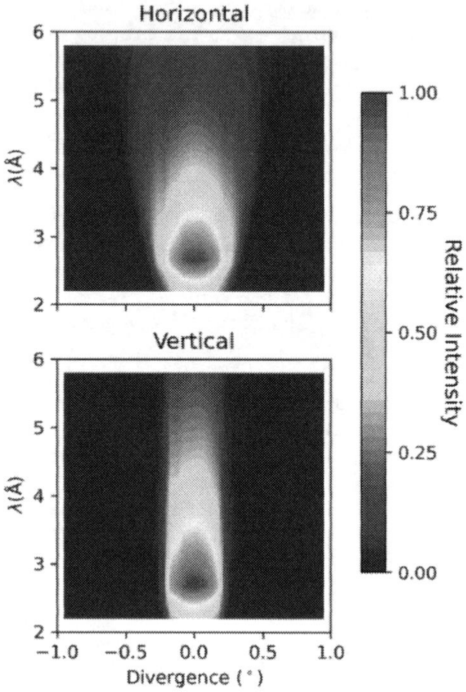

Fig. 9. Divergence distribution as a function of wavelength of the neutron beam at BIFROST before the polarizer.

Fig. 9 shows the divergence distribution as a function of wavelength in both horizontal and vertical direction before the polarizer. It can be argued in the same way as in MIRACLES that it is more convenient to have the mirrors of the cavity vertically arranged and therefore $\theta_{div,max} = 0.2°$. In addition to that, we will use $\lambda_c = 2.0$ Å, in order to obtain the required polarization at the operating wavelength range.

Table 3. Parameters of the simulated polarizers for BIFROST

m	θ_{blade}(°)
3.5	0.50
3.8	0.55
4.0	0.61
4.2	0.64
4.4	0.67
4.6	0.72

Table 3 shows the parameters of the simulated polarizers where the angle of the blades is proportional to the *m* of the coating. In an analogous way, we will also simulate the systems with $\theta_{blade} \pm 0.02°$ for a given value of m in order to observe the dependence with the angle of the blades.

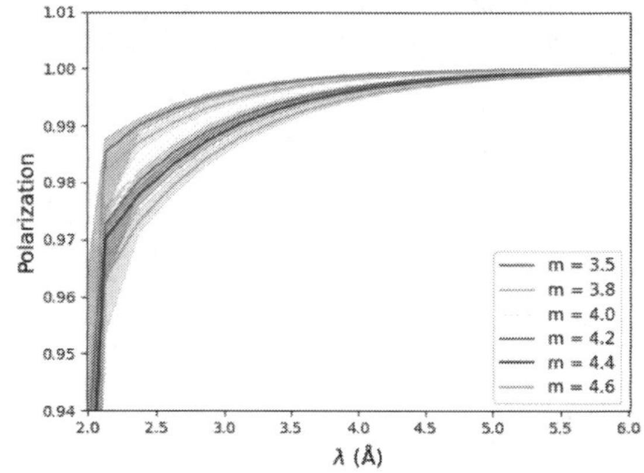

Fig. 10. Polarization at the sample position as a function of wavelength for the simulated polarizers in BIFROST described in table 3. The shady areas show the variation of the polarization modifying the nominal θ_{blade} shown in table 3 in a range of ± 0.02°.

Fig. 10 shows the polarization as a function of the wavelength for the polarizers described in table 3. Polarization is within requirements in all the simulated polarizers. Even varying θ_{blade} in the range of ± 0.02° the polarizers also fulfil the performance requirements.

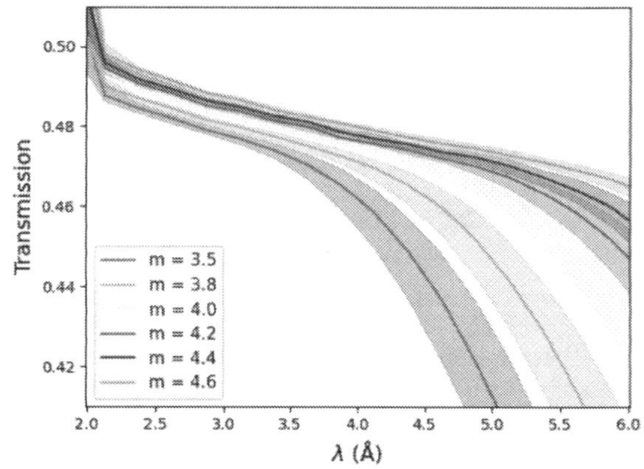

Fig. 11. Corrected transmission at the sample position as a function of wavelength for the simulated polarizers in BIFROST described in table 3. The shady areas show the variation of the transmission modifying the nominal θ_{blade} shown in table 3 in a range of ± 0.02°.

Fig. 11 shows the transmission as a function of the wavelength for the studied polarizers for BIFROST. It follows the same trend as in fig. 7 where the polarizers with higher *m* coating have a high transmission over a longer wavelength range

In the case of BIFROST, despite the strong equivalence with MIRACLES, it is also important to take into account other factors like the cost, as the *m* of the coatings is higher than in the former case of MIRACLES. All the studied polarizers obtain the required polarization, but the transmission is different. The polarizer with the highest transmission over the whole operating wavelength range (between 2 – 6 Å) is the one with the highest *m*, but the ones with $m \geq 4.0$ have little differences in transmission of a maximum 4 %, and therefore a good compromise between cost and

https://doi.org/10.1051/epjconf/202328603008

performance could be the one with $m = 4.2$. However, this choice should be reviewed when official quotations with the manufacturers are taken into account, together with the available budget.

3.3 Polarizer for instrument ODIN

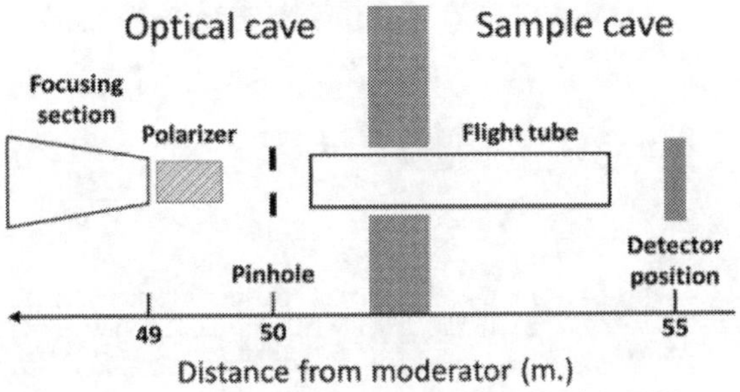

Fig. 12. Sketch of the location of the polarizer of in the ODIN instrument cave.

The case of ODIN is completely different and the polarizer has to be placed in the cave. Fig. 12 shows a sketch of the cave where the polarizer is placed. The cave is divided in two parts. The Optical cave has the end of the guide system and the pinhole that adjusts the beam shape before going to the sample (in this case the 3 x 3 cm^2 pinhole has been chosen). The flight tube after the pinhole creates vacuum to minimize neutron losses This communicates with the Sample cave, where the sample and the detector will be placed at varying distances. The most convenient location to place the polarizer is at the end of the guide and the length will be limited at 50 cm at most in order to avoid clashes with the pinhole. The enclosing around the polarizer mirrors is absorbing.

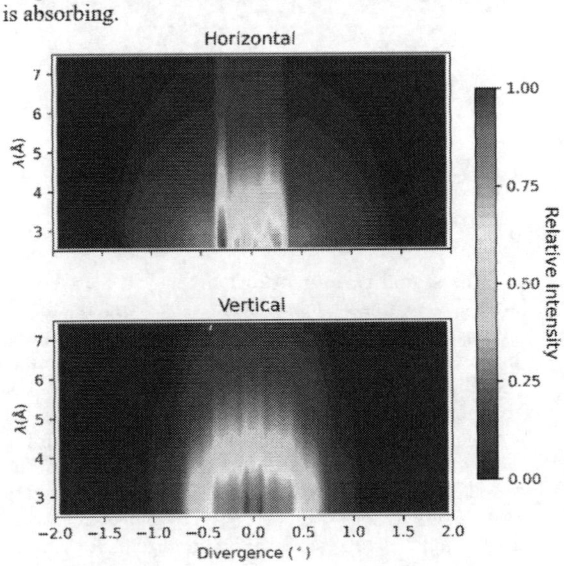

Fig. 13. Divergence distribution as a function of wavelength of the neutron beam at ODIN before the polarizer.

As in the other cases, fig. 13 shows the divergence distribution as a function of the wavelength. As the beam is after a focusing section of the guide, the beam has a larger divergence than in the former cases.

With the current multichannel v-cavity geometry, observing the trends of fig. 3, there wouldn't be any value of θ_{blade} available for such a high divergence and equation (2) would be negative, which makes no physical nor mathematical sense. An alternative solution would be to simply polarize part of the beam, and therefore use different lower values of $\theta_{div,max}$ for the calculation of θ_{blade}.

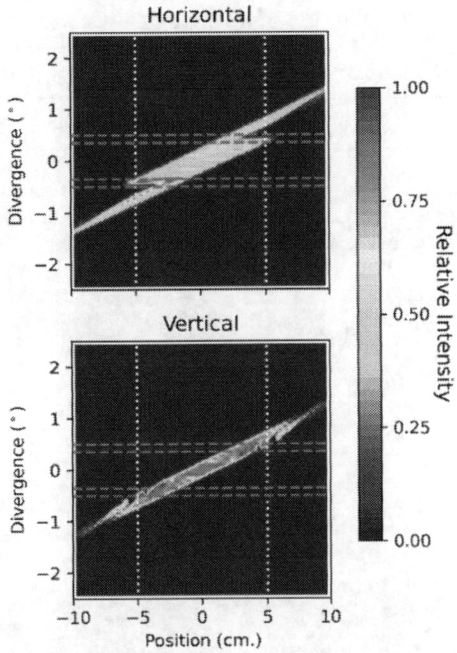

Fig. 14. Acceptance diagrams in both dimensions of the neutron beam at the detector position without polarizer. The greydashed lines denote the different values of $\theta_{div,max}$ taken for the design of the simulated polarizers (see table 4) and the white dotted lines show the spatial limits of the detector.

In fig. 14 we can can observe the acceptance diagrams in both directions of the unpolarized beam (that is, without placing any polarizer) integrated over the operating wavelength range where the white dotted lines are the spatial limits of the detector and the grey dashed lines are the values of $\theta_{div,max}$ chosen for the designs.

As can be seen, if only part of the beam is polarized, the higher divergence part of the beam that is not polarized by the polarizer would still degrade the polarization at the sides of the detector. However, taking into account that the critical angle of the coatings is wavelength dependent, we can ensure high levels of polarization by choosing a value of λ_c in eq. (2) lower than the lower limit of the operating wavelength range (2.5 Å). This way, we could ensure that the whole phase space within the detector area reaches a high level of polarization. In our case we have chosen $\lambda_c = 2.0$ Å. This strategy works better in the vertical direction, where the phase space has an approximate shape of a rhomboid. In the horizontal direction we can observe some contributions in higher divergence where this strategy would be more difficult to apply and the results would show a higher degree of degradation in the sides.

 https://doi.org/10.1051/epjconf/202328603008

Therefore, the blades of the polarizer will be arranged vertically.

Table 4. Parameters of the simulated polarizers for ODIN

m	$\theta_{div,max}$ (°)	θ_{blade} (°)	Length (m)	channels
4.0	0.35	0.45	0.47	4
4.0	0.50	0.30	0.47	6
4.5	0.35	0.55	0.39	4
4.5	0.50	0.40	0.43	5

Table 4 shows the parameters of the simulated polarizers for ODIN. In this case we use a compact multichannel arrangement for the simulations.

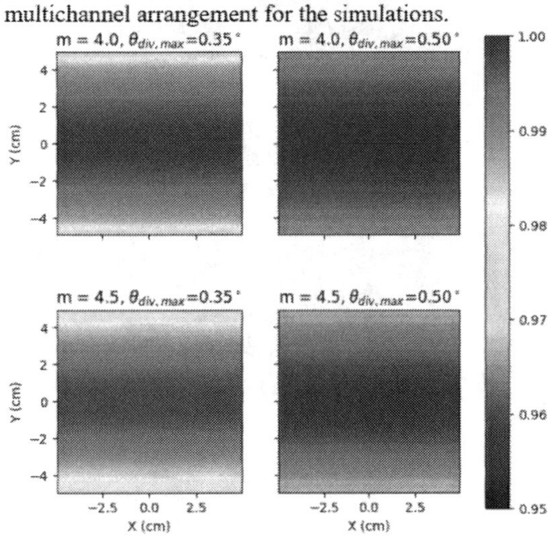

Fig. 15. Spatial distribution of polarization at the detector position of the different simulated polarizers for ODIN described in table 4.

The spatial distribution of the polarization in the detector is also important to analyse in order to ensure that the whole detector area has the required polarization. Fig. 15 shows the spatial distribution of the polarization integrated over the operational wavelength range for ODIN (2.5-7.5 Å). In all cases the total polarization is over 95 %, as required. In all cases, the highest polarization is in the center of the beam and it decreases when it goes towards the borders.

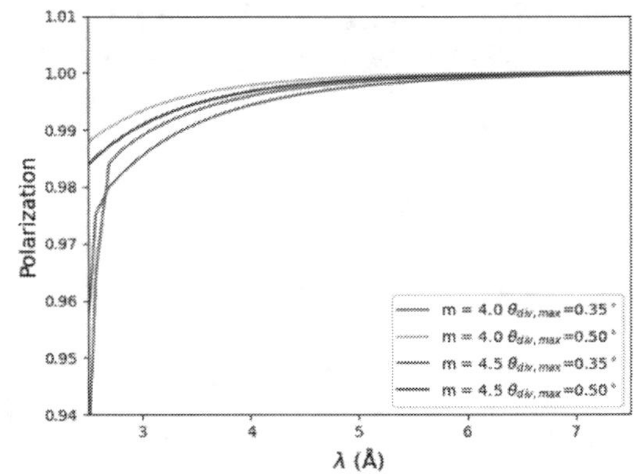

Fig. 16. Polarization at the sample position as a function of wavelength for the simulated polarizers in ODIN described in table 4

Fig. 16 shows, as in the former cases, the polarization averaged over the whole detector area as a function of the wavelength. In all cases, the polarization is within the required value.

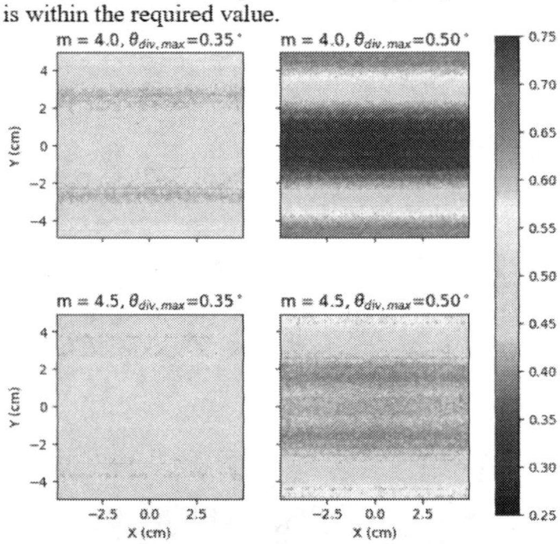

Fig. 17. Spatial distribution of the uncorrected transmission (see the text to see how it is calculated) at the detector position of the different simulated polarizers for ODIN described in table 4.

The spatial uniformity of the beam in the detector area is also important to study as, in general, beam uniformity is an important condition for imaging beamlines. It is important to analyze if the polarizer introduces additional inhomogeneities to the beam. For that purpose, we can use the ratio between the with the chosen polarizer and the beam without the polarizer. Fig 17 shows the spatial distribution of transmission in the simulated cases integrated over the operating wavelength range without applying the effect of Si absorption. There is are larger inhomogeneities created when increasing $\theta_{div,max}$ and decreasing m (and therefore, decreasing θ_{blade}). It also appears transmissions over 0.50 in some cases, which should be analyzed in further studies.

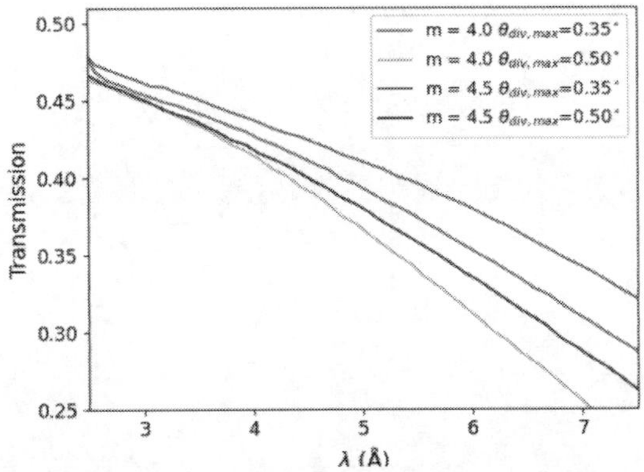

Fig. 18. Corrected transmission at the sample position as a function of wavelength for the simulated polarizers in ODIN described in table 4.

Fig. 18 shows the transmission at the detector position for the simulated devices. In all the simulated polarizers, the transmission is high for shorter wavelengths and the transmission decreases as a function of wavelength. In some cases, the transmission at the longer wavelength range can be lower than 0.30. It is also important to point out that transmission is better with increasing m but decreasing $\theta_{div,max}$.

Unlike the former cases, the results obtained for ODIN do not lead to an obvious technical choice to be implemented. Although the required polarization is obtained with a sufficient degree of spatial uniformity (figs. 15-16), the transmission is strongly dependent of the wavelength (fig. 18).

In addition to that, homogeneity of the beam is also important, and fig. 17 shows strong inhomogeneities in some of the cases that are not acceptable to conduct an experiment. It is also important to point out that, as the incoming beam is more divergent and the Si absorption is very dependent on the path through it, there may be additional inhomogeneities that are not taken into account in the simulation.

Despite not being able to find a conclusive solution for ODIN, the results indicate very important trends to find a more convenient solution. A higher m in the coating and a higher θ_{blade} in the polarizer produces a higher transmission and a higher uniformity in the beam arriving to the detector.

Taking into account these trends, a higher m than the ones analyzed in the study would improve the performance, leading to important increases in cost. In that case, other solutions would need to be explored in further studies, like for example a fan arrangement where each channel has a different angle or a logarithmic spiral geometry of the mirrors [26].

4 Conclusions

This study presents the conceptual design of the supermirror polarizers for three instruments at ESS: MIRACLES (backscattering spectroscopy), BIFROST (indirect geometry spectroscopy) and ODIN (imaging). All the three instruments have different requirements and constraints that have to be taken into account (see table 1). In all cases, it is studied the performance of polarizers in transmission geometry, as they perform better in the required operating wavelength range (table 1, column 3).

The cases of MIRACLES and BIFROST are very similar, as can be observed in figs. 4 and 8. The proposed solution places the polarizer in the guide system before the focusing section that both have. The optimal parameters have been found that ensures a high polarization and a high transmission (i. e., a higher flux at the sample position).

In the case of ODIN (fig. 12), the main constraint is that the polarizer has to be placed in the cave, where the beam is more divergent. Even if the polarization obtained at the detector position is within the requirements, transmission is lower than the other cases and strongly dependent on the wavelength. In addition to that, beam inhomogeneities have to be taken into account to find the most convenient solution. The trends in the results show that a higher m in the coating and a higher angle of the blades (θ_{blade}) lead to a higher transmission and a more uniform beam. This would lead to a solution with high costs pushing the technological limits of polarizing supermirror coating. Therefore, other solutions with different mirror arrangements will be explored in further studies.

References

1. W. G. Gavin, *Polarized Neutrons* (Clarendon Press, Oxford, 1988)
2. F. Mezei, C. Pappas, Th. Gutberlet eds., *Neutron spin echo spectroscopy* (Springer Berlin, Heidelberg, 2003)
3. J. Stahn, D. Clemens, Appl. Phys. A **74**, S1532 (2002)
4. K. H. Andersen, D. Jullien, A. K. Petoukhov, P. Mouveau, F. Bordenave, F. Thomas, E. Babcock, Physica B **404**, 2652 (2009), G. L. Jones, T. R. Gentile, A. K. Thompson, Z. Chowdhuri, M. S. Dewey, W. M. Snow, F. E. Wietfeldt, Nucl. Instrum. Methods Phys. Res. A **440**, 772 (2000).
5. P. Courtois, Physica B **267-268**, 363 (1999)
6. K. H. Andersen, D. N. Argyriou *et al.*, Nucl. Instrum. Methods Phys. Res. A, **957**, 163402 (2020)
7. N. Tsapatsaris, R. E. Lechner, M. Markó, H. N. Bordallo, Rev. Sci. Instrum. **87**, 085118 (2016)
8. M. Klausz, K. Kanaki, T. Kittelmann, R. Toft-Petersen, J. O. Birk, M. A. Olsen, P. Zagyvai, R. J. Hall-Willton, J. Appl. Cryst., **54**, 263 (2021)
9. M. Strobl, Phys. Procedia, **69**, 18 (2015)
10. F. J. Villacorta, D. Martin Rodriguez, M. Bertelsen, H. N. Bordallo, Quantum Beam Sci., **6**, 3 (2022)
11. K. Nemkovski, R. Bewley, V. Garcia Sakai, G. J. Nilsen, A. Perrinchon, I. Silverwood, EPJ Web Conf. **272**, 02004 (2022)

12. P. Böni, C. Schanzer, M. Schneider, Nucl. Instrum. Methods. Phys. Res. A **966**, 163858 (2020)
13. M. Dawson, I. Manke, N. Kardjilov, A. Hilger, M. Strobl, J. Barnhart, New J. Phys. **11**, 043013 (2009)
14. P. Willendrup, and K. Lefmann, J. Neutron Res. 22, 1-16 (2020).
15. P. Willendrup, and K. Lefmann, J. Neutron Res. 23, 7-27 (2021).
16. W.- T. Lee, J. Hagman *et al.*. *This issue.*
17. T. Krist, C. Lartigue, F. Mezei, Physica B **180-181** 1005 (1992)
18. R. R. Gainov, F. Mezei, J. Fuzi and M. Russina, Nucl. Instrum. Methods. Phys. Res. A **930**, 42-48 (2019).
19. T Bigault, G Delphin, A Vittoz, V Gaignon and P Courtois., J. Phys. Conf. Ser. **528**, 012017 (2014)
20. https://www.ncnr.nist.gov/resources/activation/
21. H. Rauch and W. Waschkowski (2003). Neutron Scattering Lengths in *ILL Neutron Data Booklet* (second edition), A.-J. Dianoux, G. Lander, Eds. Old City Publishing, Philidelphia, PA. pp 1.1-1 to 1.1-17.
22. V. F. Sears (2006). *Scattering lengths for neutrons* In Prince, E. Ed. *International Tables for Crystallography Volume C: Mathematical, Physical and Chemical Tables* Kluwer Academic Publishers, pp 444-454.
23. C. Klauser, T. Bigault, P. Böni, P. Courtois, A. Devishvili, N. Rebrova, M. Schneider and T. Soldner, Nucl. Instr. Meth. A **840**, 181-185 (2016)
24. C. D. Dewhurst, Nucl. Instr. Meth. A **683**, 16-23 (2012)
25. K. H. Andersen, M. Bertelsen, L. Zanini, E. B. Klinkby, T. Schönfeldt, P. M. Bentley and J. Saroun, J. Appl. Cryst. **51**, 264–281 (2018).
26. J. Stahn and A. Glavic, J. Phys.: Conf. Ser. **862**, 012007 (2017)

EPJ Web of Conferences **286**, 03009 (2023) https://doi.org/10.1051/epjconf/202328603009
ECNS 2023

Novel idea of neutron polychromator and application for reflectometry and spectroscopy

Norifumi L. Yamada*
Institute of Materials Structure Science, High Energy Accelerator Research Organization (KEK), Ibaraki 305-0801, Japan

Abstract. Historically, two methods have been used to determine the wavelength of neutrons: (i) a time-of-flight method that separates the velocity of pulsed neutrons by the flight time; and (ii) a method utilizing Bragg reflection by a monochromator, such as a single crystal or multilayer mirror. The former cannot be applied to electromagnetic waves because the light velocity is constant and independent of the wavelength, whereas "polychromators" such as prisms and gratings, which separate wavelengths via chromatic dispersion, are typically used in the infrared to soft X-ray range. Although polychromators require collimated beams to separate wavelengths with sufficient resolution, this aspect does not affect laser and synchrotron light because they are naturally collimated. Herein, we propose a novel idea of a neutron polychromator utilizing an elliptical multilayered mirror that can be applied to a wide beam with a large beam divergence. In addition, examples of reflectometer and spectrometer applications are presented.

1 Introduction

Historically, two methods have been used to determine the wavelength of neutrons: the time-of-flight (TOF) method, which separates the velocity of pulsed neutrons by the flight time, and a method utilizing Bragg reflection by a monochromator, such as a single crystal or multilayer mirror. The former is extremely useful for obtaining the spectrum of a white beam, i.e., the wavelength dependence of the intensity; however, it cannot be applied to electromagnetic waves because the light velocity is constant and independent of the wavelength. Hence, "polychromators" such as prisms, gratings, and bent crystals are typically used in the infrared to X-ray range.

A prism is the simplest polychromator, in which the relationship between the angles of incidence θ_{in} and extraction θ_{out} can be described as a function of wavelength λ, as follows:

$$\cos\theta_{in} = n\cos\theta_{out},$$

$$n = \sqrt{1 - \frac{\lambda^2\rho}{\pi}} \sim 1 - \frac{\lambda^2\rho}{2\pi} \quad (\lambda^2\rho \ll 1),$$

where n and ρ are the refractive index and scattering length density of the prism, respectively. Based on this relation, the λ dependence of the intensity can be acquired by measuring the intensity of a beam spreading out from the prism with various θ_{out} when a white beam with incident angle θ_{in} is introduced on the prism surface. However, if the beam passes through the prism in this geometry, the abovementioned method cannot be applied to X-rays because the beam is attenuated in the transmission geometry.

A grating with reflection geometry is a typical polychromator used for soft X-rays, in which the beam is not required to pass through a material. Here, the relationship between θ_{in} and θ_{out} satisfying Bragg's law of the line and the space patterns of the grating can be expressed as

$$d(\cos\theta_{in} - \cos\theta_{out}) = m\lambda,$$

where d is the repeat distance of the grating and m is the order of the Bragg peaks. Based on this relation, the λ dependence of the intensity can be acquired by counting the intensity based on θ_{out}, as in the case of a prism. As the value of d ranges from submicrons to several microns, this method cannot be applied to hard X-rays because λ is extremely small to separate, and the critical angle for the total reflection is extremely small as well.

Hard X-rays can be applied to a bent crystal [1] based on Bragg's law.

$$2d\sin\theta_{in} = m\lambda \; (\theta_{in} = \theta_{out}).$$

In this method, the distribution of θ_{in} depends on the position on the crystal because the crystal has curvature. If the beam is focused at the sample position, then it spreads to the detector position, depending on the position where the beam is reflected at the crystal. Consequently, the X-ray wavelength depends on the detection position, as in the case of the aforementioned methods.

The principle of these polychromators is applicable to neutrons; however, these polychromators are barely used for neutrons. This is attributable to the fact that they require "collimated beams" to suppress the distribution of θ_{in}, which worsens the λ resolution. In the cases of laser and synchrotron light, this feature does not impose a significant effect because the beam is naturally collimated. However, this results in a considerable

* Corresponding author: norifumi.yamada@kek.jp

decrease in the intensity of neutrons because they naturally diverge from the source. In other words, polychromators can be applied to instruments that utilize a collimated beam, such as a reflectometer. To the best of our knowledge, the only full-scale neutron instrument that uses a polychromator is the RAINBOWS reflectometer, which is equipped with a prism to capture the spectrum of narrow and collimated white neutrons reflected at a sample surface [2]. However, this restriction is not suitable for other instruments and is the biggest barrier hindering the use of polychromators in neutron instruments.

To overcome the problem, the author proposes the novel concept of a neutron polychromator comprising an elliptical multilayered mirror that can be used for neutrons with a large beam divergence. To fully exploit the polychromator with a higher order reflection and high wavelength resolution, a novel "crystallic multilayered mirror" is proposed. Furthermore, some ideas for neutron instruments utilizing a polychromator are presented: (i) a reflectometer at a continuous source, and (ii) a quasi-elastic neutron spectrometer at a pulsed source.

2 Concept of Novel Polychromator

2.1 Optics

Figure 1 shows a schematic illustration of the novel polychromator proposed herein, which comprises an elliptical focusing mirror coated with a multilayer having a periodic thickness and a position-sensitive detector (PSD). A virtual source is placed at the focal point of the mirror, and the PSD is placed at the other focal point. The image of the source is transferred to the PSD at the same magnification when the mirror is placed at the center of the focal points. In this polychromator, the incident angle on the mirror depends on the position of the virtual source. This indicates that neutrons propagating from a point closer to the mirror on the virtual source are reflected with a smaller incident angle at the mirror and are transferred to a point closer to the mirror on the PSD. Because the wavelength satisfying the Bragg reflection depends on the incident angle, the spectrum of the white beam can be observed on the detector as a function of the detection position. This is the essential idea of the novel polychromator with an elliptical mirror, which is based on the simple Bragg's law and allows a diverged beam from the virtual source to be used.

Prior to this study, the combination of elliptic mirror and multilayer as a polychromator for neutron reflectometry was proposed by Ott *et al.* [3]. In the study, they performed a Monte Carlo simulation with the optics that the beam size at the source was limited but a large focusing mirror was used. Because the incident angle to the mirror gradually increases with the distance from the center of the ellipse, the wavelength satisfying the Bragg condition also increases in the same manner. On the other hand, the wavelength dependence on the position is assumed to be corrected by controlling the layer thickness to compensate the wavelength satisfying the Bragg condition for the polychromator in this study. This correction allows us to use the monochromatic beam with the large beam divergence, while the beam divergence of a limited wavelength is very narrow before the correction. As a result, the polychromatic beam of the corrected polychromator with a wide virtual source can increase the beam flux compared to that of the previous work.

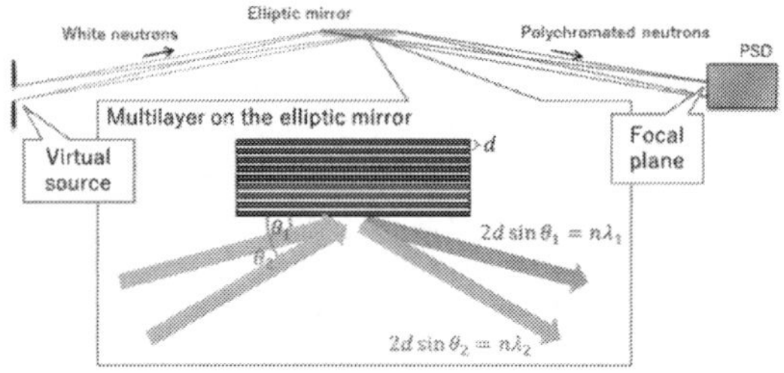

Fig. 1. Schematic illustration of the novel polychromator comprising an elliptic focusing mirror and PSD.

2.2 Crystallic multilayer

The wavelength resolution of the polychromator shown above is limited by two factors: the spatial resolution of the image on the detector, which is primarily governed by the detector resolution and the slope error of the focusing mirror; and the width of the Bragg reflection of the multilayer on the focusing mirror. The former can be reduced to less than 1% by applying the most recent technology, such as focusing mirrors via ultraprecise machining [4-6] and employing scintillation detectors [7]. Meanwhile, the width of the Bragg reflection of multiple layers of typical mirrors such as Ni/Ti remains at approximately several percent because no attempt has been made to obtain better resolutions, presumably because of the low demand for high-resolution monochromators composed of multilayers. However, polychromators require higher resolutions than monochromators as shown later. Hence, a novel method for improving the wavelength resolution of multilayers is proposed herein.

A typical multilayer mirror for non-polarized neutrons comprises Ni ($\rho = 9.4 \times 10^{-4}$ nm^{-2}) and Ti ($\rho = -1.9 \times 10^{-4}$ nm^{-2}) layers, which allows high reflectivity to be achieved using only a few of those layers owing to the high contrast in the scattering length density, i.e., the refractive indexes between the layers. On the other hand, the effective number of layers contributing to Bragg reflection is limited because of the high reflectivity, which implies that neutrons satisfying the Bragg condition cannot reach the deep layers. As the peak width of the Bragg peak decreases with an increase in the effective number of layers, the multilayer with a large stacking number composed of materials providing less reflectivity in one layer unit is expected to exhibit sharp Bragg peaks because of improved crystallinity. Hereinafter, multilayers with high crystallinity are referred to as "crystallic multilayer."

To demonstrate the concept of the crystallic multilayer, we estimated the reflectivity using the Motofit software [8] based on Parratt's recursive method [9]. To decrease the reflectivity in one unit layer, Ge ($\rho = 3.6 \times 10^{-4}$ nm^{-2}) and Ti with asymmetric layer thickness were combined to reduce the contrast in the refractive index and enhance the tunneling of neutrons by thinning Ti layers while maintaining the periodic distance at 20 nm. The solid line in Fig. 2 shows the result of the reflectivity estimation as a function of momentum transfer along the depth direction, in which the layer thicknesses of Ge and Ti were 18.8 and 1.2 nm, respectively, and the number of unit bilayer was 2000 layers. The resolution evaluated from the full width at half maximum of the Bragg peak was 0.7% at the first peak and decreased significantly with the increasing order of the peaks, reaching 0.004% while maintaining a reflectivity of approximately 100% at the 10th peak, as shown by the circles on the right axis.

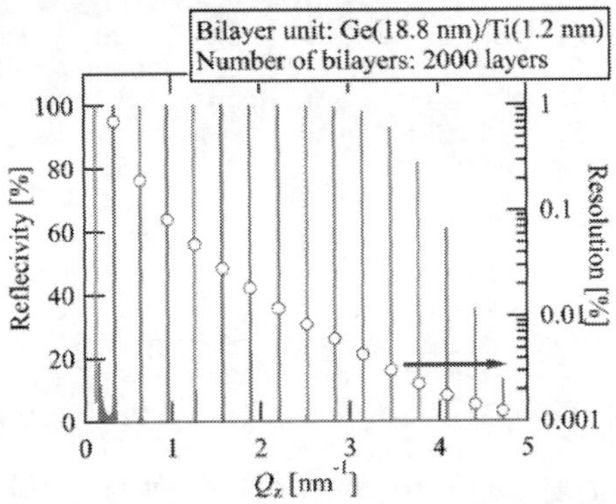

Fig. 2. Ideal reflectivity of crystallic multilayer composed of Ge/Ti bilayer (solid line, left axis) and resolution evaluated from full width at half maximum of Bragg peak (circles, right axis).

The results above are only an estimation, and the effect of microscopic roughness at the interfaces, the macroscopic inhomogeneity over the mirror, and the difference in the incident angle with respect to the position in the ellipse of the polychromator should be considered. Therefore, further research is required to realize such high resolutions by creating a multilayer with low roughness to suppress reflectivity loss and off-specular scattering, and by implementing a well-controlled thickness to compensate for the distribution of the incident angle. These ideas will be considered in future studies as the purpose of the current study is to propose a conceptual design for a novel polychromator.

3 Possible Neutron Scattering Instruments Using the Polychromator

3.1 Neutron Reflectometer

The concept of a polychromator was discussed in the previous section. Next, the possible designs for neutron scattering instruments are presented.

The design of a neutron reflectometer with a polychromator as a continuous source, which is one of the simplest designs, is considered. The principle of the instrument is extremely simple: neutrons collimated in the direction of reflection are introduced into a sample surface or interface with a small incident angle, and neutrons reflected at the sample with a reflection angle that is the same as the incident angle are counted using a detector. Because the beam is collimated and its path is finely defined, polychromators are applicable to reflectometers at a relatively low cost. Prism reflectometry has already been realized using a RAINBOWS reflectometer with a continuous source at the ILL, in which neutrons are polychromated in the direction of the reflection angle using a prism [2]. This enables us to perform rapid measurements using intense white neutrons without intensity loss using the TOF method. However, the position sensitivity in the direction of the reflection angle is diminished, which prevents us from correcting the incident angle distribution caused by beam divergence or wavy surfaces and the reflection angle caused by off-specular scattering. Also, the concept of the REFocus reflectometer using a polychromator consisting of an elliptic mirror and multilayer monochromator is proposed. This reflectometer can accept a polychromatic beam with a wide incident angle to measure the reflectivity in a wide momentum transfer range at the same time. On the other hand, by using the wide incident angle, the specular signal is disturbed by the surface waviness and the off-specular signal is buried in the specular signal.

Figure 3 illustrates the design of the neutron reflectometer with the novel polychromator for solving the problems of the reflectometers above. A sample is placed immediately in front of the elliptical mirror and reflected neutrons are polychromated in the width direction, which is insensitive to the reflection angle, not in the direction of the reflection angle. This configuration enables us to independently evaluate the wavelength of neutrons by their position on the x-axis on the detector coordinate and the sum of the incident and reflection angles of neutrons by their position on the y-axis. In other words, the degree of freedom in the x-axis position is utilized to determine the wavelength. This causes intensity loss as compared with when the RAINBOWS geometry is used, in which a homogeneous white beam propagates in the x-axis direction. However, the new geometry of the polychromator offers a significant advantage, i.e., it allows more information concerning the incident and reflection angles to be obtained. This is beneficial to the application of reflectometry to various samples and conditions for correcting the incident and reflection angles, as mentioned previously. For example, an incident beam with a large beam divergence can partially compensate for intensity loss.

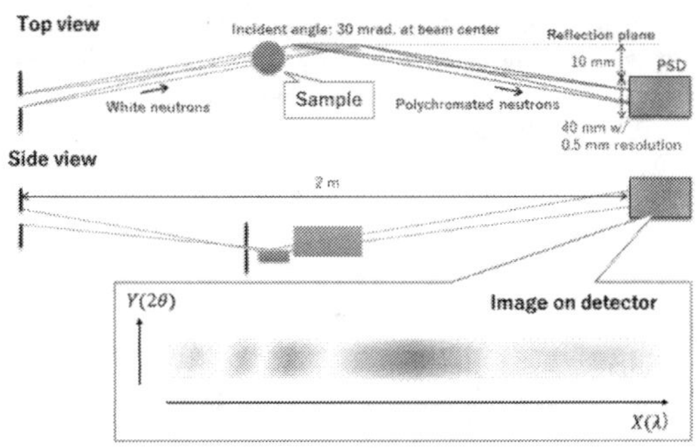

Fig. 3. Design of neutron reflectometer utilizing the novel polychromator proposed herein.

Next, the performance of the reflectometer with a polychromator is estimated. The wavelength separated by the polychromator is expressed as

$$\lambda_m = \frac{2d\sin\theta_{in}}{m} \sim \frac{2dx}{mL},$$

where x is the distance from the detection position to the reflection plane at the polychromator center, L is the flight distance from the polychromator to the detector, and the subscript m of λ represents the order number of the Bragg peak. The image on the detector is blurred because of the spatial resolution of the detector and the slope error of the focusing mirror; here, the detector resolution is dominant if the focusing mirror developed by the author and collaborators is adopted [6,7]. Therefore, the wavelength resolution, $\Delta\lambda$, of the detector can be described as

$$\frac{\Delta\lambda}{\lambda} \sim \frac{\Delta x}{x},$$

where Δx is the spatial resolution of the detector. To simplify the calculation, the wavelength resolution arising from the width of the Bragg peaks is disregarded.

Figure 4 shows the wavelength and its resolution based on the detection position under the assumption that the parameters of the optics and detector in Figs. 2 and 3 are used, and neutrons with a wavelength of less than 0.4 nm are cut using the critical angle of the total reflection of a mirror, since the use of wavelengths below the Maxwellian peak of a neutron source is inefficient for neutron reflectometers. The primary peak includes the wavelength from 0.4 to 2.0 nm (solid red line) with a resolution of 2.5% to 0.5% (dashed black line), which are typical values for a neutron reflectometer. However, higher-order peaks emerged as the incident angle to the mirror increased, where the wavelength depends on the order of the Bragg peak and the detection position (solid lines), whereas the resolution is independent of the order (dashed line). A simple method to remove the contamination arising from the higher-order peaks is by using hardware, such as a mirror that filters short wavelengths based on the total reflection. Another approach is to subtract the higher-order peaks as follows:

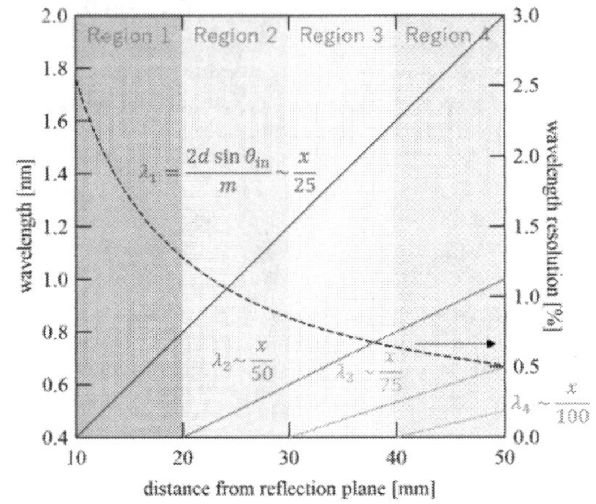

Fig. 4. Wavelength (solid lines, left axis) and resolution (dashed line, right axis) obtained by the novel polychromator for neutron reflectometry, in which the wavelength depends on the order of the Bragg peak, whereas the resolution is independent of the order.

As mentioned previously, only the primary peak was observed at a low incident angle to the polychromator (Region 1). As the incident angle increased, a second peak appeared in Region 2. The intensity of the second peak, $I(\lambda_2)$, can be evaluated using that of the primary peak in Region 1 if the normalizing factor is known. Similarly, the intensities of the third and fourth peaks in Regions 3 and 4, respectively, can be subtracted. Hence, the signal at x, $I(x)$, can be expressed as

$$I(x) = I(\lambda_1) + \sum_{m=2}^{4} a_m I(\lambda_m/m),$$

where a_m is the normalizing factor of the m-th peak, and the second term can be subtracted to evaluate $I(\lambda_1)$.

Finally, the intensity gain of the polychromator optics is evaluated and compared with that of the TOF method. The duty cycle of the TOF method is related to the wavelength resolution at the shortest wavelength as follows:

$$Duty\ cycle = \frac{\Delta t}{t_{max} - t_{min}} = \frac{\Delta\lambda}{\lambda_{max} - \lambda_{min}},$$

where t is the TOF; and the subscripts "max" and "min" represent the maximum and minimum wavelengths, respectively. Therefore, the duty cycle providing the same wavelength band and resolution as that of the polychromator is 0.625%. To compare the beam intensities, the product of the beam width, divergence, and duty cycle is evaluated. If the neutron density in phase space is uniform, then the product is proportional to the neutron intensity.

Table 1 lists the parameters for intensity comparison. To compare the flux at the same wavelength and resolution, a pixel of 0.5 mm for the shortest wavelength at x = 10 on the detector was extracted for the polychromator optics, which is the same as the spatial resolution. Because the distance between the mirror and detector is 1 m, the incident angle to the mirror is 10 mrad at x = 10. If the length of the mirror is 300 mm, then it functions as a slit of width 3 mm, and the divergence from the pixel to the mirror is 3 mrad. To maintain the same optics as the polychromator optics, a slit of width 3 mm is placed instead of a mirror at the same position. The divergence from one slit to the other at the virtual source of the mirror is 40 mrad. The products of the parameters for each optics indicate that the polychromator optics possess a gain factor of 2, although this is only an estimation.

Table 1. Comparison of parameters related to intensity estimation. Products of parameters indicate that the polychromator optics possess a gain factor of 2.

	Duty cycle	Beam width	Divergence
Polychromator	100%	0.5 mm *on detector*	3 mrad. *to mirror with 300 mm length*
TOF	0.625%	3 mm *on a slit*	40 mrad. *to slit at virtual source*

3.2 Quasi-Elastic Neutron Scattering (QENS)

In the case of the reflectometer shown above, a polychromator is used to capture a spectrum of white beams under the assumption that no energy transfer occurs between the sample and neutrons. On the other hand, the dynamics of the sample can be obtained if the energy of the incident neutron is known. Here, an idea for a QENS instrument is presented as an example of neutron spectroscopy.

Figure 5 shows a schematic illustration of the QENS spectrometer utilizing the novel polychromator. The design is based on an inverse geometry spectrometer with a pulsed neutron source, such as the IRIS [10] and OSIRIS [11] at the ISIS, BASIS [12] at the SNS, and DNA [13] at the J-PARC MLF. Whereas a monochromator crystal is used as an analyzer in the conventional spectrometers described above, a novel polychromator is used in this study. For polychromating scattered neutrons in a sample, one focal point of the mirror is placed on the sample, and the other is placed on the PSD. In the conventional crystal, only the dynamics around the energy satisfying the Bragg condition of the crystal can be observed. Although the use of the polychromatic beam by a monochromator crystal are proposed [14,15], the energy band is very limited (less than ±10% from the center value). On the other hand, a wide energy band can be utilized in the polychromator, as will be discussed later.

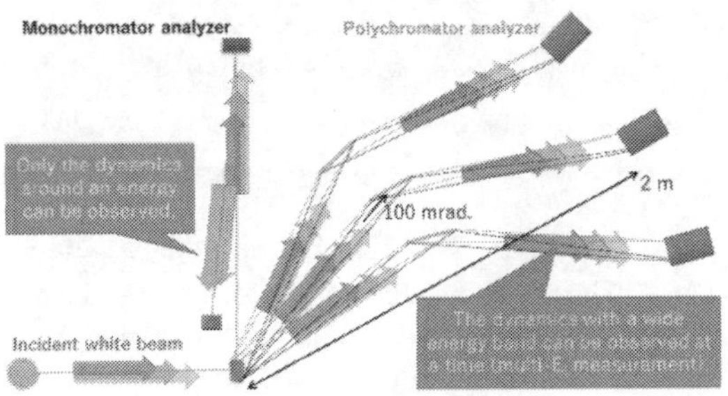

Fig. 5. Design of QENS spectrometer utilizing the novel polychromator based on an inverse-geometry spectrometer with a pulsed neutron source.

Figure 6 (a) shows the wavelength and resolution of the polychromator. The design of the polychromator differs from that of the neutron reflectometer, in that the angle of incidence that for reflectometry is 30 mrad at the center, whereas that of the QENS is 100 mrad. Combining the TOF enabled us to distinguish the higher-order reflections. Consequently, the wavelength from 0.4 to 2.0 nm was available without any gap by utilizing 10 Bragg peaks (shown as solid colored lines), in which the resolution was slightly higher than 0.3% at $x = 80$ and approximately 0.2% at $x = 120$ (shown as dashed line). Because the wavelength resolution arising from the pulse width was not considered, the value obtained was worse than those yielded by high-resolution instruments such as BASIS and DNA, i.e., 0.08% at $\lambda = 0.63$ nm based on Si(111) reflection. However, the important specification of the QENS spectrometer is not the wavelength resolution but the energy resolution, which limits the detection of slight changes in the width of a dynamic structure factor, which is described via a Lorentzian as a function of energy transfer for a relaxation mode. Because neutrons with longer wavelengths (i.e., lower energy) are available in the polychromator, a better energy resolution can be realized, even with the worse wavelength resolution.

To quantitatively evaluate the specifications of the QENS instrument, the relationship between the wavelength and energy of neutrons is used to estimate the energy resolution as follows:

$$E = \frac{h^2}{2m_n\lambda^2},$$

$$\frac{\Delta E}{E} = 2\frac{\Delta\lambda}{\lambda},$$

where m_n is the mass of a neutron, h is Planck's constant, and ΔE is the energy resolution. In addition, the energy window E_{window}, which limits the detection of significant changes in the width of a dynamic structure factor, are evaluated using as follows:

$$E_{window}(\lambda_m) = \frac{E(\lambda_{m+1}) - E(\lambda_m)}{2} - \frac{E(\lambda_{m-1}) - E(\lambda_m)}{2} = \frac{E(\lambda_{m+1}) - E(\lambda_{m-1})}{2}.$$

Figure 6 (b) shows the energy resolution (shown as solid colored lines) and window (shown as dashed colored lines) as a function of wavelength. The energy resolution is proportional to λ^{-2} or λ^{-3} when λ is scanned with m or x. The value of approximately 25 μeV at 0.4 nm decreased as λ increased to 1 μeV at 2.0 nm. The result shows that the energy resolution of 1 μeV is comparable or less than those yielded by BASIS and DNA. Notably, the resolution estimated in this study was based only on the detector resolution; in other words, the effects of the Bragg peak width of the multilayer and pulse width, which affect the TOF resolution, were not considered. However, the results demonstrated the potential of the polychromator for high-resolution QENS. Similar to the energy resolution, the energy window is proportional to λ or λ^{-2} when λ is scanned with m or x, and the value of approximately

1000 μeV at 0.4 nm decreased as λ increased to 200 μeV at 2.0 nm. The former value is comparable to that of the wide-window mode of DNA, and the latter value is comparable to that of the wide-window mode of BASIS.

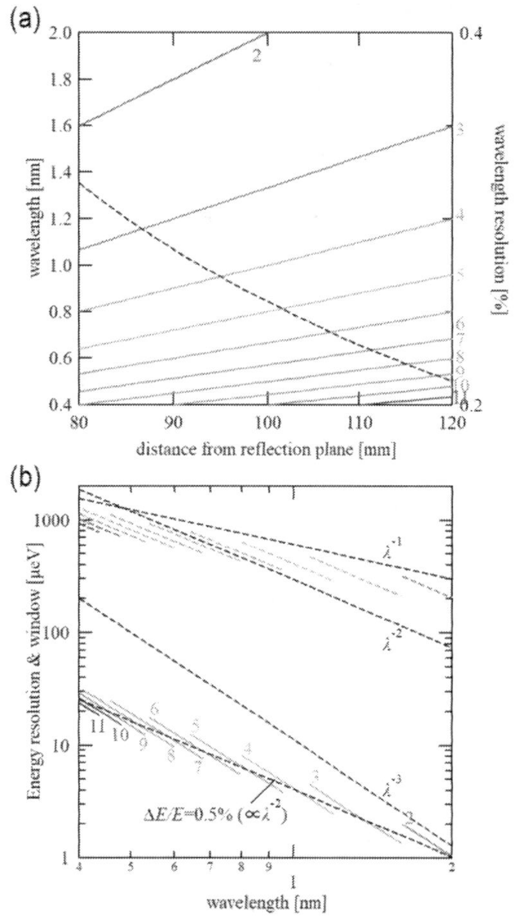

Fig. 6. (a) Wavelength (solid lines, left axis) and resolution (dashed line, right axis) yielded by the novel polychromator for QENS; (b) energy resolution (solid colored lines) and window (dashed colored lines) achieved by the polychromator design.

Therefore, we can conclude that the QENS spectrometer using the polychromator enables us to consider a wide energy range of white cold neutrons from 0.4 to 2 nm. Energy transfer with different energy resolutions can be measured simultaneously using different orders of reflection. This may facilitate peak width analysis for evaluating the width of energy transfer by considering the effect of resolution. Regarding the solid angle, a gap of 36 mm was required at the entrance of the analyzer mirror for a mirror length of 300 mm, where the maximum incident angle was 120 mrad. This resulted in a maximum coverage of the solid angle of 95% (36/38) under the assumption that analyzer mirrors with a thickness of 2 mm were placed around the sample, such as in the case of a radial collimator.

4 Conclusion

In this study, a novel polychromator comprising multiple layers on an elliptical mirror and a high-resolution PSD was proposed. One of the focal points of the mirror was placed on a virtual source, and the other was placed on the PSD. The spectrum of the white beam was observed on the detector as a function of the detection position owing to changes in the incident and reflection angles of the neutrons on the mirror. As the wavelength resolution of the polychromator is governed by the width of the Bragg peak, a "crystallic multilayer" with a large stacking number composed of materials providing less reflectivity in one layer unit was proposed to increase the crystallinity for sharp Bragg peaks. For the application of the polychromator, a neutron reflectometer with a continuous source and a QENS spectrometer with a pulsed source were designed, and their performances were estimated. The neutron reflectometer enabled us to measure the reflectivity over a wide momentum transfer range using a wide wavelength band, as in the TOF method. The QENS spectrometer encompassed an extremely wide energy range, i.e., from micro-electron-volts by extremely cold neutrons of λ = 2.0 nm, to meV by cold neutrons of λ = 0.4 nm. For the latter case, the proposed method demonstrated significant potential for improving the data quality via the application of higher-order reflections at the polychromator mirror. This is because the dynamic structure factor in an extremely wide energy transfer and momentum transfer space can be measured under different energy resolutions simultaneously. However, this is merely an estimation, and further research and development are required to demonstrate the actual performance of the polychromator in the future.

1. T. Matsushita *et al.*, Appl. Phys. Lett. **92**, 024103 (2008)
2. R. Cubitt *et al.*, J. Appl. Cryst. **51**, 257 (2018)
3. F. Ott and A. Menelle, Nucl. Instrum. Methods Phys. Res. Sect. A **586**, 23 (2008)
4. S. Takeda *et al.*, Opt. Express **24**, 12478 (2016)
5. T. Hosobata *et al.*, Opt. Express **25**, 20012 (2017)
6. T. Hosobata *et al.*, Opt. Express **27**, 26807 (2019)
7. F. Nemoto *et al.*, Nucl. Instrum. Methods Phys. Res. Sect. A **1040**, 166988 (2022)
8. A. Nelson, J. Appl. Cryst. **39**, 273 (2006)
9. L. G. Parratt, Phys. Rev. **95**, 359 (1954)
10. D. M. y Marero *et al.*, Physica B **150**, 276 (2000)
11. C. J. Carlile and M. A. Adams, Physica B **182**, 431 (1992)
12. E. Mamontov *et al.*, Neutron News, **19**, 22 (2008)
13. K. Shibata *et al.*, JPS Conference Proceedings **8**, 36022 (2015)
14. J. O. Birk, Rev. Sci. Instrum. **85**, 113908 (2014)
15. R. Bewley, Rev. Sci. Instrum. **90**, 075106 (2019)

Development of a large-area curved Trench-MWPC ^{3}He detector for the D16 neutron diffractometer at the ILL

Jean-Claude Buffet[1], *Viviana* Cristiglio[1], *Sylvain* Cuccaro[1], *Bruno* Demé[1], *Bruno* Guérard[1], *Julien* Marchal[1,*], *Jérôme* Pentenero[1], *Nicolas* Sartor[1], and *Jules* Turi[1]

[1]Institut Laue Langevin, 71 Avenue des Martyrs CS 20156, 38042 Grenoble Cedex 9, France

Abstract. The D16 instrument is a versatile cold-neutron diffractometer at the ILL. It has benefited from a number of upgrades over the years, such as the installation of a large-area ^{3}He Multi-Wire Proportional Chamber (MWPC) in 2011. This detector has provided a resolution of 1 mm x 1 mm over an area of 32 cm x 32 cm. After 12 years of operation, it was replaced by a new curved detector which covers a wider solid angle while maintaining a high angular resolution. Its 86° horizontal angular coverage makes it possible to perform time-resolved experiments with a large q-range. This new detector is based on the Trench-MWPC detector technology developed at the ILL. In the D16 Trench-MWPC, 6 modules are mounted side by side in an ^{3}He-filled curved vessel. Each module consists of 192 cathode blades positioned every 2 mm, and 192 anode wires spaced by 1.5 mm. The radius of curvature of the cathode blades is 1150 mm, providing a parallax-free resolution of 0.075°, horizontally along the 86° angular coverage of the 38 cm high detector. The various steps of the fabrication and mechanical inspection of the D16 Trench-MWPC detector modules and pressure vessel are presented, as well as experimental results obtained during the characterisation of the detector with neutrons.

1 Introduction

The D16 instrument is a versatile cold-neutron diffractometer which has evolved from the first biological membrane diffractometers from the early seventies, and has benefited from a number of upgrades over the years, such as the installation of MILAND (MIllimeter resolution Large Area Neutron Detector [1]), a large-area ^{3}He Multi-Wire Proportional Chamber (MWPC) in 2011. This detector provided a resolution of 1 x 1 mm^2 over an area of 32 x 32 cm^2. After 12 years of operation, the decision was made to replace it by a curved detector covering a wider solid angle while maintaining a high angular resolution.

Several neutron research facilities host one or more neutron diffraction instruments with a large high-resolution 2-dimensional curved ^{3}He-filled detector. These gaseous detectors are based on the well-known MWPC design [2]. A detector with 120° angular coverage was developed in 2002 at BNL by Fried et al. [3] and two examplars of this detector design were fabricated. The first BNL-120 detector was installed at the protein crystallography beam-line at the LANSCE spallation source of the Los Alamos National Laboratory [4] and was moved in 2018 to the Wide Angle Neutron Diffractometer (WAND^2) at the High Flux Isotope Reactor, Oak Ridge National Laboratory [5]. Another BNL-120 is operating in the Wombat High-Intensity Powder Diffractometer at ANSTO's Opal Reactor [6]. A large curved detector with 110° angular coverage was also fabricated by Moon et al. for single-crystal neutron diffraction studies at the research reactor HANARO [7]. During a collaboration between FRM II and PSI detector groups, the BNL detector design inspired the development of two detectors : one in operation on the DMC cold neutron diffractometer at PSI and another one under commissioning for the ERWIN new powder diffractometer at FRM II. The main difference between the FRM/PSI design and the BNL design concerns the read-out electronics. The BNL detector uses the charge-division technique whereas FRM/PSI have implemented individual channel read-out to enhance the counting rate of their detector. At the ILL, a large curved detector which covers 120° with a resolution of 0.19° horizontally has been in operation on the D19 diffractometer since 2006 [8]. More recently, a new large area-curved detector with 120° angular coverage and 0.15° resolution horizontally, similar to the D19 detector in dimensions, but with a higher count-rate capability, was developed at the ILL, and is now under commissioning on the XtremeD instrument [9]. The BNL and the FRM/PSI detectors are operated with a permanent gas cleaning system connected to the pressure vessel. At the ILL, both D19 and XtremeD detection modules have been fabricated with low-outgasing materials which allow the detectors to be operated without any permanent gas cleaning system. However, the detection modules in the XtremeD detector are based on a new type of MWPC design: the Trench-MWPC recently introduced by the ILL [10].

The D16 detector is also based on the Trench-MWPC detector technology. This technology is described in more detail in the next section. The design and the fabrication of the D16 Trench-MWPC detector are also presented, as

*e-mail: marchalj@ill.fr

well as detector characterisation results obtained with an AmBe neutron source. Finally, some information regarding the installation of the detector on the D16 instrument, and data obtained with a neutron beam are provided.

2 Trench-MWPC concept

The Trench-MWPC electrode structure used in the new large-area curved D16 detector is illustrated in figure 1. This concept was initially proposed for XtremeD detector at the ILL [9] and the advantages of this approach were demonstrated on a prototype detector which is described in reference [10].

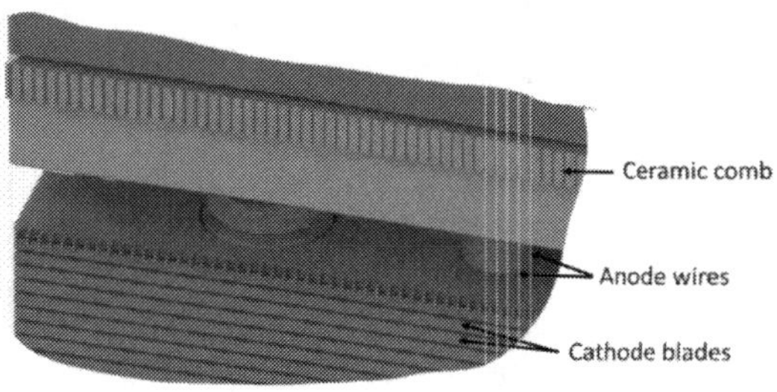

Figure 1. Schematics of the Trench-MWPC electrode design used in D16 detector. In this 3D view, the row of teeth along cathode electrode blades are shown only on the first blade. Anode wires are stretched along the trenches formed by the cathode blades.

In the Trench-MWPC design, cathode electrodes (providing the Y coordinate of neutron interactions in the detector) take the shape of a stack of curved aluminium blades. A row of teeth is present along the edge of each blade so that this stack of electrically isolated toothed blades forms trenches along which anode wires (providing the X coordinate of neutron interactions) are stretched. This Trench-MWPC configuration is used as the electron avalanche multiplication electrode structure inside the ^{3}He-filled pressure vessel. In the Trench-MWPC there is no need for a cathode electrode wire plane on top of the anode wires (used in standard MWPCs to counterbalance the electrostatic attraction of the anode wires by the bottom cathode electrode plane). In the Trench-MWPC, the electrostatic stability of the wires is provided by the cathode blade teeth surrounding the anode wires and produces a uniform gain multiplication factor even along 40 cm long wires. The absence of a cathode electrode wire on top of the anode wires leads to simple maintenance and simple assembly procedures. Moreover, neutron detection events produce electric charges which are confined within the trenches and induce pulses on a limited number of electrodes (only 1 or 2 anode wires and 1 or 2 cathode blades) increasing the signal-to-noise ratio per channel, which allows for an operation with low electron multiplication gain. In addition to the low-gain, the presence of cathode trench side-walls surrounding anode wires improves the collection of slow ions, produced during the gas avalanche process, and also contributes to a reduction of space-charge effects which usually limit the maximum local count rate of MWPCs. Another advantage is the low high-voltage difference required between anodes and cathodes (< 1000 V).

3 Detector design

The mechanical design of the D16 detector is adapted from the XtremeD detector design. The main difference between the two detectors lies in their geometrical characteristics. The height of the active area of the D16 detector is 38 cm whereas it is 32 cm for XtremeD. The horizontal angular resolution is twice as high for the D16 detector (0.075° instead of 0.15°), however the angular coverage is higher for XtremeD detector (120°) than for the D16 detector (86°).

Figure 2. 3D view of XtremeD detector design showing the pressure vessel and 3 of the 6 Trench-MWPC detection modules.

The high angular resolution of the D16 detector is provided by the longer radius of curvature of the cathode blades (115 cm for D16 compared to 76 cm for XtremeD) and a smaller space between anode wires (1.5 mm for D16 compared to 2 mm for XtremeD). The smaller spacing and higher length of D16 anode wires make mechanical precision requirements in cathode blades and anode wires positioning more stringent than they are for the XtremeD detector; the specifications of D16 detector are listed in Table 1.

Angular coverage	Horizontally : 86° Vertically : 19° (38 cm high)
Angular resolution	Horizontally : 0.075° Vertically : 0.1°
Radius of curvature	115 cm
Number of pixels	221 184
Conversion depth	26 mm
Detection efficiency	67% at 4.5 Å
Local count rate	50 kHz per read-out channel
Global count rate	1 MHz

Table 1. D16 Trench-MWPC detector specifications.

The detector consists of a curved pressure vessel machined from aluminium inside of which 6 detection modules are mounted side by side, as illustrated in figure 2. Each detection module is composed of 192 cathode blades, separated by ceramic spacers. The 29 cm long cathode blades, shown in figure 6, are curved and a series of 192 teeth have been cut-out along them, spaced out by 1.5 mm. In each module, 192 gold-plated tungsten-rhenium anode wires are stretched perpendicularly to the cathode blades along the trenches formed by the teeth of the cathode blades. From a mechanical point of view, the overall detector is therefore composed of 1152 anode wires and 1152 cathode blades. Cathode blades located at the same height in two adjacent modules are electrically connected together in order to reduce the number of signals to read-out via the high-voltage feedthroughs at the back of the detector vessel. Therefore, from a signal read-out point of view, the detector is composed of 3 double-modules, each with 192 cathode blades and 384 anode wires, resulting in a total of 1728 channels to read-out.

The read-out of anode and cathode electrode signals is performed by the front-end electronics, located in an enclosure mounted to the back of the detector vessel. High-voltage signals are AC-coupled to pre-amplifier boards via 54 decoupling boards, each with 32 channels. The pre-amplifier boards are developed by the ILL and are common to all detectors requiring an individual electrode signal read-out. The boards are based on a FET pre-amplifier with low-noise and low-power consumption characteristics, followed by two filtering stages, a gain stage and a pulse amplitude discrimination circuit. Analogue pulses with amplitude above a selectable global threshold produce logic pulses, which are sent to 8 NIM-format FPGA-based pulse processing electronics modules. These FPGA-boards are the main parts of the back-end read-out electronics set-up, together with a VME-based acquisition module receiving processed data via an optical fibre link.

The signal processing performed by FPGA boards is identical to the one described for XtremeD in reference [9]: a one-dimensional hit cluster analysis performed separately on x-coordinate anode wires and y-coordinate cathode blades followed by xy coincidence detection. The xy coincidence window implemented in D16 detector signal processing is 250 ns long and applies to each double-module independently. The specific processing to avoid count loss in regions located at the junction between modules, described in the XtremeD detector paper [9], is also implemented in the D16 detector. However, in the D16 detector, this processing concerns only the junction between double-modules since adjacent cathode blades are connected together and read-out as a single electrode in double-modules (as explained in the previous section).

4 Detector fabrication

The D16 detector required precision machining and alignment techniques for the fabrication of the Trench-MWPC detection modules as well as heavy-duty machining of 2.5 m diameter forged aluminium rings, used to fabricate the pressure vessel parts. The D16 detector is shown in figure 3 before the closing of the two parts of the pressure vessel: the vessel flange with its 6 detection modules mounted side-by-side and the vessel itself with a large metallised polyimide film drift-electrode stretched across the internal surface of the vessel window. This large electrode is biased negatively in order to generate an electrical field sufficient to drive the electrons created after neutron interactions in the gas towards the amplification region of the Trench-MWPC. This amplification region consists of the positively biased anode wires stretched across the electrically grounded cathode blades.

Figure 3. Photograph of the D16 detector before the closing of the pressure vessel, showing the detector modules mounted side by side on the vessel flange.

Figure 4. D16 detector vessel during pressure testing. Position sensors are used to control the elastic deformation of the vessel front window.

Guard electrodes are mounted on each side of the detector. These electrodes consist in polyimide circuits with copper tracks connected by resistors and are used to shape the electrical field between the negative voltage of the drift electrode and the 0V potential of the cathode blades. In addition, aluminium plates, polarised at a specific high-voltage, are mounted on the top and bottom of each detection module, to serve the same purpose of avoiding image distortion caused by a non-uniform drift field.

The detector vessel is filled with 2.4 bar of ^{3}He to provide a detection efficiency of 67% at 4.5Å and 4.6 bar of

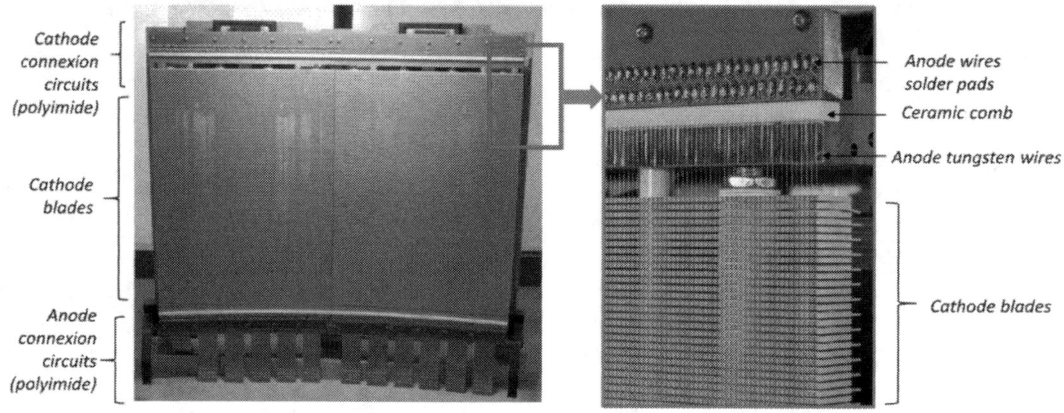

Figure 5. Photograph of a double-module of D16 Trench-MWPC detector (left) with a close-up view of the anode wires stretched along the cathode toothed blades (right).

$ArCO_2$ as stopping gas to provide a spatial resolution of 1.5 mm corresponding to the wire anode pitch.

4.1 Detector vessel

The pressure vessel assembly of the D16 detector, shown in figure 4, was machined in aluminium alloy EN AW-5083 forged rings. The vessel was designed to operate at a maximum internal pressure of 8 bar (absolute pressure). The entrance window of the vessel has a thickness of 10.2 mm and the deformation at the centre of the window is 1.4 mm under a differential pressure of 7 bar. The gas tightness between the curved flange and the curved vessel itself is provided by a double O-ring gasket in FPM Fluoroelastomer material, and 54 custom-designed multi-pin HV feedthroughs with a knife edge on a stainless steel body are directly bolted onto the vessel flange.

Various inspections of the raw material and a pressure test of the final vessel were performed to ensure that the vessel complies with the A2 module of the European directive on pressure equipment, and it has received CE marking. The main mechanical characteristics of the pressure vessel are listed in Table 2

Angular opening	98.5°
Height	735 mm
Internal radius	1086 mm
External radius	1259 mm
Weight	420 kg
Window thickness	10.2 mm
Internal volume	90 litres

Table 2. D16 Trench-MWPC detector specifications.

4.2 Detection modules

The detection modules are assembled by pairs as shown in figure 5(a) because the cathodes located at the same height of the two adjacent modules have to be connected together electrically at the back of the modules in order to reduce the number of read-out channels. With this configuration, the dead-time introduced by the xy coincidence detection window of 250 ns applies in parallel to 3 double-modules rather than 6 individual modules.

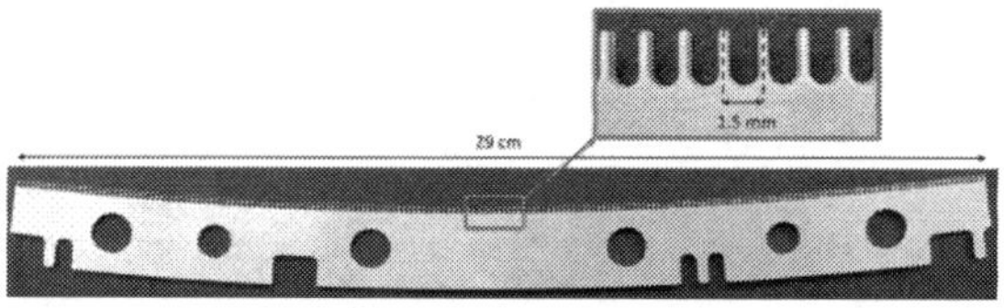

Figure 6. Photograph of an aluminium cathode blade used for D16 detector with a close-up view on the teeth machined along the blade.

A photograph of one of the 1152 aluminium cathode blades of the detector is shown in figure 6. Each blade is 29 cm long with an opening angle of 14.4°. The close-up view in figure 6 shows the row of 0.4 mm wide teeth machined by spark erosion along the blade. The mechanical frame holding the stack of 192 blades in each module was manufactured at the ILL workshop since the mechanical precision of these parts is crucial so as to avoid any deformation of the overall assembly which would otherwise lead to an error in anode centring along the trenches. The precision in the vertical positioning of each blade is obtained by using corrective spacers with a thickness of 0.05 mm or 0.02 mm above and below the nominal thickness of the 1 mm-thick ceramic spacers used to isolate the cathode blades electrically from each other. The maximum error in the vertical positioning of the cathode blades is of 0.4 mm across the 6 modules of the detector.

The exact positioning of anode wires inside the cathode trenches is of great importance in the Trench-MWPC design. Error in anode wire centring leads to a non uniform electron multiplication gain since the deformation of the wire due to an unbalanced electrical field produces variations in the anode wire/cathode tooth distance across the detector. Precision in anode centring is achieved by MACOR combs used to position each wire along a module as shown in figure 5. The spark-erosion technique used to produce the cathode blades also contributes to the overall uniformity of anode wire/cathode tooth distances. The maximum error in wire centring inside the 1.1 mm-wide trench is 50 microns across the whole module. However, on two modules, a manufacturing error on the MACOR combs resulted in a lateral shift of the wires by 70 microns between the two halves of each module. This systematic error produces a difference in amplification gain of only 10% between the two half-modules and is not visible on 2D neutron images.

5 Detector test with an AmBe neutron

The detector was tested in the lab with an AmBe neutron source in order to determine the optimum operation voltages. At a drift voltage $V_d = -750\ V$, the anode voltage was increased while recording the neutron count-rate over the detector to produce the counting curves shown in figure 7.

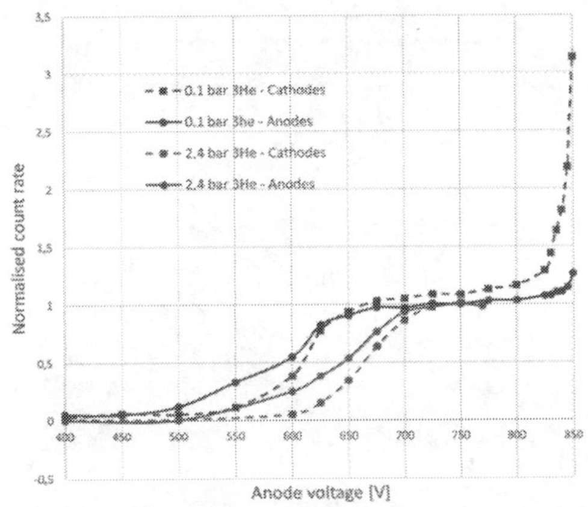

Figure 7. Counting plateau recorded over the whole detector at V_d=-750 V for two different pressures of ^{3}He (100 mbar and 2.4 bar) and a pressure of $ArCO_2$ of 4.6 bar.

These counting plateaux were obtained by recording the number of pulses above the discriminator thresholds as a function of anode wire high-voltage. The discriminator thresholds were adjusted above the electronics noise : at -350 mV for the anode channels and at a slightly higher value of 400 mV for the cathode channels due to the higher capacitance of the cathodes blades compared to the anode wires.

These curves can be used to identify the anode voltage corresponding to the centre of the neutron counting plateau which is 775 V for an ^{3}He pressure of 2.4 bar and an $ArCO_2$ pressure of 4.6 bar corresponding to the standard operating conditions of the D16 detector. This rather low operating voltage for a MWPC is due to the close proximity between anode wires and cathode blades teeth in the Trench-MWPC design: anode wires are located at 1.5 mm from the bottom of the 2.2 mm-deep, 1.1 mm wide cathode trenches.

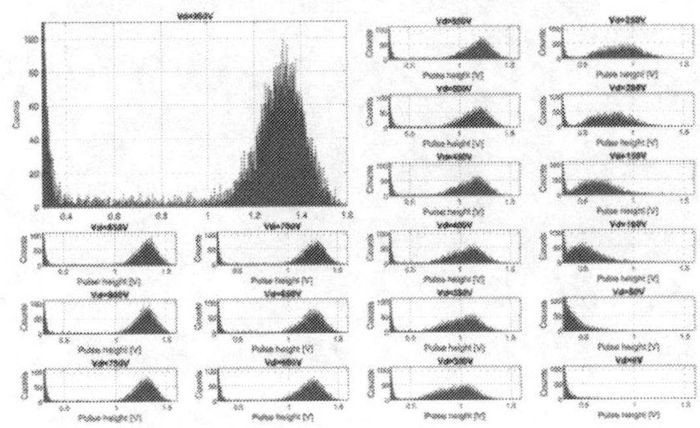

Figure 8. Anode-wire-summed-signal pulse height spectra as a function of HV applied to the drift electrode.

The optimum drift voltage was determined by varying the drift voltage at a constant anode voltage $V_a = 750\ V$. The pulse-Height-Spectra (PHS) recorded on the sum of the analogue output signals from one of the 32-channels pre-amplifier boards are shown in figure 8. These data show that amplification gain remains constant and that charge collection is complete for drift voltages V_D<-500V. For V_D>-500V, the shapes of the PHS indicate that the signal duration is longer than the shaping time of the preamplifiers (pulse FWHM of 600 ns).

The uniformity of the detector response was investigated by using the B_4C mask shown in figure 9(a) mounted on the detector front window. Two-dimensional images obtained by xy coincidence detection as described in the previous section were acquired over several hours in order to obtain sufficient counting statistics with the AmBe source. The raw image of the B_4C mask, the flat-field image (without mask) and the flat-field-corrected mask image are shown in figure 9(b). The presence of regions with 5% sensitivity reduction are visible in the raw data. These regions with lower sensitivity are due to the drift electrode not being perfectly stretched across the internal surface of the detector front window. Some ripples appearing across the large curved drift electrode when a high-voltage is applied to the copper surface create some gaps between the electrode and the entrance window of the vessel. These gaps, filled with the $^3He/ArCO_2$ gas mixture, stop a small fraction of the neutrons without generating the corresponding charge in the drift region, therefore causing sensitivity variation across the detector active area. These sensitivity variations are suppressed by a flat-field correction, as illustrated in figure 9(b) with the AmBe source. On the D16 instrument various samples such as Vanadium can be used to produce the sensitivity map required for flat-field correction.

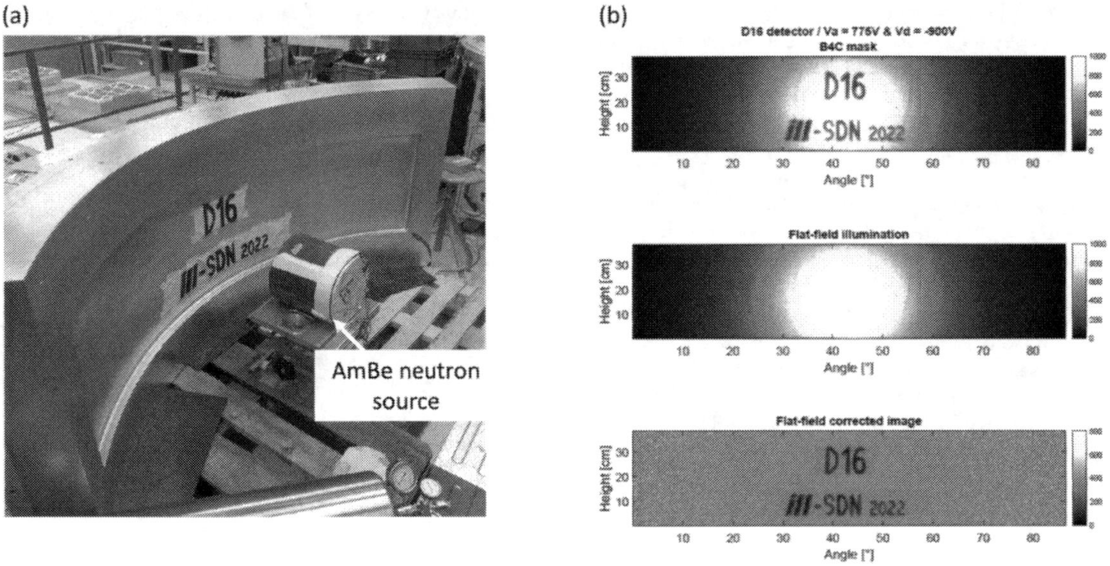

Figure 9. (a) B4C mask (neutron attenuator) mounted on the detector entrance window; (b) Mask image, flood-field image and flood-field corrected image of the mask obtained with an AmBe neutron source.

6 Detector commissioning on the D16 instrument

The previous MILAND detector (32 x 32 cm^2 MWPC detector with a 1x1 mm^2 resolution) was dismounted from the D16 instrument at ILL in Oct 2021 and a new instrument platform was installed on D16 during the ILL long reactor shutdown period in 2022. This platform comprises the following elements:

- A granite table on air-pads: to move the whole platform between two positions which correspond to two different take-off angles of the monochromator (83° for 4.5 Å or 115° for 5.6 Å neutron beam wavelength with a vertically bending HOPG monochromator).
- An helium-filled chamber: to reduce neutron attenuation and scattering between the sample and the detector. Helium circulates constantly in the chamber, maintaining an overpressure of 20 mbar above ambient pressure.
- The large-area curved MWPC-trench detector described in the previous sections.
- A motorised support frame: used to scan both the detector and the helium chamber to cover a wide Q-range of diffracted data corresponding to an angular range between -43° and 133°.
- A motorised beamstop: located in a 1 cm-thin air-gap between the helium chamber aluminium window and the detector front window.

The new platform is still compatible with previous sample environments including controlled temperature and humidity chambers, cryostats, furnaces, magnets, cryomagnets, high pressure cryostats, high pressure cells and an automatically positioned, temperature controlled, horizontal sample changer.

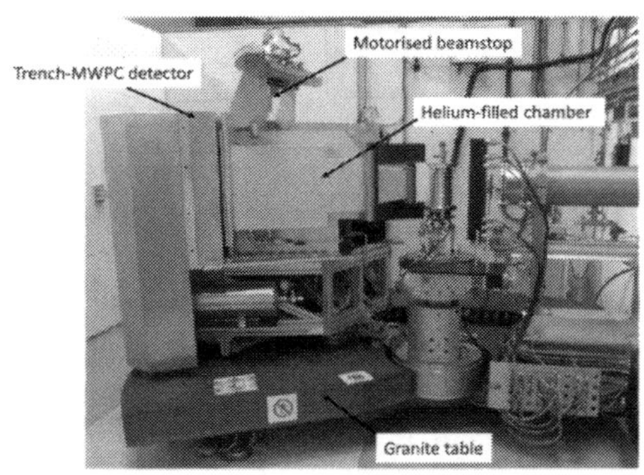

Figure 10. Photograph of the Trench-MWPC detector installed on the new platform of the D16 instrument.

After the restart of the ILL reactor in March 2023, flat-field data and diffraction data from various samples were acquired at a neutron wavelength of 4.5 Å. These results, obtained with the neutron beam, were used to optimise the detector data processing (1D pulse clusters processing and xy coincidence analysis) at a high count rate. Figure 11 shows the diffracted data recorded from a YIG (Yttrium Iron Garnet) crystal powder sample with the new Trench-MWPC detector at 4.5 Å. The Q-range covered by the new detector in one position is 4.5 times higher than the one achievable with MILAND, the previous detector installed on D16, demonstrating the new capabilities of the

 https://doi.org/10.1051/epjconf/202328603010

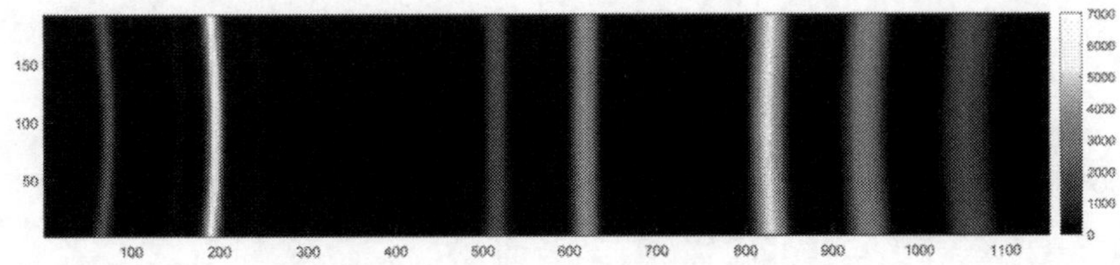

Figure 11. Neutron diffraction image produced by a YIG (Yttrium Iron Garnet) powder sample with the new Trench-MWPC detector during its commissioning on D16.

instrument for dynamic diffraction experiments over wide Q-ranges.

7 Conclusion

A new large-area curved detector is now operational on the D16 cold-neutron diffractometer instrument at ILL. This ^{3}He detector is based on the Trench-MWPC detector technology developed at the ILL. The Trench-MWPC design requires simple maintenance procedures, low operating voltages, low-electron multiplication gain operation and reduces space-charge effects due to the slow drift of ions produced during electron multiplication in the detector. The new curved D16 detector, a 38 cm high detector, offers a continuous horizontal angular coverage of 86° with an angular resolution of 0.075°. The 221 184 pixels of this high-granularity detector produce high-quality neutron diffraction images. The D16 instrument platform was upgraded to take full advantage of this new detector and to allow dynamic neutron diffraction experiments over a wide Q-range.

8 Acknowledgements

The project was funded by the Endurance phase-2 program at the ILL. The authors are thankful to the technicians of the science division at ILL who took an active part in every step of the project: Olivier Aguettaz, Michel Bonnaud and Pascal Mouveau. We would also like to acknowledge the expertise of Luc Didier and Gilles Pastrello of ILL design office during the initial phases of the project, and Adeline Robert, for the Finite-Element-Modelling of various mechanical components of the platform and detector. Finally, the electronics read-out set-up of the detector presented in this paper owes much to the work of Martin Platz, Patrick Van-Esch and Kevin Ollivier at the ILL Instrument Control Service.

References

[1] B. Gebauer and B. Guerard, Neutron News **20**, 1 (2009)

[2] G. Charpak et al., Nucl. Instr. Meth. **62**, 262 (1968)

[3] R.B. Knott et al, Nucl. Instr. Meth. Phys. Res. A **392**, 62-67 (1997)

[4] J. Fried et al., Nucl. Instr. Meth. Phys. Res. A **478**, 415-419 (2002)

[5] M.D. Frontzek et al., Physica. B: Condensed Matter **551**, 464-467 (2018)

[6] A.J. Studer et al., Physica. B **385-386**, 1013-1015 (1997)

[7] M.-K. Moon, Nucl. Instr. Meth. Phys. Res. A **717**, 14-20 (2013)

[8] A.W. Hewat, Physica B **385-386**, 979-984 (2006)

[9] D. Barkats et al., Proc. IEEE NSS-MIC conf. 2021 (2021)

[10] J. C. Buffet et al., JINST **12**, C12009 (2017)

The TAS-IN8 upgrade: towards the limit of a three-axis spectrometer performance

Andrea PIOVANO and *Alexandre* IVANOV*

Institut Max von Laue – Paul Langevin, 71 avenue des Martyrs, CS 20156, 38042 Grenoble Cedex 9, France

Abstract. IN8 is a high-flux three-axis thermal neutron spectrometer designed to measure inelastic neutron scattering on single crystals in a wide energy and momentum transfer range. It is the highest thermal flux spectrometer worldwide with the monochromatic flux at the sample up to 10^9 neutrons/cm^2/s where users can perform demanding experiments not possible anywhere else. In recent years, the IN8 spectrometer has undergone several major upgrades that have further extended the capabilities of the instrument to the limits achievable with the standard setup. The first part of the upgrade was centred on the replacement of the monochromator assembly, now consisting of four double focusing reflecting crystal faces with enhanced luminosity. The ensuing upgrade was to replace the entire secondary spectrometer. As the result, the new spectrometer, called *Thermes,* profits from a compact design and it is well shielded against ambient experimental background. All the incorporated modifications have given the instrument an unprecedented and unique luminosity combined with an optimal signal-to-noise level and a remarkable configuration flexibility for a broad range of experimental requirements.

1 Introduction

Three-axis spectroscopy (TAS) is a well-known and widely used technique for the study of excitations in condensed matter. An introduction to the TAS technique can be found in many textbooks and an ample review on the principles as well as technical aspects is given in Ref.[1].

TAS is the technique of choice whenever it is desirable to explore specific regions in the excitations phase space of a material under study defined by their momentum **Q** and energy E. The relevant quantities for a TAS instrument are luminosity, resolution and signal-to-noise ratio. Since the inelastic neutron scattering signal is intensity limited even at the neutron sources with highest brilliance, a spectrometer ''flexibility'' is an important issue: it means being able to trade in the most practical way resolution against luminosity.

The recent upgrade of the thermal three-axis spectrometer IN8 at the Institut Laue-Langevin (ILL) [2,3] aimed to optimise its experimental capacities. In the following we will summarise the characteristics of the new instrument and highlight its enhanced performance in terms of flexibility and luminosity. We will discuss its scientific impact and present the prospective steps to be taken for further improving the instrument.

2 Primary spectrometer

In order to boost the performance of the primary spectrometer we have designed and implemented a new monochromator assembly (Fig.1 and 2), making use of experience in the engineering of multi-crystal arrays with remotely controlled two-dimensional (horizontal and vertical) variable focusing accumulated over the last decades.

The new device includes four crystal faces that can be alternatively set in Bragg reflection: mosaic pyrolytic graphite and copper crystals with reflecting planes PG002 and Cu200, as well as elastically bent perfect silicon crystals [4,5] with reflecting planes Si111 and Si311 both eliminating the second order diffraction harmonics.
The mosaic crystal faces are composed of 121 individual crystals each, organized in 11 horizontal rows and 11 vertical columns for a total active surface of about 290 x 200 mm^2 (horizontal x vertical). The two silicon crystal faces, instead, are built as a combination of 11 cassettes, piled up in the vertical direction for a total active surface of 230 x 200 mm^2 (horizontal x vertical) and filled in with stacks of thin silicon blades assembled to be bent in the horizontal plane within the elastic bending limits. A system of precise camshafts is designed for each face in order to incline individual crystals of the mosaic crystal faces or the cassettes of the silicon faces and bend the silicon blades in these cassettes. With only two

* Corresponding author: aivanov@ill.fr

motor/encoder pairs per crystal face, we achieve both vertical and horizontal bending, creating the desired approximation of a chosen ideal focusing surface. This functionality enables independent adjustments of the bending radius in each direction.

FIG. 1. General view of the new monochromator assembly with the PG002 crystal face in working position. Unexposed faces are hidden by neutron shielding.

FIG. 2. The copper crystal face during the cycle of changing over between different working faces

Adding the fourth Si311 face permitted enhancement of the instrument performance at low energy transfers and provided additional flexibility for high-resolution experiments [6]. We found a possibility to include the fourth reflecting face through the original design of back-to-back arrangement of the two pairs of the reflecting faces with a smart lifting and face exchange system. The duration of the changing cycle between different faces varies from a minute to a few minutes scale depending on the crystal face to be put in the working position: exchange between silicon and mosaic faces involves an additional vertical translation (Fig.2). The collected experience in face exchange cycles demonstrates high stability and reliability of the exchange system with well reproducible settings after bringing a chosen face to the working position so that no additional alignment is required.

The actual compact structure provided additional space to extend the lateral dimensions of each crystal face by more than 20% enabling a better use of the available white beam divergence. The design of the assembly allows for dismounting of any of the 4 faces (for maintenance *etc.*) and any combination of them so that the remaining subset of the faces can continue to work. Each reflecting face is thoroughly protected with neutron absorbing materials (Cd, sintered boron carbide with ^{10}B enrichment) behind the crystals and around the whole face in order to reduce parasitic scattering from the construction materials, cables, electrical connectors *etc.* and to decrease their residual activation after exposure to the neutron beam. For the same purpose, the exchange between reflecting surfaces in the course of experiment without closing the neutron beam from reactor is fulfilled at a special rotation position of the whole monochromator drum when the entrance window for the white beam is shut with a dedicated local beam-stop.

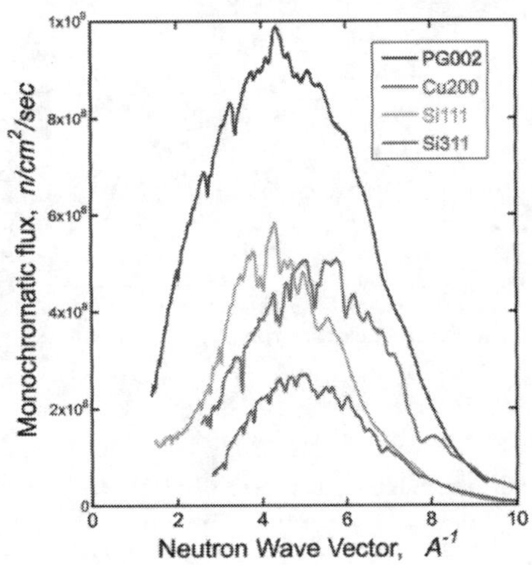

FIG. 3. The flux of monochromatic neutrons at the sample position for all 4 faces of the new IN8 monochromator. The take-off angle on the monochromator can be varied in the range 10-90 degrees.

https://doi.org/10.1051/epjconf/202328603011

The new monochromator outperforms the previously used device with 3 crystal faces [3]. Flux measurements (Fig.3) indicate an increase in the monochromatic flux at the sample position by 40 to 60% depending on the reflecting face and the wavelength range compared to the previous primary spectrometer. Correspondingly, the background conditions have been improved by about 20 to 40% through the careful choice of neutron shielding components. An additional set of Soller slit collimators in white and monochromatic beam with ^{10}B-enriched coating and better stability in the intense radiation field enhances the instrument flexibility in high-resolution configurations. The change of collimators and open tunnel in white beam is controlled remotely. The monochromator performance has been confirmed after the recent 'in-pile' beam-tube exchange, now with a slightly reduced diameter of its extremity nose in the heavy water moderator of the ILL high-flux reactor.

3 Secondary spectrometer

With the aim to keep the IN8 to the highest standards of efficiency and as the reference instrument among high-flux thermal TAS instruments, we have engaged in another project of developing a new single-detector secondary spectrometer called *ThermES* (as abbreviated from Thermal Excitation Spectra, Fig.4).

Fig. 4. The brand-new *Thermes* spectrometer in the IN8 zone during the first working cycle, here at an experiment with a vertical field superconducting cryomagnet.

Thermes has a compact design and optimized neutron shielding to minimise all parasitic background not coming from the sample or passing away from the principal optical path. While most of the background in an IN8 experiment is typically stemming from the sample itself and its nearest environment (cryostat *etc.*), the signal-to-noise ratio could be anyway improved due to thoroughly arranged neutron shielding environment. An additional gain in high-resolution setup is brought in by the new enriched boron coated slit collimators and by thorough design of their neutron shielding. This permitted to eliminate major sources of background leaking into the zone around the analyser crystals.

The analysers in use (Fig.5) are presented by the double-focussing devices built from mosaic crystals of pyrolytic graphite (reflection plane PG002) and copper (reflection plane Cu200) and also elastically bent perfect silicon crystals (reflection plane Si111). The mechanical design and implementation of neutron beam focussing in the analysers is similar to that of the corresponding monochromator faces. The size of active reflection area on the analysers is about 180 x 140 mm^2 (horizontal x vertical). The smaller dimensions of the analysers respect the closer distance from the analysers to the sample than from the sample to the monochromator so that the resolution contributions from both energy selective units match each other. Changing over between different analysers is done manually, removing one and installing the other one at place. Accuracy in the manual re-positioning is sufficiently high with no change in translation, tilt and focussing alignment.

Fig. 5. Double-focusing analysers of the *Thermes*: pyrolytic graphite (PG002, left), copper (Cu200, centre) and elastically bent silicon (Si111, right) crystals.

The classical single-detector configuration of *Thermes* is enriched with a diaphragm positioned next to the Helium-3 counter. This diaphragm permits variation of the sensitive area of the neutron counter and helps optimising the signal-to-noise ratio in the experiments with diverse sample sizes.

The new spectrometer proposes a larger dynamical range with respect to the previous setup, especially towards high values of momentum transfer or high scattering angles, now possible up to at least 120 degrees in both scattering directions. This is an advantage for lattice dynamics studies, now again highly demanded at IN8-*Thermes* along with research in magnetism.

The new compact design opens route for the use of the highest available magnetic fields previously restricted due to insufficient space required to reduce attraction power from stray fields resulting in considerable forces acting on the superconducting magnet.

The secondary spectrometer is equipped with two oriented pyrolytic graphite filters each of about 5 cm in thickness. As a rule, one is positioned in the scattering beam in order to suppress high-order harmonics scattering from the monochromator and/or analyser crystals that can create a substantial parasitic signal, in particular in combination with mosaic crystal faces. The second filter can be installed when required in order to enhance the filtering power. The particular combination of silicon monochromator together with silicon analyser creates a specifically "clean" set-up where the filters can be removed. This compensates inevitable reduction in total reflected intensity and offers comfortable freedom in the chosen neutron energy that can be different from discrete values imposed by the filter.

Due to its modular structure the *Thermes* secondary spectrometer can be replaced with the *FlatCone* multi-analyser spectrometer [7]. The sample table remains basically unchanged in the undertaken instrument upgrades.

4 Available sample environment and programmed improvements

We operate two dedicated liquid-He ("wet") cryostats in the temperature range 1.5-320K. One is the standard ILL orange cryostat with the bore for inserting sample with a diameter of 49 mm. The other cryostat offers enlarged bore of 69 mm for relatively bulky sample arrangements combined from several co-aligned single crystals. Both cryostats are equipped with multiple capillary He pumping system working with the regulated "cold-valve" and correspondingly powerful heaters permitting a relatively fast cooling/heating cycle (about 1-1.5 hour for the whole temperature range).

A particular sample environment that is available on *Thermes* includes the *VacBox* cryostat designed to work with the background reducing diaphragms maximally approached to the sample. The adjustable diaphragms and 'in-box' (that is in the vacuum-pumped neutron path) rotating tunnels are made from the boron-containing materials and are positioned only a few centimeters away from the sample. This configuration is optimized for relatively small samples (less than 10 mm in any direction) when the background conditions are especially demanding. In the *VacBox* the sample stick is rotating independently of the cryostat using an additional rotation stage in a way similar to that applied to the vertical field cryomagnets. This equipment is compatible with the *Goniostick* sample alignment device developed at the ILL [8].

Most of the other sample environment equipment available at ILL can be used on *Thermes* such as standard cryo-furnaces, high-field cryomagnets (currently the maximum authorized field is at 11 Tesla), high-pressure in the Paris-Edinburg press (at ambient and low temperature) with pressures at sample up to 100 kbar, high-temperature furnaces (up to 1600 °C).

We are considering further improvements to be implemented in the near future as the most necessary evolution of the instrument. One of them is a set of remotely controlled compact non-magnetic diaphragms operated in the presence of the cryomagnet's stray fields. The other is a 'dry' cryostat (a closed cycle refrigerator) combined with a cradle permitting fast reorientation of the sample crystal plane without removing the sample from the cryostat.

We are studying the potential advantages of working with a cold beryllium filter in combination with the pyrolytic graphite analyser in view of high-resolution studies of vibration dynamics of protonated molecules. The feasibility of Brillouin neutron scattering experiments is being reviewed with the new set of fine slit collimators and a dedicated vacuum box to reduce air scattering. A more distant project of mechanical velocity selector for *Thermes* awaits for outcome from the other thermal neuron spectrometers that now begin working with such a delicate and budget demanding equipment.

5 Scientific Impact

The new spectrometer started working at full capacity after commissioning and has been operating for the user community for about two years so far. Studied materials range from complex oxides with specific magnetic and structural properties to various novel superconductors, metals and semiconductors with peculiar electron spectra. Corresponding publications begin to appear [6,9-12].

Here we want to highlight particular success of experiments where accomplished developments had a direct impact on the quality of the collected data.

Matsuura *et al* [10] have explored the phonon renormalization effects accompanying the '6 K anomaly' in the quantum spin liquid candidate κ-(BEDT-TTF)$_2$Cu$_2$(CN). Such a system is extremely difficult to synthesize in the form of large single crystals therefore a set of smallish crystals of only 26 mg total mass was assembled for this study. The improved signal-to-noise performance of the new spectrometer finally allowed identification of the desired effect based on this capability that remains unique to IN8-*Thermes* spectrometer.

Xie et al [11] studied a Kagome lattice compound CsV_3Sb_5 that exhibits intertwined charge density wave (CDW) order and superconductivity. The experiments on *Thermes* with the sample built from 400(!) co-aligned single crystals demonstrate that the CDW order in this compound is associated with a static lattice distortion and a sudden hardening of an optical phonon below the ordering transition. The results indicate the effect of wave vector dependent electron-phonon coupling is playing an important role in the CDW order.

These and other examples of the stable and reliable operation ensure fruitful outcome in the ILL user program.

6 Conclusions

We have optimized the IN8 spectrometer for the easiness and comfort of operation, faster change of configurations and maintenance to minimize dead times. IN8-*Thermes* at ILL with unrivalled performance continues to supply the highest monochromatic flux among all thermal three-axis spectrometers worldwide. The instrument is dedicated to inelastic neutron scattering measurements with energy transfers in the range of from one to about one hundred meV, it is used to investigate thermal spectrum range of magnetic excitations and lattice vibrations. Its high incident flux offers a wide range of flexibility for optimizing the intensity *versus* resolution conditions and permits investigations of small samples and weak inelastic response not possible elsewhere. The enhanced performance, reliability and flexibility of IN8-*Thermes* constitute a major step forward to the successful functioning for the ILL user program. The present instrument state together with planned improvements ensures the instrument operation in the years to come. Current information about the instrument and additional details of the available configurations and operation can be found on the instrument webpage [13].

We acknowledge the unvaluable help from all the ILL services whose cooperation made it possible to realize the whole project. We particularly thank Franck Charpenay and Stephane Roux for their technical and engineering assistance as well as all helpful discussions, advices and recommendations from our colleagues Jiri Kulda and Martin Boehm.

References

1. G.Shirane, S.M.Shapiro, J.M.Tranquada, Neutron Scattering with a Triple-Axis Spectrometer, Cambridge University Press, Cambridge (2002).
2. A.Hiess, R.Currat, J.Saroun, F.J.Bermejo, Physica B **276-278**, p.91-92 (2000).
3. A.Hiess, M.Jimenez-Ruiz, P.Courtois, R.Currat, J.Kulda, F.J. Bermejo, Physica B **385-386**, p.1077-1079 (2006).
4. P.Mikula, J.Kulda, P.Lukas, M.Ono, J.Saroun, M.Vrana, V.Wagner, Physica B **283**, p.289-294 (2000).
5. J.Kulda, J.Saroun, NIMA **379**, p.155-166 (1996).
6. T.Tejsner, A.Piovano, A.Tutueanu, A.T.Rømer, B.O.Wells, J.Grivel, M.Boehm, L.Udby, Phys. Rev. B **101**, p.100504(R) (2020).
7. M.Kempa, B.Janousova, J.Saroun, P.Flores, M.Boehm, F.Demmel, J.Kulda, Physica B **385–386**, p.1080–1082 (2006).
8. E.Lelièvre-Berna, E.Bourgeat-Lami, J.Allibon, P.Decarpentrie, J.-P.Gonzales, J.Halbwachs, J.Ollivier, B.Ouladdiaf, J. Neutr. Res. **19**, p.27–32 (2017).
9. A.S.Sukhanov, Y.V.Tymoshenko, A.A.Kulbakov, A.S.Cameron, V.Kocsis, H.C.Walker, A.Ivanov, J.T.Park, V.Pomjakushin, S.E.Nikitin, I.V.Morozov, I.O.Chernyavskii, S.Aswartham, A.U.B.Wolter, A.Yaresko, B.Büchner, D.S.Inosov, Phys.Rev.B, **105**, p.134424(1-13) (2022).
10. M.Matsuura, T.Sasaki, M.Naka, J.Müller, O.Stockert, A.Piovano, N.Yoneyama, M.Lang, Phys.Rev.Res., **4**, p.L042047(1-6) (2022).
11. Y.F.Xie, Y.K.Li, Ph.Bourges, A.Ivanov, Z.J.Ye, J.-X.Yin, M.Zahid Hasan, A.Y.Luo, Y.G.Yao, Zh.W.Wang, G.Xu, P.Ch.Dai, Phys.Rev.B, **105**, p.L140501(1-6) (2022).
12. K.J.Neubauer, F.Ye, Y.Shi, P.Malinowski, B.Gao, K.M.Taddei, Ph.Bourges, A.Ivanov, J.-H.Chu, P.Ch.Dai, NPJ Quantum Mat., **7**, p.112(1-8) (2022).
13. www.ill.eu/users/instruments/instruments-list/in8/description/instrument-layout

Recent studies of highly oriented pyrolytic graphite and hybrid graphite-silicon monochromator systems

Andreas K. Freund[1*], *Dawn* Krencisz[2], *Mike* Crosby[2], *Changyong* Chen[2], *Brian* Kozak[2], *Pavol* Mikula[3], and *Gergely* Farkas[3]

[1] Consultant, 40 Rue Auguste Poirson, 33000 Bordeaux, France
[2] Momentive Technologies, 22557 Lunn Rd, Strongsville, Ohio, OH 44149, USA
[3] Nuclear Physics Institute, ASCR, 250 68 Řež, Czech Republic

Abstract The excellent performance of highly oriented pyrolytic graphite (HOPG) as monochromator, analyzer and filter material for neutron instrumentation was discovered more than 50 years ago. Even though HOPG has been in use for a long time, there is still room for improvements regarding its optimized use in neutron beam conditioning. The features of a novel concept called POSI based on careful x-ray characterization of the HOPG crystals and on proven mounting and focusing technologies show that the efficiency of HOPG monochromators can be increased while decreasing complexity and cost. Moreover, by combining HOPG and bent perfect silicon crystals the functionality of separate single monochromators and analyzers can be expanded to a wider range of applications provided by dual systems or bichromators.

1 Introduction

The superiority of highly oriented pyrolytic graphite (HOPG) as a material for neutron beam conditioning is well established. When comparing both the nuclear scattering properties and the structural data with those of potentially competing materials, only diamond and beryllium show similar performances as neutron monochromators and analyzers. The figures of merit normalized to that of HOPG are 1.2 for diamond and 0.96 for beryllium, followed by 0.55 for lead, 0.51 for silicon, 0.40 for niobium, 0.35 for germanium and 0.26 for copper, respectively [1]. This theoretical approach assumes that these materials would be available as ideal mosaic crystals featuring a mosaic spread of typically 0.2° to 1°.

Based on these predictions considerable efforts to produce beryllium and diamond mosaic crystals with uniform defect structures were undertaken in the past. They did not succeed because of the unfavourable defect and mechanical properties of these materials. By the way, the Greek root of the word diamond signifies the "*untameable*". Thus a reproducible monochromator production on a large scale could not be envisaged. The development of niobium crystals strained by diffusion of deuterium had to be abandoned, too. Nowadays, plastically deformed copper, germanium and silicon mosaic crystals are applied in addition to HOPG. Copper is preferred for the short wavelength range because of its small unit cell while germanium and silicon single crystals permit to suppress second order reflections. Their mosaic distribution is relatively uniform. It can be made anisotropic which is an advantage, in particular for increasing the focusing efficiency in vertical (sagittal) direction where the smaller mosaic spread diminishes the beam widening.

As an alternative to mosaic crystals, nearly perfect, bent silicon crystals were introduced successfully as monochromators, see *e.g.* refs. [2-4]. Elastic bending permits to focus the reflected neutron beam and creates lattice deformations that increase the integrated reflectivity per unit volume. The diffraction process can be well described by a modified dynamical theory. To obtain sufficient intensity a thickness of about 10mm is required. To prevent breakage when bending silicon blades, these monochromators are made of stacks containing typically ten blades of 1mm thickness. Their neutron rocking curve width is smaller than that produced by HOPG which results in higher spectral resolution at the expense of integrated intensity. On the other hand, the five times bigger thickness of the silicon stacks leads to a widening of the monochromatic beam that increases the focal spot size. However, the global performance of bent silicon crystals is still quite high. An advantage of silicon is the possibility to use asymmetric reflections in both reflection and transmission geometry. The latter is applicable only if there is no material in the beam trajectory behind the stack.

The specific properties of mosaic HOPG and bent silicon crystals make them complementary. Therefore, many instruments employ several monochromators. Their number and the materials depend on the spectral resolution and the neutron wavelength range. Usually three- or four-face monochromators are mounted on turrets located inside the monochromator housing. As an example, the three axis spectrometer IN8 at the Institute Laue-Langevin uses a triple device consisting of double focusing HOPG and copper mosaic crystal systems plus a stack of bent silicon crystals with fixed horizontal and variable vertical curvature [5].

*Corresponding author: kafreund8@gmail.com

In a scenario called serial beam multiplexing, two (or more) instruments are placed one behind the other on the same neutron source, for example on a guide. Here the downstream instrument is fed by the neutron beam exiting the upstream monochromator. Therefore the transmission of the upstream device should be as high as possible. For example, this scheme was applied at the USANS instrument Kookaburra at ANSTO [6]. The HOPG crystals composing the double focusing premonochromator were soldered on a perfect silicon support plate of toroidal shape.

A very elegant design of a different type of double focusing monochromator was reported for a triple axis spectrometer at NIST [7, 8]. Variable vertical curvature was achieved by buckling specially shaped, vertically aligned aluminium support plates carrying the HOPG crystals. Horizontal focusing is obtained by rotating the support plates around a vertical axis.

A still different system was fabricated recently by the company Positechnics, France, for the texture diffractometer HETU at the China Mianyang research reactor, CMRR [9-11]. A fixed horizontal curvature was generated by soldering Momentive ZYA-N grade HOPG crystals on faceted perfect silicon support plates made by the company Siliciumbearbeitung Andrea Holm GmbH, Germany. These plates were mounted on a mechanical device allowing for variable focusing in vertical direction by rotation about horizontal axes.

In this paper we outline recent attempts to further perfect the neutron monochromator technology. We propose hybrid schemes regarding the application of HOPG combined with bent perfect silicon crystals. These compact double or dual HOPG-silicon systems also minimize the intensity loss of the non-diffracted, throughgoing neutron beam. We begin by briefly recalling the scattering properties of HOPG and by describing the x-ray characterization of the HOPG crystals delivered by Momentive Technologies that are beneficial for optimizing composite HOPG systems.

2 Neutron scattering properties of PG

The scattering properties of imperfect crystals depend on the radiation used. They are calculated theoretically by the kinematical approximation applied to the so-called mosaic model assuming that the diffracting crystal can be regarded as a homogeneous assembly of independently scattering volumes called mosaic blocks that represents the defect structure of real crystals.

These crystal domains are considered to be perfect inside and small enough so that dynamical diffraction effects can be neglected. But to avoid beam widening and particle size broadening of the diffraction pattern they should not be too small as well. The orientation of these mosaic blocks is supposed to follow a Gaussian distribution function. For several reasons all these conditions characterizing the so-called "ideal mosaic crystal" are never fulfilled by a real, non-perfect crystal containing defects such as dislocations that are most frequently arranged in small angle grain boundaries. The main deviations from the model are caused by non-uniformities of the spatial and angular strain distributions. The experimental rocking curves shown in fig. 1 can be matched by a double mosaicity, a deviation from a purely Gaussian angular mosaic block distribution and by the effect of primary extinction due to a mosaic block size exceeding what is called the primary extinction thickness. This effect decreases the peak reflectivity.

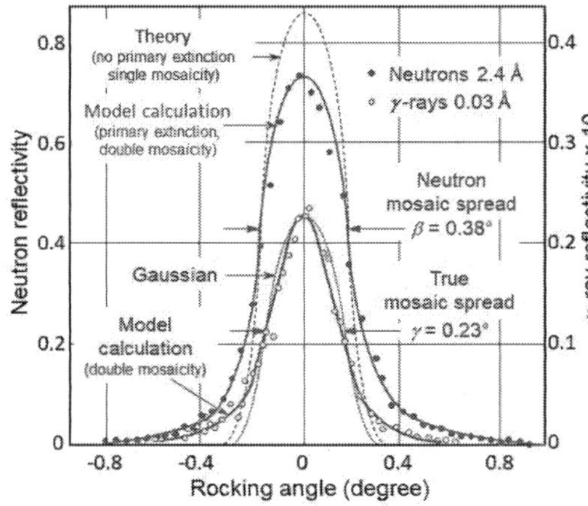

Fig. 1. Neutron and γ-ray rocking curves of a HOPG-ZYA grade sample [11].

When attempting to predict the monochromatic flux obtained by mosaic crystals, for example by using computer simulations, it is important to consider the fact that there is a difference between what we call the *true mosaic spread, γ,* that can be measured almost extinction-free by high-energy γ-rays, the *x-ray mosaic spread, ξ,* that is the FWHM of the x-ray rocking curve, and the wavelength dependent *neutron mosaic spread, β,* that corresponds to the FWHM of the neutron rocking curve. The empirical correlation of these three quantities is displayed in fig. 2.

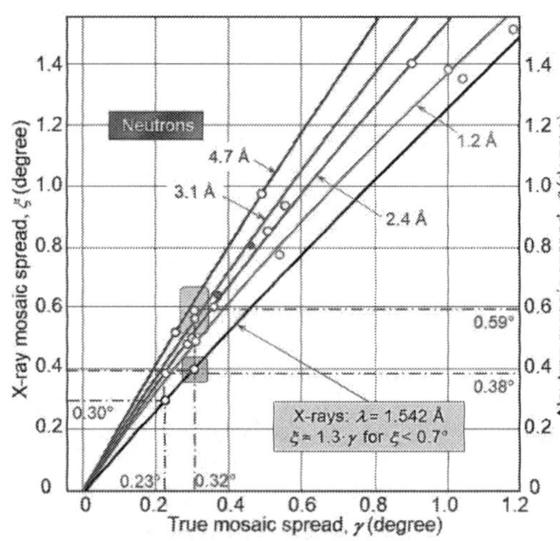

Fig. 2. Correlation between the true mosaic spread measured with short wavelength (0.03Å) γ-rays and the FWHM of x-ray and neutron rocking curves, HOPG 002-reflection, depending on the grade of the samples [11].

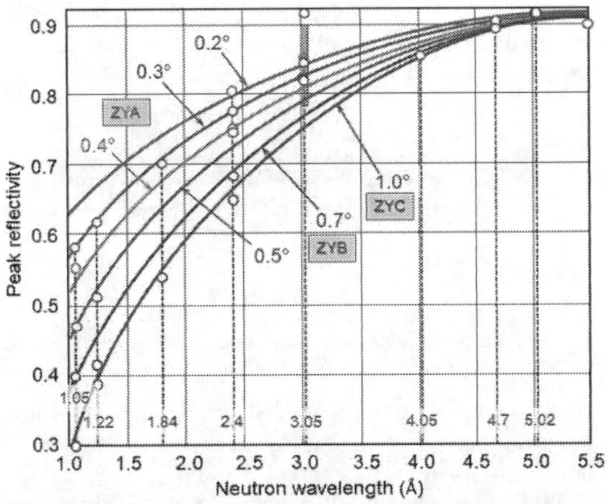

Fig. 3. Experimental peak reflectivities of Momentive ZYA, ZYB and ZYC grade HOPG as a function of wavelength and true mosaic spread for a crystal thickness of about 2 mm [11].

For example, a crystal characterized by x-rays of 1.542 Å wavelength producing a 0.40° wide rocking curve exhibits a true mosaic spread of 0.32° when recorded by diffraction of γ-rays of 0.03Å wavelength in transmission geometry. The corresponding mosaic spread for neutrons of wavelength 3.1Å is 0.59°, almost twice the true mosaic spread. The curves shown in figs. 2 and 3 are averages of results obtained for many samples. The range of measured values is indicated by the shaded areas. At certain wavelengths competing HOPG reflections are excited that influence the width and the peak reflectivity. Thus the smoothened curves represent approximations. However, these data can be used for resolution and flux estimations based on computer simulations. For more details see ref. [11] and the literature cited therein.

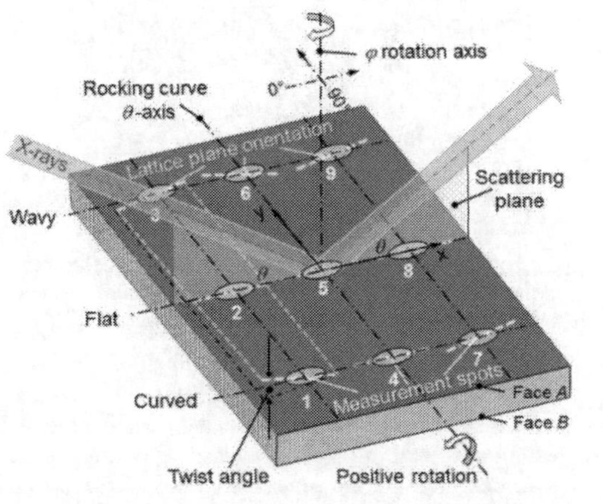

Fig. 4. Diagram of x-ray tests performed by Momentive Technologies to determine the mosaic spread and the lattice plane orientation of HOPG typically 20x20mm² in size.

Detailed x-ray diffraction tests of HOPG crystals proved to be very useful and sufficient for optimizing position and orientation of the pieces in the matrix of composite monochromator systems. By scanning a typically 1mm² wide x-ray beam across the surface of the sample with high spatial resolution, the topology of the lattice plane orientation is recorded in much more detail than with neutron beams. As shown below in fig. 4, three types of lattice plane orientation changes are observed: a wavy behaviour, a flat shape and a uniform curvature. The crystals are studied in two azimuthal positions and on both faces. The data obtained allow us to identify the concave face to be turned towards the incident neutron beam and to select the best samples to be placed in the center of the composite system.

3 Focusing devices

A drawback of HOPG is its strong mechanical rigidity and its fragility that prevent the creation of variable crystal curvature by elastic bending. Therefore, the majority of today's single and double focusing devices consist of typically 20x20mm² wide, more or less flat pieces. They are mounted on mechanical systems where each piece can be aligned individually by a mechanical lever-cam system to achieve fixed or variable meridional and/or sagittal focusing, see the photograph in fig. 4. To protect the mechanical parts situated behind the crystals against neutron and γ-ray irradiation, the crystal supports are covered by a neutron absorbing material, for example B_4C. Then the non-diffracted quasi-white neutron beam is blocked and cannot feed other instruments situated downstream that otherwise could use neutrons exiting from the same source.

Fig. 5. Conventional double-focusing monochromator composed of HOPG mosaic crystals bonded on individually orientable supports. (Courtesy: Positechnics).

In an attempt to achieve dynamical focusing by a variable curvature of the HOPG crystals, the bendability of Momentive grade ZYA crystals was studied as a function of their thickness and curvature. To achieve a minimum radius of 1.5m without breakage the thickness had to be inferior to 0.1mm, but even then a non-zero risk of damage after several bending and flattening cycles remained [12]. The results also showed that the elasticity limit was sample dependent. Therefore, the idea of bending thin HOPG blades for dynamical focusing was not followed up.

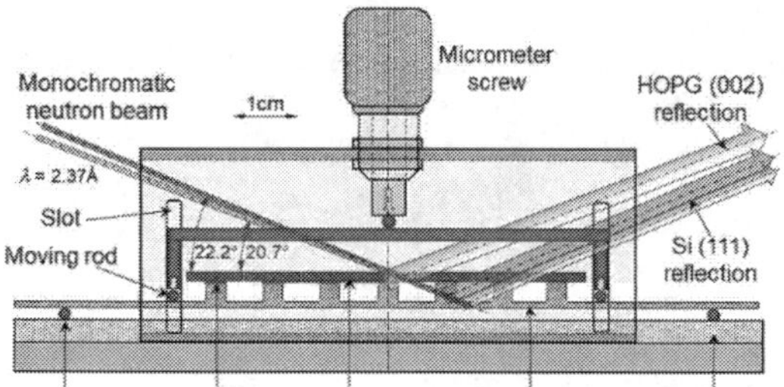

Fig. 6. Schematic view of the four-point bending device for studying the POSI system and the neutron diffraction geometry used for the tests [12].

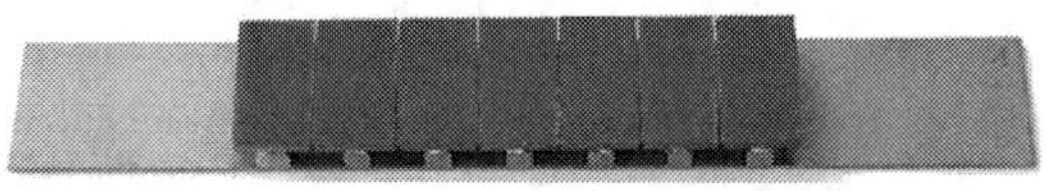

Fig. 7. A long HOPG crystal indium bonded on a ribbed silicon blade and then sliced into 9mm wide segments.

An alternative design was examined where flat HOPG crystals are soldered onto a bendable, ribbed silicon plates, supplied by Siliciumbearbeitung Andrea Holm GmbH, Germany. This concept is called POSI, an acronym for **p**yrolytic graphite **o**n **si**licon. The use of ribs decreases the amount of stress on the indium bonds created by bending the plates and thus avoids the risk of detachment. Mechanical and both x-ray and neutron diffraction tests using the device shown in fig. 6 confirmed that this new type of monochromator with precise, tunable curvature can be safely used in neutron instrumentation [12]. A photograph of the assembly is presented in fig. 7.

Similar to the design described in refs. [7] and [8] the POSI device is mechanically less complex than the conventional scheme. In comparison with the buckling technique of aluminum blades the use of silicon perfect crystals as support material reduces the background produced by Bragg scattering. The four-point bending design does not require the complex shape of the support needed for buckling.

The POSI design has the same advantage of transmitting the non-diffracted neutron beam with a minimum loss to the instruments located further downstream on the same beam tube or neutron guide that can be arranged according to an optimal sequence [13]. The length reduction of the individual HOPG pieces could improve the uniformity of the global curvature. This would improve the monochromator efficiency for thermal neutrons and under high resolution conditions where the beam divergence and the width of the neutron rocking curves are small.

But this is not the end of the story. A material called glassy carbon commercialized under the name Sigradur, well characterized with neutrons [14], will replace the perfect silicon support plates. Then the acronym POSI stands for 'pyrolytic graphite on Sigradur' too. When compared to silicon, in addition to the advantages of smaller attenuation and less thermal diffuse scattering, this amorphous material does not produce parasitic Bragg scattering. Sigradur type G plates can be obtained from the company Hochtemperatur-Werkstoffe GmbH, Thierhaupten, Germany.

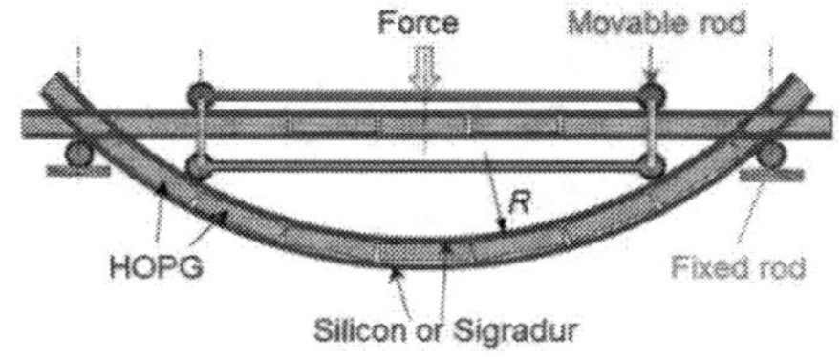

Fig. 8. Four-point bending principle acting on a sandwich of HOPG crystals and silicon or glassy carbon plates.

The system shown in fig. 8 is presently studied. It does not use any bonding material. To ascertain a uniform curvature very precise dimensioning of the HOPG pieces is required. In view of the varying shapes and dimensions of these pieces this condition is likely to represent a challenging technical difficulty.

4 Dual and double monochromators

The POSI technique permits to integrate individual HOPG and silicon monochromators into single devices. For example, by adding a ribbed Si crystal carrying the HOPG pieces or a Sigradur-HOPG sandwich on top of a stack of Si blades, using *e.g.* the device shown in fig. 9, a bichromator, is formed.

Fig. 9. Conventional double-focusing silicon monochromator with variable curvature composed of bent stacks of typically 10 perfect crystals each. (Courtesy: Positechnics).

Such a system extracts simultaneously two monochromatic neutron beams of different flux and spectral resolution out of a white incident beam; see the example shown in fig. 10. Depending on the degree of asymmetry and on the Bragg reflection of the silicon crystals they focus neutrons of the same or of different wavelength(s) with respect to the neutrons scattered by the HOPG crystal. The beams produced by dual reflections can be received by one or two sample(s), either simultaneously or one after the other. If no obstacles are situated behind the crystals, neutrons diffracted in symmetric or asymmetric transmission geometry of the silicon crystals can be used too.

As an illustration of possible hybrid scenarios we present the application of the POSI design principle to a monochromator actually requested for the ANDES stress diffractometer that is planned to be installed at the LAHN facility presently under construction in Bariloche, Argentina (https://www.lahn.cnea.gov.ar/).

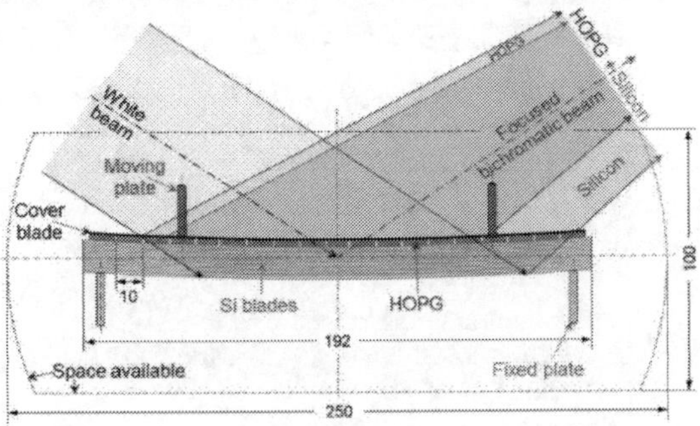

Fig. 10. Top view of one stack of a hybrid HOPG-Silicon monochromator proposed for the instrument ANDES focusing two neutron beams of slightly different wavelengths simultaneously on the same sample position. The dimensions are mm.

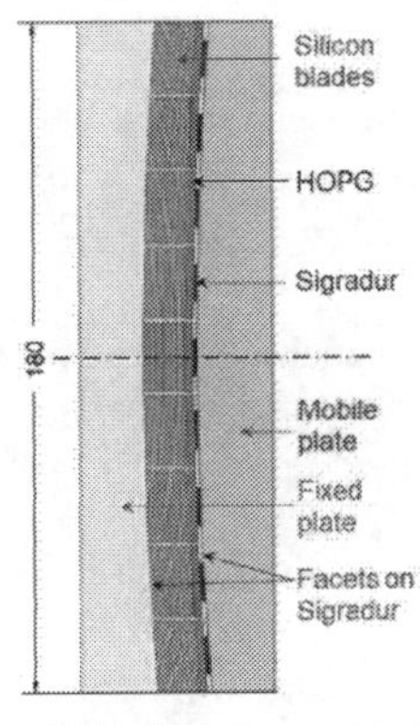

Fig. 11. Side view of the dual monochromator composed of 9 stacks shown in fig. 10.

Figs. 10 and 11 show schematically the proposed hybrid monochromator consisting of stacks of perfect silicon blades carrying 19 HOPG crystals each. The HOPG pieces are covered by Sigradur plates to spread out the local pressure of the force acting on the HOPG crystals and to uniformize the curvature approximated by the crystal polygon. The dimensions were adapted to the relatively small space inside the monochromator vessel of the ANDES device. Instead of the rods shown in figs. 7 and 8, this four-point bending system uses faceted plates made of Sigradur whose shapes are designed for fixed vertical focusing, see fig. 11. The attenuation of the incident and reflected neutron beams traversing these plates is very small.

Two POSI scenarios and a dual system are presented in fig. 12. In line with the specifications, in all three scenarios the constant vertical radius of curvature corresponds to the central wavelength used. The horizontal focusing distance can be kept constant when changing the Bragg angle by adapting the curvature of the HOPG-silicon stacks.

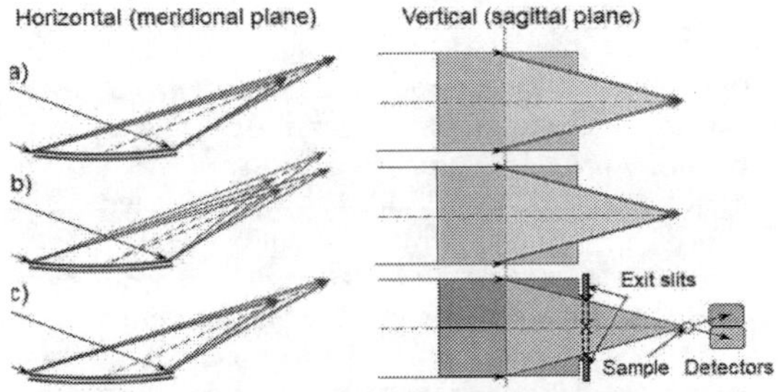

Fig. 12. Three horizontal (left) and vertical (right) geometries of a dual POSI monochromator (a, b) and of a double HOPG-silicon device (c) for two radii of meridional curvature, respectively.

a) Reflected by the HOPG crystals and by the silicon crystals in symmetric geometry a dual beam containing two wavelengths is focused simultaneously on the sample. An example of this option is shown in fig. 10. Here the HOPG (002)- and the silicon (111)-reflections were chosen. Because their *d*-spacings are quite similar, the two corresponding diffraction peaks of the sample are close to each other.

b) When using asymmetric reflection geometries the lattice planes of the silicon monochromator are not parallel to the HOPG lattice planes and the sample will be exposed successively to the two beams by moving it to the respective focal positions. If the sample is wide enough it can intercept both beams.

c) The upper half of the monochromator consists of HOPG crystals, whereas the lower part is composed of silicon stacks. This double monochromator scenario allows us to choose between three possibilities that can be selected by an appropriate positioning of the slits at the monochromator entrance or exit.

i) By shielding the lower half of the double monochromator, the sample receives neutrons only from the HOPG crystals. After reflection by the sample they are recorded by the lower detector.

ii) By shielding the upper half of the double monochromator, the sample receives neutrons only from the Si crystals. After diffraction by the sample they are received by the upper detector.

iii) When the slits are fully open, the sample receives neutrons from both the HOPG and the silicon crystals. The corresponding neutrons scattered by the sample are recorded separately by the upper and lower detector, respectively.

A simultaneous recording of the sample response to the high resolution/low intensity beam and the low resolution/high intensity beam, respectively, is possible for a specific scattering geometry. Such a configuration can be interesting for time-dependent experiments, especially when studying irreversible processes in real-time. An interesting feature of scenario c), case ii), is the availability of a 'clean' beam reflected by the Si crystals only. This is not possible for the options a) and b) where the dual wavelength beam focused on the sample may be contaminated by some parasitic Bragg reflections of the HOPG crystals other than the (002) reflection.

5 Summary

The neutron diffraction properties of highly oriented pyrolytic graphite crystals and their application in focusing monochromator and analyzer systems are briefly recalled. The detailed x-ray characterization carried out routinely by Momentive Technologies appears to be sufficient to optimize such composite systems as a function of the shape and the mosaicity of the HOPG crystals. The relation between the intrinsic mosaic spread of this material, the widths of x-ray diffraction curves and the wavelength dependence of neutron rocking curves are emphasized. These data can be useful for computer simulations of instrument performance.

Recent bendability studies of HOPG showed that elastic bending of this material to achieve dynamical focusing of neutron beams is unrealistic. A novel technique called POSI that is different from the presently used mechanical focusing devices has been tested successfully where approximately flat HOPG pieces are bonded on the ribs of flexible silicon plates. This technique has the advantage that the non-diffracted, quasi-white beam is transmitted with only small attenuation to instruments situated downstream on the same neutron guide. Compared to the properties of aluminum and silicon as support materials the use of bendable plates made of glassy carbon would still increase the transmissivity of POSI monochromators and decrease the background intensity. When holding the HOPG crystals between two glassy carbon blades, any glue or solder might become unnecessary. This possibility is presently being studied.

The complementarity of HOPG and bent perfect silicon crystals suggests their use in hybrid systems where both materials are integrated into compact bichromator devices. Neutron scattering experiments can then be conducted, either simultaneously or successively, under high flux-low resolution and low flux-high resolution conditions. The POSI technique is a perfect tool to realize such scenarios. An example is given describing an application to a strain analysis diffractometer that is presently in the design status.

Because of their increased functionality, reduced size and complexity, bichromators are cost-effective and thus attractive from both ecological and economical viewpoints. Similar considerations may apply to crystal analyzer systems. Frontend monochromator multiplexing using different materials might be interesting when correlated with backend analyzer multiplexing techniques presently applied or proposed [15, 16].

The choice of the most suitable dual or double monochromator option for a given experimental study is determined by the best compromise between a multipurpose scenario and a highly specific mode of instrument operation. The hybrid concept as presented in this paper is meant to be a first step towards an alternative kind of monochromator design. We trust that it will find useful applications.

The neutron diffraction experiments were carried out at the CANAM instrument of NPI CAS Řež installed the CICRR infrastructure, which is financially supported by the Czech Ministry of Education and Culture - project LM2023041.

References

1. A.K. Freund, J. Appl. Cryst. **42**, 36 (2009).
2. P. Mikula, J. Kulda, P. Lukáš, M. Ono, J. Šaroun, M. Vrána and V. Wagner, Physica **B 283**, 289 (2000).
3. M. Popovici, A. Stoica, C. Hubbard, S. Spooner, H. Prask, T. Gnaeupel-Herold, P. Gehring and R. Erwin, Proc. SPIE 4509, **21** (2001).
4. P. Mikula, M. Vrána, M. Furusaka, V. Wagner, Y.-N. Choi, M.-K. Moon, V. Em and C.-H Lee, Nucl. Instrum. Meth. Phys. Res. **A 529**, 138 (2004).
5. https://www.ill.eu/users/instruments/instruments-list/in8/description/instrument-layout.
6. C. Rehm, A. Brûlé, A.K. Freund and S. Kennedy, J. Appl. Cryst. **46**, 1699 (2013).
7. S.A. Smee, P.C. Brand, D.D. Barry, C.L. Broholm and D.K. Anand, Nucl. Instrum. Meth. Phys. Res. **A 466**, 513 (2001).
8. S.A. Smee, J.D. Orndorff, G.A. Scharfstein, Y. Qiu, P.C. Brand, C.L. Broholm and D.K. Anand, Appl. Phys. **A74** [Suppl.], S255 (2002).
9. A.K. Freund, unpublished report to CMRR (2021).
10. B. Wang, S. Zhong, H. Lin, J. Li, Z. Yang, A. Goukassov, H. Zhang and G.-A. Sun. Submitted to J. Appl. Cryst., March 2023.
11. A.K. Freund, H. Qu, X. Liu, M. Crosby, C. Chen, J. Appl. Cryst. **55**, 247 (2022).
12. A.K. Freund, C. Chen, M. Crosby, G. Farkas, B. Kozak, D. Krencisz, X. Liu, P. Mikula, Nucl. Instrum. Meth. Phys. Res. **A 1048**, article 16802 (2023), in print, available on the internet.
13. D.R.F. Mildner, M. Arif and S.A. Werner, J. Appl. Cryst. **34**, 258 (2001).
14. R.L. Capelletti, T.J. Udovic, H. Li and R.L. Paul, J. Appl. Cryst. **51**, 1323 (2018).
15. F. Groitl, D. Graf, J. Okkels Birk, M. Markó, M. Bartkowiak, U. Filges, C. Niedermayer, C. Rüegg and H.M. Rønnow, Rev. Sci. Instrum. **87**, 035109 (2016).
16. R. Toft-Petersen, R. Georgii, M. Schneider, N. Nishiki and P. Böni, Nucl. Instrum. Meth. Phys. Res. **A 977**, 164341 (2020).

Where Does an Enzyme Reside in a Bicontinuous Structure?

Stefan Wellert[1,*], *Sandra* Engelskirchen[2], *Thomas* Hellweg[3], and *Olaf* Holderer[4,**]

[1]Department of Chemistry, Technische Universität Berlin, Berlin, Germany
[2]Department of Chemistry, Universität Stuttgart, Stuttgart, Germany
[3]Department of Chemistry, Physical and Biophysical Chemistry, Bielefeld University, Bielefeld, Germany
[4]Jülich Centre for Neutron Science (JCNS) at Heinz Maier-Leibnitz Zentrum (MLZ), Forschungszentrum Jülich GmbH, Garching, Germany

Abstract. Using enzymes as catalysts in biochemical processes requires to bring them into close contact with the substrates to be processed. For oil-soluble substrates and water soluble enzymes this might be achieved by bringing the water and oil phase in a bicontinuous microemulsion into close contact. In this contribution we review two possible scenarios of how the enzyme and the interface influence each other. Small angle neutron scattering (SANS) and neutron spin echo spectroscopy (NSE) act as a microscope to look into the details of the interfacial region of microemulsions.

1 Introduction

Microemulsions as thermodynamically stable mixtures of oil, water and surfactant, can provide a very unique environment for enzymatic catalysis. The thermodynamically stable compartmentalization of polar and non-polar phases by a large internal interface in the range of m^2/ml inspired the use of microemulsions as reaction media for a series of catalytic reactions, e.g. [1]. Among them, enzyme-catalyzed reactions are of particular interest. Domains of water and oil are only nanometers apart from each other, and provide therefore a very elegant way of bringing water soluble enzymes and often unpolar substrates together [2–4]. Depending on their macromolecular structure, the enzymes can accumulate either within the aqueous phase or directly at the amphiphilic interface. It can be assumed that an enzyme without enrichment at the interface has no influence on the mechanics, whereas an interfacially active enzyme should lead to changes in the mechanical properties in a concentration-dependent manner. Corresponding effects have already been observed with the addition of amphiphilic polymers (boosting).

Additives to microemulsions on the other hand can modify the location of phase boundaries in the phase diagram [5–7]. Mainly the elasticity of the surfactant layer separating oil and water is modified by incorporation of molecules [8] or co-surfactants [9, 10]. Neutron small angle scattering (SANS) and neutron spin echo spectroscopy (NSE) are scattering techniques which can provide a valuable insight into these systems structured on nanometer length scales, by combining structural information from SANS [11] and dynamic information about fluctuations from NSE [7, 12]. The influence of enzymes has been investigated with SANS and NSE for two different systems [13, 14]. This contribution reviews briefly how additives modify microemulsions, and how scattering techniques can scrutinize their enzyme induced changes in nanoscale structure and motion. The focus is thereby on the addition of proteins as bio-catalysts to microemulsion. Besides different interactions between enzymes and surfactant layers of the microemulsion, also the domain size of the microemulsion itself influences the physical properties of the system. The question of the relevance of density fluctuations or breathing modes in microemulsions will be addressed.

*e-mail: s.wellert@tu-berlin.de
**e-mail: o.holderer@fz-juelich.de

2 Microemulsions and Enzymes

Two main aspects of enzyme containing microemulsions will be briefly reviewed. On the one hand, the elastic properties of the surfactant membrane of microemulsions, described by the bending rigidity κ and saddle splay modulus $\bar{\kappa}$ relate the structure of the microemulsion to its dynamics. The addition of enzymes to the water phase of the emulsion might influence both, structure and dynamics depending how and where the enzymes reside. This is illustrated in figure showing in sketch a) the different possible locations and in b) an SEM measurement of the bicontinuous structure. [13].

2.1 Microemulsions as Compartments and Reaction Media

Microemulsions form two types of compartments, droplets (oil-in-water or water-in-oil) and bicontiuous structures. Here we focus only on the bicontiuous case. In bicontinuous microemulsions the two immiscible liquids (such as oil and water) are stabilized by a surfactant and are dispersed into each other, forming continuously connected but separated networks of both phases. The key advan-

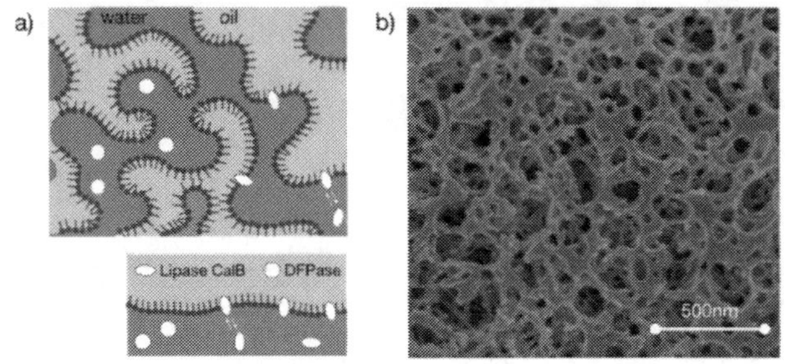

Figure 1. a) Top: Scenarios of enzyme – membrane interactions in microemulsions. Enzymes can be purely dissolved in water, rather unaffected by the presence of the lipid or surfactant interface, or they can interact and adsorb (for some residence time) at the interface. Bottom: When residing at the interface the presence of the enzyme can influence the bending elasticity of the interface and therefore the thermally activated fluctuations.
b) Freeze fracture SEM micrograph of a bicontinuous microemulsion structure which serves as the host medium in the discussed experiments.

tage is their enormous interfacial area, which facilitates mass transfer across the interface enhancing reaction constants of the reaction kinetics. Additionally, the bicontinuous structure also provides a highly stable environment for the reactants, potentially preventing unwanted side reactions or product degradation, since products and reactants are mainly separated via the interface. For example, the activity of DFPase remained high also in the complex environment of a mircoemulsion [13].

Bicontinuous microemulsions have been successfully employed in a wide range of chemical reactions, including organic synthesis, enzymatic reactions, and catalysis[15–17]. The potential of bicontinuous microemulsions in various fields such as pharmaceuticals, materials science, and biotechnology makes them a promising area of research for future applications.

2.2 Bicontinuous Microemulsions

As mentioned before, the bicontinuous structure resembles a sponge. The structural parameters, that is the domain size D_{TS} and the correlation length ξ_{TS} are obtained from fit of SANS measurements with the well-known Teubner-Strey formula [11]:

$$I_{\text{TS}}(Q) = \frac{8\pi\langle\nu^2\rangle/\xi_{TS}}{p^2 - 2Q_{\text{max}}^2 Q^2 + Q^4} \quad (1)$$

with $p^2 = (2\pi/D_{\text{TS}})^2 + 1/\xi_{\text{TS}}^2$ and $Q_{\text{max}}^2 = (2\pi/D_{\text{TS}})^2 - 1/\xi_{\text{TS}}^2$. These measurements are performed in "bulk contrast", where D_2O and hydrogenated oil provide different scattering length densities for the two main domains in the sample.

2.3 Bicontinuous Microemulsion Membrane Dynamics

The Helfrich free energy is typically the starting point for considerations about the membrane structure and dynamics [18]. It reads

$$F_{el} = \int dS \left[\frac{\kappa}{2}(c_1 + c_2 - 2c_0)^2 + \bar{\kappa} c_1 c_2\right] \quad (2)$$

where the principal curvatures c_1 and c_2, the mean curvature c_0 and the bending rigidities κ and $\bar{\kappa}$ are related.

Here, the bending rigidity κ determines the energy required to deviate the interfacial film from the spontaneous curvature c_0 at the given minimum of the free energy. The saddle-splay modulus $\bar{\kappa}$ gives the energy cost for saddle-splay deformations [19]. The bending elasticity constant is highly sensitive to any changes in topology and interfacial composition. And it is directly accessible by scattering in an easy way. A softening of the interfacial film is observed when undulations are thermally activated in the order of k_BT. This leads to a length scale dependent modification of the bare bending rigidity κ_0, the so-called renormalized bending elasticity constant:

$$\frac{\kappa_{\text{SANS}}}{k_BT} = \frac{\kappa_{0,\text{SANS}}}{k_BT} - \frac{3}{4\pi}\ln\left(\frac{D_{\text{TS}}}{2l_s}\right) \quad (3)$$

where the renormalization term contains the menbrane thickness l_s and therefore $D_{\text{TS}}/2l_s$ is the membrane volume fraction.

This quantity is connected to the structural length scales inside the bicontinuous structure by [20]:

$$\frac{\kappa_{\text{SANS}}}{k_BT} = \frac{10\pi\sqrt{3}}{64}\frac{\xi_{\text{TS}}}{D_{\text{TS}}} \quad (4)$$

Simulations [21] in combination with measurements of the phase behavior, SANS and NSE experiments [22] revealed that the bending rigidity obtained with SANS are a combination of κ and $\bar{\kappa}$, $\kappa_{0,SANS} = (a_1\kappa_0 + a_2|\bar{\kappa}_0|)$. Also the renormalization term in Equation 3 is modified and reads $(3a_1 + 10/3|a_2|)/(4\pi)\ln(D_{TS}/2l_s)$ with $a_1 = 0.19$ and $a_2 = -0.84$ [22]. These corrections have been applied for the values of the bending rigidity from SANS measurements discussed here.

While the main structural features are obtained by fits with the Teubner-Strey structure factor [5, 11] as described above, the height fluctuations of the interface layer, governed by the film elasticity, are obtained in the framework of the Zilman-Granek model [12] from NSE mea-

https://doi.org/10.1051/epjconf/202328604001

surements. The contrast applied here is typically "film contrast", with only the interface layer made up from hydrogenated components, water and oil are deuterated (although at higher scattering vectors also "bulk contrast" samples provide the interface fluctuations in NSE). For relatively stiff membranes the intermediate scattering function from NSE can be approximated by a stretched exponential function, for softer membranes some modifications are required, e.g. by numerical integration of the correlation function [7]. But also the approach with the stretched exponential function results in the correct length scale dependence (i.e. the relaxation rate $\Gamma \propto q^3$) and the right curve shape of S(q,t) with a stretching exponent of $\beta = 2/3$. NSE experiments provide therefore the bending rigidity determined from height fluctuations in a rather direct way, while also SANS provides elastic properties from the ratio of the involved length scales of domain size and correlation length, which includes length scale dependent renormalization effects and also includes contributions from the saddle splay modulus [21, 22].

3 Experimental Aspects

3.1 The Microemulsion Systems

The results discussed in this contribution were obatined from scattering experiments with two microemulsion systems. A bicontinuous microemulsion from the ternary system pentaethylene glycol monodecyl ether($C_{10}E_5$)-D_2O/NaCl(4 wt.%)-d(18)-octane was used. Details of the phase behavior of the pure system and the system with additional lipase calB can be found in the literature [14].
The amphiphilic component of the second microemulsion system consists of a sugar surfactant ($C_{8/10}G_{1.3}$) and a medium chain alcohol (pentanol). In this case, instead of the temperature the alcohol content is used to tune the curvature of the amphiphilic film and to control the phase behavior. The quaternary system cyclohexane-D_2O-$C_{8/10}G_{1.3}$-pentanol resembles all structural features of a ternary system using the pentanol content as tuning parameter.[13]
All samples were prepared near the X-point of the respective phase diagram. The position of the phase boundaries for each sample was visually determined prior to the measurements. All bicontinuous samples used in the experiments were prepared at an oil-to-water ratio α=0.5. The surfactant concentration γ in case of the ternary system was γ=0.15 irrespective of the enzyme concentration. The temperature was 35.5° C for the blank microemulsion going down to 24.5° C when the enzyme concentration was raised up to 100mg/ml. In case of the quaternary system, the surfactant concentration was at γ=0.21 with a concentration δ=0.06 of pentanol serving as cosurfactant in the amphiphilic component.

3.2 The Enzymes lipase CalB and DFPase

Lipases show the technically interesting tendency to adsorb to oil-water interfaces which leads to a restructuring of the enzyme at the interface and in fact activates the enzyme (at least the wild type).
The enzyme diisopropyl fluorophosphatase (DFPase) from

Protein	conc. (mg/ml)	Surfactant	Oil phase
CalB	0	$C_{10}E_5$	octane
CalB	10	$C_{10}E_5$	octane
CalB	50	$C_{10}E_5$	octane
CalB	100	$C_{10}E_5$	octane
DFPase	0	$C_{8/10}G_{1.3}$/pent.	cyclohexane
DFPase	5	$C_{8/10}G_{1.3}$/pent.	cyclohexane
DFPase	10	$C_{8/10}G_{1.3}$/pent.	cyclohexane

Table 1. Sample composition of the investigated microemulsions.

the squid Loligo vulgaris belongs to a group of enzymes that efficiently detoxify highly toxic organophosphorus (OP) compounds, which act as irreversible inhibitors of acetylcholinesterase [23]. DFPase has a remarkable thermal stability and is very tolerant toward organic solvents which makes it an excellent candidate for an catalytic additive inside a microemulsion. Moreover, DFPase can routinely be produced by heterologous expression in E. coli.
The lipase from Candida antarctica B (CalB) is from a broad class of enzymes which have the potential of acting as catalyzers for organic reactions with often only poorly water soluble substrates, which makes such microemulsion systems with the proximity of water and oil phases a very attractive environment.
Experimental details related to the expression, purification and characterization of the enzymes and their introduction with different concentration into the buffered aqueous phases of the microemulsions were previously described in the literature [13, 14].
The enzyme concentrations used vary widely and were mainly chosen with regard to the technically desirable quantities in each case.

3.3 Measurement of Membrane Fluctuations

Neutron spin echo spectroscopy (NSE) uses spin encoding and decoding of the neutron spin velocity by a large number of spin precessions in a magnetic field before and after the scattering in the sample [24]. This decouples the energy resolution from the wavelength spread $d\lambda/\lambda$ of the neutron beam and allows to use a relatively broad wavelength band (typically 10-20 %) by providing a very high energy resolution. The actually measured signal is the Fourier transform of the scattering function from the energy domain to the time domain, the intermediate scattering function $S(Q,\tau_{NSE})$. A high energy resolution is equivalent to a large Fourier time obtained at the NSE. For a Lorentzian line broadening with a FWHM Γ in energy space, the Fourier transform leads to an exponential decay with relaxation time τ [ns] = 1.317 / Γ[μeV] (see e.g. [25]). The intermediate scattering function is the time correlation function in reciprocal space, in the case of bicontinuous microemulsion this reflects the height fluctuations of the surfactant membrane on the relevant NSE length scales.

The NSE bending rigidity is obtained from the full integration of the Zilman-Granek model [7, 12]:

$$S(Q,\tau_{\text{NSE}}) \propto \int_0^1 d\mu \int_0^R dr\, rJ_0\left(Qr\sqrt{1-\mu^2}\right) \exp\left(-\frac{k_B T}{2\pi\kappa}Q^2\mu^2 \int_{k_{\min}}^{k_{\max}} \frac{dk}{k^3}\left(1 - J_0(kr)e^{-\omega(k)t}\right)\right) \quad (5)$$

with the Bessel function $J_0(kr)$ and the integration limits of the membrane undulations, k_{min} and k_{max}, which correspond to the maximum and minimum undulation wavelength respectively present in the microemulsion membrane.

4 Results

4.1 On the Analysis of Intermediate Scattering Functions in the Case of a Bicontinuous Microemulsion

Two main dynamical processes may occur in the time and length scale window of typical NSE experiments. On larger length scales, density fluctuations or "breathing modes" might be present, which show a typical Q^2 dependence of the relaxation rate similar to simple diffusion. On shorter length scales, the Q^3 behaviour and the stretched exponential decay of the Zilman-Granek model dominates [22, 26]. Density fluctuations may well be observed with dynamic light scattering [9]. Their contribution can be included into the data modelling by an additional diffusive Q^2-dependent decay, a path which has been followed in Ref. [13]. If one assumes, both modes of motion to be decoupled from each other, the result is an additive superposition of their contributions. However, if the breathing motion and interfacial fluctuations depend on each other in the time and size range accessible with a NSE measurement, their contributions must be considered accordingly.

$$S(Q,\tau_{NSE})/S(Q,0) = exp(-\Gamma_{col}\tau_{NSE})(A + (1-A) exp(-(\Gamma_u\tau_{NSE})^{\beta})) \quad (6)$$

The two relaxation rates here are the collective relaxation rate Γ_{col}, which is related to a diffusive Q^2 dependent decay, and the undulation relaxation rate of a membrane patch, Γ_u. Plotting the resulting values of Γ_u vs. Q^3 yields the bending elastic constant κ_{NSE} according to

$$\Gamma_u = 0.025\gamma_\kappa \left(\frac{k_B T}{\kappa_{NSE}}\right)^{1/2} \frac{k_B T}{\eta_{eff}} Q^3 \quad (7)$$

Here, η_{eff} is the effective solvent viscosity of the fluid surrounding the fluctuating membranes and γ_κ is a numerical parameter.
Taking a look at the microemulsions used there and fitting the dynamics for a series of scattering vectors simultaneously either purely with the full Zilman-Granek model with numerical integration from Ref [7], or with the additional diffusive contribution. Figure 2 shows the data with this simultaneous fits. Only including the diffusion allows to reproduce the decay of the intermediate scattering functions correctly, the diffusion constant $D_{col}=\Gamma_{col}/Q^2$ with

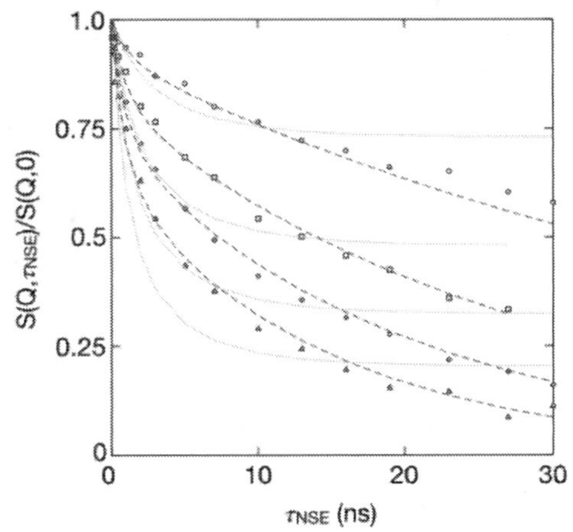

Figure 2. Normalized intermediate scattering functions of a pure film contrast microemulsion, used for the measurements on the microemulsions dynamics in case of addition of DFPase in Ref. [13]. The blue symbols mark the data points for a set of four Q values (0.05 Å^{-1} to 0.24 Å^{-1}), the dashed line shows the fits to the data points calculated with the fully dynamic structure factor including a diffusive mode while the solid lines in green show the result where only the undulation mode is visible in the time and length scale window of NSE (Γ_{col} = 0 in Eq. 6)

this simultaneous fit is $1.49 \pm 0.02 \times 10^{-1}$ cm^2/s, very similar to the 1.75×10^{-1} cm^2/s reported in Ref. [13] from DLS measurements. The form of Eq. 6 allows for multiplicative diffusive motion (which is visible on top of the membrane undulations, i.e. the undulations follow additionally the density fluctuations), or decoupled to some extend (by the factor A).

The microemulsions used with CalB in Ref. [14] had a significantly larger domain size than the sugar surfactant based microemulsion shown in Fig. 2. The reason for the difference is most probably the different domain size and hence different length scale where density fluctuations play a role. A smaller domain size shifts the region where density fluctuations are relevant into the window accessible by NSE. Length scale (or q-)dependent evaluation is then required and possible, and the diffusive contribution can be deduced, either directly from NSE experiments (provided the data quality and q-t-range is sufficient), or by an additional DLS measurement.

An interesting theoretical problem remains in the prediction at which domain sizes the signature of the breathing modes has to be included into the analysis. Up to now, only the actual analysis process can answer this question for the respective investigated system.

4.2 On the Effect of the Enzymes on the Membrane Fluctuation Dynamics

Two experiments will be compared here: the microemulsion structure and dynamics of the protein DFPase in a sugar surfactant based microemulsion, as reported in [13], and the enzyme CalB in a nonionic surfactant based microemulsion [14].

From the protein point of view, there is one major difference which is worth a direct comparison. The bending elasticity measured with NSE and SANS should both depend in the same way on the protein concentration.

Figure 3 compiles the bending rigidities determined with NSE and SANS for the two systems from Refs. [13] and [14]. The observed κ values are within the observation window constant with protein concentration for the DFPase, but not equal between NSE and SANS. For CalB, the NSE bending rigidity slight increases, while the one determined from the structure via SANS slightly decreases. The slight decrease from the structural determination is interpreted similar to the effect of adding homopolymers to a bicontinuous microemulsion [27] ("anti-boosting effect"). It is obtained from SANS, which measures a snapshot of the fluctuating structure, which is then averaged over many "snapshots" during the time of the measurement. NSE on the other hand provides insight into the local shape fluctuations on nanosecond time scales and provides therefore a different view on the elasticity. Adsorption may result in a stiffening of the membrane visible in the increase in κ from NSE. This behaviour might reflect the temporary adsorption of the lipase at the interface and a kinetic exchange equilibrium with the lipase residing in the aqueous phase. This is in contrast to the DFPase mainly dissolved in the water phase. The bending rigidities of the DFPase containing microemulsion differs significantly between NSE and SANS. The pure $\kappa_{0,NSE}$ shows this behaviour with the full integration of the Zilman-Granek model (which takes into account the undulation spectrum of the membrane) as well as the stretched exponential simplified model (which might take a too broad undulation spectrum, often corrected for with an adapted viscosity of the surrounding medium). The reason for the difference might be the rather medium efficient technical grade sugar surfactant, which requires a high surfactant concentration and a locally very croweded environment. The additional distance to the X-point in the phase diagram might change the contribution of κ and $\bar{\kappa}$ to the SANS bending rigidity, or in a simple picture the time averaged structure seems to be rather disordered compared to the high energy needed to bend the membrane. The role of the average viscosity which affects the diffusive as well as the undulation modes in the microemulsion is more complex in such technical grade high concentration microemulsions and need further attention. A more detailed view into this question requires additional experiments.

5 Conclusion

Two bicontinuous microemulsion systems with enzymes introduced in the aqueous phase are revisited and compared. The structure and dynamics measured with SANS and NSE provide interesting insights and details about the effect of the enzyme on the elastic properties of the surfactant membrane of the microemulsion. The enzyme can either be mainly dissolved in the water phase with no significant influence on the membrane elasticity as in the case of DFPase which lacks an amphiphilic character, or it can

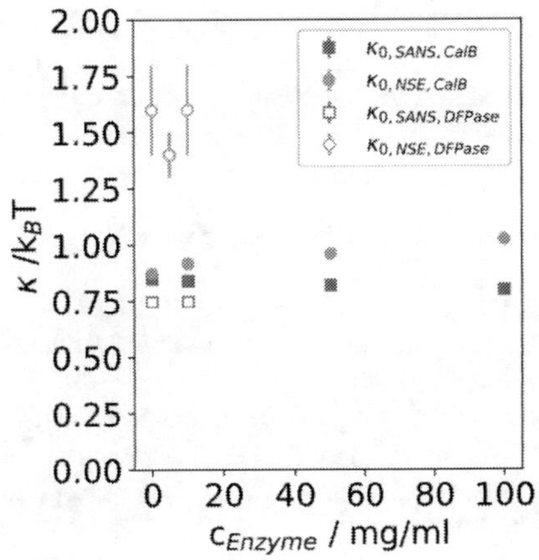

Figure 3. Comparison of bending rigidities from DFPase and CalB determined from structural data (SANS) and from membrane height fluctuations (NSE). Data are from Refs. [14] and [13].

modify the bending elasticity as in the case of CalB which can undergo conformational changes at a water-oil interface, where NSE reveals on nanosecond time scales a stiffening of the amphiphilic membrane, while SANS with its time averaged measurement shows the opposite effect. It has been suggested that the residence time of the protein CalB at the surfactant interface is the origin of this result. The reason why the DFPase does not change κ significantly, for example similar to the reduction in κ observed upon homopolymer addition into microemulsion phases as in Ref. [6], is a bit speculative. It may be that having the not surface active component only in the water phase is not sufficient for affecting the conformational space of the membrane and thus reducing κ. However, since also a combined effect of κ and $\bar{\kappa}$ could lead to this results, future measurements on both bending elasticity constants in droplet microemulsions could provide additional insights.

A second observation between the two systems under investigation is the relevance of breathing modes observable in the membrane dynamics depending on the characteristic size of the domains of the microemulsion. With smaller structures, the importance of breathing modes observed on the length scales accessible with NSE increases. The coupling of breathing modes with the undulation modes might be also a topic of further study.

Further experiments might shed light on the interfacial activity of enzymes in microemulsions and on the influence of the structural length scales of the microemulsion on its dynamics as well as on possible enzymatic activities. Such experiments can in future support theoretical considerations to predict the potential interfacial activity of enzymes.

Acknowledgement

This work is based upon experiments performed at the J-NSE instrument operated by Forschungszentrum Jülich at the Heinz Maier-Leibnitz Zentrum (MLZ), Garching, Germany.

References

[1] B. Orlich, R. Schomäcker, Enzyme Microb. Technol. **28**, 42 (2001)

[2] M. Subinya, A.K. Steudle, B. Nestl, B. Nebel, B. Hauer, C. Stubenrauch, S. Engelskirchen, Langmuir **30**, 2993 (2014)

[3] K. Homberg, Adv. Colloid Interf. Sci. **51**, 137 (1994)

[4] S. Serrano-Luginbühl, K. Ruiz-Mirazo, R. Ostaszewski, F. Gallo, P. Walde, Nature Rev. Chem. **2**, 306 (2018)

[5] C. Frank, H. Frielinghaus, J. Allgaier, H. Prast, Langmuir **23**, 6526 (2007)

[6] O. Holderer, H. Frielinghaus, D. Byelov, M. Monkenbusch, J. Allgaier, D. Richter, Journal of Chemical Physics **122** (2005)

[7] M. Mihailescu, M. Monkenbusch, H. Endo, J. Allgaier, G. Gompper, J. Stellbrink, D. Richter, B. Jakobs, T. Sottmann, B. Farago, The Journal of Chemical Physics **115**, 9563 (2001)

[8] H. Klemmer, H. Frielinghaus, J. Allgaier, M. Ohl, O. Holderer (2017), Vol. 862, ISSN 17426596

[9] S. Wellert, H.J. Altmann, A. Richardt, A. Lapp, P. Falus, B. Farago, T. Hellweg, The European Physical Journal E **33**, 243 (2010)

[10] H. Klemmer, J. Allgaier, H. Frielinghaus, O. Holderer, M. Ohl, Colloid and Polymer Science **295** (2017)

[11] M. Teubner, R. Strey, The Journal of Chemical Physics **87**, 3195 (1987)

[12] A.G. Zilman, R. Granek, Physical review letters **77**, 4788 (1996)

[13] S. Wellert, B. Tiersch, J. Koetz, A. Richardt, A. Lapp, O. Holderer, J. Gäb, M.M. Blum, C. Schulreich, R. Stehle et al., European Biophysics Journal **40** (2011)

[14] S. Engelskirchen, S. Wellert, O. Holderer, H. Frielinghaus, M. Laupheimer, S. Richter, B. Nestl, B. Nebel, B. Hauer, Frontiers in Chemistry **8**, 613388 (2021)

[15] M.J. Schwager, K. Stickdorn, R. Schomäcker, Chem. Rev. **95**, 849 (1995)

[16] K. Homberg, Eur. J. Org. Chem. **2007**, 731 (2007)

[17] K.M. Larsson, P. Adlercreutz, B. Mattlasson, J. Chem. Soc. Faraday Trans. **87**, 465 (1991)

[18] W. Helfrich, Zeitschrift für Naturforschung A **33**, 305 (1978)

[19] S. Safran, Advances in physics **48**, 395 (1999)

[20] P. Pieruschka, S. Safran, S. Marcelja, Physical Review E **52**, 1245 (1995)

[21] M. Peltomäki, G. Gompper, D.M. Kroll, The Journal of chemical physics **136**, 134708 (2012)

[22] O. Holderer, H. Frielinghaus, M. Monkenbusch, M. Klostermann, T. Sottmann, D. Richter, Soft Matter **9** (2013)

[23] M. Blum, F. Löhr, A. Richardt, H. Rüterjans, J.C.H. Chen, J. Am. Chem. Soc. **128**, 1275012757 (2006)

[24] F. Mezei, ed., *Neutron Spin Echo*, Vol. 128 (Springer Berlin Heidelberg, 1980), ISBN 978-3-540-10004-1

[25] F. Mezei, C. Pappas, T. Gutberlet, eds., *Neutron Spin Echo Spectroscopy*, Vol. 601 (Springer Berlin Heidelberg, 2003), ISBN 978-3-540-44293-6

[26] M. Monkenbusch, O. Holderer, H. Frielinghaus, D. Byelov, J. Allgaier, D. Richter, Journal of Physics Condensed Matter **17** (2005)

[27] D. Byelov, H. Frielinghaus, O. Holderer, J. Allgaier, D. Richter, Langmuir **20** (2004)

Open problems in liquids dynamics: the role of neutron scattering

Eleonora Guarini[1*], *Gianmarco* Masini[1], *Ubaldo* Bafile[2], *Milva* Celli[2], *Daniele* Colognesi[2], *Alessandro* Cunsolo[3], *Luisa* Scaccia[4], *Alessio* De Francesco[5], and *Ferdinando* Formisano[5]

[1]Dipartimento di Fisica e Astronomia, Università degli Studi di Firenze, I-50019 Sesto Fiorentino, Italy
[2]Consiglio Nazionale delle Ricerche, Istituto di Fisica Applicata "Nello Carrara", I-50019 Sesto Fiorentino, Italy
[3]Department of Physics, University of Wisconsin-Madison, Madison, WI 53706, United States
[4]Dipartimento di Economia e Diritto, Università di Macerata, I-62100 Macerata, Italy
[5]CNR-IOM & INSIDE@ILL c/o Operative Group in Grenoble and Institut Laue Langevin, F-38042 Grenoble, France

Abstract. We review recent inelastic neutron scattering experiments aimed at investigating still open issues in the microscopic dynamics of liquids. It is shown that the interpretation of experimental results is put on solid ground by the application of modern methods of analysis and lineshape modelling which ensure the fulfillment of fundamental physical properties that the spectra must obey. This last condition becomes crucial to avoid overinterpretations of the genuine information conveyed by scattering data, especially when studying weak signals in the dynamic structure factor. Moreover, we highlight the different roles that neutron data presently play compared to molecular dynamics simulations depending on the nature of the sample, including the case of quantum liquids. In particular, we show how neutron measurements remain an indispensable benchmark in assessing the present capabilities of classical and quantum simulation methods. We also mention the potential of statistical methods, such as Bayesian inference, when applied to neutron data analysis and the opportunity they provide in establishing the spectral features without arbitrary assumptions on the model lineshape.

1 Introduction

Inelastic neutron scattering (INS) is a master technique for experimental studies on the dynamic behaviour, at the nanometre and picosecond scales, of various classes of liquids. Topical examples are given by the pioneering works on the longitudinal collective excitations (sound waves) in liquid metals as Pb and Rb [1,2], which opened the way to the numerous investigations on metallic liquids carried out up to present times [3,4]. In particular, these simple monatomic liquids were more recently taken as reference systems for studies of shear waves through the search for possible low frequency excitations in the dynamic structure factor $S(Q,\omega)$ [5], which is accessed by both INS and inelastic x-ray scattering (IXS) techniques when the probe exchanges a momentum $\hbar\mathbf{Q}$ and an energy $E = \hbar\omega$ with the sample, $\hbar$ being the reduced Planck constant.

However, the true realm of neutron scattering is liquid H_2, together with its heavier counterpart D_2. Across the years, both these fluids attracted much interest when investigating general aspects of the single-particle (self) dynamics of an incoherent scatterer like H_2 [6-9], or quantum effects on either the self [10,11] or the collective properties [12-14] of these systems. In fact, the low molecular mass of H_2 and D_2 and the low temperatures at which they are in the liquid phase lead to evident manifestations of quantum behaviour, as further detailed in Sect. 3. Moreover, H_2 and D_2 are the most used moderating materials for the production of low-energy (about 1 meV) neutrons. In this respect, the design of new generation hydrogen-based cold neutron sources requires the availability of reliable databases of the scattering law taking explicitly the quantum nature of these fluids into account. Simultaneously, it demands an efficient mapping of the whole kinematic (Q,ω) range for best estimates of the response function. Such requirements unavoidably call for simulation methods capable of providing reliable dynamical data on H_2 and D_2, since experiments will never cover, in reasonable times, the innumerable kinematic conditions contributing to a total cross section calculation, with varying incident neutron energy. Available semiclassical methods for simulating the dynamics of H_2 and D_2, like ring polymer molecular dynamics (RPMD) [15] and two versions of the Feynman-Kleinert (FK) approach [16,17], urge then an experimental check, and neutrons are the best probe for this purpose.

Here we review recent INS investigations of both the mentioned classes of liquids, with specific interest in the methods of analysis. In particular, we first address the case of liquids which can be assumed to behave classically, like molten metals, with special focus on the still debated claim that the experimental $S(Q,\omega)$ of some metals [18,19] bears the signature of low-frequency transverse-like excitations. As a matter of fact, a

* Corresponding author: guarini@fi.infn.it

dependence on the methods of analysis emerges, so the state of the art of reliable approaches for model fitting to the data is discussed in some detail.

The remainder of the paper deals with the case of a quantum liquid like D_2, while discussing the present need for absolute scale comparisons between neutron data and quantum simulations probing the translational dynamics through the centre of mass dynamic structure factor $S_{CM}(Q,\omega)$ of the liquid. Indeed, the effectiveness of quantum calculations of the microscopic dynamics has been poorly tested until present, hindering the possible use of existing quantum simulation methods in the mentioned applications related to cold neutron production. The benchmark of neutron measurements evidences the still lacking performance of the semiclassical approximations adopted in each of the available quantum simulation techniques.

2 Transverse dynamics of liquid metals

In the 1970s, simulation studies of specific time autocorrelation functions [20] have shown that dense fluids can sustain shear wave propagation at sufficiently small wavelengths (around and below 0.5 nm, approximately). The specific functions we are referring to are the transverse current autocorrelation function $C_T(Q,t)$ and the velocity autocorrelation function $Z(t)$. The former is given by

$$C_T(Q,t) = \langle \mathbf{j}_T^*(\mathbf{Q},0) \cdot \mathbf{j}_T(\mathbf{Q},t)\rangle / (2N), \quad (1)$$

with $\mathbf{j}_T(\mathbf{Q},t) = \mathbf{j}(\mathbf{Q},t) - \mathbf{j}_L(\mathbf{Q},t)$, where the current of an N particle system is $\mathbf{j}(\mathbf{Q},t) = \sum_\alpha \mathbf{v}_\alpha(t) \exp[i\,\mathbf{Q} \cdot \mathbf{R}_\alpha(t)]$, $\mathbf{R}_\alpha(t)$ and $\mathbf{v}_\alpha(t)$ denoting the position and velocity of the α-th particle at time t, and $\mathbf{j}_L(\mathbf{Q},t)$ is the longitudinal current, i.e. the projection of $\mathbf{j}(\mathbf{Q},t)$ on the wavevector $\mathbf{Q}$. The angular brackets in Eq. (1) stand for the statistical average in the canonical ensemble [5]. The velocity autocorrelation function is defined by

$$Z(t) = \sum_\alpha \langle \mathbf{v}_\alpha(0) \cdot \mathbf{v}_\alpha(t)\rangle / N. \quad (2)$$

In particular, the spectrum $Z(\omega)$ of Eq. (2) can be considered, for a liquid, as a sort of equivalent of the phonon density of states (DoS) in solid state physics. Unfortunately, neither $Z(\omega)$, nor the spectrum of the transverse current autocorrelation function, $C_T(Q,\omega)$, can actually be measured. Indeed, past determinations of $Z(\omega)$ by incoherent neutron scattering [6-8] revealed to be demanding. INS and IXS experiments enable the determination of the space and time Fourier transform of the microscopic density autocorrelation function, i.e. of the dynamic structure factor $S(Q,\omega)$ mentioned in the Introduction. However, $S(Q,\omega)$ is a longitudinal quantity by definition [5], therefore detectability of transverse-like contributions to this function has often been objected as a matter of principle [21]. Nonetheless, more than ten years ago, at least two papers analyzing IXS data on two molten metals [18,19] reported the presence of an additional low-frequency component in the measured $S(Q,\omega)$. In both works, the same phenomenological model was fitted to the experimental spectra. In particular, the fit function was modelled by adding one or two (depending on Q) damped harmonic oscillator (DHO) doublets [22] to a central Lorentzian, the latter accounting for relaxation processes in an effective way. However, such a modelling disregards the fact that the global fit function has a divergent second frequency moment [5,22] in place of the finite theoretical value k_BTQ^2/M, where k_B is Boltzmann constant, T the temperature, and M the atomic mass of the fluid. As we will show, the physical consistency of the adopted models can instead be decisive for a correct interpretation of the experimental results.

At present, there are two possibilities to limit the arbitrariness of fit-based procedures. The first is based on the availability of an exact theory for the functional form of autocorrelation functions and corresponding spectra [23-25], where known constraints can be easily imposed. In this case, one can choose a few tentative models guided by physical considerations, then enforce at least the most important sum rules that $S(Q,\omega)$ must obey, and finally check which model provides the best fit quality with simultaneous minimization of the number of free parameters.

The other possibility is to control model fitting to the data on a statistical basis, without imposing *a priori* the number of excitations. At present, we use simple phenomenological models in a complex algorithm exploiting Bayes theorem to estimate the posterior probabilities, conditional to the experimental data, of the various parameters, including the number of excitations itself [26]. The information from the posterior distributions permits to control, in a statistical sense, what the data actually support.

Both the theoretically and the probabilistically grounded approaches for the modelling of the scattering signal are briefly described in the next subsections.

2.1 Exponential Expansion Theory (EET)

The theory states that any normalized autocorrelation function $c(t)/c(0) = \langle A(0)\,A(t)\rangle / \langle A(0)^2\rangle$ of a generic dynamical variable $A(t)$ can be represented by a series of exponential terms, called *modes*, with generally complex amplitudes I_j and frequencies z_j, i.e.:

$$c(t)/c(0) = \sum_j I_j \exp(z_j\,|\,t\,|). \quad (3)$$

Consequently, the spectrum is a sum of so-called generalized Lorentzian lines

$$c(\omega)/c(t=0) = \sum_j (-I_j\,z_j) / [\pi(\omega^2 + z_j^2)]. \quad (4)$$

In particular, if I_j and z_j are real, we are dealing with relaxation processes with decay constant $z_j < 0$, which contribute with genuine central Lorentzians to the spectrum. If I_j and z_j are complex, pairs of conjugate modes are present in the series and account for damped oscillatory components in the correlation function due to collective excitations characterized by a damping $\mathrm{Re}(z_j) < 0$ and a frequency $\mathrm{Im}(z_j)$. Such complex pairs correspond to two distorted Lorentzian lines in the spectrum, centred at the nonzero frequencies $\pm\,\mathrm{Im}(z_j)$, respectively. As far as the sum rules are concerned, normalization is guaranteed by enforcing the condition $\sum_j I_j = 1$. In addition, constraints are imposed on the odd time derivatives of $c(t)$ at $t = 0$ in the form

$$[d^{\,p}c(t) \,/\, dt^{\,p}]_{t=0} = \sum_j I_j \, z_j^{\,p} = 0, \text{ for odd } p \qquad (5)$$

which ensure the correct behaviour of the function at the time origin and are equivalent to require the finiteness of the even spectral moments in the frequency domain.

Luckily, most correlation functions of interest in liquids dynamics are accurately described by a small number of modes, meaning that the main dynamical processes are actually few. The scheme of the modes may of course change with varying the thermodynamic state and the probed length scale (Q). At each wavevector and thermodynamic state, the best model is chosen on the basis of the corresponding fit quality and of some general knowledge on liquids' behaviour.

In what follows, three models complying with the general EET will be considered for $S(Q,\omega)$. The simpler one is a generalized hydrodynamics (GH) triplet with Q-dependent parameters [5]. It is a single-excitation model consisting of one central Lorentzian, accounting for thermal diffusion and structural relaxation in an effective way, plus one doublet of distorted Lorentzians representing the longitudinal sound waves propagating in the fluid. The model spectrum is constrained to have a finite second frequency moment. The second modelling is the viscoelastic (VE) lineshape [22], which differs from the GH one as regards the relaxation phenomena (two central Lorentzians in place of a single one) and the obeyed sum rules. The VE model again foresees a single oscillatory component due to longitudinal collective excitations and its spectral moments are constrained to converge up to the fourth one. Finally, a two-excitation model, labelled 2C to mean "two complex pairs" [4], is considered in those cases where also shear waves have set in in the liquid. Relaxation phenomena are represented by an effective central Lorentzian and constraints are enforced to ensure, at least, a finite fourth frequency spectral. Detailed-balance asymmetry must be applied to the above models before comparison with experiment.

2.2 Bayesian inference

As demonstrated in previous works by some of us [26,27], Bayesian methods can be successfully applied to achieve a minimally biased and probabilistically grounded modelling of experimental spectra. The inferential approach relies on the use of a Markov Chain Monte Carlo algorithm endowed with a Reversible Jump (RJ) option [28] which enables the space of free parameters to contain also the number k of excitations, and identifies the model with the highest posterior probability evaluated according to Bayes theorem, i.e., conditionally on the experimental outcome. As mentioned above, thus far the code foresees the use of simple phenomenological models for $S(Q,\omega)$, consisting of an effective central Lorentzian plus a number k of DHO doublets. Details of the Bayesian inference algorithm can be found in Ref. [26]. It is important to note that when the number of possible models (numbered by k) is a parameter itself, a Bayesian analysis naturally includes the so-called Occam's razor principle or "lex parsimoniae", which states that between two models providing an equally good account of some evidence, the one containing the lower number of parameters is always to be preferred. Thus, overparametrizations are automatically avoided within the Bayesian approach [27].

In tackling the detectability of a second (transverse) excitation in the experimental $S(Q,\omega)$, we are specially interested in the posterior distribution of the parameter k, for which we assumed a uniform prior distribution (i.e., all models are considered to be *a priori* equally probable) in order to let the inferential process be driven uniquely by the experimental evidence. An example of the results of such a statistical analysis is discussed, along with EET-based ones, in the next section.

2.3 INS and simulation results for Au and Ag

Here we summarize the main results of two investigations carried out on liquid Au [3] and Ag [4] by both INS and *ab initio* molecular dynamics (AIMD) simulations. Both metals were studied experimentally in the so-called neutron Brillouin scattering (NBS) regime, i.e., at the rather low wavevectors ranging from a few inverse nanometres (e.g., 4 nm^{-1}) to values slightly exceeding $Q_p / 2$, where $Q_p \approx 26$ nm^{-1} is the position of the maximum in the static structure factor $S(Q)$ of both metals. Since liquid metals are characterized by a high sound velocity c_s (i.e., 2568 m/s in Au and 2790 m/s in Ag) the rather energetic beam (incident energy $E_0 \approx 80$ meV, energy resolution HWHM = 1.5 meV, corresponding to 2.3 rad ps^{-1} in ω) of the small angle BRISP spectrometer at the Institut Laue Langevin (ILL) [29,30] was required to span the mentioned Q range in both cases. At the same time, the experimental work was paralleled by AIMD calculations (details are given in Refs. [3] and [4]) to check the capability of *ab initio* methods in reproducing the neutron results and, in the case of satisfactory comparisons, use the simulations to extend the study of the dynamics to higher wavevectors.

The first experiment was performed on Au. An example of the experimental $S(Q,\omega)$ at the highest Q of the measurements is given in Fig. 1.

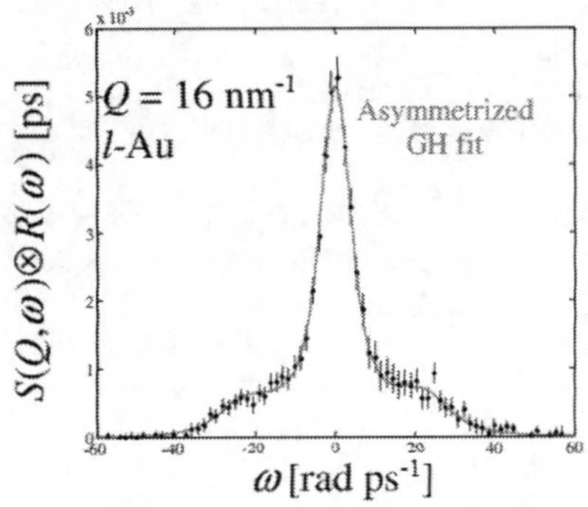

Fig. 1. Experimental (resolution broadened) $S(Q,\omega)$ of liquid Au (dots with error bars) and GH fit (red solid curve).

An EET analysis of the spectra showed that a simple GH triplet (see red curve in Fig. 1) was more than sufficient to obtain an accurate description of the data at all Q values. Thus, no evidence of transverse excitations was found from the experimental spectra of liquid Au. Nonetheless, the very good agreement between neutron data and simulations (see Fig. 2) justified an EET analysis of the simulated $S(Q,\omega)$ in the wide range 4 nm^{-1} < Q < 70 nm^{-1} probed by AIMD. In this case, a GH modelling was found to be insufficient, while spectra were perfectly described by a VE lineshape. An example of the fit quality is given in Fig. 3.

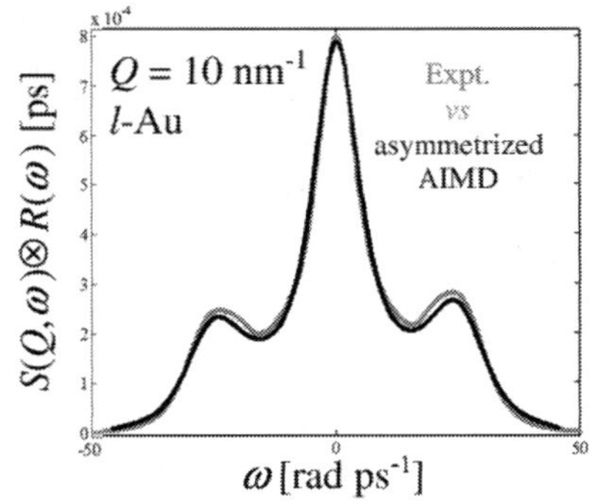

Fig. 2. Simulated $S(Q,\omega)$ at a representative Q value, taking asymmetry and experimental resolution into account (black curve). The red curve is the GH fit to the experimental data.

The VE fits to the simulations thus permitted to better resolve the central peak, but again only a single excitation could be detected. In summary, signs of shear waves turned out to be absent in both the experimental and simulated $S(Q,\omega)$ of liquid gold.

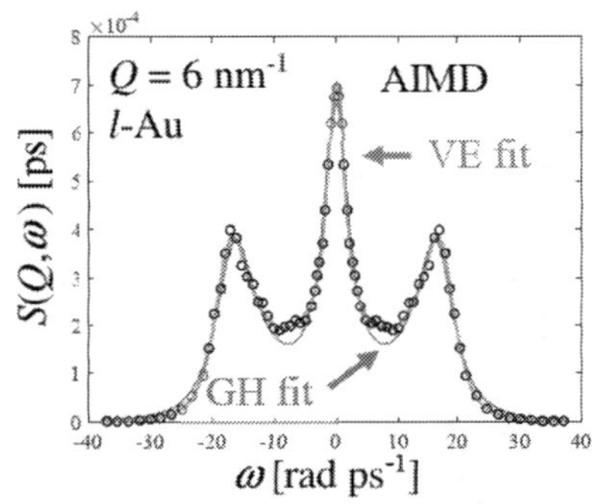

Fig. 3. Simulated $S(Q,\omega)$ of liquid Au (black circles). The GH fit (red thin curve) is clearly inaccurate, while the VE modelling (pale blue curve) provides a high fit quality.

However, shear waves clearly propagate in this liquid as witnessed by the maximum in the DoS shown in Fig. 4(b), and by the evolution, as Q grows, of a low-frequency maximum in $C_T(Q,\omega)$ (see Fig. 4(c)). The dispersion relation we were able to determine from both experiment and simulations contains instead only the longitudinal branch (Fig. 4(a)).

A study similar to that on Au was later carried out on liquid Ag, again by using the BRISP spectrometer and performing parallel AIMD calculations. As far as the measurements are concerned, a GH modelling was the only justified within the accuracy of the data (see Fig. 5), thus providing no evidence of a second excitation in the neutron spectra, like in the case of gold.

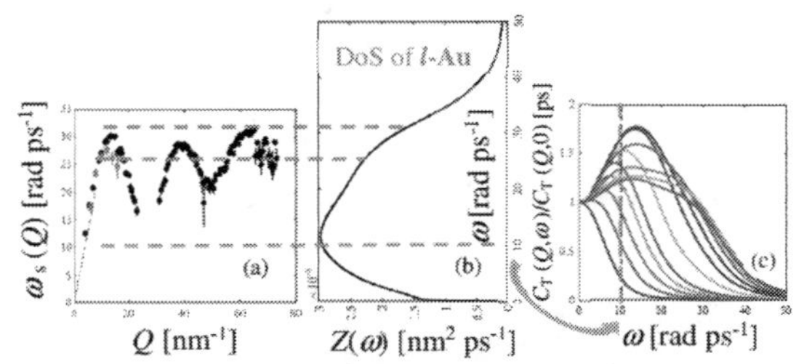

Fig. 4. (a) Dispersion relation obtained from NBS (red dots) and AIMD data (black dots); the dashed black line is the hydrodynamic law $\omega_s = c_s Q$. (b) DoS of liquid Au from AIMD. (c) Normalized $C_T(Q,\omega)$ at Q values ranging from 4.0 (blue monotonic curve) to 25.5 nm^{-1} (broad purple curve). The magenta dashed lines, containing the maxima of $\omega_s(Q)$, highlight the frequency band where the DoS shows the typical shoulder due to longitudinal modes. The green dashed curve, marking the frequency around which the maxima of $C_T(Q,\omega)$ evolve in panel (c), is shown to correspond to the frequency of the maximum in the DoS, the latter owing to the weakly dispersive transverse modes that were not detected from $S(Q,\omega)$, thereby leaving a missing branch in panel (a).

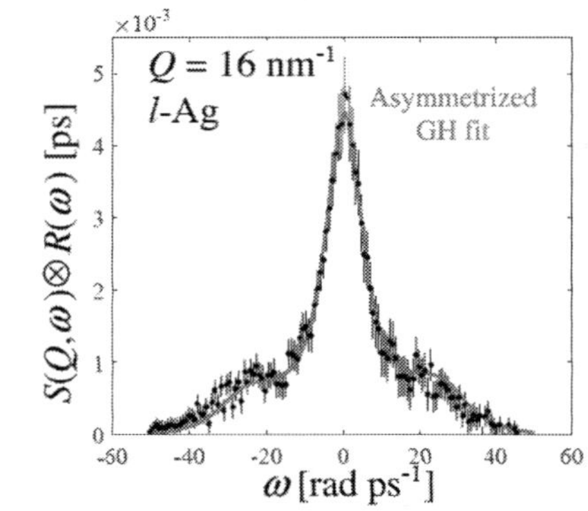

Fig. 5. Experimental (resolution broadened) $S(Q,\omega)$ of liquid Ag (dots with error bars) and GH fit (red solid curve).

By contrast, when we analyzed the AIMD simulations, we found that above Q = 15 nm^{-1} a VE lineshape became insufficient, while a 2C model (see Sect. 2.1) provided a very accurate description of the simulated spectra, as shown in Fig. 6.

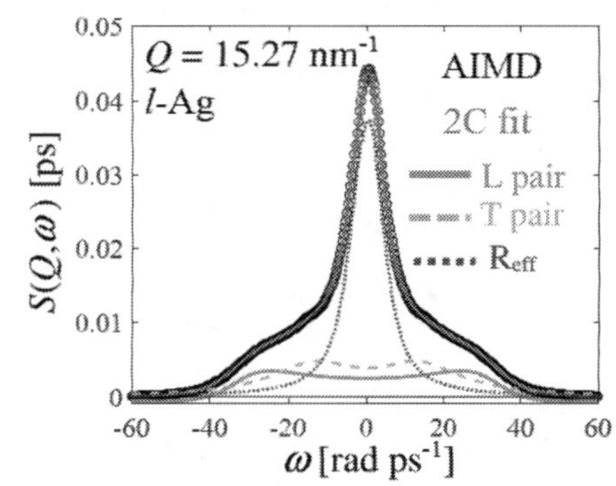

Fig. 6. AIMD $S(Q,\omega)$ of liquid Ag (circles) and 2C fit (red solid curve). The spectral components are also shown.

https://doi.org/10.1051/epjconf/202328604002

The 2C fits to the AIMD data of Ag thus enabled the determination of a second, low-frequency, branch $\omega_t(Q)$ in the dispersion relation displayed in Fig. 7. However, when ω_s starts to decrease, the fits become unstable and the low-frequency modes are badly determined. Conversely, transverse-like modes are clearly detected whenever ω_s and ω_t have considerably different values.

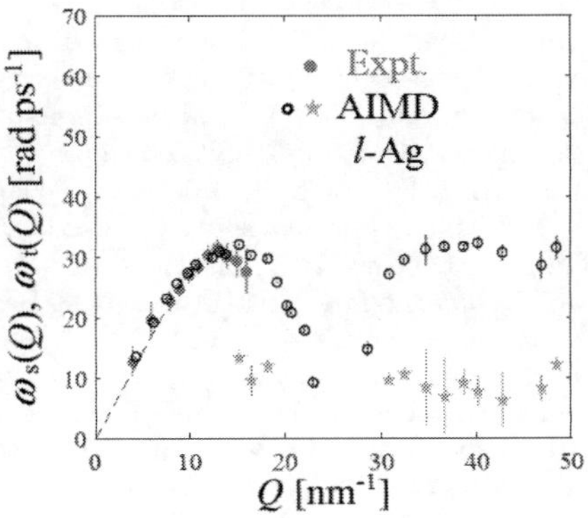

Fig. 7. Dispersion relation of liquid Ag. Red dots and black circles represent the longitudinal branch as obtained from experiment and simulations, respectively. The transverse branch (green stars) could be determined only from the AIMD results at Q values where ω_s and ω_t are sufficiently different.

Given the ambiguous situation between experimental and simulation results for liquid Ag, we revisited the neutron data by means of the second statistical route based on Bayes' theorem and briefly described in Sect. 2.2. As mentioned, we are particularly interested in the posterior distribution of the parameter k, conditional to the experimental data set at hand (globally denoted as Y). The posterior distribution of the number of excitations is reported in Fig. 8 at both low and high values of the measured Q range.

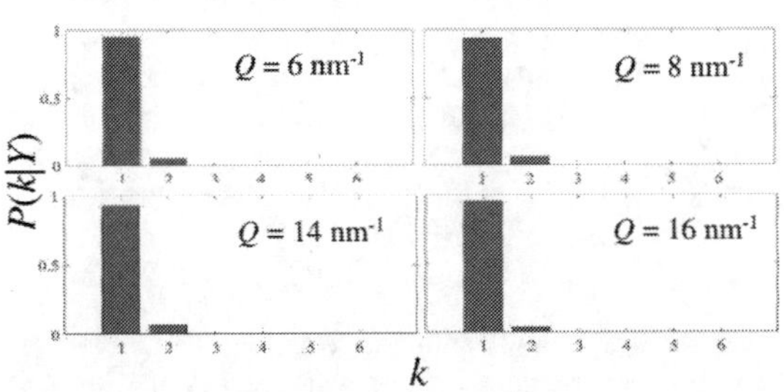

Fig. 8. Conditional posterior distributions of the number k of inelastic components in the experimental $S(Q,\omega)$ of liquid Ag at selected Q values.

Clearly, the algorithm privileges the one-excitation case, confirming the undetectability of transverse modes in the available neutron data. Moreover, very well-shaped unimodal posteriors were obtained for all the parameters [31] indicating the high reliability of the fit results which provided, within the errors, the same longitudinal dispersion curve of Fig. 7. Therefore, different ways to control the plausibility and coherence of model fitting to the data, like the EET and the Bayesian approaches, provide the same results.

To further verify the output of the inferential analysis, it is possible to switch off the RJ algorithm and check the specific posterior distributions pertaining, separately, to the k = 1 and k = 2 cases. Figure 9 shows the corresponding posterior distributions for the undamped frequency $\Omega_s = (\omega_s^2+\Gamma_s^2)^{1/2}$ of sound waves (Γ being the damping), along with that of shear waves Ω_t present only in the k = 2 case. It is seen that in the two-excitation case a less symmetric distribution is obtained for Ω_s and, more importantly, a broad and nearly flat distribution pertains to Ω_t. The second excitation is therefore completely undetermined, confirming the RJ-on result where k is a free parameter.

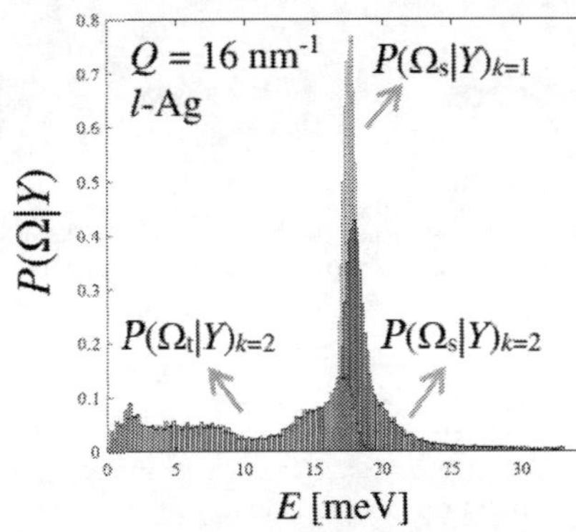

Fig. 9. Posterior distributions for the undamped frequency in the one- (cyan) and two-excitation (green for the transverse and pink for the longitudinal) cases, specifically obtained by switching off the RJ algorithm option.

Finally, to better understand why experiments did not reveal a second excitation in $S(Q,\omega)$, we performed a rather stringent test on the Ag neutron data. First, we compared the performance of the EET 2C model with the GH one of Fig. 5. Figure 10 shows that at both reported Q values the second complex pair of the 2C lineshape is characterized by a negligible amplitude.

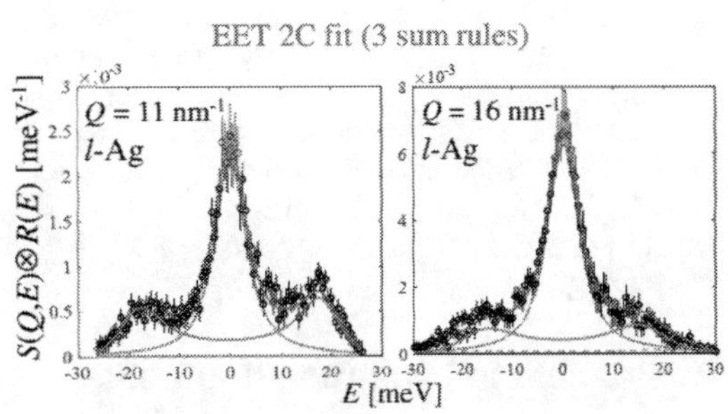

Fig. 10. Fits of the 2C lineshape to the experimental $S(Q,\omega)$ of liquid Ag. The second complex pair of the model is the dashed green curve hardly distinguishable from the zero axis.

Moreover, we found for this inelastic component an unreasonably large inelastic shift and an error on the damping doubling the damping value itself, meaning that such a parameter is undetermined. Therefore, according to *lex parsimoniae*, there are no reasons to choose the 2C result (8 parameters) in place of the GH one (5 parameters) which provides an identical global fit curve.

Then, we analyzed the neutron spectra in the phenomenological fashion adopted in other works [18,19], without any constraint except normalization. Figure 11 puts in evidence quite an embarassing situation, resembling very much that of Fig. 1 of Ref. [18] or of Ref. [19].

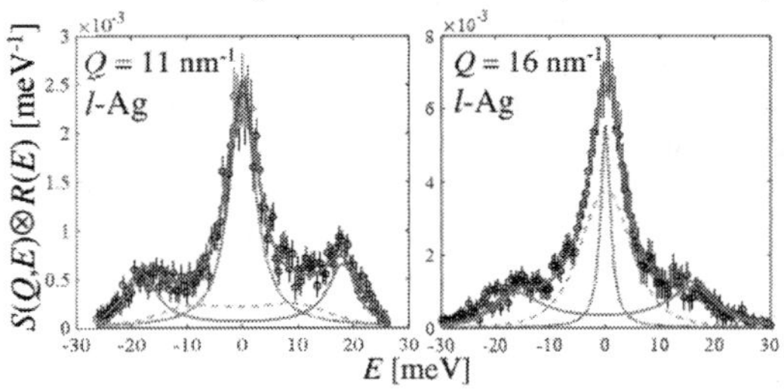

Fig. 11. Fits of the neutron data on liquid Ag using the unconstrained phenomenological model specified in the figure title.

The previous example demonstrates that the use of constrained models (i.e., obeying at least the most important sum rules) can completely change the results and the deduced physical properties. Indeed, uncontrolled fit procedures, escaping either physical or statistical consistency criteria are likely prone to biases of confirmation. Moreover, the above test shows that the undetectability of low-frequency contributions in appropriate modellings of the experimental $S(Q,\omega)$ is not due to the limited resolution or to the scattering of the data-points, but is merely *a matter of analysis*.

As a final remark regarding this discussion on high-temperature classical fluids, we can conclude that the experimental observation of shear waves is not yet assessed for either system, and still remains out of reach of the present spectroscopic techniques. In fact, convincing indications that $S(Q,\omega)$ bears also the signature of the transverse dynamics were only found by means of rather recent, quantitative EET-studies of simulation results [4]. On the other hand, neutron data on these classical systems showed the effectiveness of *ab initio* simulation methods, which open the way to extended investigations of the dynamical behaviour of simple classical liquids, without limitations in wavevector.

3 "Boltzmann" quantum liquids: D_2

Liquid hydrogen and deuterium represent an intermediate case within the few systems displaying quantum behaviour (He, H_2, D_2, and Ne). In fact, they are so-called Boltzmann quantum fluids, where, differently from the case of helium, exchange effects due to particle indistinguishability are supposed to be negligible in comparison to particle delocalization, and Boltzmann statistics can still be assumed to hold. Therefore, distinct trajectories in phase space can still be defined, and molecular dynamics methods applied, although with obvious differences from a purely classical treatment. This is the reason why H_2 and D_2, can be considered as *moderate* quantum fluids, since only single-molecule delocalization is actually responsible for a nonclassical behaviour, as suggested by the limited values [32] reached by the de Broglie thermal wavelength $\Lambda = h / (2\pi M k_B T)^{1/2}$ even close to their respective triple points.

The characteristic length scale considered for comparisons with Λ at a certain temperature is the mean interparticle distance $l = n^{-1/3}$, with n being the number density. At all relevant liquid densities and temperatures of H_2 and D_2, the condition $\Lambda < l$ holds [32], implying that overlap of the spatial wave functions of two adjacent molecules, and consequently quantum exchange, does not occur on average, so that quantum statistics need not be invoked. Nonetheless, Λ can reach values of the order of the molecular size, giving anyway rise to quantum delocalization effects that must be accounted for in some approximate, semiclassical way.

A recent review of the available semiclassical approximations for calculations of the dynamic structure factor of moderate quantum fluids, like RPMD [15], FK-Linearized Path Integral (LPI) [16] and FK-Quasi Classical Wigner (QCW) [17], can be found in Ref. [14].

On the experimental side, neutron measurements on these liquids have been few, and often reporting the double differential cross section in arbitrary units (including the rotational contributions) [12,13] rather than the centre of mass dynamic structure factor $S_{CM}(Q,\omega)$ in absolute ones. Experimental knowledge of the translational dynamics would instead be very helpful for an in-depth verification of the mentioned simulation methods, enabling or not their possible use in applications. These reasons induced us to perform a neutron measurement on liquid D_2, accompanied by the RPMD, FK-LPI and FK-QCW calculations detailed in Ref. [14], all relying on the isotropic Silvera and Goldman intermolecular potential [33].

The experiment was carried out on the BRISP spectrometer in the standard configuration used also for the metallic samples. After corrections for background, attenuation and multiple scattering, the single scattering intensity of a diatomic homonuclear molecule can be schematized as

$$I^{(1)}(Q,E) = C\,\{(k_1 / k_0)\,[u(Q)S_{CM}(Q,E)+J(Q,E)]\}\otimes R(E), \quad (6)$$

where C is a normalization factor and k_1/k_0 is the ratio of scattered to incident neutron wavevector. In Eq. (6), $u(Q)$ is the intermolecular cross section depending only on the coherent scattering length of D, while $J(Q,E)$ is an intramolecular term accounting for the rotational structure and which depends also on the incoherent scattering length of D. Both $u(Q)$ and $J(Q,E)$ can be confidently calculated within well-known approximations complying, in the case of $J(Q,E)$, with the asymmetry requirements of quantum spectra [14]. Finally, $R(E)$ is the instrument energy resolution function.

The quantity we are interested in is the quantum $S_{CM}(Q,E)$, which is related to the Kubo (symmetric) dynamic structure factor $S_{CM,K}(Q,E)$ through the well-know asymmetry factor $\beta E / [1\text{-exp}(-\beta E)]$, with $\beta = (k_B T)^{-1}$. Therefore, one can model $S_{CM,K}(Q,E)$ by means of a classical lineshape and perform an overall fit to $I^{(1)}(Q,E)$ according to Eq. (6) and to the asymmetry condition quoted above. Again, a GH model for $S_{CM,K}(Q,E)$ was found to provide an accurate description of the data, as shown in Fig. 12(a) for a representative Q value. The various components of the fit function can be better appreciated in the zoom of Fig. 12(b) and are detailed in the caption.

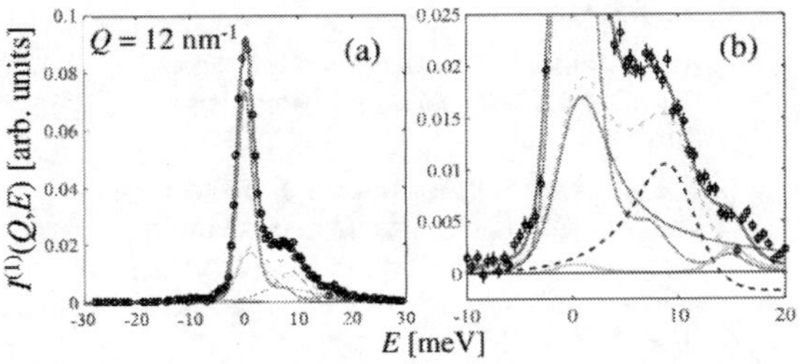

Fig. 12. (a) Single scattering intensity of liquid D_2 (black circles with error bars) and global fit curve (red solid line) according to Eq. (6). (b) Zoom of panel (a): the magenta dot-dashed curve is the intramolecular component, with visible rotational lines; the dotted and dashed black curves are the elastic and inelastic components which sum up to give, within a normalization factor, the asymmetric GH-based $S_{CM}(Q,E)$ (cyan dashed curve); the green solid curve accounts for a small amount of H_2 likely present in the sample.

The fits to $I^{(1)}(Q,E)$ then provided the sought-for absolute scale and resolution-free $S_{CM}(Q,E)$. The latter is compared, at 9 nm^{-1}, with the various simulation results in Fig. 13. Similar results were obtained at other wavevectors [14].

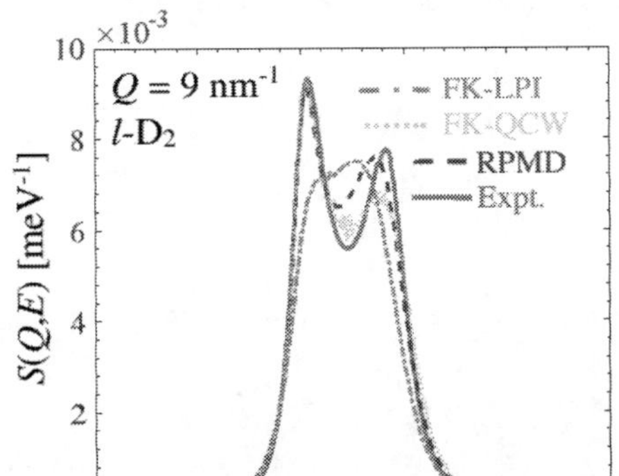

Fig. 13. Resolution-free quantum dynamic structure factor of liquid D_2 in absolute units, as obtained from the fits to the neutron data (red solid curve) and from different semiclassical approximations used in available quantum simulation techniques, each one specified in the legend and detailed in Ref. [14].

The comparisons in Fig. 13, needless to say, are quite unsatisfactory independently of the attempted simulation method. Neutron data visibly show much more marked collective excitations. Moreover, simulations do not agree even among themselves. Despite the clear loss of important details of the dynamical structure, it is anyway valuable that the absolute scale of neutron and simulation data is the same.

In order to compare frequencies and damping coefficients derived from experiment and simulations in more detail, we carried out also a thorough EET analysis of the RPMD and FK-QCW results. All simulation outputs were found to be accurately described by a VE model at the investigated Q values. The corresponding longitudinal dispersion curve and wavevector dependence of the damping coefficient of liquid D_2, are finally shown in Fig. 14.

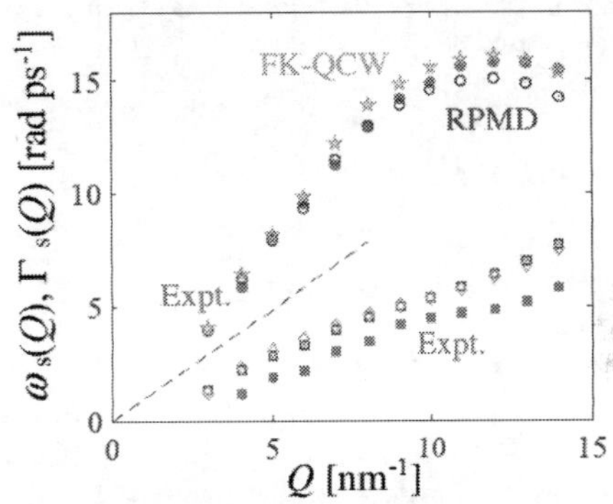

Fig. 14. Experimental (red full circles), RPMD (blue empty circles), and FK-QCW (green empty stars) dispersion curve of sound modes. The dashed black line is the hydrodynamic behaviour $c_s Q$, with $c_s = 0.984$ nm/ps. The damping coefficient is displayed with red full squares, blue empty squares, and green empty diamonds for experiment, RPMD and FK-QCW, respectively.

Figure 14 shows that, although the dispersion curves do not agree within the errors at some Q value, this property is reasonably captured by simulations, especially by the FK one. Conversely, the striking feature in the plot is the smaller damping deduced from the measurements.

The long lifetime of collective excitations is considered to be one of the main dynamical manifestations of quantum behaviour [34]. Apparently, the present simulation techniques do not fully grasp a salient feature of quantum liquids dynamics. Of course, we cannot exclude with absolute certainty some systematic errors in the (demanding) neutron data analysis. However, the disagreement among simulation results suggests that the semiclassical approximations adopted in each technique should anyway be improved. Indeed, while the dispersion curve is acceptably accounted for by computations, which is anyway an achievement, an important property of cold liquids, i.e., the damping of sound waves, is largely missed and systematically overestimated.

4 Final remarks

The examples reported in this work show that the degree of accuracy with which simulations of $S(Q,\omega)$ are able to reproduce experimental data is very different for classical and quantum liquids: it is very high in the first case and still rather low in the second.

As a general result, we demonstrated that unconstrained analyses of $S(Q,\omega)$ can be deceptive, while use of the EET enables very accurate and physically-grounded descriptions of both experimental and simulation data.

The future of course lies in the implementation of EET-based models within algorithms exploiting Bayesian inference, so to maximize both physical and statistical consistency of fit results and avert possible biases in the analysis.

We acknowledge the BRISP spectrometer at the ILL, no longer operational, which was the state of the art instrument for neutron Brillouin scattering at thermal energies: a neutron technique of extreme importance for studies of dense liquids dynamics. Regretfully, no equivalent instrument exists at present in the world, which is a deadweight loss.

References

1. J. R. D. Copley and J. M. Rowe, Phys. Rev. Lett. **32**, 49 (1974).
2. O. Söderström, J. R. D. Copley, J.-B. Suck, B. Dorner, J. Phys. F: Metal Phys. **10**, L151 (1980).
3. E. Guarini *et al.*, Phys. Rev. B **88**, 104201 (2013) and references therein.
4. E. Guarini *et al.*, Phys. Rev. B **102**, 054210 (2020).
5. U. Balucani and M. Zoppi, Dynamics of the Liquid State (Clarendon, Oxford, 1994).
6. Chr. Morkel, Chr. Gronemeyer, W. Gläser, J. Bosse, Phys. Rev. Lett. **58**, 1873 (1987).
7. C. Morkel and C. Gronemeyer, Z. Phys. B **72**, 433 (1988).
8. P. Verkerk, J. Westerweel, U. Bafile, L. A. de Graaf, W. Montfrooij, I. M. de Schepper, Phys. Rev. A **40**, 2860 (1989).
9. M. Celli *et al.*, Phys. Rev. B **84**, 140510 (2011).
10. M. Zoppi, D. Colognesi, M. Celli, Eur. Phys. J. B **23**, 171 (2001).
11. E. Guarini *et al.*, Phys. Rev. B **93**, 224302 (2016).
12. F. J. Bermejo *et al.*, Phys. Rev. B **47**, 15097 (1993).
13. F. J. Bermejo *et al.*, Phys. Rev. Lett. **84**, 5359 (2000).
14. E. Guarini *et al.*, Phys. Rev. B **104**, 174204 (2021).
15. I. R. Craig and D. E. Manolopoulos, J. Chem. Phys. **121**, 3368 (2004).
16. J. A. Poulsen, G. Nyman, P. J. Rossky, J. Chem. Phys. **119**, 12179 (2003).
17. K. K. G. Smith, J. A. Poulsen, G. Nyman, P. J. Rossky, J. Chem. Phys. **142**, 244112 (2015).
18. S. Hosokawa *et al.*, Phys. Rev. Lett. **102**, 105502 (2009).
19. V. M. Giordano and G. Monaco, PNAS **107**, 21985 (2010).
20. D. Levesque, L. Verlet, J. Kürkijarvi, Phys. Rev. A **7**, 1690 (1973).
21. J.-B. Suck, Cond. Matter Phys. **11**, 7 (2008).
22. U. Bafile, E. Guarini, F. Barocchi, Phys. Rev. E **73**, 061203 (2006).
23. F. Barocchi, U. Bafile, M. Sampoli, Phys. Rev E **85**, 022102 (2012).
24. F. Barocchi and U. Bafile, Phys. Rev E **87**, 062133 (2013).
25. F. Barocchi, E. Guarini, U. Bafile, Phys. Rev E **90**, 032106 (2014).
26. A. De Francesco, E. Guarini, U. Bafile, F. Formisano, L. Scaccia, Phys. Rev. E **94**, 023305 (2016).
27. A. De Francesco, A. Cunsolo, L. Scaccia, Bayesian approach for X-ray and neutron spectroscopy. In *Inelastic X-ray scattering and X-ray Powder Diffraction Applications* (Chap. 2) 26 (IntechOpen 2020).
28. P. J. Green, Biometrika **82**, 711 (1995).
29. D. Aisa *et al.*, Nucl. Instr. Meth. A **544**, 620 (2005).
30. D. Aisa *et al.*, J. Non-cryst. Solids **352**, 5130 (2006).
31. A. De Francesco, U. Bafile, A. Cunsolo, L. Scaccia, E. Guarini, Sci. Rep. **11**, 13974 (2021).
32. M. Zoppi, M. Celli, A. K. Soper, Phys. Rev. B **58**, 11905 (1998).
33. I. F. Silvera and V. V. Goldman, J. Chem. Phys. **69**, 4209 (1978).
34. H. H. Glyde, Excitations in Liquid and Solid Helium, (Clarendon Press, Oxford, 1995).

Neutron scattering study on the structure-property relationship of radiation-grafted proton exchange membranes

Yue Zhao[1*], *Kimio* Yoshimura[1], *Akihiro* Hiroki[1], *Aurel* Radulescu[2], and *Yasunari* Maekawa[1]

[1]Department of Advanced Functional Materials Research, Takasaki Institute of Advanced Quantum Science, National Institutes for Quantum Science and Technology (QST), Watanuki-machi 1233, Takasaki, Gunma, 370-1292, Japan
[2]Forschungszentrum Jülich GmbH, Jülich Centre for Neutron Science @ MLZ, Lichtenbergstraße 1, D-85747 Garching, Germany

Abstract. The partial scattering function (PSF) analysis through the contrast variation small-angle neutron scattering technique was used to determine the exact structure of the hydrated radiation-grafted proton-exchange membranes, made of poly(styrene sulfonic acid)-grafted poly(ethylene-co-tetrafluoroethylene) (ETFE-g-PSSA) with a high ion exchange capacity of 2.5 mmol/g. The membrane was treated as a three-component system composed of ETFE base polymer, PSSA graft polymer, and absorbed water. The analysis on PSF self-terms gave the exact structure of individual components and that on PSF cross-terms explored the correlation between two components to establish their locations. The characterization was performed in multiple length scales, and the mechanistic insights into membrane conductivity and structure correlations were provided.

1 Introduction

Radiation-grafted proton exchange membranes (PEMs), made of poly(styrenesulfonic acid)-grafted poly(ethylene-co-tetrafluoroethylene) (ETFE-g-PSSA), are promising alternatives to Nafion® membranes for electrochemical applications such as electrodialysis and fuel cells. They offer the advantages of a potentially low-cost fabrication technique, and the adaptability of both improving membrane ionic conductivity by PSSA graft polymers and maintaining the excellent mechanical/thermal properties of ETFE base polymer [1-3]. To the current stage of research, challenges remained to overcome are the relatively low proton conductivity of ETFE-g-PSSA under reduced relative humidity and long-term stability, which need thorough understanding of structure-property relationships [1-9].

It is known that in hydrated ETFE-g-PSSA PEMs, the sulfonic acid (SA, $-SO_3H$) groups absorb water and form hydrophilic ionic channels that phase segregate from the hydrophobic polymer matrix. Thus, the proton conductivity is strongly controlled by the morphology and connectivity of ion channels [4-9]. As one of the most essential techniques to understand the nano-scaled structure of a hydrated material, small-angle neutron scattering (SANS) technique has been employed in the previous studies. However, the conventional scattering analysis is based on the scattering intensity profile that contains contributions of all components in the system such as the hydrophobic polymer, hydrophilic polymer, ions, and water molecules, and fails to provide the detailed structure of the individual components. This undesirable original data problem was recently solved by partial scattering function (PSF) analysis, which is the quantitative decomposition of a series of intensity profiles obtained through contrast-variation SANS (CV-SANS) experiments [10-12]. PSF analysis was applied to polymer nanocomposites in early studies [10], and recently was developed by us for Nafion and ETFE-g-PSSA PEMs with moderate ion exchange capacity (IEC) of 1.0 ~ 2.0 mmol/g [11,12].

In this article, we extended the application of this PSF analysis to determine the detailed structure of ETFE-g-PSSA PEMs with a high IEC of 2.5 mmol/g (denoted as ETFE-g-PSSA_2.5). The unique capability that PSF analysis provides to understanding structure correlations can result into new insights on the role of the polymer micro-/nano- structure and water in the emergence of the ion conduction. This in turn can help in the design of high-performing PEMs for a wide range of energy conversion applications.

2 Experimental details

2.1 Materials

The ETFE base film with a thickness of 50 μm (mass density, d_{BP} = 1.75 g/cm^3, crystallinity = 0.32) were purchased from Asahi Glass Co. Ltd, Japan. ETFE-g-PSSA_2.5 membrane with a grafting degree (GD) of 60% and an IEC of 2.5 mmol/g was prepared according to our previous report [13]. The water uptake (WU) was estimated to be ~65% at room temperature, which is defined by $WU = \frac{W_{wet}-W_{dry}}{W_{dry}} \times 100\%$, with W_{dry} and W_{wet} being the weight of the dry and water-equilibrated membrane, respectively. The conductivity of the

* Corresponding author: zhao.yue@qst.go.jp

membrane was 0.12 S/cm at 60 °C in liquid water measured by electrochemical impedance spectroscopy.

2.2 CV-SANS measurements

CV-SANS measurements were performed on KWS-2 SANS diffractometer operated by Juelich Centre for Neutron Science (JCNS) at the neutron source Heinz Maier–Leibnitz (FRM II reactor) in Garching, Germany [14]. The incident neutron beam was monochromatized with a velocity selector to have an average wavelength (λ) of 5 Å with a wavelength resolution of 20%. Measurements were performed on ETFE-g-PSSA_2.5 PEMs equilibrated in H_2O/D_2O mixtures with eight different volume fractions of D_2O (f_{D2O}= 100, 90, 80, 70, 60, 50, 40, 0%) at room temperature. The scattering patterns were collected using a two-dimensional scintillation detector and circularly averaged to obtain scattering intensity profiles, $I(q)$s, as a function of scattering vector, q. The final $I(q)$s were corrected for the background, the electronic noise, detector sensitivity, and incoherent scattering.

3 Results and discussion

3.1 Theoretical development of PSF analysis for ETFE-g-PSSA_2.5 PEM

The ETFE-g-PSSA_2.5 PEM is partitioned into separate components of ETFE base polymer (BP, $-CH_2CH_2CF_2CF_2-$), graft polymer (GP, $-C_8H_7-SO_3H$), and water (W). The scattering length density (SLD) of each component is calculated using the chemical structure and mass density being 2.74 and 1.96 ($\times 10^{10}$ cm^{-2}) for BP (b_{BP}) and GP (b_{GP}), respectively [15]. SLD of water (b_W) varies as a function of f_{D2O} of the water mixture as

$$b_w = b_{D2O} f_{D2O} + b_{H2O}(1 - f_{D2O}) \quad (1)$$

where b_{D2O} and b_{H2O} are the SLD of D_2O and H_2O being 6.34 and −0.56 ($\times 10^{10}$ cm^{-2}), respectively.

According to incompressibility assumption, $I(q)$ can be described by three PSF self-terms $S_{ii}(q)$ as follows,

$$I(q) = (b_{BP}-b_{GP})(b_{BP}-b_W)S_{BP\text{-}BP}(q) + (b_{GP}-b_{BP})(b_{GP}-b_W)S_{GP\text{-}GP}(q) + (b_W-b_{BP})(b_W-b_{GP})S_{W\text{-}W}(q) \quad (2)$$

where S_{ii} is PSF self-term of the i component (i = BP, GP and W), representing its structural information.

CV-SANS experiments are performed with m different contrast by using H_2O/D_2O mixtures, the obtained $I(q)$s are a group of linear equations of Eq (2). Thus, three PSF self-terms of $S_{BP\text{-}BP}(q)$, $S_{GP\text{-}GP}(q)$ and $S_{W\text{-}W}(q)$ on the right side of Eq. (2) can be mathematically determined through the simultaneous equations of $I(q)$ [11, 12].

The PSF cross-term S_{ij} ($i \neq j$) contains information about the interaction between the i and j components and can be deduced from S_{ii} using Eqs (3)–(5) below.

$$S_{BP\text{-}GP} = 1/2(S_{W\text{-}W} - S_{BP\text{-}BP} - S_{GP\text{-}GP}) \quad (3)$$

$$S_{GP\text{-}W} = 1/2(S_{BP\text{-}BP} - S_{GP\text{-}GP} - S_{W\text{-}W}) \quad (4)$$

$$S_{BP\text{-}W} = 1/2(S_{GP\text{-}GP} - S_{BP\text{-}BP} - S_{W\text{-}W}) \quad (5)$$

3.2 Results of PSF analysis

$I(q)$ profiles are plotted as a function of q in Figure 1 for representative f_{D2O} in the CV-SANS measurements. They show typical scattering features in that a small-q upturn and two scattering maxima in three q-regions. In region I at $q < 0.12$ nm^{-1}, the small-q upturn follows a power-law relationship with an exponent of about −2.4. In Region II at $0.12 < q < 1.7$ nm^{-1}, a shoulder-like scattering peak appears at $q_1 \sim 0.17$ nm^{-1}, corresponding to a d-spacing ($=2\pi/q_1$) ~ 36.9 nm. In Region III at $q >$ 1.7 nm^{-1}, the second scattering peak, so-called "ionomer peak", appears at q_2 = 2.6 nm^{-1}, with a d-spacing ($=2\pi/q_2$) of 2.4 nm.

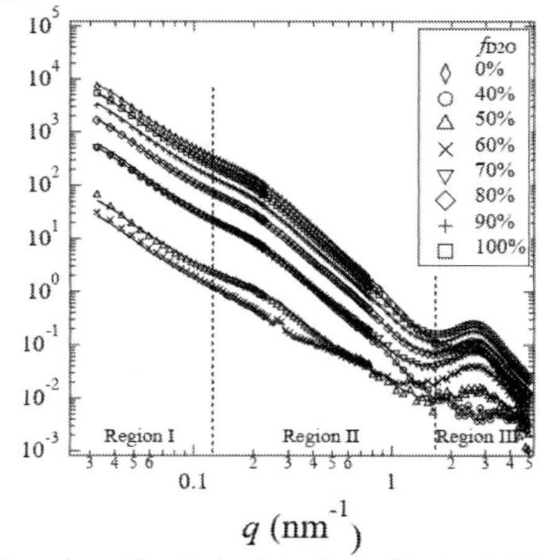

Fig. 1 Experimental scattering intensity profiles (symbols) and the reconstructed intensity profiles (solid lines) of the fully hydrated ETFE-g-PSSA_2.5 PEMs equilibrated in water mixtures of D_2O and H_2O with different ratios.

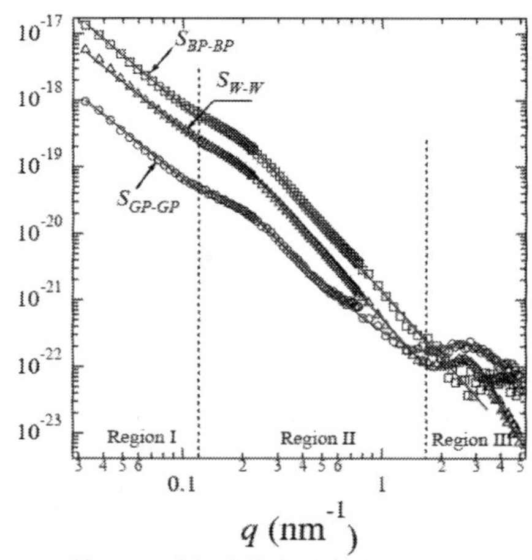

Fig. 2 PSF self-terms of the fully hydrated ETFE-g-PSSA_2.5 PEMs (symbols) and the best-fitted results (solid lines).

The self-terms of PSFs including $S_{BP\text{-}BP}$, $S_{GP\text{-}GP}$, and $S_{W\text{-}W}$ were calculated based on section 3.1 and plotted as a function of q in Figure 2. In Region I, all S_{ii} exhibit an upturn with a power-law exponent of about −2.4. In Region II, all S_{ii} shows a shoulder-like scattering maximum with a centre position close to the peak observed in $I(q)$ profiles in Figure 1. In Region III, $S_{GP\text{-}GP}$ and $S_{W\text{-}W}$ show a peak with a centre position close to

https://doi.org/10.1051/epjconf/202328604003

the ionomer peak, and the peak in S_{GP-GP} is broader than that in S_{W-W}. On the contrary, S_{BP-BP} in Region III is flat, revealing ETFE BP is structureless at this length scale. With these three S_{ii}, all reconstructed $I(q)$ profiles (solid lines) using Eq (2) via back-substitution are well matched to the experimental profiles (symbols) in Figure 1, evidencing the correctness of S_{ii}.

As mentioned in section 3.1, S_{ii} reflects the concrete structure of the *i* component, including its arrangement and phase-separated morphology in the membrane. Since S_{ii} show similar features to the previously studied ETFE-g-PSSA PEMs with an IEC of 2.0 mmol/g (denoted as ETFE-g-PSSA_2.0) [12], we reasonably propose the same fitting functions here to interpret S_{ii}: Mass fractal (MF) structure model to describe the small-*q* upturn in Region I, the unified Guinier-exponential function (GE model) to fit the shoulder peak in Region II, and Teubner-Strey (TS) model to describe the local structure in Region III. However, it should be noted that the scattering origin for the two membranes ETFE-g-PSSA_2.0 and ETFE-g-PSSA_2.5 are different. It is the aggregation of individual GP nanodomains in the hydrophobic BP matrix that makes the structure pattern of ETFE-g-PSSA_2.0, while the structure of the ETFE-g-PSSA_2.5 is originated from the aggregation of hydrophobic BP nanodomains in the GP matrix. This difference is evidenced by taking account of the volume fraction of BP in the hydrated PEMs, ϕ_{BP},

$$\phi_{BP} = \frac{\frac{1}{d_{BP}}}{\frac{1}{d_{BP}}+\frac{GD/100}{d_{GP}}+\frac{WU/100(1+GD/100)}{d_W}} \quad (6)$$

where d_x (x=BP, GP and W) is the mass density of x being approximately 0.55 and 0.3 for ETFE-g-PSSA_2.0 and ETFE-g-PSSA_2.5, respectively.

Thus, S_{ii} of ETFE-g-PSSA_2.5 is the sum of the three models given by

$$S_{ii}(q)=C_1S_{MF}(q)+C_2S_{GE}(q)+C_3S_{TS}(q)+C_B \quad (7)$$

where C_1, C_2, and C_3 are the fitting constants, and C_B is the constant background. The first term $S_{MF}(q)$ is expressed as

$$S_{MF}(q) \propto \frac{\Gamma(D_f-1)}{(1+q^2\xi^2)^{(D_f-1)/2}} \times \frac{\sin[(D_f-1)\tan^{-1}(q\xi)]}{q\xi} \quad (8)$$

which is the MF function with the gamma function (Γ), MF dimension (D_f), and the upper cutoff length of the MF structure (ξ) being roughly estimated as ~150 nm according to the previous report using the Ultra-SAXS method [13]. The second term, $S_{GE}(q)$, is proposed to describe the structure of irregularly shaped individual BP nano-domains as expressed below.

$$S_{GE}(q) \propto \exp\left(\frac{-q^2R_g^2}{3}\right) + B\left\{\frac{[\mathrm{erf}(\frac{kqR_g}{\sqrt{6}})]^3}{q}\right\}^P \quad (9)$$

where R_g is the radius of gyration of individual BP particle, erf(x) is the error function of x, *k* is an empirical constant equal to 1.06. *P* ($3 < P < 4$) describes the particle's surface fractal dimension, and *B* is a constant prefactor. The third term, $S_{TS}(q)$, is the scattering function of the TS model describing bicontinuous-shaped domains with short-range order as below:

$$S_{TS}(q) \propto \frac{8\pi d^4}{\varepsilon[16\pi^4+8d^2\pi^2(\varepsilon^{-2}-q^2)+d^4(\varepsilon^{-2}+q^2)^2]} \quad (10)$$

where *d* is the mean separation distance between two domains determined from the peak position, q_m ($d = 2\pi/q_m$), and ε is considered as the dispersion of *d* (inversely proportional to the peak width). A smaller ε resulted from a broader peak indicates a more disordered bicontinuous structure. The best-fitted curves obtained using Eq. (7) are shown together with S_{ii} in Figure 2, and all the fitting parameters are listed in Table 1.

Table 1. Fitting parameters in Eq (7).

Models	MF	GE	TS
S_{ii}	ξ(nm) D_f	R_g(nm) P	d(nm) ε(nm)
S_{BP-BP}	150 2.4	10.0 3.3	- -
S_{GP-GP}	150 2.4	10.0 3.1	2.2 0.9
S_{W-W}	150 2.4	10.0 3.1	2.3 1.2

The PSF cross-terms S_{ij} ($i \neq j$) through Eqs.(3)–(5) reflect cross-correlation between the components *i* and *j*. The positive and negative sign of S_{ij} reveals the interaction force between *i* and *j*, denoting attractive and repulsive interactions, respectively. The cross-terms of PSFs as a function of *q* are shown in Figure 3. S_{BP-GP} and S_{BP-W} are always negative, whereas the sign of S_{GP-W} is opposite at *q* lower and higher than 1.5 nm^{-1} as shown in the inset, respectively.

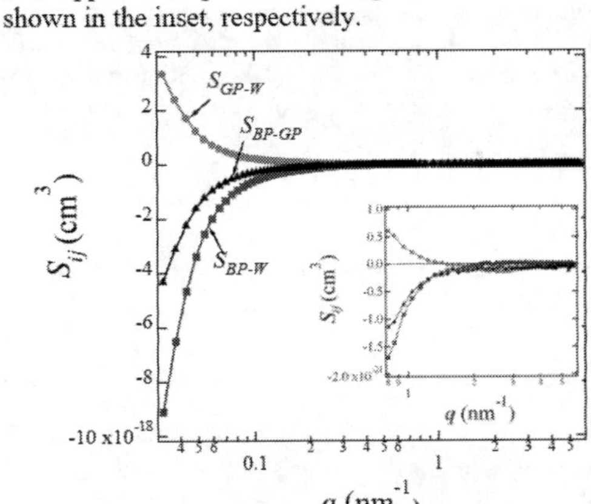

Fig. 3 PSF cross-terms of the fully hydrated ETFE-g-PSSA_2.5 PEMs. Enlarged plots at $q > 1$ nm^{-1} in the inset.

3.3 Visualization of three-component domains in the hydrated ETFE-g-PSSA_2.5 PEM

3.3.1 Large scale structure

The MF model describes a structure consists of self-similar polymer particles within a spatial range. Based on the analysis of S_{ii} in Region I, the schematic picture of the large-scale structure is shown in Figure 4 (a). Note that BP hydrophobic domains are clearly phase-separated from hydrophilic domains made of GP and water, as evidenced by the repulsive interactions suggested by the negative sign of S_{BP-GP} and S_{BP-W}. All components show a similar mass fractal structure with a dimension of $D_f = 2.4$, indicating a dense packing of BP nano-domains in space, close to percolating networks. Instead, GP and water domains form the compensative matrix according to Babinet's principle, therefore, S_{GP-GP} and S_{W-W} shows the same MF structure feature as S_{BP-BP} does. In addition, the positive S_{GP-W} indicates that water is closely attached to GP due to the hydrophilic nature of SA groups. Therefore, water domains coordinatively move with GP domains.

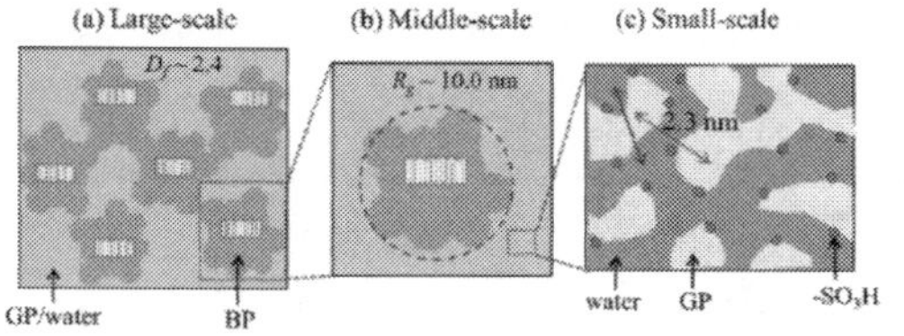

Fig. 4 Schematic of the hierarchical structure of the individual components in the fully hydrated ETFE-g-PSSA_2.5 PEMs at (a) large-scale; (b) middle-scale; and (c) small-scale.

3.3.2 Middle scale structure

The GE model analysis in Region II shows all domains have a characteristic R_g of ~10.0 nm, which is the average size of the individual building blocks for each component to form the MF structure in Region I. Since GP and water domains coordinatively move, we reasonably conclude that BP nano-domains with a size of $R_g = 10$ nm are distributed in the homogeneous matrix of GP and water. The power-law relationship in the high-q-regime in Region II allows for the estimation of P, which is an indicative of the surface roughness of the domain. P is in the range of 3.1-3.3, smaller than the typical sharp Porod surface ($P = 4$), indicating a rough domain surface. Based on the above discussion, a schematic of the structure in Region II is shown in Figure 4(b).

3.3.3 Small scale structure

The detailed local structure in a GP/water domain is explained by TS model analysis of S_{GP-GP} and S_{W-W} that shows both components have bicontinuous-like structure with similar d value (2.2–2.3 nm) as schematically shown in Figure 4(c). S_{GP-W} is negative in this region with a molecular length level < 3.7 nm (q > 1.7 nm^{-1}), indicating a repulsion between GP and water probably due to the strong repulsive force between PS polymer and water. This result is in good agreement with what we observed in ETFE-g-PSSA_2.0 PEM, and much different from Nafion membrane where the interaction between polymer side chain and water are positive. This considerable difference results in the low swelling of ETFE-g-PSSA_2.5 PEM but high conductivity under low relative humidity conditions, as compared to Nafion.

The detailed structure evolution of ETFE-g-PSSA_2.5 at multiple length scales illustrates the power of PSF analysis to gain insight into structure-property relationships in disordered and bi-continuous systems.

4 Conclusions

We applied PSF analysis to gain quantitative knowledge of the role of each component in the entire structure of fully hydrated radiation-grafted ETFE-g-PSSA_2.5 PEMs by CV-SANS experiments. Our results suggested three-component domains consisting of ETFE base polymer, PSSA graft polymers and water. PSF self-terms analysis revealed the detailed structure of each component, whereas the cross-terms gave the correlation between two components, leading to the location determination of the components. The entire structure patterns of the hydrated PEM were constructed in Figure 4. PSF analysis provides mechanistic insights concerning structural correlations over a range of length scales, from micro-/nano-meter to molecular scale, which offers the structural guidelines in the design of high-performing PEMs for a wide range of energy conversion applications.

References

1. M. M. Nasef, Chem. Rev., **114**, 12278 (2014).
2. J. A. Horsfall, K. V. Lovell, Fuel Cells, **1**, 186 (2001).
3. A. Aricò, V. Baglio, P. Creti, A. Di blasi, V. Antonucci, J. Brunea, A. Chapotot, A. Bozzi, J. Schoemans, J. Power Sources, 123, 107 (2003).
4. S. A. Gürsel, L. Gubler, B. Gupta, G. G. Scherer, in Fuel Cells I (Springer, Berlin, Heidelberg, 2008).
5. S. Balog et al., Macromolecular Chemistry and Physics 211, 635-64 (2010).
6. K. Mortensen et al., Journal of Polymer Science Part B-Polymer Physics 46, 1660 (2008).
7. V. Sproll et al., Macromolecules 49, 4253 (2016).
8. S. Balog et al., Polymer 53, 175 (2012).
9. S. Balog et al., Journal of membrane science 383, 50 (2011).
10. H. Endo, D. Schwahn, H. Colfen, J. Chem. Phys. **120**, 9410 (2004).
11. Y. Zhao, K. Yoshimura, T. Motegi, A. Hiroki, A. Radulescu, Y. Maekawa, Macromolecules, **54**, 4128 (2021).
12. Y. Zhao, K. Yoshimura, S. Sawada, T. Motegi, A. Hiroki, A. Radulescu, Y. Maekawa, Macromolecules, **55**, 7100 (2022).
13. T. D. Tap, S. Sawada, S. Hasegawa, K. Yoshimura, Y. Oba, M. Ohnuma, Y. Katsumura, Y. Maekawa, Macromolecules, **47**, 2373 (2014).
14. A. Radulescu, V. Pipich, H. Frielinghaus, M. S. Appavou, Journal of Physics: Conference Series, **351**, 012026 (2012).
15. R. J. Roe, Methods of X-ray and neutron scattering in polymer science (Oxford Uni. Press, New York, 2000).

Lamellar diffraction from lipid bilayers on MIRA, a triple axis spectrometer at the MLZ

Christopher J. Garvey[1*], *Markos* Skoulatos[1], and *Robert* Georgii[1]

[1] Technische Universitat Munchen, Forschungsneutronenquelle Heinz Maier-Leibnitz FRM II and Physik Department E13, Lichtenbergstr. 1, 85747 Garching, Germany

Abstract. Diffraction used in conjunction with molecular deuteration provides a model independent means to examine detailed structural and compositional information of model and real biological membranes in the lamellar phase. The technique provides specific information of localization of molecules and smaller units with respect to the unit cell. Deuteration is used to provide specific labelling and to provide phasing for the crystallographic reconstruction. Typical data sets consist of a series of lamellar diffraction peaks, usually collected under conditions of 3 contrasts of the water gas phase. Accurately integrated the diffraction peaks can be used for a Fourier reconstruction of the composition of the lamellar unit cell in real space. Each diffraction peak which can be integrated for 3 contrasts contributes to an additional Fourier term in the reconstruction and optimization of the number of peaks enhances the spatial resolution of the crystallographic reconstruction. Here we report on the use of the flexibly configurable cold triple-axis spectrometer, MIRA, at the Heinz Maier-Leibnitz Zentrum (Garching, Germany) for investigations of different lamellar systems using this approach. The data sets are acquired from lamellar stacks in an sample environment with humidity and temperature control with exceptionally low background. We discuss further enhancements of the instrument and sample environment which will provide information on the composition and equilibration of the sample.

1 Introduction

The low intrinsic curvature of most cell membranes ensure that highly oriented bilayers are excellent model systems for membrane biophysics. For this reason specular reflectivity[1] and lamellar diffraction[2, 3], which both provide compositional information in a direction normal to the lipid bilayer, are important tools to reveal the time averaged locus of molecules and their compositional fragments within the bilayer (Figure 1). This understanding is an important paradigm for a range of biophysical and biochemical problems but also forms an important and direct experimental reference point for computational membrane biophysics[4].

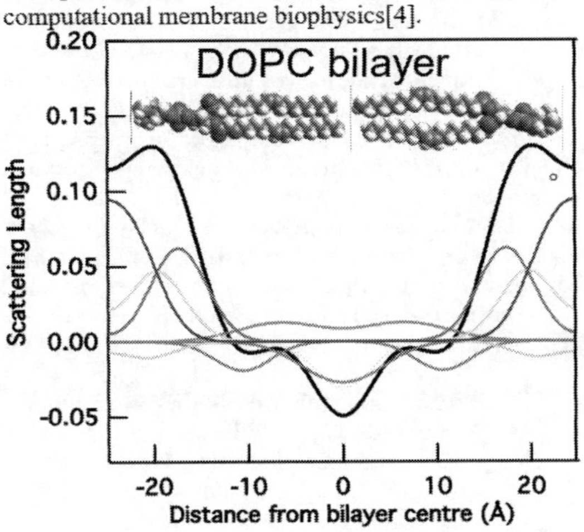

Fig. 1. Two 1,2-Dioleoyl-sn-glycero-3-phosphocholine (DOPC) molecules forming a bilayer are shown at the top, giving a reference for the hydrocarbon tails, headgroups and water. A scattering length density profile determined from diffraction data. This profile (thick black line) has been decomposed into individual molecular fragments: 1. terminal methyl group (yellow); 2. –(CH2)7- group (magenta); 3. double bond (orange); 4. –(CH2)7- groups (grey); 5. glycerol (red); 6. phosphatidyl (light blue); 7. choline (green); and 8. water (dark blue).

Here we describe the use of lamellar neutron diffraction in the context of the problems in membrane biophysics to find the average location of molecules in the bilayer structure shown in Figure 1. It is also important to place this technique in the context of the complimentary neutron technique, specular neutron reflectivity. Both techniques are used to determine compositional information of the bilayer structures, however the demands of beamtime and samples are quite different. Both techniques can make use of deuterated molecules, which provided certain caveats are observed[5], provide a convenient means of labelling specific aspects/regions of the SLD profile shown in Figure 1.

Bilayer sample preparation for neutron reflectivity is extremely demanding. The real-space analysis is achieved by modelling of composition slabs, requires deconvolution the structure of interest from the overall scattering length density profile. The quality of data is largely limited by the ability to prepare molecularly smooth samples, a single bilayer, over the area of the reflected neutron beam (~cm^2) with appropriate contrast over a liquid sub-phase. The information content, and thus uniqueness of SLD modelling from reflectivity data, can be improved by selective labelling with deuterium, or contrast variation with the water phase or magnetic methods[6]. While sample preparation can be a fundamental limitation of the technique the success of

* Corresponding author: christopher.garvey@tum.de

 https://doi.org/10.1051/epjconf/202328604004

neutron reflectometers is evident in the number of instruments and their facility support.

Lamellar neutron diffraction utilises multiple stacks of bilayers self-assembled on a solid substrate and data is analysed by a direct reconstruction of the scattering length density profile[7]. Cylindrical geometry of membranes may also be usefully exploited[8]. Samples are placed in an environment of controlled humidity. This environment is used for gas phase exchange and control of the contrast of the sorbed water between bilayers. The ease of control of humidity and ability to study dehydration processes has enabled successful studies related to cryobiology and anhydrobiology[9-11]. Understanding of molecular packing of lipid molecules which form important barrier in the skin's stratum corneum layer has been also been enabled by the technique[12, 13]. When samples consist of whole plasma membranes biophysically important aspects of the composition profile may be revealed[14, 15]. More generally the technique has revealed high resolution and important details of the packing of many small molecules into self-assembled bilayer stacks.

The Fourier reconstruction of the unit cell (Figure 1) is made from the variation of peak intensity with contrast which recovers that phase information that is lacking from simple diffraction peak intensity measurements. The mathematical formalism for the Fourier reconstruction of SLD is that same as that adopted elsewhere[3]. The structure factor contribution, *f(h)*, of diffraction peak of order h, is

$$f(h) = \sqrt{I(h)A(h)\sin\theta} \qquad (1)$$

Where *I(h)* is the measured peak integral, *A(h)* is the adsorption correction and is the Lorentz correction at angle θ. Each measured *f(h)* and its phasing is a contribution in the Fourier series to the overall variation of the scattering length density in the unit cell, $\rho^*(z)$ in the real space co-ordinate, *z*, per lipid molecule:

$$\rho^*(z) = \rho_0{}^* - \left(\frac{2}{dk}\right)\sum_{h=1}^{h_{max}} \cos\left(\frac{2\pi hz}{d}\right) f(h) \qquad (2)$$

where $\rho_o{}^*$ is the product of the average scattering length density per unit cell and the area per lipid molecule, *d* is the repeat spacing of the unit cell and *k* is an instrumental scaling constant. Determination of the phasing of each *f(h)* is made from plots of $\pm\sqrt{f(h)}$ versus % D_2O.

2 Methods and Materials

2.1 Sample Preparation

Orientated lamellar stacks with good registry with the substrate surface are most easily prepared by spraying all sample components from a suitable solvent mixture[9, 10] using a commercial airbrush on to a quartz microscope slide. Other sample strategies may be adopted depending on the amount of available material and/or a suitable solvent systems[7]. The area of sample should be of the order cm^2, containing ~ 5 mg of materials. Thicker samples, while diffracting more strongly, particularly for lipids in the gel phase[13], provide challenges in terms of formation of well ordered lamellae, equilibration times for D_2O exchange and efficient use of neutron beamtime. After initial deposition solvent can be removed in a vacuum oven, the layer can be annealed towards improved lamellar order with the high humidity, overnight at high humidity (e.g. 97 % RH) and/or temperature at above gel to fluid transition. Sample quality, particularly the number of lamellar reflections orders and mosaic spread may be monitored with laboratory X-ray diffraction in grazing incidence or Bragg-Brentano geometry (Figure 2). Typically a mosaic spread of the lamellar phase ± 0.05° may be expected.

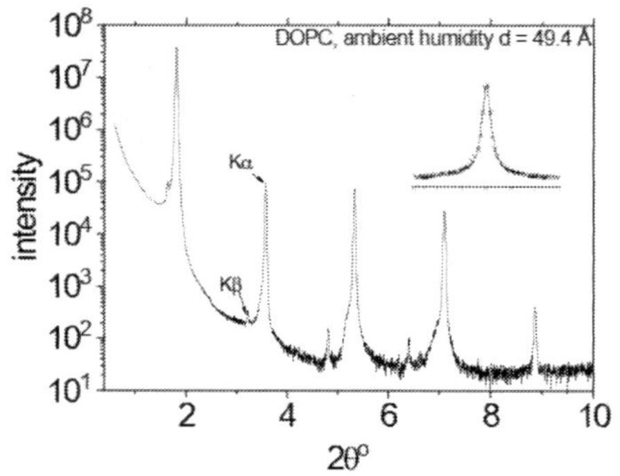

Fig. 2. Laboratory X-ray diffraction from successfully prepared sample of DOPC bilayers in ambient conditions showing well resolved Bragg peaks and the clear effects of the polychromaticity of Cu X-ray source, Kα and **Kβ** lines (labelled). The inset, shows a rocking curve around the first order diffraction peak.

2.2 MIRA – As two circle cold diffractometer and sample environment

The cold triple axis spectrometer MIRA (now known as LaDiff), can be utilized in elastic mode where the analyser axis is still in place but tuned to the same energy as the monochromator. This enables a much cleaner beam, with a very low background, and is ideal for measurement of normalized diffracted intensity from thin lipid lamellar films on flat substrates. The low background allows the measurement of many orders of lamellar diffraction, and most importantly accurate integration of the peak intensity. Each peak integrated for the 3 contrasts contributes an extra term to the Fourier series of equation 2. Samples are mounted in an environment with controlled humidity and temperature. Two systems are available for humidity and temperature control (10- 60°C):

1. Thermostated cylindrical aluminium cans which are equipped with a reservoir for a saturated salt solution. The humidity generated is determined by the salt identity. A range of salts is available with varying in the humidity and temperature dependence of the humidity[16].
2. 3D-printed metal cells also offer temperature control. This adaptation of the flexiProb[17] sample environment offers control of humidity and D_2O/H_2O contrast by use of a flow through system. There also

https://doi.org/10.1051/epjconf/202328604004

exists the possibility of introducing other volatile molecules to samples for gas phase adsorption with this system.

Using the sample environments above, a typical experiment consists of the measurement of peak integrals at 3 contrasts on the same sample for comparison of the peak intensities. For this reason it is critical to position the sample identically for each contrast and samples are typically cycled through the 3 contrasts in the one of 3 duplicate sample cans with locating pins ensuring identical sample positioning with respect to the incident neutron beam.

The triple-axis spectrometer MIRA has access to an appropriate range of scattering vectors to interrogate the integral intensity of each diffraction peak with variation of the contrast of the sorbed water phase. Additionally it has the ability to significantly reduce the background signal from the sample[18]. After the sample is first aligned with a laser illuminating the sample position along the direction of the neutron beam an iterative procedure between a θ/2θ scan and θ scan on the first order Bragg peak is used to align the sample and calibrate the θ/2θ geometry via an offset (Figure 3).

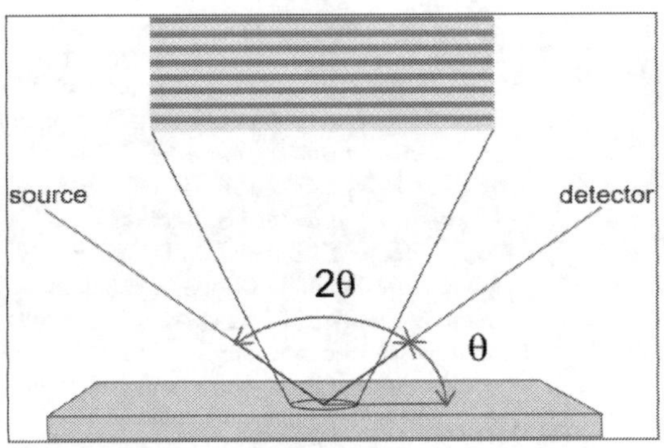

Fig. 3. Top schematic view of the measurement geometry with respect to the sample, a bilayer stack with excellent registry to the quartz substrate. With the source fixed with respect the sample, the θ/2θ measurement is a achieved by detector movement around the sample (2θ), and sample rotation, θ.

A typical measurement series consists 3 complete scans at contrasts of 50, 20 and 8 % (null water) D_2O in H_2O. Depending on the number of observable peaks, with each higher order peaks measured with 3 contrasts contributing an additional Fourier term to the reconstruction, measurement times have varied between 2-8 hours. While higher order peaks, which require longer acquisition times, are clearly desirable in order to avoid truncation artefacts of the Fourier series (equation 2) care must be exercised that these artefacts/features are smaller than features of interest in the SLD profile.

An important consideration is the equilibration time for samples. The time for equilibration between contrast changes is of the order several hours, but can be longer for samples in the gel phase.

3 Results

Figure 4 shows a data typical set from MIRA (DOPC 75 % relative humidity, 100% D_2O, 25°C) utilising 5 Å neutrons in the θ/2θ geometry. Together with the low background the instrument configuration is sufficient to obtain 5 orders of lamellar diffraction from DOPC in the lamellar fluid phase in approximately 3 hours for each contrast. The accessible minimum angle is sufficient for accurate integration of the first order peak. The maximum angle for the scan measured was determined by the available counting time. In the case of the profile shown as part of Figure 1 the profile has been reconstructed from 5 Bragg peaks at 3 different contrasts. This profile is consistent with those determined in the literature[19].

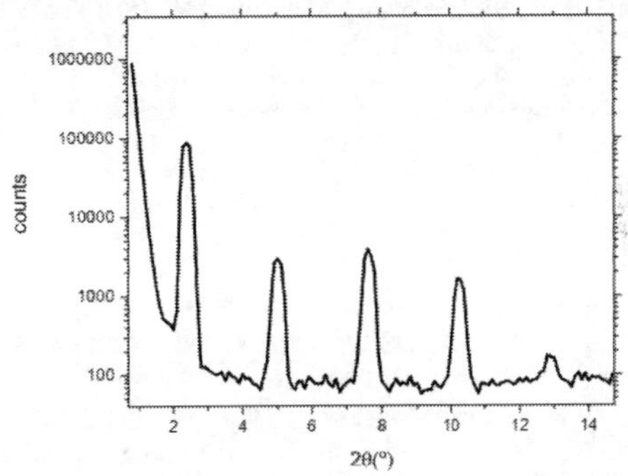

Fig. 4. Neutron diffraction from a oriented lamellar sample of DOPC on a quartz substrate at 25°C, 75 % relative humidity in 100 % D_2O. The humidity was produced by a saturated solution of NaCl in D_2O.

The overall profile was deconvolved into different Gaussian functions representing the average position of the molecule or molecular fragment and the contribution of each fragment to the scattering length density profile. The Fourier reconstruction and deconvolution of the resulting profile were achieved using specific macros for the program IgorPro (Wavemetrics, Oswego, USA) written by Dr Thomas Hauβ (Helmholtz Zentrum Berlin).

4 Conclusions and Outlook

Neutron diffraction is a technique which produces model independent information which is quite similar to that produced by the modelling of neutron reflectometry. However the technique is suitable for quite different problems in membrane biophysics, particularly those at reduced and variable hydrations, complex but ordered biological systems and lipid mixtures.

MIRA and its sample environment are quite suitable for making quantitative lamellar diffraction measurements. Sample preparation facilities and initial characterization are provided onsite at MLZ.

Current developments around the instrument include:

1. An *in situ* Raman measurement system. This development is aimed at quantify gas phase adsorption into the sample and details on the packing of molecules into the membrane stack.
2. An upgrade to the detector on MIRA (LaDIFF). The current point detector of MIRA is being replaced with detector bank around the sample. The 2D detector will be optimized for a lower background and will enable measurements of data sets which include sample rocking curves in measurement times currently unfeasible with a point detector.
3. Development of python based in house open source software for data processing, Fourier reconstruction and deconvolution of SLD profiles.

We would like to acknowledge the contributions of Dr Thomas Hauβ (Helmholtz Zentrum Berlin) for the IgorPro macros used for SLD profile reconstruction (available on request), Dr Ben Kent (School of Chemistry, University of New South Wales) and Professor Gary Bryant (RMIT University) to this work.

References

[1] Koenig B W, Krueger S, Orts W J, Majkrzak C F, Berk N F, Silverton J V and Gawrisch K 1996 Neutron Reflectivity and Atomic Force Microscopy Studies of a Lipid Bilayer in Water Adsorbed to the Surface of a Silicon Single Crystal *Langmuir* **12** 1343-50

[2] Wiener M C and White S H 1991 Fluid Bilayer Structure Determination by the Combined Use of X-Ray and Neutron-Diffraction .2. Composition-Space Refinement Method *Biophysical Journal* **59** 174-85

[3] Wiener M C and White S H 1991 Fluid bilayer structure determination by the combined use of x-ray and neutron diffraction. I. Fluid bilayer models and the limits of resolution *Biophysical journal* **59** 162-73

[4] Pan J, Cheng X, Monticelli L, Heberle F A, Kucerka N, Tieleman D P and Katsaras J 2014 The molecular structure of a phosphatidylserine bilayer determined by scattering and molecular dynamics simulations *Soft Matter* **10** 3716-25

[5] Bryant G, Taylor M B, Darwish T A, Krause-Heuer A M, Kent B and Garvey C J 2019 Effect of deuteration on the phase behaviour and structure of lamellar phases of phosphatidylcholines – Deuterated lipids as proxies for the physical properties of native bilayers *Colloids and Surfaces B: Biointerfaces* **177** 196-203

[6] Lakey J H, Paracini N and Clifton L A 2022 Exploiting neutron scattering contrast variation in biological membrane studies *Biophysics Reviews* **3** 021307

[7] Tristram-Nagle S 2007 *Methods in Molecular Biology*, ed A M Dopico (Totowa: Humana Press Inc.) pp 63-75

[8] Baumann A, Denninger A R, Domin M, Demé B and Kirschner D A 2022 Metabolically-incorporated deuterium in myelin localized by neutron diffraction and identified by mass spectrometry *Current Research in Structural Biology* **4** 231-45

[9] Bremer A, Kent B, Hauss T, Thalhammer A, Yepuri N R, Darwish T A, Garvey C J, Bryant G and Hincha D K 2017 Intrinsically Disordered Stress Protein COR15A Resides at the Membrane Surface during Dehydration *Biophysical Journal* **113** 572-9

[10] Kent B, Hauss T, Deme B, Cristiglio V, Darwish T, Hunt T, Bryant G and Garvey C J 2015 Direct Comparison of Disaccharide Interaction with Lipid Membranes at Reduced Hydrations *Langmuir* **31** 9134-41

[11] Kent B, Hunt T, Darwish T A, Hauss T, Garvey C J and Bryant G 2014 Localization of trehalose in partially hydrated DOPC bilayers: insights into cryoprotective mechanisms *J. R. Soc. Interface* **11** 6

[12] Schmitt T, Lange S, Sonnenberger S, Dobner B, Demé B, Langner A and Neubert R H H 2019 The long periodicity phase (LPP) controversy part I: The influence of a natural-like ratio of the CER[EOS] analogue [EOS]-br in a CER[NP]/[AP] based stratum corneum modelling system: A neutron diffraction study *Biochimica et Biophysica Acta (BBA) - Biomembranes* **1861** 306-15

[13] Schroeter A, Stahlberg S, Školová B, Sonnenberger S, Eichner A, Huster D, Vávrová K, Hauß T, Dobner B, Neubert R H H and Vogel A 2017 Phase separation in ceramide[NP] containing lipid model membranes: neutron diffraction and solid-state NMR *Soft Matter* **13** 2107-19

[14] King G I, Stoeckenius W, Crespi H L and Schoenborn B P 1979 The location of retinal in the purple membrane profile by neutron diffraction *Journal of Molecular Biology* **130** 395-404

[15] McCaughan L and Krimm S 1980 X-RAY AND NEUTRON-SCATTERING DENSITY PROFILES OF THE INTACT HUMAN RED-BLOOD-CELL MEMBRANE *Science* **207** 1481-3

[16] Greenspan L 1977 Humidity Fixed Points of Binary Saturated Aqueous Solutions *Journal of Research of the National Bureau of Standards - A. Physics and Chemistry* **81A** 89-95

[17] Widmann T, Kreuzer L P, Kühnhammer M, Schmid A J, Wiehemeier L, Jaksch S, Frielinghaus H, Löhmann O, Schneider H, Hiess A, Klitzing R v, Hellweg T and Müller-Buschbaum P 2021 Flexible Sample Environment for the Investigation of Soft Matter at the European Spallation Source: Part II—The GISANS Setup *Applied Sciences* **11** 4036

[18] Georgii R, Weber T, Brandl G, Skoulatos M, Janoschek M, Mühlbauer S, Pfleiderer C and Böni P 2018 The multi-purpose three-axis spectrometer (TAS) MIRA at FRM II *Nuclear Instruments and Methods in Physics Research Section A: Accelerators, Spectrometers, Detectors and Associated Equipment* **881** 60-4

[19] Wiener M C, King G I and White S H 1991 Structure of a Fluid Dioleoylphosphatidylcholine Bilayer Determined by Joint Refinement of X-Ray and Neutron-Diffraction Data .1. Scaling of Neutron Data and the Distributions of Double-Bonds and Water *Biophysical Journal* **60** 568-76

Dynamics of polymer electrolyte with LiTFSI via Quasi-Elastic Neutron Scattering

Hui Wang[1], *Naresh C.* Osti[2], *Jürgen* Allgaier[1], *Marcella Cabrera* Berg[3], *Rene* Halver[1], *Eugene* Mamontov[2], *Godehard* Sutmann[1, 4], *Yuya* Doi[5], *Nicolas* Bucher[6], *Takeshi* Egami[7], *Stephan* Förster[1], and *Michael* Ohl[1, 7]

[1]Research Center Jülich, 52428 Jülich, Germany
[2]Neutron Scattering Division, Oak Ridge National Laboratory, Oak Ridge, TN 37831, USA
[3]Department of Physics, University of Regina, Saskatchewan S4S 0A2, Canada
[4]Interdisciplinary Center for Advanced Materials Simulations (ICAMS), Ruhr-University Bochum, 44801 Bochum, Germany
[5]Department of Materials Physics, Nagoya University, Nagoya, Aichi 4648603, Japan
[6]Varta Microbattery GmbH, 73479 Ellwangen, Germany
[7]Department of Materials Science and Engineering, Univ. Tennessee, Knoxville TN. 37996, USA

Abstract. Most lithium batteries offer a wide range of applications. However, safety issues are still an unresolved issue for several applications. To solve the safety issue of Li-ion batteries, solid polymer electrolyte is a promising candidate to replace commercial liquid electrolyte. A 4-arm star poly(ethylene oxide) polymer with LiTFSI salt as an electrolyte was studied. The dynamics of this polymer were explored with the Quasi-Elastic Neutron Scattering technique. Furthermore, the influence of temperature and Li salt concentration on the polymer dynamics was investigated. The dynamics of the polymer ends of the arm show much higher flexibility than the core parts making those types of polymers attractive for further studies in battery research.

Li-ion batteries (LIBs) play a key role in our everyday lives as a component of electronic devices such as mobile phones and electric vehicles [1, 2]. They are also employed as high-density energy storage source in various industries, such as power plants, military equipment, and aerospace. In spite of their wide use, LIBs suffer from some shortcomings, such as the low boiling point of the commercial liquid electrolyte, which poses safety risks, including leaks, burning, and explosions [3, 4]. Solid polymer electrolytes (SPEs) stand out as promising potential replacements due to their high energy density, lightweight, and flexibility and attract a lot of research interests [5, 6]. These batteries could greatly improve the way we power our devices [7]. However, the ionic conductivity of SPEs is still not ideal, which is one of the significant obstacles to their wide application [8]. An assumption was proposed that the polymer segmental, backbone motions, and Li-ion charge transport are closely correlated [5]. Do et al. observed a strong coupling between conductivity and the relaxation of the polymer electrolyte [9], verifying the assumption experimentally. They found that cages are formed with about 5-6 EO monomer units surrounding the Li–ion and the cage decay time is about 5 ns. They assumed that the flexibility of the PEO chain reduces the cage decay time resulting in an increase in Li–ion transport. In this paper, the dynamics of the 4-arm star poly(ethylene oxide) (4armPEO) and the effect of LiTFSI will be investigated with Quasi-Elastic Neutron Scattering (QENS).

The 4armPEO (M_w=13.8 kg/mol, *PDI*=1.01), as illustrated in Fig. 1, was synthesized by polymerizing ethylene oxide (EO) (Sigma-Aldrich, $\geq$ 99.9%) and ethylene oxide-d4 (Cambridge Isotope Laboratories, 98% D) with partially potassium metalated pentaerythritol as initiator in anhydrous DMSO in our deuteration laboratory [10]. With the experience of our previous studies [11], we surmise that Li+ is attracted by the ether-O of the core part and tends to be trapped in the area close to the center carbon atom. To study the dynamics of the core and the hand parts of the polymer (as shown in Fig. 1 (a)) separately, the partially deuterated samples were also synthesized. The arm lengths are about 72 EO monomer units and, when fully stretched, are about 200 Å long. For the deuteration, either the first 36 EO units close to the core (core-deuterated 4armPEO, M_w=12.8 kg/mol, *PDI* = 1.05) or the last 36 EO units close to the end (hand-deuterated 4armPEO, M_w=13.9 kg/mol, *PDI* = 1.07) were selected. Additionally, Lithium bis(trifluoromethanesulfonyl)imide (LiTFSI) was added to the 4armPEO as the source of Li+. QENS was conducted on original and deuterated 4armPEO samples with varying amounts of LiTFSI (as listed in Table 1) using a backscattering silicon spectrometer (BASIS) of the Spallation Neutron Source at Oak Ridge National Laboratory, USA [12]. Samples were sealed in a flat plate Aluminum sample holder and measured under He environment. The energy of the Neutron beam is 2.08 μeV (λ=6.4 Å). The sample was heated to 400 K, equilibrated for 10 minutes, and then cooled to 380 K, 360 K, 340 K, 300 K, 280 K, and 260 K for measurement (Fig. 1(b)). The instrument was set up to get a reflection from Si(111) analyzers panel covering the Q range of 0.3 $Å^{-1}$ to 1.9 $Å^{-1}$ and the energy transfer window of -120μeV to 120 μeV with an instrumental energy resolution of 3.5 μeV. The resolution was measured separately from each sample at 20 K.

Table 1. Sample List

Sample Name	Deuterated part of 4armPEO	Proportion of LiTFSI
4armPEO	No	0
4armPEO + 20% LiTFSI	No	20 w.t.%
4armPEO + 40% LiTFSI	No	40 w.t.%
Core Deuterated 4armPEO	Core part	0
3Core Deuterated 4armPEO + 20% LiTFSI	Core part	20 w.t.%
Hand Deuterated 4armPEO	Hand part	0
Hand Deuterated 4armPEO + 20% LiTFSI	Hand part	20 w.t.%

The QENS spectra of 4armPEO with 0%, 20%, and 40% LiTFSI are shown in Fig. 2 (a), (b), and (c). The QENS spectra below 320 K are sharp and intense, whereas peaks collected above 320 K exhibits considerable broadening. There is a discontinuity at 340 K of 4armPEO with 0% and 20% LiTFSI, but not in 4armPEO with 40% LiTFSI. Meanwhile, DSC measurements reveal melting peaks at around 320 K for 4armPEO with 0% and 20% LiTFSI, while there is no prominent peak for 4armPEO with 40% LiTFSI (Fig. 3). It became evident that with the increase of the Li – salt content, the melting peak is suppressed, making the system softer. In addition, higher Li–salt content makes a better battery system that can function at lower temperatures. Because of the higher scattering cross section, H atoms from 4armPEO contribute most of the Neutron scattering signal. Therefore, the QENS data and the DSC results suggest a significant change in the dynamics of the polymer after melting in the samples with 0% and 20% LiTFSI content.

The QENS spectrum comprises of the elastic–incoherent structure factor (EISF) and the quasi-elastic scattering. The spectra were fitted using delta and Lorentzian functions convoluted with the instrument resolution measured from the same sample at 20K (Fig. 4). Below the melting temperature, the QENS spectra could be fitted using a single Lorentzian, which can be assigned to a localized motion of Hydrogen atoms in 4armPEO (Model I, Equation 1). Above the melting temperature, two Lorentzians, one for fast and another for slow dynamics (Mode II, Equation 2), were needed. Among them, the narrower component captures the slower diffusion process of 4armPEO.

$$S(Q,\omega) = A(Q)\delta(\omega) + B(Q)L(Q,\omega) \quad \text{(Mode I)} \quad (1)$$

$$S(Q,\omega) = A(Q)\delta(\omega) + B(Q)L_{fast}(Q,\omega) + C(Q)L_{slow}(Q,\omega) \quad \text{(Mode II)} \quad (2)$$

The EISF and the half width at half maximum (HWHM) were obtained from the fitting of the QENS spectrum. The EISFs of 4armPEO below the melting temperature (320 K) do not show a strong decay suggesting a strong elastic neutron scattering over the length scales from appr. 3 Å to 20 Å. This is attributed to the relatively slow motion of the H atoms of 4armPEO below the melting temperature. However, the EISFs above the melting temperature drop to 0.2.This reduction in EISF indicates that more than 80% of the elastic scattering was

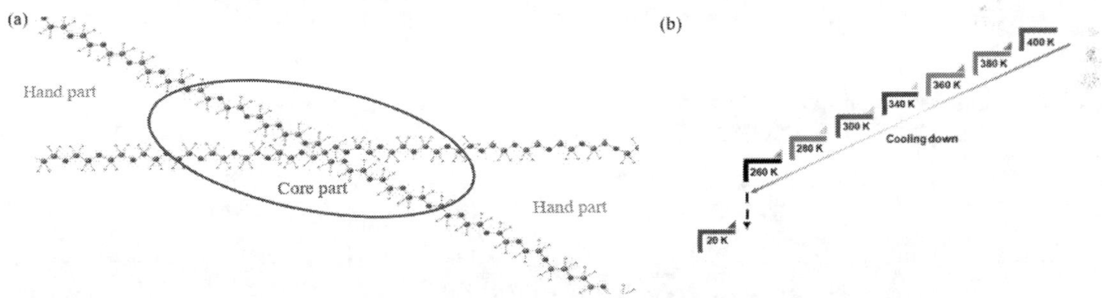

Figure 1. Snapshot of a Molecular structure of (a) 4armPEO, red balls represent oxygen atoms, the black balls represent carbon atoms, and the small white balls represent hydrogen atoms; (b) Temperature steps at which QENS data were collected during the cooling process. Note, in Fig.1a the core part and hand part consist just of about 3-5 monomer units for better illustration of the principal idea. In our samples the core part and hand parts when deuterated where about 36 monomer units each.

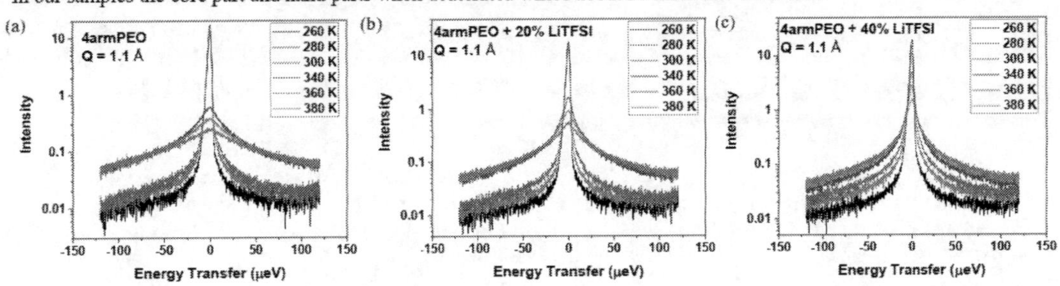

Figure 2. Representative QENS patterns for (a) 4armPEO, (b) 4armPEO+20%LiTFSI, and (c) 4armPEO+40%LiTFSI samples at a Q = 1.1 Å^{-1}.

 https://doi.org/10.1051/epjconf/202328604005

transformed into inelastic scattering after the melting temperature. The H atoms move at a relatively low frequency below 320 K. Hence, the corresponding HWHMs from the Lorentzian were small, below 15 μeV, as shown in Fig. 5(b). While, QENS signals get broader after melting with HWHMs reaching 80 μeV (Fig. 5 (c)).

Assuming the plateau region in Fig. 5(b) is limited to a maximum Q = 1.25 Å^{-1} we can calculate the proper confinement radius r_{conf} according to the Volino-Dianoux confined diffusion model [13, 14] with $r_{conf} = \pi / Q$ =2.5 Å. This is in agreement to the literature and additional MD simulations we performed. It would lead to a spherical confinement volume of about 62 Å^3.

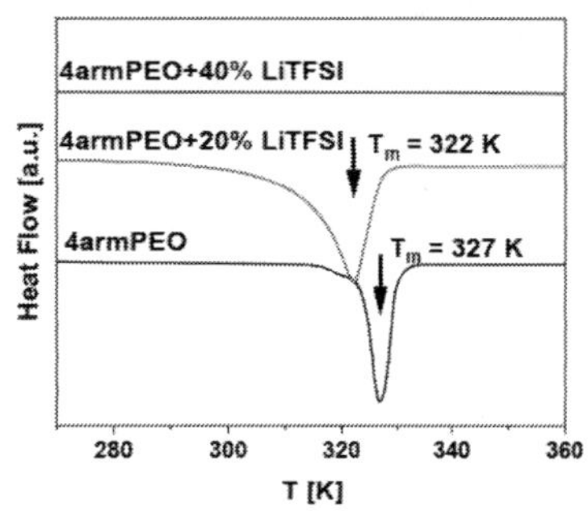

Figure 3. DSC profile for 4armPEO, 4armPEO+20%LiTFSI, and 4armPEO+40%LiTFSI samples.

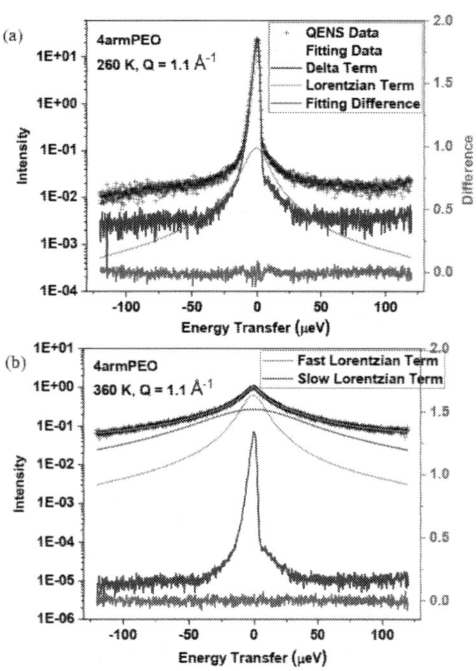

Figure 4. QENS spectra together with the model fit of 4armPEO at Q=1.1 Å^{-1} (a) from a single component fit below the melting temperature 320 K and (b) using two components fit above 320 K.

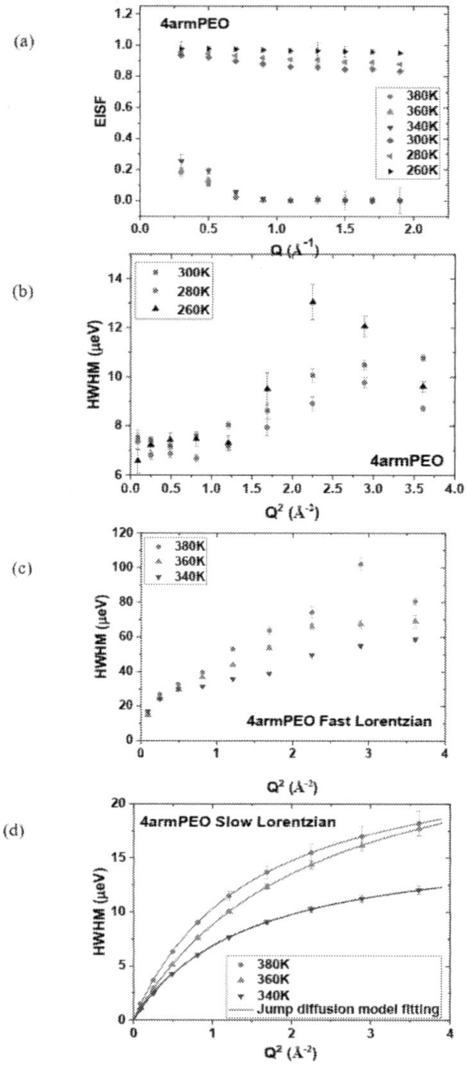

Figure 5. Q dependence of fitting parameters of 4armPEO. (a) EISF; (b) HWHM of the QENS spectra below 320 K; (c) and (d) are the HWHMs of the fast and slow components QENS spectra collected above 320K.

A comparison of the HWHMs of the broad component (Fig. 6) indicates that the adding LiTFSI slows down the chain dynamics. Furthermore, the narrower component (Fig. 5 (d)) observed after melting was fitted using the jump-diffusion model to obtain the diffusion coefficient, *D*, of this dynamics process (Equation 3).

$$\Gamma = \frac{DQ^2}{1+DQ^2\tau} \tag{3}$$

Table 2 shows that the diffusion coefficients of 4armPEO increase with temperature as expected. However, the addition of LiTFSI leads to a decrease in diffusivity values. Theoretically, the core-deuterated polymer should have a higher diffusivity than the hand-deuterated polymer, as the hand part of 4armPEO is more flexible than the core part. However, QENS results

 https://doi.org/10.1051/epjconf/202328604005

show that the D values of the core-deuterated and hand-deuterated polymers are similar, likely due to the long arm of 4armPEO (with around 72 EO monomers on each arm), which smears out the effect of the constrained core part. It is unclear why the D values for the partially deuterated polymers are consistently smaller than that of 4armPEO. Still, it may be related to the larger molecular weight of the deuterated 4armPEO. Also, when adding Li salt, the diffusion coefficient decreases (Fig. 8). Here, Li – ions bind with the oxygen in the 4armPEO and reduce mobility. The addition of more of the Li – ions is found to restrict the mobility of the polymer chain further.

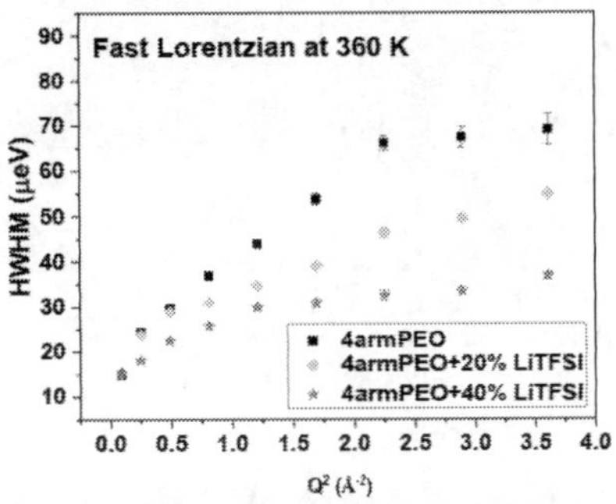

Figure 6. A comparison of HWHM of the fast Lorentzian component for 4armPEO, 4armPEO+20%LiTFSI, and 4armPEO+40%LiTFSI samples at 360 K.

Table 2. Diffusion coefficient of different samples at temperatures above T_m

Sample	D at 340 K (10^{-10} m²/s)	D at 360 K (10^{-10} m²/s)	D at 380 K (10^{-10} m²/s)
4armPEO	1.76±0.03	1.93±0.03	2.59±0.05
4armPEO+20%LiTFSI	1.07±0.02	1.60±0.03	1.86±0.04
4armPEO+40%LiTFSI	1.02±0.03	0.88±0.02	1.17±0.02
Core Deuterated 4armPEO	1.43±0.03	1.71±0.04	1.90±0.02
Core Deuterated 4armPEO+20%LiTFSI	0.91±0.02	1.35±0.02	1.49±0.03
Hand Deuterated 4armPEO	1.25±0.02	1.53±0.03	1.70±0.04
Hand Deuterated 4armPEO+20%LiTFSI	0.93±0.02	1.23±0.02	1.45±0.02

With increasing temperature, the thermal activation triggers, not just for the polymer chain but also for the Li – ions, making them move faster and consequently raising the diffusion coefficient (Figures 7 and 8). In addition, the hands-deuterated 4armPEO has a smaller diffusion coefficient than the core deuterated at all temperatures and concentrations (Fig. 7). The core part moves slower than the hand part, which is what we expected to achieve.

The diffusion coefficients of the slow process follow the Arrhenius behavior. The activation energy, E_a, of this slow process is depicted in Fig. 7 and Fig. 8. The activation energy for the 4armPEO system is 10 mJ/mol, which agrees with the values in the literature [15]. Table 3 demonstrates that the activation energy increases with the addition of LiTFSI, indicating that the polymer chain motions get more restricted by the ions. Additionally, the activation energy for the core-deuterated and hand-deuterated samples is similar due to the long chains of the four arms in 4armPEO. The overall trend of the activation energy is consistent with that of the diffusion coefficients showing the restrictions of the chain dynamics with increasing Li – salt concentration due to the engagement of more of the Li-ions. Also, the hand part of the chain shows higher flexibility (diffusion coefficients), which is of great advantage for Li-ion transportation in batteries.

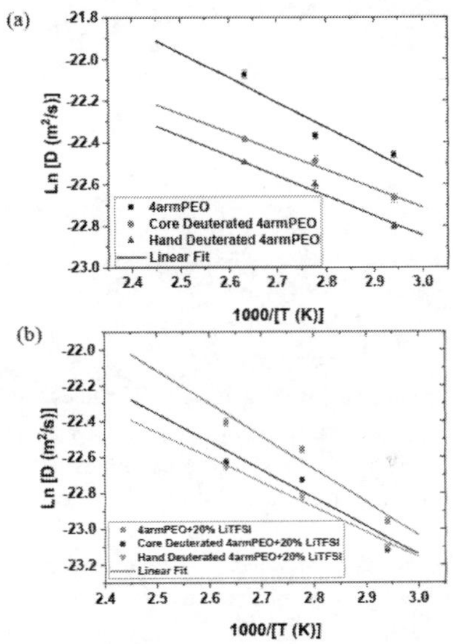

Figure 7. Temperature dependence of diffusivity showing an Arrhenius behavior.

Table 3. Activation energy of slow dynamics of different samples at temperatures above T_m

Sample	E_a (kJ/mol)
4armPEO	9.94±3.79
4armPEO+20%LiTFSI	15.27±3.55
4armPEO+40%LiTFSI	16.2
Core Deuterated 4armPEO	7.46±0.68
Core Deuterated 4armPEO+20%LiTFSI	13.05±5.05
Hand Deuterated 4armPEO	7.97±0.77
Hand Deuterated 4armPEO+20%LiTFSI	11.59±1.47

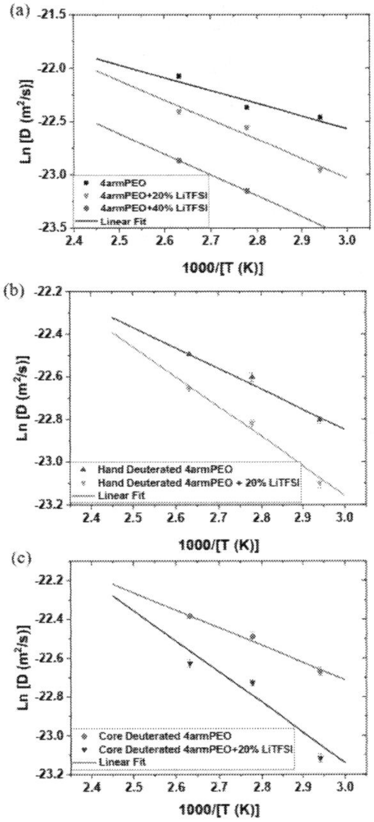

Figure 8. Temperature dependence of diffusivity showing an Arrhenius behavior.

To summarize, the present study investigates dynamics of 4armPEO and the influence of LiTFSI. Below the melting temperature, the H atoms move relatively slow resulting in the strong elastic neutron scattering. Above the melting temperature, two dynamics processes were observed: a faster-localized and a long-range slow process. The relative fast motions of H atoms cased the significant inelastic neutron scattering (EISF below 0.2). All dynamics processes observed here become faster with increasing temperature as expected. The addition of LiTFSI softens 4armPEO and contributes to its room temperature application. However, it also slows down the polymer dynamics by forming bond with ether-O atoms in the polymer chain. Furthermore, this study also indicates that the hand part is always more flexible than the core part, which improves the mobility of Li+. This study provides valuable insights into the dynamics of 4armPEO and highlights the role of LiTFSI in its performance. These findings have important implications for the development of advanced polymer electrolytes in energy storage applications.

Research at the BASIS of the ORNL's Spallation Neutron Source was sponsored by the Scientific User Facilities Division, Office of Basic Energy Sciences, U.S. Department of Energy.

References

[1] J.W. Choi, D. Aurbach, Nat. Rev. Mater. **1**, 4 (2016).
[2] V. Etacheri, R. Marom, R. Elazari, G. Salitra, D. Aurbach, Energy Environ. Sci. **4**, 3243 (2011).
[3] O.W. Sheng, C.B. Jin, J.M. Luo, H.D. Yuan, H. Huang, Y.P. Gan, J. Zhang, Y. Xia, C. Liang, W.K. Zhang, X.Y. Tao, Nano Lett. **18**, 3104 (2018).
[4] X. Zhang, T. Liu, S.F. Zhang, X. Huang, B.Q. Xu, Y.H. Lin, B. Xu, L.L. Li, C.W. Nan, Y. Shen, J. Am. Chem. Soc. **139**, 13779 (2017).
[5] F. Croce, G.B. Appetecchi, L. Persi, B. Scrosati, Nature (London) **394**, 456 (1998).
[6] A. M. Stephan, K. S. Nahm, Polymer **47**, 5952 (2006).
[7] F. B. Dias, L. Plomp, J. B. J. Veldhuis, J. Power Sources **88**, 169 (2000).
[8] R. J. Chen, W. J. Qu, X. Guo, L. Li, F. Wu, Mater. Horizon. **3**, 487 (2016).
[9] C. Do, P. Lunkenheimer, D. Diddens, M. Götz, M. Weiß, A. Loidl, X. G. Sun, J. Allgaier, M. Ohl, Phys. Rev. Lett. **111**, 018301 (2013).
[10] J. Allgaier et al. unpublished data.
[11] H. Wang et al. unpublished data.
[12] E. Mamontov, K. W. Herwig, Rev. Sci. Instr., **82**, 85109 (2011).
[13] F. Volino and A. Dianoux, Mol. Phys., **41**, 271–279 (1980).
[14] A. J. O'Malley, M. Sarwar, J. Armstrong, C. R. A. Catlow, I. P. Silverwood, A. P. E. York and I. Hitchcock, Phys. Chem. Chem. Phys., **20**, 11976-11986 (2018).
[15] V. G. Sakai, J. K. Maranas, I. Peral, J. R. D. Copley, Macromolecules, **41**, 10 (2008).

The high-Q static scattering of 3-methyl pyridine/D_2O mixtures without and with antagonistic salt

Henrich Frielinghaus[1,*], *Purushottam S.* Dubey[1], *Eunjoo* Shin[2], *Mary* Odom[3], *Piotr* Zolnierczuk[3], *Baho* Wu[1], *Olaf* Holderer[1], *Theresia* Heiden-Hecht[1], *Jan V.* Sengers[4], and *Stephan* Förster[1]

[1]Forschungszentrum Jülich GmbH, Jülich Center for Neutron Scattering JCNS-4 at MLZ, Lichtenbergstrasse 1, 85747 Garching, Germany
[2]Korea Atomic Energy Research Institute, 111, Daedeok-daero 989beon-gil, Yuseong-gu, Daejeon, 34057, Republic of Korea
[3]Neutron Sciences Directorate, Oak Ridge National Laboratory (ORNL), POB 2008, 1 Bethel Valley Road, Oak Ridge, TN 37831, USA
[4]Institute for Physical Science and Technology, University of Maryland, College Park, Maryland 20742, USA

Abstract. Here we focus on the high-Q small-angle neutron scattering where we observed deviations from an ideal power law Q^{-2}. From theory, this deviation Δ arises from the critical correlation-function exponent η_d in d dimensions. The investigated systems were 3-methyl pyridine/D_2O without and with antagonistitc salt. They display the critical behavior of a 3d and 2d Ising system, respectively. In the first case, the value of Δ matches the ideal value η_3 well, but in the latter case the value of Δ is affected by the two confined dimensions *and* the third dimension.

1 Introduction

We have studied the critical dynamics of the system 3-methyl pyridine (3MP) / D_2O without and with antagonistic salt [1]. We found a 3d- and 2d-Ising behavior in the static structural measurements (also firstly found in [2, 3]) and a characteristic critical behavior for the dynamics [1]. At large length scales the hydrodynamic diffusion and at small length scales the critical diffusion was observed. While at large length scales whole domains are diffusing, at small length scales an exchange between domains occurs. The two behaviors have a footprint of a dynamic critical exponent $z = 0.063$ and virtually 0, respectively. More important is the difference in the measured bare viscosities between the 3d and 2d cases that also affects the amplitudes for critical dynamics. The 2d confined liquid displays the well-known lubrication effect that goes in hand with the lower bare viscosities.

The understanding of the confinement is based mainly on the theory of Onuki [4–6]. A sketch of the situation is displayed in Fig. 1. The charges mainly form layers and the different ions are found in the domains of 3MP and D_2O. They also serve as wavefronts for the charge density waves and are responsible for the 2d confinement of 3MP/D_2O. The solvation effect described by Onuki means that a rather sharp boundary between 3MP and D_2O domains at this interface leads to strong enrichments, whereas in the other directions the transitions between domains are rather soft. On the other hand, the system without salt is assumed to be a rather ideal 3d Ising system.

*e-mail: h.frielinghaus@fz-juelich.de

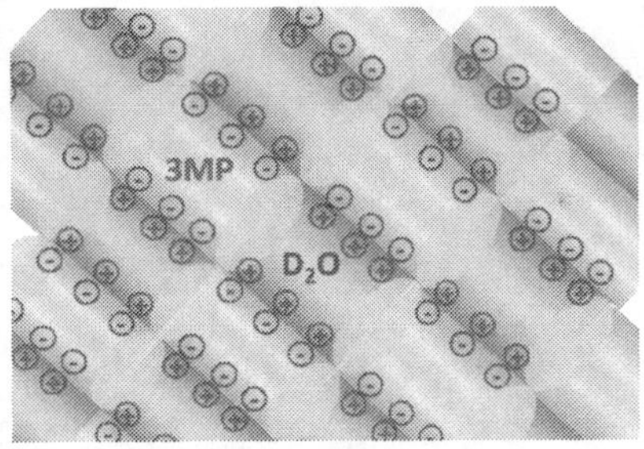

Figure 1. A sketch of the arrangement of the domains of 3MP and D_2O and the ions of the antagonistic salt. The charges are aligned in planes throughout the sample and serve as wavefronts of the charge density waves. The solvation effect is exaggerated by the blue and orange colors (water and 3MP) in terms of a fast decay towards mixing indicated by the light brown color.

By evaluating the high-Q small-angle scattering we look at the phase boundaries just with an orientational average. We first evaluate the asymptotic power laws of the small-angle scattering and then discuss the obtained exponents. Then we summarize and discuss the findings in the context of existing theories.

2 Theory

The classical theory for critical fluctuations in three dimensions is the Ornstein-Zernike theory [7–9], which yields a very simple formula for the scattering, i.e., the

Table 1. Summary of the critical exponent η_d in d dimensions.

dimensionality d	3	2	1
η_d	0.033	$\frac{1}{4}$	$\frac{1}{2}$
citation	[10]	[11]	[12]

correlation function in reciprocal space:

$$S(Q) = \frac{S(Q=0)}{1+\xi^2 Q^2} \tag{1}$$

The real space correlation function then is obtained by a Fourier transform after:

$$G(r) = \frac{1}{4\pi\rho R^2}\frac{\exp(-r/\xi)}{r} \tag{2}$$

Here ρ is the density, and R^2 is the second moment of the direct correlation function $C(r)$ [7]. This approach is widely used and works well for small Q when only the correlation length is extracted from scattering experiments. However, at large Q, there could be deviations from this ideal behavior as Fisher pointed out [7]. He generalized the real space correlation function by including a correction with the critical exponent η_d. The real-space correlation function then reads:

$$G(r) \sim \frac{\exp(-r/\xi)}{r^{d-2+\eta_d}} \tag{3}$$

The general case for d dimensions is covered here. Of course also η_d depends on the number of dimensions. A list of values for η_d can be found in Table 1. We tabulate the classical values that have been established for 3 to 1 dimensions [10–12]. They compare quite well with the values obtained from a Renormalization-Group expansion that depends on $\epsilon = (4-d)$ according to $\eta_d = 0.050\epsilon^2$ [13]. Fisher then proposed a simple corrected formula for the scattering function in three dimensions that reads [7]:

$$S(Q) = \frac{S(Q=0)}{(1+\xi^2Q^2)^{1-\eta_3/2}} \tag{4}$$

Similarly, one could write a scattering function for two dimensions (that is exact for $\eta_2 = 0$):

$$S(Q) = \frac{S(Q=0)}{\left(\sqrt{1+\xi^2Q^2}\right)^{1-\eta_2}} \tag{5}$$

One has to be cautious with eqs. 4 and 5 because they give the impression to be exact over the entire Q-range. At small Q, the corrections of η_d are small in the sense that they are nearly invisible (an analytic analysis is presented in the Appendix). At large Q, the asymptotic scaling is correct and then reflects the low-r dependency of eq. 3. However, the prefactor might change and is then different from the low-Q behavior. The medium Q-range is actually not very well known, and usually needs more delicate discussion. This overall mismatch is more pronounced the bigger the exponent η_d is.

For completeness, we summarize the scattering law that was applied to the $3MP/D_2O$ system with antagonistic salt. The ions were taken into account and one arrives at [2]:

$$S(Q) = \frac{S(Q=0)}{1+\left\{1-\kappa^2/\left(1+\lambda_D^2Q^2\right)\right\}\cdot\xi^2Q^2} \tag{6}$$

This function was derived from the real-space correlation function of Onuki [4, 5]. It employs the Debye length $\lambda_D = 39.25$Å as a fixed parameter, and the parameter κ. When $\kappa < 1$ the system is more like a two-component system, and $\kappa > 1$ includes the amphiphilicity of the antagonistic salt with a preferred wavelength of the charge density waves (see [2]). For the current manuscript, we are only interested in the asymptotic Q^{-2} scattering that starts at $Q > 10^{-3}$Å^{-1} where the correlation length ξ is determined and further on might show deviations from the ideal Q^{-2} behavior.

3 SANS data evaluation at high Q

The experiments have been described in the main paper [1]. We just mention that the small-angle neutron scattering (SANS) data have been collected at the Hanaro reactor in the Republic of Korea. Examples for the $3MP/D_2O$ mixtures are displayed in Fig. 2. While at small Q we can extract the critical behavior, we now focus on the large-Q scattering. We applied the following behavior for the macroscopic cross section (i.e., the calibrated intensity) as a function of the scattering vector Q:

$$\frac{d\Sigma}{d\Omega}(Q) = \frac{A}{Q^{2-\Delta}} + b_{\text{background}} \tag{7}$$

The amplitude A was kept as a free parameter and was not connected to the forward scattering. The exponent Δ describes the difference from the ideal scattering law; and $b_{\text{background}}$ describes the incoherent background (also a free parameter). In the following we focus on the exponent Δ

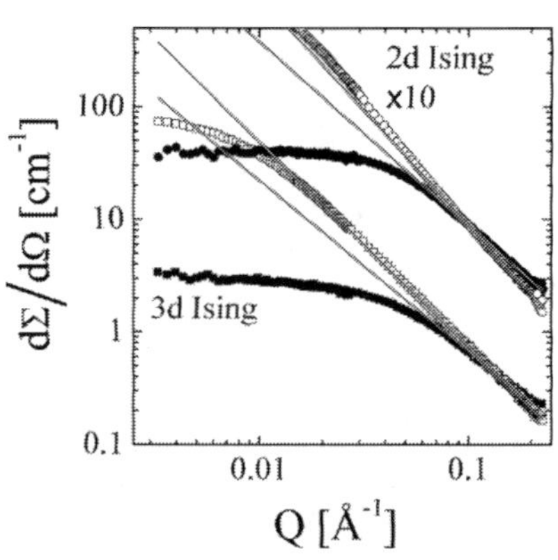

Figure 2. The macroscopic cross section (with background subtracted) as a function of the scattering angle Q for the system $3MP/D_2O$ (3d Ising system) and $3MP/D_2O$ with antagonistic salt (2d Ising system, multiplied by 10). The solid symbols indicate a temperature of 20°C and the open symbols the highest temperature of 36°C and 40°C, respectively. The asymptotic high-Q scattering is fitted at $Q > 0.104$Å^{-1} (red lines).

only as the deviation from the ideal behavior. We applied this in the 3d and 2d Ising case in the same manner independent on the dimensionality. The parameter Δ is shown as a function of temperature in Fig. 3. We can see that the temperature dependence is quite linear for temperatures below 32°C (or 34°C). These values are also considerably higher than what would be expected for 3 dimensions, i.e., η_3. The reason for the decay towards higher temperatures is the importance of the incoherent background. We cannot exactly motivate a reason for a linear behavior and so we consider this as an experimental observation. At even higher temperatures there are four data points that are considerably higher than the linear approach. Here, we consider the values to originate from the intermediate Q-scattering where extrapolations become difficult. From the linear extrapolations to $T = T_C$, the critical temperature of 36.9°C (42.9°C), we obtain the exponent $\Delta = 0.043 \pm 0.04$ (-0.292 ± 0.026) at the critical temperature where background issues should be minimal. For the 3d Ising system our result is consistent with the expectation for the critical exponent η_3, except that the statistical errors are nearly larger than the value itself. The negative Δ of the 2d system indicates an incorrect estimate of the ideal asymptote. So we would like to propose the following formula for high Q:

$$S(Q) \sim \frac{1}{\left(\sqrt{1+\xi^2Q^2}\right)^{1-\eta_2}} \cdot \left|\frac{\sin(QD)}{QD}\right|^{2-\eta_1} \tag{8}$$

This reflects the critical behavior in the 2 dimensions between the charge density wave fronts (first term), and the sharply defined domains in the third dimension. We also assume a criticality in the direction of the third dimension using η_1. The factorization of two formfactors (critical fluctuations in two dimensions, i.e. eq. 5, and a compact object in the third dimension) does either hold for a separation of length scales [14] or for large polydispersities as in our case. Due to the polydispersity of the domian sizes D in the third dimension, we can replace the oscillatory part $\sin^2(QD)$ by $\frac{1}{2}$ [15]. So, we would correct the original Δ into:

$$\Delta_{\text{corr}} = 3 - 2 + \Delta = 0.708 \pm 0.026 \approx \eta_2 + \eta_1 = \frac{3}{4} \tag{9}$$

For the values of η_d we refer to Table 1. So, we can state that in the high-Q behavior a strong influence of the one dimension along the propagation of the charge density waves is visible. The contribution of η_1 is stronger than from the other two dimensions in the lateral directions. This is surprising because the criticality of the structural static measurements (at low Q) and the critical dynamics do not display any influence of this separated third dimension. We argued [1] that the finite viscosity that applies for the two lateral dimensions manages to avoid the third dimension for shorter times (when one could speculate about the Pomeau divergence [16]) but at later stages the molecules enter the third dimension when the charge density waves come into play. In the experiments with strong contrast between 3MP and D_2O the whole influence of the charge density waves appears rather low. So it is even more interesting that the third dimension also comes into play at the high-Q static scattering.

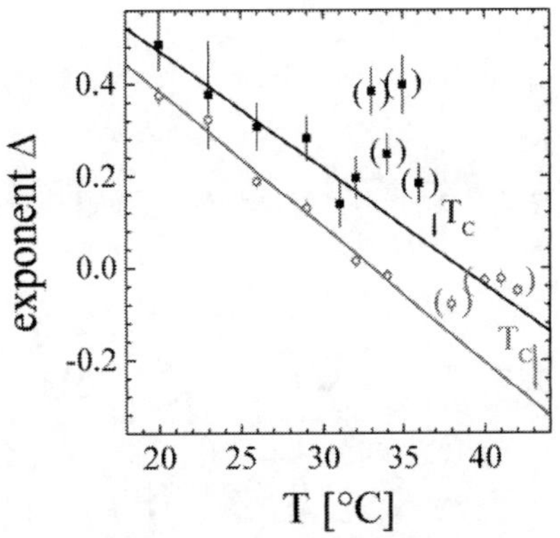

Figure 3. The deviation of the exponent from the ideal value 2 arising from the high-Q SANS power law scattering. This deviation Δ is plotted as a function of temperature. From smaller temperatures towards higher temperatures the trend seems to be linear and then suddenly breaks off to higher values (in brackets). The limit of the linear trend at the critical temperature T_C is determined to be compared to the theoretical exponents η_d.

The one-dimensional Ising criticality is a quite interesting subject. For instance many critical exponents seem to diverge when approaching the dimensionality one [17] (and references herein). For instance the critical exponent ν goes with $\nu = \epsilon^{-1} - \frac{1}{2} + \frac{1}{2}\epsilon + O(\epsilon^2)$ with ϵ being the difference to the ideal one-dimensional case [18]. Thus it is surprising that the exponent η_1 stays finite. A more detailed study of our system would possibly give new insight about other critical exponents in one dimension.

4 Summary

We analyzed the high-Q SANS patterns of 3MP/D_2O mixtures without and with antagonistic salt. Here, we determined the effective exponent from the scattering data that deviates from the ideal exponent of 2. The deviation of this exponent to the ideal case of 2 was then plotted as a function of temperature. We observed a linear behavior at lower temperatures while at elevated temperatures the medium Q-range behavior does not match anymore. The exponent was then extrapolated towards the critical temperature T_C. These values were then compared to the theoretical values of the exponent η_d. For the 3d Ising case the extrapolation matches the ideal 3d-case. For the 2d Ising case we had to derive the ideal power law with an exponent 3, and the deviation Δ_{corr} is affected by the two dimensions where the domains are confined to and the last, third dimension, in which the charge density waves propagate. The last influence is surprising because the charge density waves seem to be not that dominating in the SANS patterns.

In the future the alignment of the charge density waves would be desireable – be it in a shear field or in an electric or magnetic field. Then, the decomposition of different

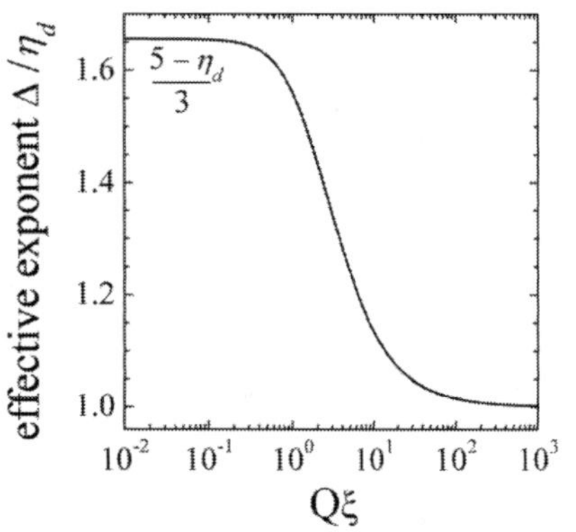

Figure 4. The effective exponent Δ as a function of the parameter $Q\xi$ normalized to the actual exponent η_d (here we assumed the 3-dimensional case, that is the only known analytic and exact solution given in eq. 10). However, for other dimensionalities we expect an extremely similar behavior.

terms would be easier and the exponent η_1 would appear isolated.

5 Appendix

From eq. 3 we could derive the scattering function in three dimensions exactly:

$$S(Q) \sim \frac{\cos(\eta_3 \arctan(Q\xi)) - \sin(\eta_3 \arctan(Q\xi))/(Q\xi)}{(1+Q^2\xi^2)^{1-\eta_3/2}} \tag{10}$$

The enumerator contains the details about the transition from small $Q\xi < 1$ to large $Q\xi > 1$. The amplitudes and effective exponents (or slopes) change between the two cases. The effective exponent Δ at a specific $Q\xi$ normalized to the actual exponent η_d is displayed in Fig. 4. It decays from a maximum at small $Q\xi$ towards the expected ratio unity at large $Q\xi$. This also means that the intensities (or the effective amplitude) at small $Q\xi$ are smaller and a faster decay towards higher $Q\xi$ is observed (most curves in Fig. 2 stay below the high-Q power law, and only for the 2d-Ising case at 40°C the deviation is reversed - possibly due to the charge density waves, see eq. 6). Thus we believe that the enumerator of eq. 10 might be seen as a general switching function between the two Q-regions also for other dimensionalities. However, for different dimensions the exact expressions are not analytic anymore.

References

[1] H. Frielinghaus, P.S. Dubey, B. Wu, M. Odom, F. Zheng, E. Shin, P. Zolnierczuk, O. Holderer, S. Förster, T. Heiden-Hecht, Phys. Rev. Res. **5**, 023053 (2023)

[2] K. Sadakane, N. Iguchi, M. Nagao, H. Endo, Y.B. Melnichenko, H. Seto, Soft Matter **7**, 1334 (2011)

[3] K. Sadakane, M. Nagao, H. Endo, H. Seto, The Journal of Chemical Physics **139**, 234905 (2013)

[4] A. Onuki, H. Kitamura, The Journal of Chemical Physics **121**, 3143 (2004)

[5] A. Onuki, The Journal of Chemical Physics **128**, 224704 (2008)

[6] D. Jung, N. Rivas, J. Harting, The Journal of Chemical Physics **150**, 064912 (2019)

[7] M.E. Fisher, Journal of Mathematical Physics **5**, 944 (1964)

[8] F. Leclercq, S. Pouget, P. Damay, in *Neutron Spin Echo Spectroscopy: Basics, Trends and Applications* (Springer, 2002), pp. 232–245

[9] H. Frielinghaus, D. Schwahn, J. Dudowicz, K.F. Freed, K. Foreman, The Journal of Chemical Physics **114**, 5016 (2001)

[10] M. Anisimov, J. Sengers, Supercritical Fluids: Fundamentals and Applications pp. 89–121 (2000)

[11] A. Pelissetto, E. Vicari, Physics Reports **368**, 549 (2002)

[12] Y.W. Dai, X.H. Chen, S.Y. Cho, H.Q. Zhou, D.X. Yao, arXiv preprint arXiv:1805.03464 (2018)

[13] G.R. Golner, Physical Review B **8**, 339 (1973)

[14] J.S. Pedersen, P. Schurtenberger, Macromolecules **29**, 7602 (1996)

[15] M. Wagener, S. Förster, Scientific Reports **13**, 780 (2023)

[16] Y. Pomeau, Physical Review A **5**, 2569 (1972)

[17] M. Novotny, Physical Review B **46**, 2939 (1992)

[18] D. Forster, A. Gabriunas, Physical Review A **23**, 2627 (1981)

Potential of environmental scanning electron microscopy and SAXS to determine structural insights of plant-based emulsions with increasing dry matter content

Theresia Heiden-Hecht[1,*], *Baohu* Wu[1], *Marie-Sousai* Appavou[1], *Stephan* Förster[2], *Henrich* Frielinghaus[1], and *Olaf* Holderer[1]

[1]Jülich Centre for Neutron Science (JCNS) at Heinz Maier-Leibnitz Zentrum (MLZ), Forschungszentrum Jülich GmbH, Garching, Germany

[2]Jülich Centre for Neutron Science (JCNS), Forschungszentrum Jülich GmbH, Jülich, Germany

Abstract.
Plant-based emulsions with increasing dry matter content show a large range of structural features from atomic to macroscopic length scales, which may be examined with scattering techniques in reciprocal space and microscopic techniques in real space. In this contribution, we focus on plan-based emulsions in terms of mesoscopic structure, and report on the impact of temperature and humidity on the structure measured via environmental scanning electron microscopy (ESEM) in real space. Small angle x-ray scattering in reciprocal space extends the knowledge on structural properties on smaller length scales at different temperature. Decreasing the humidity for the ESEM experiments revealed structural properties of different products. Temperature decrease from room temperature to 5 °C showed emerging crystalline peaks during SAXS measurements.

1 Introduction

Plant-based emulsions have a huge potential in substituting classical milk products [1]. A lot of research focuses on the implementation of novel plant proteins, and investigates the potential usage in plant-based emulsions with increasing dry matter content like milk drinks, fermented or heat-induced yoghurt or cheese like products [2, 3].

Imitating the texture and structure of dairy products is a challenge [4]. The texture and structure of emulsion products with different dry matter content is often analyzed via rheology, tribology and microscopy. Common microscopy methods in use are: transmission electron microscopy [5] and scanning electron microscopy [6, 7], both methods are used for studying casein micelles, confocal laser scanning microscopy [8, 9]. However, most of these methods have the disadvantage that an invasive sample preparation (freezing, cutting, substitution of water, staining, dye labelling) is required. Environmental scanning electron microscopy (ESEM) offers possibilities to investigate a range of samples in high resolution without invasive sample preparation, for example in Ref. [10] for a model cheese system. Varying pressure, temperature and with this combination also the relative humidity opens new possibilities in studying the morphology of food products on length scales below 1 μm to 1 mm. With small angle x-ray scattering (SAXS), complementary information in reciprocal space are accessible from the μm range down to atomic length scales (of the order of 0.1 nm)[11]. Also here, a rather natural environment with control parameters such as temperature are accessible.

This contribution shows some examples of how real space information obtained with environmental scanning electron microscopy (ESEM), complemented by reciprocal space information from small angle x-ray scattering (SAXS), can provide insight into plant-based emulsions and food products with different dry matter. Series of images with decreasing humidity in ESEM and scattering curves at two relevant temperatures are presented and discussed. The last section shows perspectives of neutron scattering to extend the range of scattering techniques for the characterization of food products [12]. Neutron scattering provides a view into these systems with different contrast and hence can reveal new structural details.

2 Methods

2.1 Environmental Scanning Electron Microscopy

The environmental scanning electron microscope (ESEM) provides the possibility to investigate surfaces with an electron beam without further treatment or coating, i.e. with almost no additional sample preparation, also for water containing samples, since a partial pressure of up to ~ 800 Pa can be tolerated, corresponding to 100% relative humidity (r.h) at temperatures of 1.5-5°C. Reducing the pressure results in water evaporation revealing surface structural information of the non-water components [13, 14].

ESEM images were obtained at JCNS MLZ using a Thermo Fischer Quattro S ESEM (Thermo Fischer Sci-

*e-mail: t.heiden-hecht@fz-juelich.de

entific, Eindhoven, Netherlands) equipped with a 500μm aperture gaseous secondary electron detector (GSED). During observation, the temperature of the sample stage was controlled with a Peltier cooling system. All observations were carried out at a constant temperature of 2-5°C of the sample stage and the temperature of the samples was equilibrated for 15 min prior to any measurement. The relative humidity was controlled by changing the water vapor pressure in the chamber. The relative humidity (r.h.) is then 100% at a pressure of 800 Pa, shown as initial picture of all series. Subsequently, the pressure is lowered in two steps, to 350 Pa, corresponding to a r.h. of 50%, further to 105 Pa (r.h. 15%) and reversed to a r.h. of about 100%.

All ESEM experiments were carried out with the following operating conditions: 10 kV beam accelerating voltage for the samples PD1 "Drink1", PD2 "Drink1", PCC and PC, 5 kV was used for sample PY to diminish beam damage, spot size 3.0, sample working distance 6.7 +/- 1.2 mm, and and 5–10 us dwell time. The corresponding sample currents (measured in the Faraday cup in high vacuum) were 80 pA and 110 pA for 5 and 10 kV respectively. Secondary electron images were acquired via the integrated imaging software (xT microscope control) at 500, 650, 1000, and 1200 magnifications.

2.2 Small angle x-ray scattering

SAXS experiments were performed by a laboratory based SAXS beamline KWS-X (XENOCS XUESS 3.0 XL) at JCNS MLZ. The X-ray source is a D2+ MetalJet (Excillum) with a liquid metal anode operating at 70 kV and 3.57 mA with Ga-Kα radiation (wavelength λ = 1.314 Å). Samples were measured in a glass capillary at ambient pressure (1.5 mm ID) that was kept at 25°C or 5°C in a temperature controlled peltier stage. The sample to detector distances from 0.1 m to 1.70 m which cover the scattering vector Q range from 0.003 to 4 Å^{-1} (Q is the scattering vector, Q=$(4\pi/\lambda)sin(\Theta)$, 2$\Theta$ is the scattering angle). The SAXS patterns were normalized to an absolute scale and azimuthally averaged to obtain the intensity profiles, and the solvent background was subtracted.

2.3 Samples

We present results on two plant-based emulsions and three plant-based products with different amounts of dry matter, a yoghurt, a cream cheese and a cheese (all plant-based). Details of the samples are provided in Ref. [15], where plant-based and dairy emulsions and fermented emulsions are compared with respect to their structural characteristics. Here only the plant-based products are in the focus of interest. Table 1 lists the names and basic state of the sample.

Short name	Sample name Ref. [15]	State
PD1	Drink 1	Plant Drink 1
PD2	Drink 2	Plant Drink 2
PY	Ferm. 1-Y	Plant Yoghurt
PCC	Ferm. 3-CC	Plant Cream Cheese
PC	Ferm. 2-C	Plant Cheese

Table 1. Samples studied with ESEM and SAXS with short names. More details on the composition of the samples can be found in Ref. [15].

3 Results and Discussion

3.1 SAXS and ESEM

Plant-based emulsions and gelated emulsions were analyzed with ESEM and SAXS, the latter with a broad Q-range up to 4 Å^{-1}.

Surface structures on micrometer length scales were observed via ESEM by decreasing the r.h. from 100% down to 50%, 15% and then back to 100%. This series of images has been taken for two emulsions (PD1 and PD2), and three types of plant-based food products (PCC, creamy; PC, cheese like and PY, youghurt like). The series of pictures are presented in Figures 1 and 2 for the liquid emulsions and Figures 3, 4, and 5 for the plant-based food products with higher amounts of dry matter. For the latter two, all steps of the r.h. series are shown, for the first three only the 100% and 15% r.h. steps.

The two types of emulsions, PD1 and PD2, showed initially a smooth surface (as expected for a liquid), with reduced humidity a surface structure became visible, best at the lowest r.h. of 15% (Figures 1, 2). This process is reversible, an increase of r.h. leads to a fully smooth surface again. The emerging surface structures are in our point of view due to evaporation of water, and opposing water capillary effects exposing the structuring phase at the surface. The mobility of the low dry matter samples probably increases just enough to smoothen the surface again at high r.h., it is therefore not considered a macroscopic phase separation due to the soft environmental changes and conditions. Due to the relatively high hydrophobicity of the system, reducing the r.h. from 100 % to 15 % does not expel the hydration shell at the interfaces and is probably the reason for a certain reversibility upon rehydration as well as the high amount of water in the sample itself. The emerging structures of PD1 at low r.h. could be salts (Potassium phosphate and sea salts present in the solution) and fat droplets. A particle size analysis of 77 particles of the 15% r.h. sample showed a mean diameter of 6 μm (with a standard deviation of 3 μm). For PD2, structuring particle like elements seem to emerge at the top part of the image with diameters of about 8 μm and a rather broad distribution, difficult to assess because partly hidden in the interface. The time constant for recovery when going back to 100 % r.h. was about 5 min for PD1 and quasi-instantaneous for PD2.

The plant-based yoghurt (PY, yoghurt like) with low dry matter (Figure 3) showed a similar low structuring at the beginning at 100% r.h., which increased slightly upon drying at 50% r.h. with no further change when reducing the r.h. to 15%. Again, this process is reversible, increasing the r.h. to 100% leads again to a completely smooth surface. The associated length scales are of the order of some 10-50 μm similar to a fractal surface.

Other in the case of the second low dry matter cream cheese like product (PCC, Figure 4), where the structure consisting mainly of globular particles of very regular size

 https://doi.org/10.1051/epjconf/202328604007

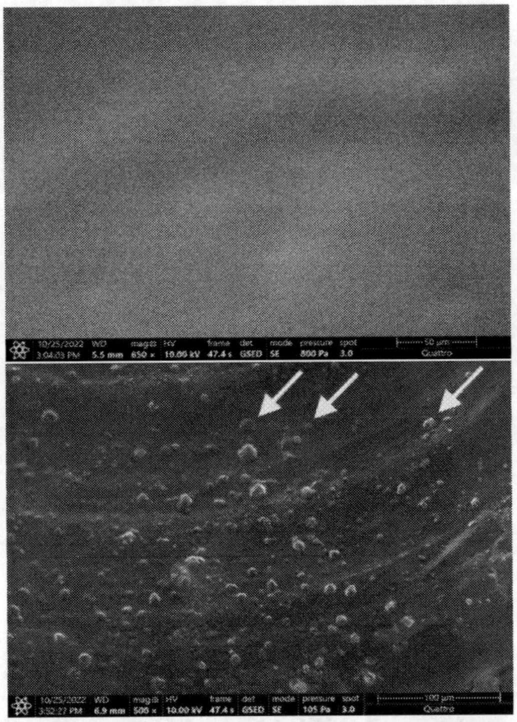

Figure 1. ESEM pictures at different relative humitidy of emulsion PD1, "Drink 1". From top to bottom, the r.h. was changed from 100% to 15%. Arrows mark some of the particles identified as most characteristic and discussed in the text.

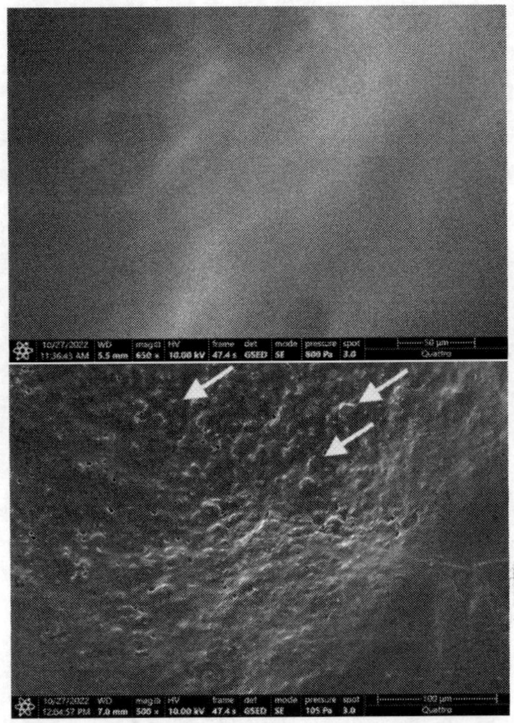

Figure 2. ESEM pictures at different relative humitidy of emulsion PD2, "Drink 2". From top to bottom, the r.h. was changed from 100% to 15%. Arrows mark some of the particles identified as most characteristic and discussed in the text.

of about 1.5 μm diameter (std dev. 0.3 μm, measured over 74 particles approximately at the same surface level in the image), which became visible upon reduction of humidity to 50% persisted also after re-hydration at the end of the experiment, probably due to the less mobile dry matter phase compared to the liquid drinks and the yoghurt like sample.

The high dry matter plant-based cheese PC 5 was much more solid and already well structured at 100% r.h. on the length scales of the ESEM experiment. It showed already at the initial 100% r.h. observation disperse globular structures of about 2-5 μm, and otherwise a rather fractal surface structure. There was no significant effect of changing the humidity.

With small angle x-ray scattering (SAXS) the three plant-based products PY, PCC and PC have been studied at two temperatures over a Q-range from 0.003-4 Å^{-1}, covering the typical small angle area with mainly a power law decay as well as the region where diffraction peaks from crystalline domains can be visible. The SAXS results of the two emulsions and the gels at room temperature have been discussed elsewhere [15] in more detail. Here the focus is on the temperature dependence of the three products. Figure 6 shows the full Q-range and a zoom into the high-Q-area of PY, PCC and PC at 25 °C and 5 °C. The power law decay is hardly affected by the temperature change, while on very small length scales a significant increase in crystallinity could be observed. The softest sample (top in Figure 6 is PY with low dry matter, as shown in the hydration change series in Figure 3, which stayed intact also when cooled down to 5 °C. PCC and PC showed an increase in crystalline features upon cooling. The Bragg peaks visible for the low (PCC) and high (PC) dry matter fermented emulsions (middle and bottom of Figure 6) at Q$\simeq 0.15 - 0.17$ Å^{-1} stem from aligned fat molecules from either coconut oil [16] for PC or from rapeseed oil for PCC (middle) [17] which is already present in a partly aligned form at 25 °C but which increases in ordering at 5 °C.

At higher Q above 1 Å^{-1} local ordering also increased with decreasing temperature significantly for PC, while for PCC it was already initially in a partly ordered state and only sharpened slightly.

The length scales covered by the SAXS experiments here, $d = 2\pi/Q$, are approximately 2 μm to 2 Å. The ESEM images shown here complement the length scale range towards larger structures. The slope of the SAXS intensities at low Q has been analyzed in Ref [15] with $I(Q) \propto Q^{-3.2}$ to $Q^{-3.9}$ indicating a surface fractal structure on these length scales below the length scales of the ESEM images, which seems quite consistent with the ESEM images at low relative humidity. While SAXS provides truly a view into the bulk structure, with the property that an ensemble average of the structures in the probed volume is represented in the scattering curves, the ESEM technique

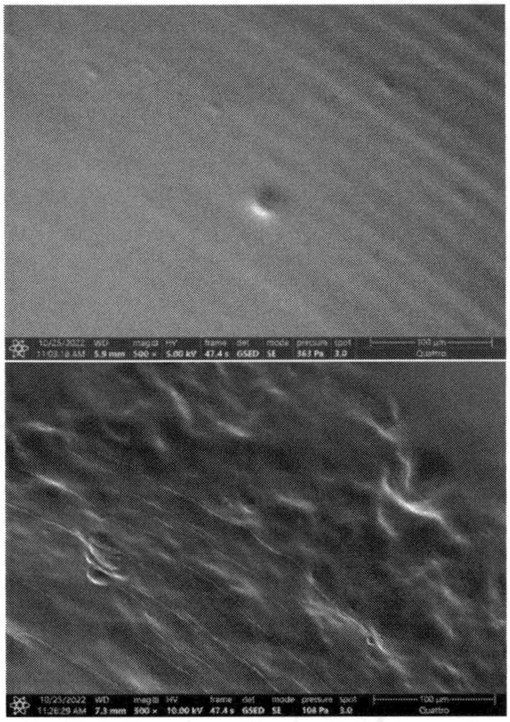

Figure 3. ESEM pictures of PY, a plant-based yoghurt with low dry matter, at r.h. of 100% (top) and 15% (bottom).

shows the surface morphology of a cut through the sample or of the liquid droplet surface, at different temperature/pressure conditions. From the images presented here it is also obvious that the 100% r.h. images often do not show any significant structuring in the cuts of the soft matter samples, but that contrast between parts with different vapor pressure may be created by reducing the relative humidity. It can therefore complement scattering information with respect to larger length scales and with insight into structuring elements emergent under vapor evaporation.

3.2 Outlook: neutron perspectives

The main advantage for neutron scattering in the area of food gels and emulsions is the possibility to investigate thick and nontransparent samples, with less radiation damage or heating as with x-rays due to the much lower kinetic energy of the neutron beam, and the possibility of varying the contrast by partially deuterating components of such multicomponent systems as the food emulsions. Small angle neutron scattering (SANS) provides a broad range of scattering vectors, and it can be extended towards low Q with double crystal diffractometers or with spin echo resolved SANS (SESANS)[18], or finally radiography for real space information, and to high Q with classical diffractometers. This also offers the possibility to access length scales from millimeters to Angstroem, but requires normally to generate model systems by partly deuterating the samples, in the simplest case only by ex-

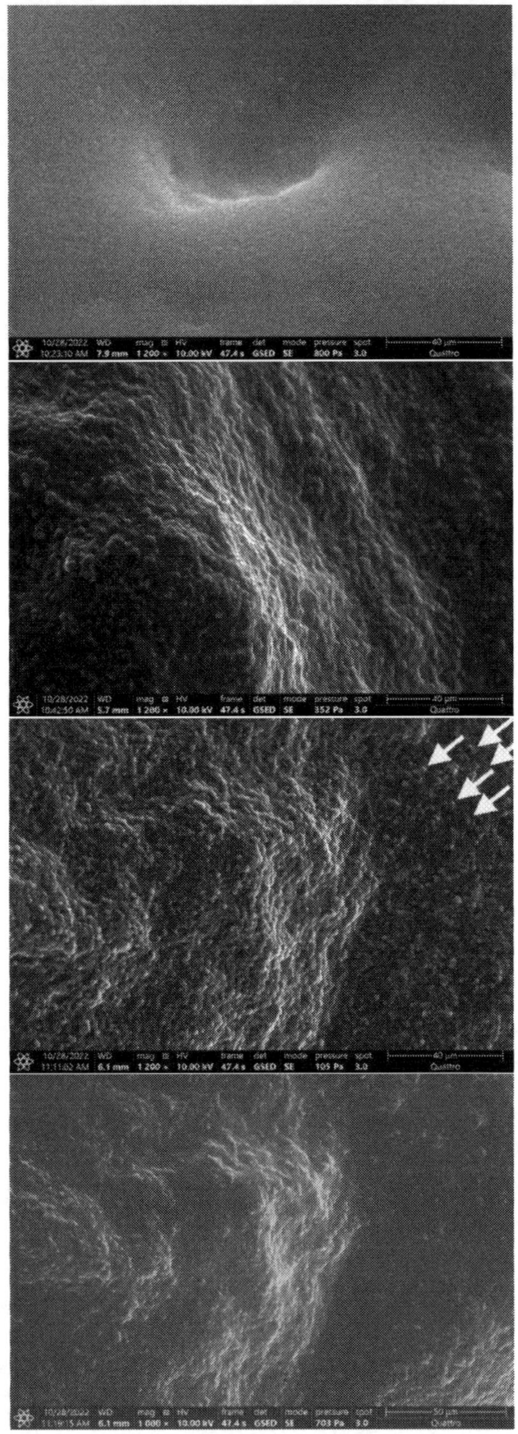

Figure 4. ESEM images of a plant-based cream cheese PCC with low dry matter at a r.h. of 100% to 50%, 15%, back to 100% (top to bottom). Arrows mark some of the particles identified as most characteristic and discussed in the text.

https://doi.org/10.1051/epjconf/202328604007

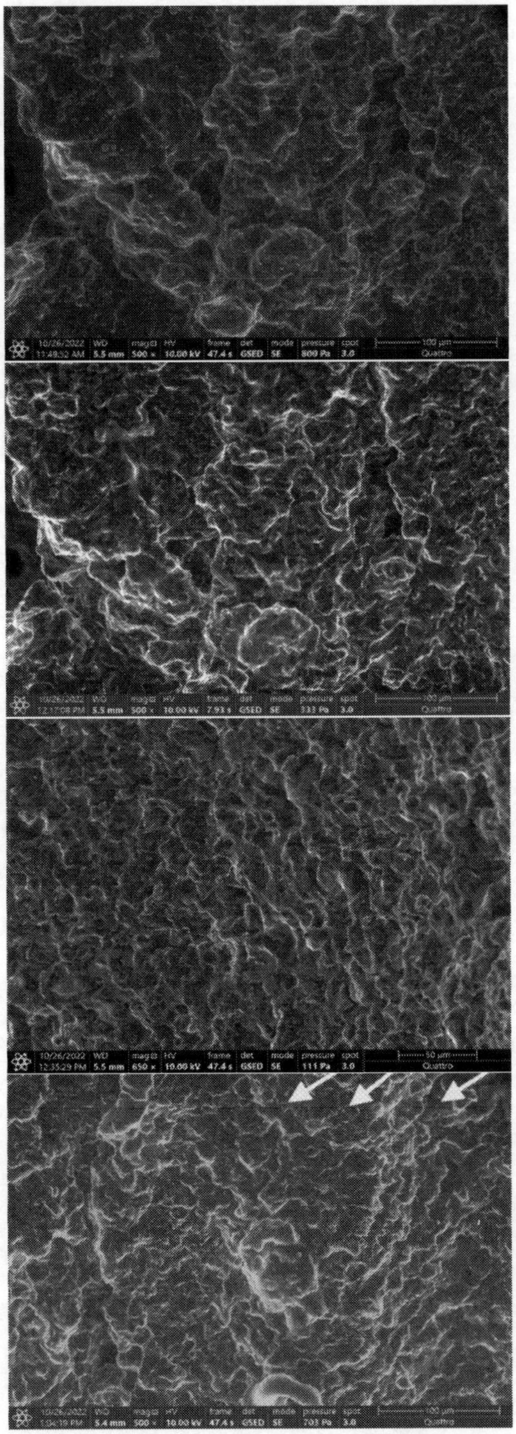

Figure 5. ESEM images of the plant cheese PC with high dry matter at a r.h. of 100% to 50%, 15%, back to 100% (top to bottom). Arrows mark some of the particles identified as most characteristic and discussed in the text.

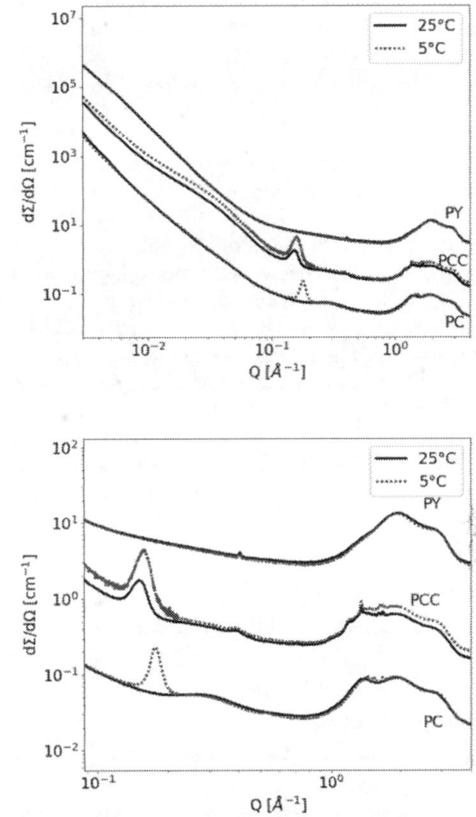

Figure 6. Top: Small Angle X-ray Scattering over a broad Q-range. From top to bottom PY, PCC, PC, measured at 25 °C and 5 °C. Bottom: zoom into the region of larger Q.

changing the water contents with heavy water, or by more advanced contrast variation methods, with deuterated oils, lipids or proteins, which would fully exploit the strength of neutron scattering. The results presented here were obtained on commercially available food products. Studying related model systems with selective deuteration and contrast variation allows to address specific questions with neutron scattering, such as details of the origin of different features from the SAXS results.

4 Conclusion

This contribution briefly illustrates how combining scattering techniques as an ensemble average in reciprocal space and microscopy such as ESEM as a high resolution technique without the need of extensive sample preparation and modification can lead to interesting information about structural features on length scales relevant for the texture, for the rheological and stability properties of gels made from plant-based food emulsions. ESEM provides the possibility to observe humid samples and to vary humidity in the sample chamber and hence observe surface structures during de-hydration and re-hydration in situ, but is mainly restricted to observations of the surface or cuts through the bulk. SAXS is a complementary technique in reciprocal space, able to observe ensemble averaged structural features from sub-nm to micrometer length scales. Crystallisation of fat molecules on length scales of nanometers as well as crystal structures possibly from salt crystals on sub-nanometer length scales could be observed. A power law decay from the larger scale structures can be observed. More details are expected from further neutron scattering studies, where the contrast between different domains and areas in the sample can be modified by using partly deuterated compounds such as deuterated water, oil, proteins or lipids. This requires to prepare comparable samples to the consumer products shown here in a controlled manner in the lab. Linking the information on this important range of length scales with macroscopic rheological properties might be the path of systematically improving food based fermented emulsions such as plant-based cheese or yoghurt. Since the study of plant-based emulsions and gels is only at its very beginning, we think that these techniques might contribute in the future significantly to further investigate and develop new products in this emerging sector.

Data availability statement

The data from the SAXS and ESEM experiments are available from the corresponding author upon reasonable request.

Acknowledgement

This work is based upon experiments performed at the ESEM and KWS-X instrument operated by Forschungszentrum Jülich at the Heinz Maier-Leibnitz Zentrum (MLZ), Garching, Germany. The Thermo Fischer Quattro S ESEM is located at the Meier-Leibnitz Zentrum in Garching and is jointly operated by Helmholtz-Zentrum Hereon and Forschungszentrum Jülich.

References

[1] W. Kim, Y. Wang, C. Selomulya, Trends in Food Science Technology **105**, 261 (2020)
[2] M. Brückner-Gühmann, E. Vasil'eva, A. Culetu, D. Duta, N. Sozer, S. Drusch, Journal of the Science of Food and Agriculture **99**, 5852 (2019)
[3] M. Brückner-Gühmann, A. Kratzsch, N. Sozer, S. Drusch, Future Foods **4**, 100053 (2021)
[4] P. Fischer, E.J. Windhab, Current Opinion in Colloid Interface Science **16**, 36 (2011)
[5] T. Kamigaki, Y. Ito, Y. Nishino, A. Miyazawa, Microscopy **67**, 164 (2018)
[6] J. Glaser, P.A. Carroad, W.L. Dunkley, Journal of Dairy Science **63**, 37 (1980)
[7] D.J. McMahon, W.R. McManus, Journal of Dairy Science **81**, 2985 (1998)
[8] J. Blonk, H. Van Aalst, Food research international **26**, 297 (1993)
[9] M. Ferrando, W. Spiess, Food Science and Technology International **6**, 267 (2000)
[10] N. Castaneda, Y. Lee, Gels **5** (2019)
[11] E.P. Gilbert, Current Opinion in Colloid and Interface Science **42**, 55 (2019)
[12] A. Lopez-Rubio, E.P. Gilbert, Trends in Food Science and Technology **20**, 576 (2009)
[13] A. Donald, Nature materials **2**, 511 (2003)
[14] D.J. Stokes, Microscopy and Analysis **26(6)**, 67 (2012)
[15] T. Heiden-Hecht, B. Wu, M.S. Appavou, S. Förster, H. Frielinghaus, O. Holderer, submitted (2023)
[16] X. Meng, Q. Pan, Y. Ding, L. Jiang, Food chemistry **147**, 272 (2014)
[17] Y. Miyagawa, K. Shintani, K. Katsuki, K. Nakagawa, S. Adachi, Food Structure **11**, 8 (2017)
[18] W.G. Bouwman, Food Structure **30** (2021)

https://doi.org/10.1051/epjconf/202328605001

The ortho-para transition, confinement and self-diffusion of H_2 in three distinct carbide-derived carbons by quasi- and inelastic neutron scattering

Riinu Härmas[1*], *Rasmus* Palm[1,2], *Miriam* Koppel[1], *Laura* Kalder[1], *Margarita* Russina[3], *Heisi* Kurig[1], *Eneli* Härk[1,3], *Jaan* Aruväli[4], *Indrek* Tallo[1], *Jan P.* Embs[5], and *Enn* Lust[1]

[1]Institute of Chemistry, University of Tartu, Ravila 14a, 50411 Tartu, Estonia
[2]Department of Applied Physics, KTH Royal Institute of Technology, SE-106 91 Stockholm, Sweden
[3]Helmholtz-Zentrum Berlin für Materialien und Energie, Hahn-Meitner-Platz 1, 14109 Berlin, Germany
[4]Department of Geology, University of Tartu, Ravila 14a, 50411 Tartu, Estonia
[5]Laboratory for Neutron Scattering and Imaging, Paul Scherrer Institute, Forschungsstrasse 111, CH-5232, Villigen PSI, Switzerland

Abstract. Microporous carbon materials are promising for hydrogen storage due to their structural variety, high specific surface area, large pore volume and relatively low cost. Carbide-derived carbons are highly valued as model materials because their porous structure is fine-tuned through the choice of the precursor carbide and the synthesis route. This study investigates H_2 adsorption in three carbide derived carbons with well-defined pores and pore size distributions with quasi- and inelastic neutron scattering methods. Concerning previous studies, a wider neutron energy transfer window is investigated, and a detailed quantitative evaluation of the graphitic structure is presented. The graphitic structure of the carbon is shown to influence the speed of the ortho-to-para transition of H_2. Namely, the ortho-para transition was the slowest in carbon derived from TiC, which also had the smallest average stacking size of graphene layers. The possibility to inhibit the ortho-para transition in cryo-adsorption devices is sought after to mitigate the evaporation of H_2 during storage. In addition, the self-diffusion of H_2 in different timescales is detected in carbon derived from Mo_2C, demonstrating the usefulness of obtaining data in a wide energy window.

1 Introduction

Currently, hydrogen storage is one of the most expensive components in implementing entire hydrogen value chains [1]. In order to increase the density of hydrogen, it needs to be stored either at high pressure (300–700 bar) or at a very low temperature (20 K). Both methods result in a significant drop in the overall energy efficiency of the hydrogen value chain. Porous materials have attracted interest as hydrogen storage materials, since these can help to alleviate the problem of the unfavourable compressibility factor of H_2 at low temperatures and high pressures. Although low temperatures and pressures higher than atmospheric pressure are still needed for H_2 storage in cryo-adsorption systems, the conditions for hydrogen storage are much milder in the case of porous materials [2]. It has been shown that the optimal conditions for using porous materials for hydrogen storage lie in the temperature range of 77–150 K, and the storage pressure is up to approximately 250 bar. At 80 K and in the pressure range from 100 to 200 bar, cryo-adsorption vessels have been shown to contain up to 20% more H_2 compared with the corresponding cryo-compression vessels, which do not contain adsorbents [3].

Carbide-derived carbons (CDCs) have been shown to be excellent model materials to analyze the effect of porous structure on performance of the carbon in specific applications (e.g. H_2 storage, charge storage in supercapacitors) [4,5]. CDCs are synthesized from inorganic carbides (e.g., TiC) through the reaction with chlorine at high temperature, typically 700–1200 °C, where the formed chloride is removed in gas flow. Thus, from the carbide structure only the chemically pure carbon, noted as CDC, is left behind. It has been previously shown that the average graphene platelet size in these types of materials is in the range of 2.3–8.2 nm [6], and the average pore shapes depend on the precursor carbide [7] and the chlorination temperature used [8]. In a previous quasi-elastic neutron scattering (QENS) study on the instrument FOCUS at PSI, the transport properties of H_2 in the pores of CDCs derived from SiC, Mo_2C and TiC was investigated in the energy range from -2 to 2 meV. As a result, the sub-nanometer pores of C-SiC were shown to contain almost immobile hydrogen (i.e. solid-like) even at a relatively high temperature of 120 K [5]. In this study, the Quasi-elastic neutron scattering (QENS) data of H_2 diffusing in the pores of CDCs derived from SiC, Mo_2C and TiC is revisited to investigate a larger transfer energy window with respect to the previous study. The aim is to

* Corresponding author: riinu.harmas@ut.ee

investigate the ortho-para transition band of H_2 and to verify if fast H_2 diffusion can be detected in the broader energy window.

2 Experiment

The precursor carbides and the chlorination temperatures of the studied carbon materials can be seen in Table 1.

Table 1. Overview of the main synthesis parameters of the studied CDCs.

	Precursor carbide	Chlor. *T* °C	Ref.
C-TiC	TiC	950	[7,9]
C-SiC	SiC	1000	[7,10]
C-Mo_2C	Mo_2C	1000	[7,11]

X-ray diffraction patterns were measured using D8 Advance diffractometer (Bruker) with Cu Kα radiation and Bragg-Brentano geometry. To obtain structural parameters, the data were fitted with the Ruland and Smarsly algorithm [12] using the CarbX software (https://www.uni-giessen.de/CarbX) [13].

QENS measurements of H_2 adsorbed in C-TiC, C-SiC, and C-Mo_2C were performed at Paul Scherrer Institute in Villigen, Switzerland, on the cold neutron time-of-flight spectrometer FOCUS [14] with incident neutron wavelength of 5.02 Å and the elastic energy resolution of 100 μeV. Data were obtained in the energy transfer range of E = −37 meV to 2 meV and scattering vector, Q, range of 0.4–2.5 Å^{-1} (corresponding scattering angle range 2θ 18–174°). A cylindrical aluminium sample holder with an inner cylinder was used for all samples. The outer diameter of the sample was 1.02 cm, and the inner diameter of the sample was 0.555 cm. Empty sample holder and vanadium measurements were performed for background subtraction and data correction.

Before the QENS measurement, all three CDCs were outgassed to remove moisture and other adsorbed species. After the outgassing process, the CDCs were transferred into the sample holder in a glovebox. The signal from the outgassed samples was measured in the temperature range of 10–120 K. H_2 with normal ortho-para content (3:1) was dosed in the sample holder at 77 K until the pressure reached the $p_{H2,load}$ value (Table 2). The adsorption equilibrium was established and, thereafter, the sample holder was hermetically sealed and, thus, disconnecting it from the gas dosing apparatus, and data were measured at different temperatures ranging from 10 to 120 K. H_2 pressure in the sample holder varied according to the reached adsorption equilibrium at each experimental temperature point as the sample holder was hermetically sealed during the experiment.

Table 2. Conditions for hydrogen loading, $p_{H2,load}$, is the H_2 pressure in the sample holder at 77 K, n_{H2}, mmol g^{-1} is the total amount of H_2 per mass of carbon in the sample holder.

CDC adsorbent	$p_{H2,load}$ bar	n_{H2} mmol g^{-1}	*w*% of H_2 in sample at 20 K
C-TiC	0.011	2.0	0.4%
C-SiC	0.011	2.4	0.48%
C-Mo_2C	0.087	1.2	0.24%

Data reduction, which included normalization with vanadium, corrections for detector efficiency and subtraction of background, was performed with the DAVE 2.4 software. [15] Binning of the scattering data were conducted in the whole energy transfer range, from dE = −37 meV to 2 meV, with step size 0.047 meV; Q was binned over the entire range as one slice. Since Q was summed over all detectors, the reported dynamic incoherent structure factors $S(Q,E)$ are noted as intensity $S(E)$, in the following text.

The intensity of signal from hydrogen, $S_{H2}(E)$, was obtained by subtracting the intensity of signal of degassed carbon, $S_C(E)$, from the total intensity $S(E)$.

$$S_{H2}(E) = S(E) - S_C(E) \quad (1)$$

To investigate the diffusion of H_2 in multiple timescales, i.e. to detect whether multiple quasi-elastic broadening signals are convoluted in $S_{H2}(E)$, the quasi-elastic broadening of $S_{H2}(E)$ was analysed. The $S_{H2}(E)$ was symmetrised based on the neutron energy gain side and fitted in the energy range from E = −30 meV to 30 meV with equation 2.

$$S_{model}(E) = A_0\,\delta(E) + A_1\,L(E,\Gamma_1) + A_2\,L(E,\Gamma_2) + y_0 \quad (2)$$

where $S_{model}(E)$ is the model scattering signal, $\delta(E)$ is the Dirac delta function describing the elastic scattering, A_0 is the fraction of elastic and A_1 and A_2 are the fractions of quasi-elastic scattering signals, L is the Lorentzian function which describes the quasi-elastic component mathematically, Γ is the half width at half maximum of the quasi-elastic component, and y_0 is the background. The characteristic times for rotational diffusion were calculated based on the Γ_1 and Γ_2 values by converting energy units (1 meV = 2.42·10^{11} Hz).

3 Results and discussion

3.1 Ortho-para transition of H_2

The rotational energy levels are given by

$$E_J = \mathrm{B}\,J\,(J+1) \quad (3)$$

where J is the rotational quantum number and B = 7.37 meV. The free quantum rotor energy of the

https://doi.org/10.1051/epjconf/202328605001

transition from the lower energy para level to the higher energy ortho level (p→o, $J = 0 \rightarrow 1$) is 14.7 meV [16,17].

The para-ortho transition band is seen when the H_2 is adsorbed strongly in the carbon material, i.e. due to adsorption, the H_2 molecule struck by the neutron is not able to diffuse freely and therefore, undergoes rotational excitation [18–20]. The detection of p→o transition band has been seen as evidence of very strongly adsorbed H_2 in the porous material [21]. The ortho to para (o→p, $J = 1 \rightarrow 0$), i.e. a reverse transition could be investigated in the energy transfer window available in this study. At room temperature, the ortho to para ratio of hydrogen is about 3:1, but para form will become prevalent at low temperatures. The spin conversion process from ortho to para H_2 is exothermic releasing 535 kJ kg^{-1} of heat, meaning that a rapid o→p conversion could result in desorption and evaporation of the adsorbed H_2 [22]. Therefore, it is of interest to stabilize the H_2 adsorbed in ortho state.

H_2 adsorbed in C-TiC, a o→p transition band near $E = -14$ meV is visible at 10 K (Fig. 1a). However, no o→p transition band is detected at low hydrogen coverages ($p_{H2,load} \leq 87$ mbar) in C-SiC and Mo_2C in the investigated temperature range (Fig. 1b-c). The pore size distributions of C-SiC and C-TiC are very similar [5], so the porous structure alone cannot explain why the o→p transition band was seen for H_2 adsorbed in C-TiC, but not for H_2 adsorbed in C-SiC and C-Mo_2C at 10 K. As the temperature is increased to 50 K, a slight quasi-elastic broadening becomes visible in the $S(E)$.

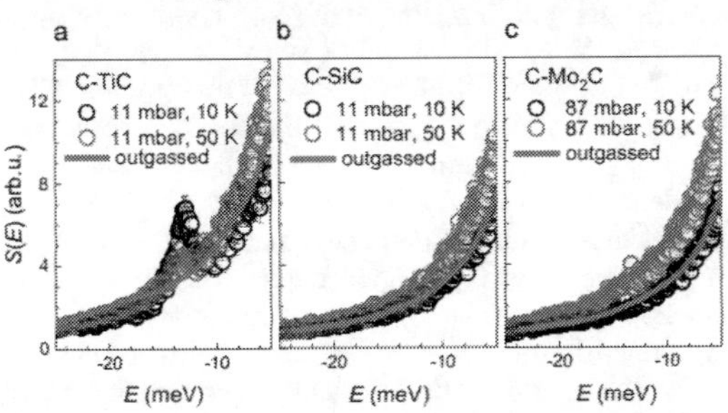

Fig. 1. The intensities, $S(E)$, in the o→p energy transfer range. (a) C-TiC at $p_{H2,load} = 11$ mbar, (b) C-Mo_2C at $p_{H2,load} = 87$ mbar and (c) C-SiC at $p_{H2,load} = 11$ mbar. The red lines correspond to the intensities of the corresponding outgassed carbons, $S_C(E)$.

The main structural aspect differentiating C-TiC from other studied carbon material is the lack of graphitic ordering. Namely, the main difference lies in the 002 reflections in the X-ray diffraction patterns (Fig. 2), which is related to stacking of graphene layers.

The average stacking size, L_c, average interlayer spacing, a_3, and the average graphene layer extent, L_a, were estimated based on the X-ray diffraction patterns (Table 3). The algorithm that was used for this analysis (See part 2) calculates the L_c and a_3 parameters based on the shape of the 002 reflection and the L_a parameter from the shape of the 10 reflection [12]. Based on these parameters, the C-TiC material shows a more disordered structure compared to the other investigated CDCs. L_c values show the most significant difference between the investigated materials, less than 1 nm for C-TiC and 12.7 nm for C-SiC. Also, the L_a value is almost two times smaller compared to that of C-SiC and C-Mo_2C.

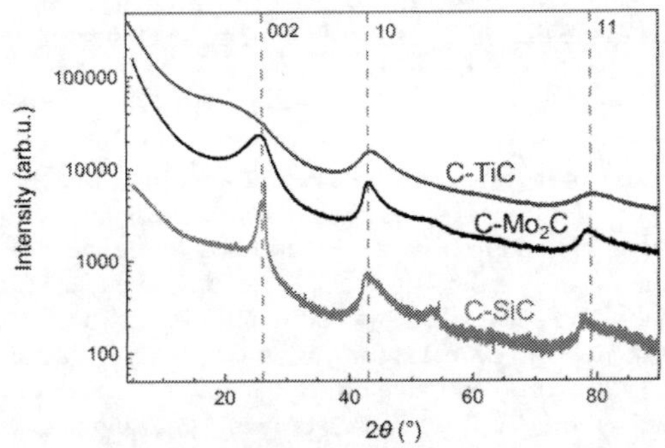

Fig. 2. The X-ray diffraction patterns of studied materials (noted in Figure). The dashed lines represent the reflections along with their assignment.

Table 3. Main structural parameters of CDCs calculated based on XRD patterns with Ruland and Smarsly algorithm [12].

CDC	L_c nm	a_3 nm	L_a nm
C-TiC	0.6	0.38	4.5
C-SiC	12.7	0.36	7.1
C-Mo_2C	2.7	0.35	8.3

L_c - average stacking size, a_3 - average interlayer spacing, L_a, average graphene layer extent

In this study, a normal mixture of ortho and para hydrogen was dosed into the carbon at 77 K, i.e. the ratio of ortho to para H_2 corresponded to 3:1. If by the time of measurement, there was a negligible amount of ortho-H_2 in the carbon, also the ortho-para transition band would not be visible as seen for C-SiC and C-Mo_2C. Namely, at low temperatures (<80 K), the thermodynamically more stable para form should become prevalent, but the carbon's structure influences the kinetics of the transition from o→p. It has been shown that depending on the carbon material, the o→p transition can be as slow as 0.4% h^{-1} [23] or as fast as ~140% h^{-1}. [24] This difference has been attributed to paramagnetic sites at the edges of graphitic planes, which can catalyse the o→p transition.

Therefore, since C-TiC is the most disordered, it should contain more edges, defects, etc., which can catalyse the o→p transition. However, since no o→p transition was seen in the more graphitic C-SiC and C-Mo_2C at 10 K, the o→p conversion must have been faster in these materials.

In addition, preferred adsorption of o-H_2 can occur in some pores. The physisorption energy is a sum of the isotropic and anisotropic terms of the adsorption potential. While the isotropic term is the same for both ortho and para species, the anisotropic term can be different. Namely, the energy of the o-H_2 rotational levels can be lowered or raised due to anisotropic term, while the para level is left unchanged. This indicates that due to the decrease in the o-H_2 energy level in specific anisotropic adsorption sites, o-H_2 can be preferably bound to these sites and conversion to p-H_2 would then be less likely [16].

Recently, in the study of Terry et al. [19], the specific adsorption of o-H_2 in carbon materials was detected. These adsorption sites were attributed to the smallest pores (pore width, w = 0.4 nm) or some specific adsorption sites with a disorder-induced curvature. So, it might be the case that for low surface coverage, H_2 in C-TiC is adsorbed in the ortho state and not convert to para form easily. Thus, despite the anticipated conversion of most of the ortho species to para species already before the start of the neutron scattering measurement, an appreciable amount of hydrogen is still left in the ortho state in the pores in C-TiC. Thus, the existence of o→p transition band proves the existence of o-H_2 still in the pores which alludes to the H_2 rotation being strongly affected by the anisotropic adsorption potential in these pores.

This explanation is also supported by the fact that the o→p band seen in C-TiC was centred at approx. dE = −13 meV (Fig.1), instead of dE = −14.7 meV seen for freely rotating H_2. This deviation refers to strong anisotropic adsorption potential in the pores of C-TiC.

3.2 Quasi-elastic broadening, i.e. H_2 diffusion

The H_2 diffusion was recently seen to occur in multiple different timescales in the pores of carbon materials. [25] This has increased the interest to investigate the quasi-elastic broadening component of $S(E)$ in a wide energy transfer window. From the investigated CDC materials, a wide temperature-dependant quasi-elastic broadening was detected in the $S_{H2}(E)$ in the C-Mo_2C material at temperatures over 25 K (Fig. 3a). This could refer to fast self-diffusion of H_2 that could occur in mesopores and be also related to multilayer adsorption.

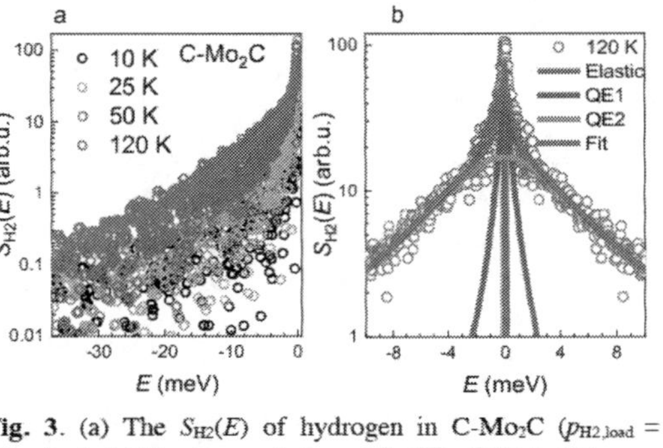

Fig. 3. (a) The $S_{H2}(E)$ of hydrogen in C-Mo_2C ($p_{H2,load}$ = 87 mbar). (b) The deconvolution of $S_{H2}(E)$ of C-Mo_2C ($p_{H2,load}$ = 87 mbar) at 120 K. QE1 is the quasi-elastic intensity related to the slower self-diffusion, and QE2 is the quasi-elastic intensity related to the faster self-diffusion of H_2.

To investigate the existence of self-diffusion at different timescales, the $S_{H2}(E)$ was fitted with equation 2. The successful fit was obtained using two Lorentzian functions in addition to the resolution function (Fig. 3b). This demonstrates that H_2 self-diffusion with speed faster than previously reported exists in C-Mo_2C.

Assuming the rotational diffusion mechanism, the characteristic times were calculated based on the half-width half-maxima of the quasi-elastic broadening components (Table 4). It can be seen how the residency times between rotational jumps decrease with the increase in temperature. Further measurements are needed to ascertain the mechanism of H_2 diffusion by investigating the Q dependence of the quasi-elastic broadening signal.

Table 4. Assuming rotational diffusion, the characteristic residency times of H_2, τ_{rot}, in the pores of C-Mo_2C was calculated.

T K	QE1, τ_{rot} ps	QE2, τ_{rot} ps
50	18.5 ± 1.5	1.0 ± 0.1
70	6.9 ± 0.2	0.5 ± 0.1
120	4.8 ± 0.2	0.5 ± 0.1

Conclusions

The graphitic structure of the porous carbon material affected the transition from ortho to para hydrogen. C-TiC was the most disordered of the studied CDC materials and the only one for which the o→p transition band of H_2 was seen at 10 K. Moreover, there were evidence of H_2 rotation being strongly affected by the anisotropic adsorption potential in the pores of C-TiC. This finding can have an impact on the practical application since it would be preferable to keep hydrogen in the ortho state. Namely, storing H_2 in the ortho form would help avoid heat generation that comes from ortho to para transition during hydrogen storage in cryo-adsorption systems.
In addition, H_2 self-diffusion in different timescales was seen in the case of C-Mo_2C at temperatures over 25 K. This gives the incentive to study the self-diffusion of H_2 in the pores of carbon material in a wider energy transfer window in the future.

Acknowledgements

The sample holder was designed by Dr. Dirk Wallacher from Helmholtz-Zentrum Berlin für Materialien und Energie Probenumgebung. Paul Scherrer Institute is thanked for beamtime on FOCUS. The authors would like to thank Marek Bartkowiak for his help during the experiment at PSI. This research was supported by the EU through the European Regional Development Fund (Centers of Excellence, TK141 20142020.4.01.15-0011 and TK117 3.2.0101e0030); European Spallation Source: Estonian Participation in ESS Instrument design, development and building and application for scientific research SLOKT12026T and SLTKT16432T and Graduate School of Functional materials and technologies receiving funding from the European Regional Development Fund in University of Tartu, Estonia. This work has been also partially supported by Estonian Research Council Institutional Research Grant IUT20e13 and personal research grants PUT55 and PUT1033.

References

1. V. Dias, M. Pochet, F. Contino, H. Jeanmart, Front. Mech. Eng., **6** (2020)
2. D.P. Broom, C.J. Webb, K.E. Hurst, P.A. Parilla, T. Gennett, C.M. Brown, R. Zacharia, E. Tylianakis, E. Klontzas, G.E. Froudakis et al., Appl. Phys. A, **122** (2016)
3. G. Petitpas, P. Bénard, L.E. Klebanoff, J. Xiao, S. Aceves, Int. J. Hydrog. Energy., **39** (2014)
4. V. Presser, M. Heon, Y. Gogotsi, Adv. Funct. Mater., **21** (2011)
5. R. Härmas, R. Palm, M. Russina, H. Kurig, V. Grzimek, E. Härk, M. Koppel, I. Tallo, M. Paalo, O. Oll et al., Carbon, **155** (2019)
6. R. Härmas, R. Palm, H. Kurig, L. Puusepp, T. Pfaff, T. Romann, J. Aruväli, I. Tallo, T. Thomberg, A. Jänes et al., *C*, **7** (2021)
7. H. Kurig, M. Russina, I. Tallo, M. Siebenbuerger, T. Romann, E. Lust, Carbon, **100** (2016)
8. R. Palm, R. Härmas, E. Härk, B. Kent, H. Kurig, M. Koppel, M. Russina, I. Tallo, T. Romann, J. Mata et al., Carbon, **171** (2021)
9. H. Kurig, T. Romann, A. Jänes, E. Lust, ECS Trans., **25** (2010)
10. M. Käärik, M. Arulepp, M. Karelson, J. Leis, Carbon, **46** (2008)
11. A. Jänes, T. Thomberg, H. Kurig, E. Lust, Carbon, **47** (2009)
12. W. Ruland, B. Smarsly, J Appl Cryst, J Appl Crystallogr, **35** (2002)
13. T. Pfaff, M. Simmermacher, B.M. Smarsly, J Appl Crystallogr, **51** (2018)
14. J. Mesot, S. Janssen, L. Holitzner, R. Hempelmann, J. Neutron Res., **3** (1996)
15. R.T. Azuah, L.R. Kneller, Y. Qiu, P.L.W. Tregenna-Piggott, C.M. Brown, J.R.D Copley, R.M. Dimeo, J Res Natl Inst Stand Technol, **114** (2009)
16. K. Fukutani, T. Sugimoto, Prog. Surf. Sci., **88** (2013)
17. I.F. Silvera, Rev. Mod. Phys., **52** (1980)
18. P.A. Georgiev, D.K. Ross, P. Albers, A.J. Ramirez-Cuesta, Carbon, **44** (2006)
19. L.R. Terry, S. Rols, M. Tian, I. da Silva, S.J. Bending, V.P. Ting, Nanoscale, **14** (2022)
20. P.A. Georgiev, D.K. Ross, A. De Monte, U. Montaretto-Marullo, R.A.H. Edwards, A.J. Ramirez-Cuesta, M.A. Adams, D. Colognesi, Carbon, **43** (2005)
21. C.I. Contescu, D. Saha, N.C. Gallego, E. Mamontov, A.I. Kolesnikov, V.V. Bhat, Carbon, **50** (2012)
22. K.T. O'Neill, S. Al Ghafri, B. da Silva Falcqão, L. Tang, K. Kozielski, M.L. Johns, Chem Eng Process, **184** (2023)
23. P.R. Kubik, W.N. Hardy, H. Glattli, Can. J. Phys, **63** (1985)
24. C.A. Swenson, J. Chem. Phys., **18** (1950)
25. M. Koppel, R. Palm, R. Härmas, M. Russina, V. Grzimek, J. Jagiello, M. Paalo, H. Kurig, M. Månsson, O. Oll et al., Carbon, **197** (2022)

In-situ Neutron Reflectometry to Determine Ge Self-Diffusivities and Activation Energy of Diffusion in Amorphous $Ge_{0.8}Si_{0.2}$

Erwin Hüger[1,2*], *Jochen* Stahn[3], and *Harald* Schmidt[1,2]

[1]Clausthal Centre for Materials Technology, Clausthal University of Technology, D-38678 Clausthal-Zellerfeld, Germany
[2]Solid State Kinetics Group, Institute of Metallurgy, Clausthal University of Technology, D-38678 Clausthal-Zellerfeld, Germany
[3] Paul Scherrer Institut, 5232 Villigen PSI, Switzerland

Abstract. Amorphous Ge-Si solid solutions are an interesting class of materials from the fundamental as well as the technological point of view. Self-diffusion of the constituents is an important process because of the inherent metastability. While self-diffusion was already examined in crystalline Ge_xSi_{1-x} ($0 < x < 1$) this is not the case for the amorphous counterparts. This work reports on Ge self-diffusivities obtained from in-situ neutron reflectometry measurements during isothermal annealing of ion-beam sputter-deposited amorphous $Ge_{0.8}Si_{0.2}$ films. The diffusivities are modified peculiarly fast with annealing time by a maximum factor of two due to structural relaxation. The diffusivities in the relaxed state are lower (higher) than in amorphous germanium (silicon). They follow the Arrhenius law and show an activation energy of (2.06 ± 0.1) eV, which equals that of amorphous germanium, but differs from that of amorphous silicon. Thus, it is concluded that the diffusion mechanism of Ge in amorphous $Ge_{0.8}Si_{0.2}$ and Ge are similar, despite of the presence of dispersed 20 at.% of Si.

1 Introduction

Ge-Si alloys are considered to improve state-of-the-art electronic devices which are still based on silicon, and to add new possibilities of performance such as optics to integrated circuits [1]. The difference in bonding distance in germanium and silicon lattices causes strain and modifies the electronic band structure with increasing electron and hole mobility [1]. Miniaturization in electronic integrated circuits is still an ongoing demand [1]. Miniaturization is correlated to material confinement with development of quantum well states in the occupied and non-occupied electronic band structures [2]. In particular, Ge-Si nanostructures are considered to provide a new scientific field regarding materials properties [1]. Ge-Si spintronics and efficient light emission even with indirect band-gap materials [1] beneficial for opto-electronic integrated circuits are examples. Amorphous nanoscaled Ge-Si is of particular interest for solar cells whose optical band gaps matching the whole solar radiation spectrum [3-5]. This work contributes to the scientific examination of Ge-Si nanostructures by the measurement of self-diffusion in thin amorphous Ge-Si films with an innovative in-situ diffusion measurement methodology based on neutrons [6].

Self-diffusion is the thermally activated migration of atoms in solids [7,8]. It is a basic transport process, important for precipitation, grain growth, oxidation, joining processes and plastic deformation [7,8] as well as for the functionality of fuel cells and batteries [9-12].

Atomic diffusion was in the past routinely investigated by radiotracer diffusion studies [7,8,13] with the disadvantage of using radioactive tracers and doing depth profiling with mechanical sectioning and limited depth resolution [13]. A further disadvantage of the radiotracer method is the lack of measuring diffusion in-situ, i.e. during the heat treatment without cooling down the sample. This is also the case for using stable tracer isotopes and depth profiling with secondary ion mass spectrometry (SIMS) [14-27].

In-situ measurement of diffusivities has the advantage of a significant reduction of the experimental time, a reduction of error limits due to lack of heating/cooling steps, and an identification of time-dependent processes [6]. In order to carry out such measurements, non-destructive methodologies are necessary.

* Corresponding author: erwin.hueger@tu-clausthal.de

Neutron reflectometry (NR) [6,25,28-52] is such a non-destructive methodology, where the neutrons do not interact with electrons and possess therefore a high penetration depth in almost all materials. For example, the neutron beam transmission through a 1 cm thick silicon wafer, Al metal, or sapphire is 99.5 %, 99.1 %, or 97.4 %, respectively [53]. This ability of neutrons allows neutron beam investigations during device operation. A typical example is the monitoring of atomic diffusion in-situ or in-operando [6,40,42-44] and in real time [6,40,42-44,46-52] during furnace operation (sample heating).

Self-diffusion investigations on amorphous Ge-Si alloys where not published until now. This is mainly due to the metastable [54] or even unstable [55] state of the amorphous structure, as predicted for amorphous silicon [54,55]. There, diffusivity has to be determined on short lengths scales in order not to average over different transient metastable states and to avoid unwanted crystallization. NR experiments are capable to measure small diffusion lengths down to 1 nm [6,28,31,38] and, hence, in transient states [31]. This study reports on in-situ self-diffusion investigations on amorphous $Ge_{0.8}Si_{0.2}$ films. The second section gives a review on recent literature results on germanium, silicon, and Ge-Si alloys. The third section describes the experimental procedure. The fourth section presents and discuss the experimentally derived in-situ NR results. A comparison to published results obtained for pure amorphous germanium and silicon will be given. The results are summarized in the fifth section.

2 Literature survey on self-diffusion in Ge, Si and Ge-Si alloys

Ge-Si is a completely miscible system [1]. Thus, Ge-Si compounds with any Ge content can be produced [1]. There is a lack of reports on self-diffusion in amorphous Ge_xSi_{1-x}. However, this is not the case for crystalline Ge_xSi_{1-x}, where self-diffusion was examined by a few research groups [13,23-25,38]. In general, the Si and Ge self-diffusivities are similar due to the valence isoelectronicity of the two species, and they decrease with increasing Si content. Figure 1 gives an overview. The activation energy of self-diffusion of dependence on the Ge content in Ge_xSi_{1-x} with x between 0 and 1 is plotted. Results were reported for Ge_xSi_{1-x} single crystals in form of epitaxial films as obtained by Zangenberg et al. [23] (green circles), and more recently by Kube et al. [24] (red dots and blue squares), as well as for polycrystalline Ge_xSi_{1-x} [13] (black filled triangles).

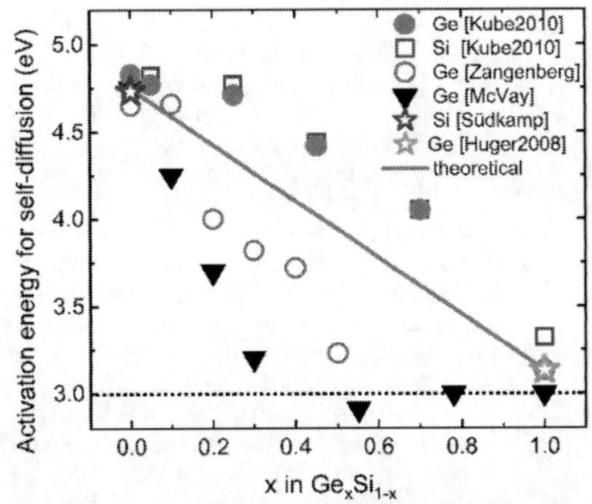

Fig. 1. Activation energy of self-diffusion in crystalline Ge_xSi_{1-x}. The references by Südkamp et al. [26] and Hüger et al. [38] provides values for Si self-diffusion for x = 0 (pure silicon) and for Ge self-diffusion for x = 1 (pure germanium) which are marked with a blue star and an orange star, respectively. The red dots and the blue squares mark the results by Kube et al. [24] for Ge and Si self-diffusion in epitaxial films. The green circles mark the results on Ge self-diffusion obtained by Zangernberg et al. [23], also on epitaxial films. The black filled triangles are for Ge self-diffusion in polycrystalline Ge_xSi_{1-x} obtained by McVay et al. [13]. The expected diffusivity behaviour according to equation (1) is marked with a red continuous line.

The activation energy of Ge self-diffusion is similar to that of Si self-diffusion for the same Ge content (results of Kube et al. [24]). The self-diffusivity examined in crystalline pure germanium can be accurately described over nine orders of magnitude by a single Arrhenius straight line [38]. A diffusion mechanism taking place by single-vacancies is postulated with an activation energy of (3.13 ± 0.03) eV for temperatures between 429 °C and 904 °C. Calculations [56] predict that the vacancy formation energy in crystalline germanium is positive and lower than in crystalline silicon. Consequently, vacancies in germanium are more prevalent than in silicon [56]. Thus, the situation for self-diffusion in crystalline pure silicon is more complicate [25,26], where contributions of self-interstitials and vacancies have to be considered. Self-diffusivities over ten orders of magnitude were determined from experiments [25]. An activation energy of self-diffusion of 4.73 eV is reported [26]. Hence, the examination of self-diffusion in Ge_xSi_{1-x} alloys is of interest also from the view point of fundamental diffusion science. It has to be clarified if a germanium-like diffusion mechanism or a silicon-like diffusion mechanism is operating predominantly in Ge_xSi_{1-x}.

For crystalline Ge_xSi_{1-x} alloys, the reported activation energy decreases with increasing Ge content x (Figure 1). As a simple approximation, the activation energy vs Ge content should obey to the Vegard's law [24]:

$$E_A(x) = (1-x)E_A^{Si} + xE_A^{Ge} \quad (1)$$

This behaviour was clearly not observed in all the literature studies (Figure 1). The results of Kube et al. [24] show an enhancement and that of the other authors a reduction of the activation energies in comparison to the behaviour predicted by equation (1) (see the red continuous line in Figure 1). For example, the single-crystalline Ge_xSi_{1-x} epitaxial films with the relative high Ge composition of x = 0.7 gives [24] an activation energy of self-diffusion of 4.05 eV which is still more close to that of pure silicon (4.73 eV) than to that of pure germanium (3.13 eV). This indicates a diffusion behaviour similar to crystalline silicon. This is not the case for the activation energy of Ge self-diffusion in polycrystalline Ge_xSi_{1-x} (Figure 1) [13], which reaches that of pure Ge already for x = 0.3 and remains constant for higher Ge contents (Figure 1). All activation energies reported by McVay lie below the straight line corresponding to equation (1), which was attributed to self-diffusion via grain boundaries in the polycrystalline Ge_xSi_{1-x} structure. The results of Zangenberg et al. [23] are older compared to that of Kube et al. [24] but are also on epitaxial Ge_xSi_{1-x} films. Their results are more similar to that of McVay et al. [13].

Investigations on self-diffusion in amorphous Ge_xSi_{1-x} were not reported until now. However, reports exist for the edge systems pure amorphous germanium [6,40] and silicon [39,27]. Recent computer calculations [54,55,57] suggest, that although the local order of amorphous silicon is close to that of crystalline silicon, the energetics of defect formation [54,55] and the diffusion mechanism [57] strongly differ. The formation energy of point defects such as vacancies, self-interstitial and dangling bonds is negative [54,55]. This means that defect production is spontaneous and does not need thermal energy in the case of the amorphous state which is contrary to the situation in the crystalline state. The self-diffusion mechanism in amorphous silicon is predicted by calculations also to be different from that in crystalline material in a way that well-defined elemental jump lengths do not exist in the amorphous network [57]. The self-diffusion proceeds by atomic bond rearrangement [57]. Experimentally, the self-diffusivities in amorphous silicon were found [27,39] to be many orders of magnitude higher than in crystalline silicon [25]. It was suggested from experimental data that self-diffusion in amorphous silicon films is mediated by local bond rearrangement leading to an activation energy of (2.70 ± 0.10) eV [27] for samples amorphisized by ion implantation or, alternatively by the migration of extended point defects leading to an activation energy of (4.4 ± 0.3) eV for samples produced by sputter deposition [39].

The energetics of defects and the diffusion mechanism in amorphous germanium is expected to be roughly similar to that in amorphous silicon. For amorphous germanium (sputter deposition), NR experiments found that the self-diffusivities are many orders of magnitude larger than in crystalline germanium [6,40] due to a lower activation energy of (2.11 ± 0.12) eV instead of (3.13 ± 0.03) eV. This indicates different diffusion mechanisms in crystalline and amorphous modifications. Local bond rearrangement is also suggested here as diffusion mechanism [6]. Within this work, the Ge self-diffusion in amorphous $Ge_{0.8}Si_{0.2}$ films is investigated and the results are compared with literature.

3 Experimental procedure

The amorphous $Ge_{0.8}Si_{0.2}$ samples in form of thin films were prepared by reactive ion-beam co-sputter deposition. Segmented targets were used to tailor the requested composition. Figure 2a,b presents photographs of the segmented targets. The thin films are produced in form of isotope [$^{73}Ge_{0.8}Si_{0.2}$ (14 nm) / $^{nat}Ge_{0.8}Si_{0.2}$ (14 nm)] × 10 multilayers (MLs). The base of the method is given in references [6,28-30,38]. The natural isotope abundance of germanium, termed ^{nat}Ge, consists of ^{70}Ge (20.5%), ^{72}Ge (27.4%), ^{73}Ge (7.8%), ^{74}Ge (36.5%), and ^{76}Ge (7.8%). The germanium target termed ^{73}Ge possesses a 95% ^{73}Ge isotope enrichment. For sputter target production, 1 mm thick polycrystalline ^{nat}Ge and ^{73}Ge disks with 20 mm diameters (MaTeck GmbH, Jülich, Germany) were bonded on a copper target holder. On top of the Ge disks, a quadrant of a 1 mm thick polycrystalline Si disk of 20 mm diameter (MaTeck GmbH, Jülich, Germany) was bonded as it is visualized in Figure 2a,b, to produce segmented targets. There is a three times larger Ge surface than silicon surface.

The depositions were performed using a sputter coater (IBC 681, Gatan, USA), equipped with two Penning sources. Ar^+ ion beams (5 kV, 180 μA) were used for film deposition. The targets were used successively without breaking the vacuum with a base pressure below 5×10^{-7} mbar. During deposition, the specimen is rotated (30 rotations per minute) and rocked (rock angle: 30° and rock speed: 15° per second) to ensure well dispersed Ge and Si atoms in the deposited film. (100) oriented, polished, nominally undoped silicon wafers (CrysTec GmbH, Berlin, Germany) were used as substrates for the MLs. For Raman spectroscopy and for energy dispersive X-ray spectroscopy (EDX) examination, the films were deposited on (0001) oriented, polished, saphire wafers (CrysTec GmbH, Berlin, Germany) and on flat Cu foils, respectively. The substrates were cleaned with isopropanol beforehand. The as-deposited multilayers have an overall thickness of 280 nm. Additional annealing was not applied to the deposited amorphous layers in the sputter chamber before removal and handling in air. According to literature, the self-heating of the sample during ion-beam sputter experiments is generally low (below 80 °C) due to the low impact energy (tens of eV) of ions which are deposited [6]. The native oxide layer of the substrates was not removed.

The multilayers were examined exclusively with non-destructive measurement techniques. X-ray reflectometry (XRR) and grazing incidence X-ray diffraction (GI-XRD) was done ex-situ with a Bruker D8 DISCOVER (Germany) diffractometer ($Cu_{K\alpha}$, 40 keV, 40 mA). The GI-XRD measurements were performed with 1° incident angle. Raman scattering was performed also ex-situ on a Bruker SENTERRA Raman microscope with a laser of 532 nm wavelength. For more information and references on Raman scattering, the reader is referred to a recent work [58]. EDX was performed ex-situ with a High-Resolution Scanning Electron Microscope (SEM, EVO 15, Zeiss). Except in-situ NR measurements, all other measurements were performed at room temperature.

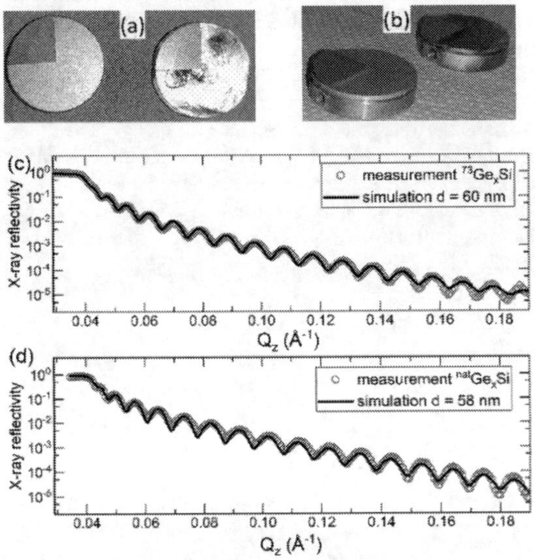

Fig. 2. a,b) top view (a) and side view (b) photographs of the segmented ^{nat}GeSi (left) sputter target and ^{73}GeSi (right) target to produce isotope ML. c,d) XRR data (circles) and corresponding simulations (Parratt32) for films deposited from the ^{73}GeSi or ^{nat}GeSi target.

NR was performed in-situ, meaning during isothermal heating of the ML sample inside a rapid thermal annealing oven (RTA AO500 MBE Komponenten, Germany) in argon gas. The NR examination was performed on the reflectometer AMOR at SINQ, Paul Scherrer Institut, Switzerland [42, 59]. The focusing Selene guide implemented there reduces the counting time considerably for small scattering vectors and specular measurements, only. Reflectometry simulations were performed using the Parratt32 software package [60].

The layer deposition rate was determined with the help of XRR measurements (Figure 2c and 2d). Thicknesses of (60 ±1) nm and (58 ±1) nm are obtained for as-deposited ^{73}Ge$_x$Si and ^{nat}Ge$_x$Si films (each 16 minutes of deposition time). Hence, the deposition rate from the ^{73}Ge$_x$Si and ^{nat}Ge$_x$Si target was (3.75 ± 0.06) nm and (3.62 ± 0.06) nm per minute, respectively, in good agreement.

4 Results and discussion

4.1 Sample characterization

EDX investigations found a relative amount of 80 at.% germanium and 20 at.% silicon (i.e., $Ge_{0.8}Si_{0.2}$) in films deposited from the ^{73}Ge$_x$Si and also the ^{nat}Ge$_x$Si target, indicating no or only minor chemical contrast within a bilayer

Figure 3 presents the measured GI-XRD data of the isotope $Ge_{0.8}Si_{0.2}$ ML (curves (a-c)) in comparison to that of isotope modulated pure Ge MLs (curves (d-f)). Both as-deposited materials are X-ray amorphous. The GI-XRD data show a lack of long-range order but the existence of local order (curves (a,d)). The heat treatment of the pure Ge ML at 418 °C for three hours and at 425 °C for one hour results in crystalline germanium (curves (e,f)). This is not the case for the amorphous $Ge_{0.8}Si_{0.2}$ ML (curves (b,c)) for even higher temperatures and longer annealing times. The XRD investigations clearly show that the presence of silicon inhibits the crystallization of the amorphous $Ge_{0.8}Si_{0.2}$ ML compared to Ge ML. Thus, the diffusion experiments performed in this work were done on X-ray amorphous $Ge_{0.8}Si_{0.2}$ layers.

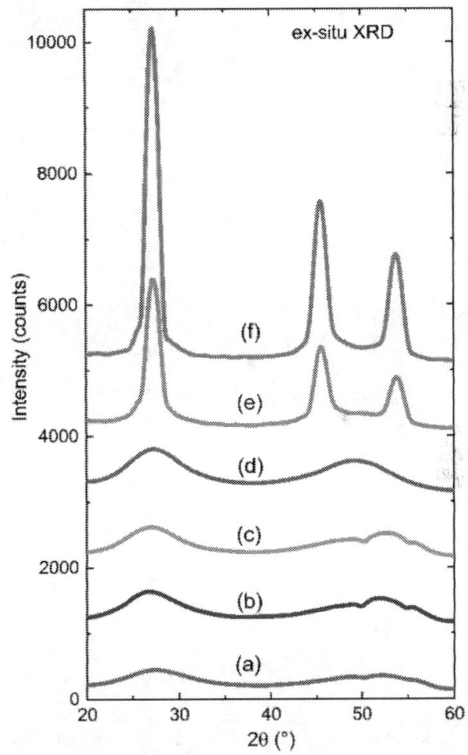

Fig. 3. GI-XRD data of the $Ge_{0.8}Si_{0.2}$ isotope ML (curves marked with (a-c)) compared to that of a pure Ge isotope ML film (curves marked with (d-f)). (a,d) As-deposited films. (b,c) After in-situ NR measurements of $Ge_{0.8}Si_{0.2}$ isotope MLs during isothermal annealing at 415 °C for 14 hours (curve (b)), and at 436 °C for 7 hours (curve (c)). (e,f) After in-situ NR measurements of pure Ge isotope MLs during isothermal annealing at 418 °C for 3 hours (curve (e)), and at 425 °C for 1 hour (curve (f)).

Figure 4 presents the Raman scattering measurements of the amorphous $Ge_{0.8}Si_{0.2}$ film (Figure 4b) in comparison to that of amorphous pure Ge film (Figure 4a) and additionally, an amorphous Si-rich $Ge_{0.25}Si_{0.75}$ film (Figure 4c). The amorphous $Ge_{0.25}Si_{0.75}$

https://doi.org/10.1051/epjconf/202328605002

film was obtained by using three instead of one quadrants of Si pieces for the sputter-target (see Figure 2a,b). The dashed vertical lines show the wave number position of Raman bands for amorphous germanium and amorphous silicon as found in literature [61-63]. Ge-Si local bonds produces Raman lines between the pure substances [61-63]. In the $Ge_{0.8}Si_{0.2}$ film (Figure 4b), there are no Si-Si bonds. The Si atoms are well dispersed into the amorphous germanium matrix. This slightly shifts the Raman emission of amorphous germanium to higher wave numbers (Figure 4b), as expected. In the case of the higher silicon content of the amorphous $Ge_{0.25}Si_{0.75}$ film (Figure 4c), there exist also Raman scattered intensity at the position of pure amorphous silicon. This means that for the $Ge_{0.25}Si_{0.75}$ film there exist Si atoms with only Si nearest neighbour atoms. Raman intensity is enhanced also between the peaks of the pure amorphous films without producing distinct peaks (Figure 4c). This means that the amorphous $Ge_{0.25}Si_{0.75}$ film possess an atomic environment with Si and Ge atoms as nearest neighbour atoms. In overall terms, the EDX, GI-XRD, and Raman measurements revealed that Ge self-diffusion examined in this work on $Ge_{0.8}Si_{0.2}$ takes place inside an amorphous matrix with well dispersed Si atoms inside a Ge network.

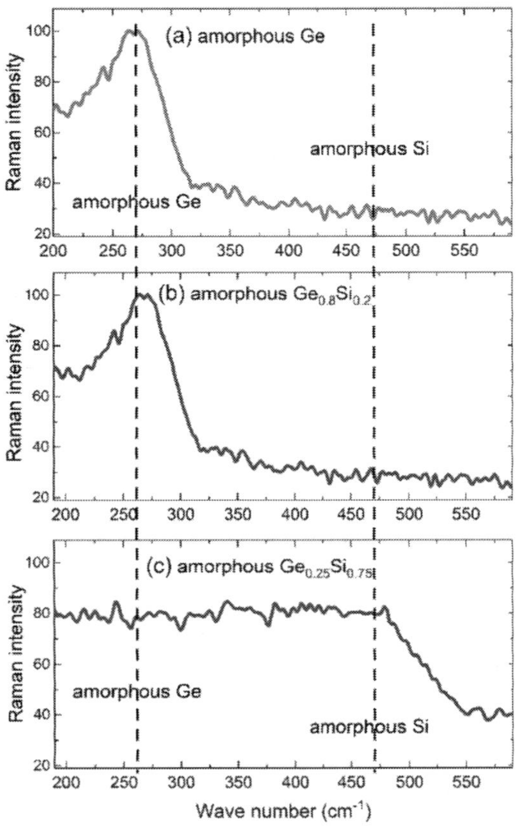

Fig. 4. Raman spectra of amorphous Ge (a), $Ge_{0.8}Si_{0.2}$ (b) and $Ge_{0.25}Si_{0.75}$ (c) films sputter-deposited on sapphire substrate.

4.2 Ge self-diffusion experiments

The Figures 5a, 6a, and 7a present three-dimensional colour maps of the in-situ measured NR data on $[^{73}Ge_{0.8}Si_{0.2}$ (14 nm) / $^{nat}Ge_{0.8}Si_{0.2}$ (14 nm)] × 10 MLs in the Rq_z^4 representation, during isothermal annealing at 397 °C, 415 °C and 436 °C, respectively. The total reflection edges are located around q_z = 0.013 Å^{-1}, and the Bragg peaks due to the Ge isotope modulation are located around q_z = 0.027 Å^{-1}. There is a Bragg peak decrease during the heat treatment. Figure 5b, 6b, and 7b plot the integrated area of the Bragg peak (termed Bragg peak intensity) of dependence on the annealing time. The integrated area is obtained by Gauss-fits to the Bragg peak. The integrated area decreases due to Ge isotope interdiffusion. Details on such experiments can be found in [6].

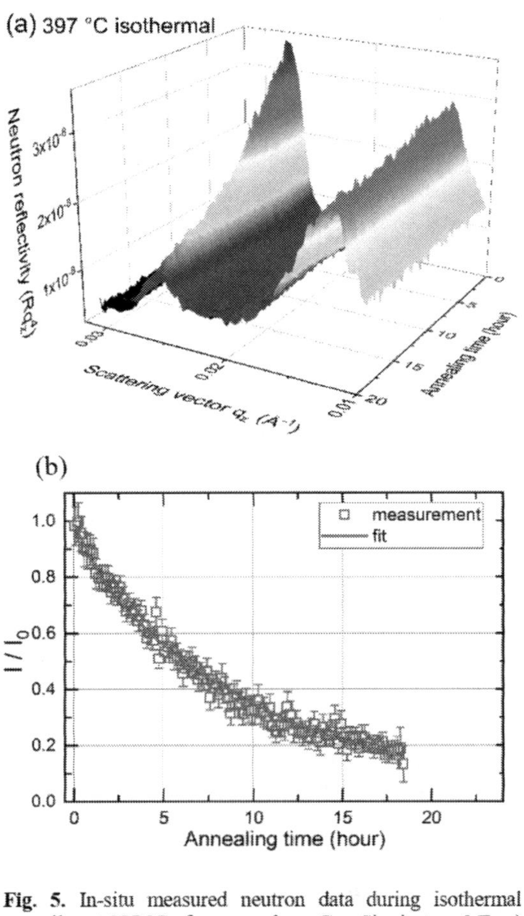

Fig. 5. In-situ measured neutron data during isothermal annealing at 397 °C of an amorphous $Ge_{0.8}Si_{0.2}$ isotope ML. a) Three dimensional color map of (Rq_z^{-4},q_z). b) Bragg peak intensity versus annealing time. The fit was performed with equation (6) with the following result D_I = (4.0 ± 3.5) ×10^{-22} m^{2}s^{-1}, D_R = (2.2 ± 1) ×10^{-22} m^{2}s^{-1}, and τ = (1.6 ± 0.3) h.

https://doi.org/10.1051/epjconf/202328605002

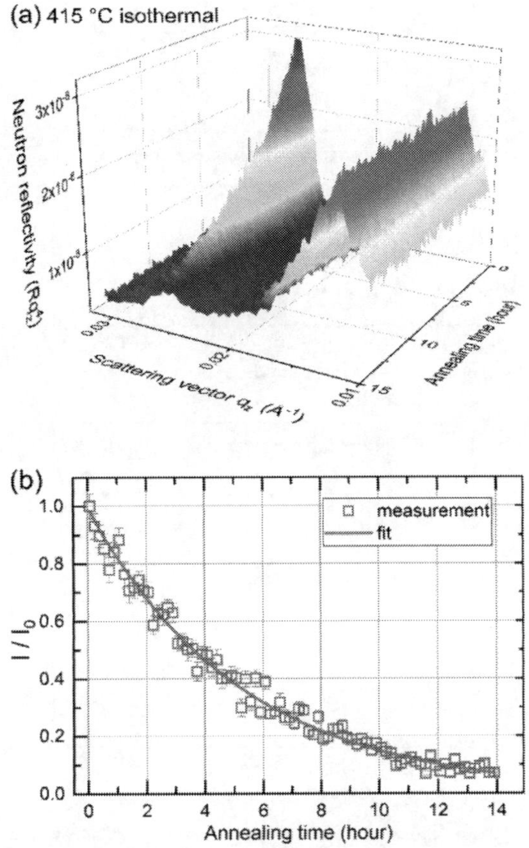

Fig. 6. In-situ measured neutron data during isothermal annealing at 415 °C of an amorphous $Ge_{0.8}Si_{0.2}$ isotope ML. a) Three dimensional color map of (Rq_z^{-4},q_z). b) Bragg peak intensity versus annealing time. The fit was performed with equation (6) with the following result $D_i = (5.5 \pm 4.5) \times 10^{-22}$ m^2s^{-1}, $D_R = (4.5 \pm 0.5) \times 10^{-22}$ m^2s^{-1}, and $\tau = (0.2 \pm 3)$ h.

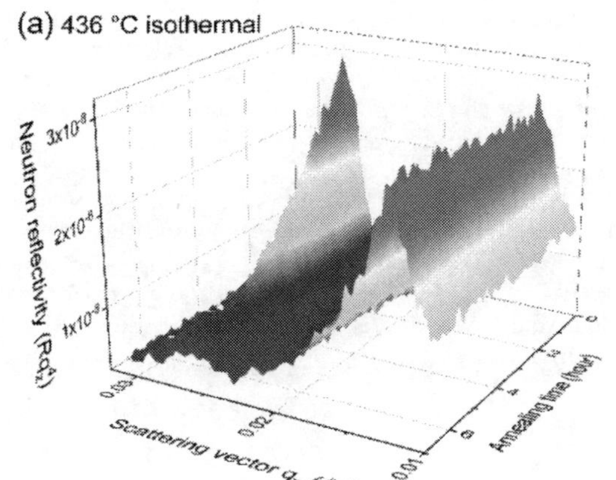

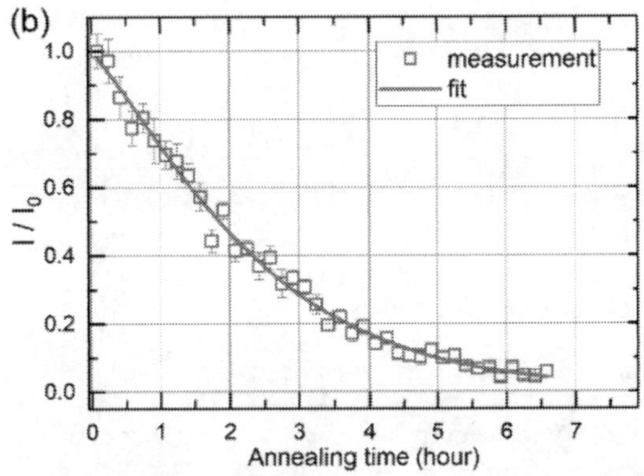

Fig. 7. In-situ measured neutron data during isothermal annealing at 436 °C of an amorphous $Ge_{0.8}Si_{0.2}$ isotope ML. a) Three dimensional color map of (Rq_z^{-4},q_z). b) Bragg peak intensity versus annealing time. The fit was performed with equation (6) with the following result $D_i = (5.5 \pm 2) \times 10^{-22}$ m^2s^{-1}, $D_R = (1.3 \pm 0.2) \times 10^{-21}$ m^2s^{-1}, and $\tau = (1.5 \pm 1)$ h.

In general, Ge self-diffusivities can be determined from the Bragg peak decrease according to the simple equation

$$D_{average} = \frac{r^2 \cdot ln\left(\frac{I_0}{I}\right)}{8 \cdot \pi^2 \cdot t} \quad , \qquad (2)$$

assuming that the diffusivity, $D_{average}$, is constant during the heat treatment. I_0 and I is the Bragg peak intensity before annealing and at the time t of the annealing process, respectively. The bilayer thickness is given by r. In case of the diffusivity changes during the annealing process, equations (3-5) have to be used instead of equation (2) [6,29]. The equations determine the average diffusivity, $D_{average}$, over the time interval t. This analysis is more appropriate for amorphous films, where the annealing process may change defects densities, e.g., frozen defects produced by the sputter-deposition process [6]. For that case, the average diffusivity for a given time interval t, is given by

$$D_{average}(t) = \frac{1}{t} \cdot \int_0^t D(t') \cdot dt' \quad , \qquad (3)$$

where $D(t)$ represents the instantaneous diffusivity at time t which can be described as

$$D(t) = D_R + (D_i - D_R) \cdot exp\left(-\frac{t}{\tau}\right) \qquad (4)$$

where D_R, D_i, and τ, are the diffusivity in the relaxed state and in the initial (t = 0) state, respectively (see references [6,29]). Integration of equation (3) (see references [6,29]) leads to the following expression

$$D_{average}(t) = D_R + (D_i - D_R)\left(\frac{\tau}{t}\right) \cdot \left(1 - exp\left(-\frac{t}{\tau}\right)\right) \qquad (5)$$

Combining relation (2) and (5) the following expression for the Bragg peak intensity decrease is obtained

$$\frac{I}{I_0} = exp\left\{-\frac{8\cdot\pi^2}{r^2}\cdot\left[D_R + (D_i - D_R)\cdot\tau\cdot\left(1 - exp\left(-\frac{t}{\tau}\right)\right)\right]\right\} \qquad (6)$$

Equation (6) was used to fit the Bragg peak decrease in Figure 4b, 5b, and 6b, with D_R, D_i, and τ as fit variables. The obtained annealing time dependence of the instantaneous diffusivities is plotted in Figure 8. For isothermal annealing at 397 °C (Figure 8a), the diffusivity decreases by a factor of 2. This is a low decrease in instantaneous diffusivity compared to other reports on amorphous samples [6,29]. For example, the diffusivity in amorphous pure Ge films decreases by almost one order of magnitude (Figure 8b). Moreover, quite surprising, the diffusivity relaxation is almost non-existent for isothermal annealing at 415 °C (Figure 8c), and even reversed (i.e., diffusivity increases during the relaxation process) for annealing at 436 °C (Figure 8d). Noticeably, the determined instantaneous diffusivities at t = 0, are almost independent of the annealing temperature.

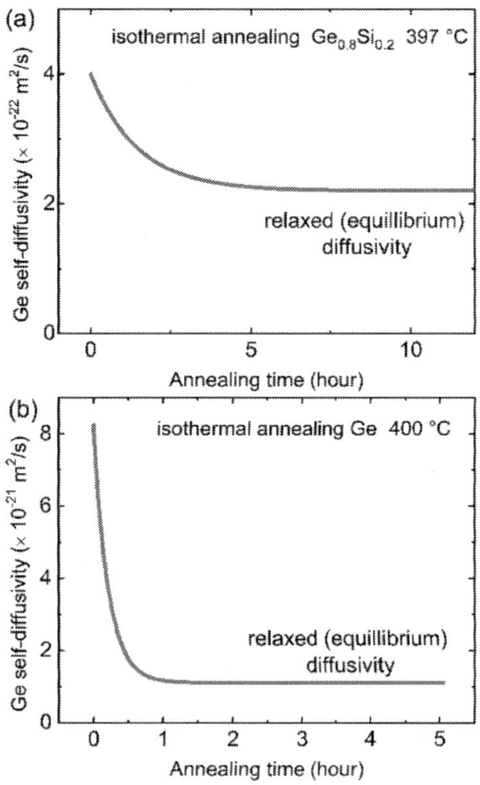

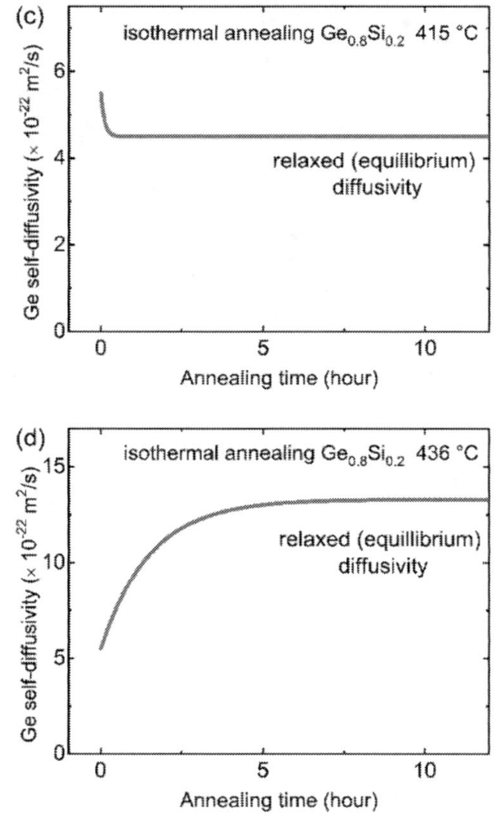

Fig. 8. Modification of the instantaneous Ge self-diffusivity in amorphous $Ge_{0.8}Si_{0.2}$ (a,c,d) and amorphous Ge (b) films as obtained by in-situ NR measurements during isothermal annealing at (a) 398 °C, (b) 400 °C, (c) 415 °C, and (c) 436 °C.

The obtained diffusivities in relaxed amorphous $Ge_{0.8}Si_{0.2}$ are plotted in Figure 9 as red filled triangles in Arrhenius representation. They obey the Arrhenius law with an activation energy of (2.06 ± 0.10) eV. For comparison, the diffusivities of relaxed amorphous Ge films are plotted as black-filled circles (with an activation energy of (2.11 ± 0.12) eV). The dotted blue line shows the Si self-diffusivities calculated from the reported activation energy of Si self-diffusion in amorphous pure silicon of (2.7 ± 0.11) eV as determined by Kirschbaum et al. [27]. The amorphous silicon was produced by Si ion-implantation and self-diffusion was investigated ex-situ after annealing the samples between 460 °C and 600 °C and cooling to room temperature afterwards. We assume that Si and Ge diffusion in amorphous $Ge_{0.8}Si_{0.8}$ is similar as also shown in Figure 1.

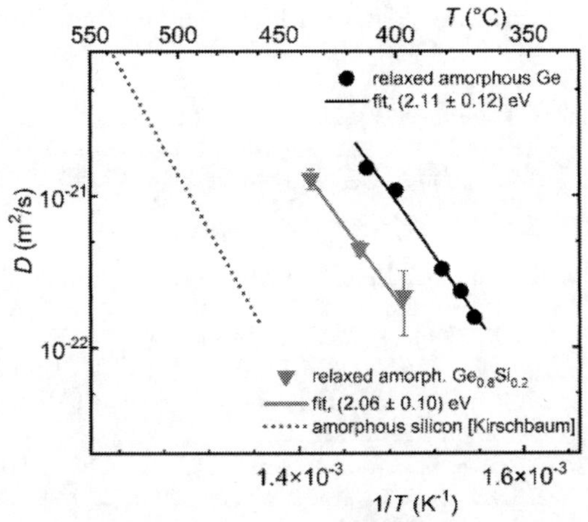

Fig. 9. Arrhenius plot of Ge self-diffusivities in relaxed amorphous $Ge_{0.8}Si_{0.2}$ films obtained during isothermal annealing and in-situ NR experiments (red triangles). Also shown are Si self-diffusivities in amorphous silicon (dotted blue line, Kirschbaum et al. [27], ion-implantation) and Ge self-diffusivities in amorphous relaxed germanium (black dots, [6], sputter deposition). Note that for sputtered amorphous silicon a higher activation energy of (4.4 ± 0.3) eV is found for temperatures above 500 °C [39].

The Ge self-diffusivities of amorphous $Ge_{0.8}Si_{0.2}$ are lower than of that of amorphous Ge by a factor of about five, while the activation energy is the same within error limits. The latter aspect indicates that the Ge self-diffusion mechanism is similar in both materials, in spite of the existence of Si-Ge bonds. The existence of Si atoms in the amorphous Ge network decreases the Ge self-diffusivities as also found for crystalline counterparts. The fact that the activation energy remains unchanged is similar to the results found for the crystalline system by Zangenberg et al. [23] and McVay et al. [13].

5 Conclusion

Literature shows that self-diffusion in crystalline silicon and crystalline germanium proceeds by different mechanisms. In crystalline germanium, literature reports self-diffusion to be mediated by single-vacancies. In contrast, self-diffusion in crystalline silicon is more complex, with contributions of self-interstitials and vacancies, but is governed predominantly by self-interstitials at high temperatures. Thus, the mixture (i.e. alloying) between germanium and silicon may cause peculiar self-diffusion. Literature also reports on self-diffusion in crystalline Ge_xSi_{1-x}, but not in the amorphous counterparts, which is presented here for the case of amorphous $Ge_{0.8}Si_{0.2}$.

The investigation of Ge self-diffusion in $Ge_{0.8}Si_{0.2}$ was performed by in-situ NR measurements during isothermal heating. All Ge self-diffusivities obtained for amorphous $Ge_{0.8}Si_{0.2}$ are smaller than that of amorphous germanium, but higher than that of amorphous silicon. The experiments revealed that the Ge self-diffusivities show a peculiar fast modification as a function of annealing time due to structural relaxation. The self-diffusivities in relaxed $Ge_{0.8}Si_{0.2}$ obey to the Arrhenius law, and have an activation energy of (2.06 ± 0.10) eV which is similar to that in relaxed amorphous germanium of (2.11 ± 0.12) eV, and different to that in amorphous silicon of (2.7 ± 0.1) [27] or (4.4 ± 0.3) [39] eV. This indicates that the Si content of 20 at.% does not change the diffusion mechanism in amorphous $Ge_{0.8}Si_{0.2}$ compared to pure Ge.

This work is based upon experiments performed at the Swiss spallation neutron source SINQ, Paul Scherrer Institut (PSI), Villigen, Switzerland. The research project has been supported by the German Research Foundation under the contract Schm1569/28-2.

References

1. Y. Shiraki, N. Usami, Edts., *Silicon–germanium (SiGe) nanostructures. Production, properties and applications in electronics*. Woodhead Publishing, Sawston, Cambridge, UK, Philadelphia, USA (2011). ISBN 978-0-85709-142-0.
2. E. Hüger, K. Osuch, *"Quantum well states in thin (110)-oriented Au films and k-space symmetry,"* Eur. Phys. J. B **37**, 149–162 (2004). https://doi.org/10.1140/epjb/e2004-00041-1.
3. M. Stutzmann, R. A. Street, C. C. Tsai, J. B. Boyce, S. E. Ready, *"Structural, optical, and spin properties of hydrogenated amorphous silicon-germanium alloys,"* J. Appl. Phys. **66**, 569-592 (1989).
4. B. Liu, L. Bai, X. Zhang, D. Zhang, C. Wei, J. Sun, Q. Huang, X. Chen, J. Ni, G. Wang, Y, Zhao, *"Light management in hydrogenated amorphous silicon germanium solar cells,"* Solar Energy Materials and Solar Cells **128**, 1-10 (2014). https://doi.org/10.1016/j.solmat.2014.05.008.
5. S. Sreejith, J. Ajayan, S. Kollem, B. Sivasankari, *"A Comprehensive Review on Thin Film Amorphous Silicon Solar Cells,"* Silicon **14**, 8277-8293 (2022). https://doi.org/10.1007/s12633-021-01644-w.
6. E. Hüger, F. Strauß, J. Stahn, J. Deubener, M. Bruns, and H. Schmidt, *"In-situ Measurement of Self-Atom Diffusion in Solids Using Amorphous Germanium as a Model System,"* Scientific Reports **8**, 17607 (2018). https://doi.org/10.1038/s41598-018-35915-1.
7. P. Heitjans, J. Kärger, *Diffusion in condensed matter. methods, materials, models*. Springer, Berlin, Germany (2005).
8. H. Mehrer, *Diffusion in solids. Fundamentals, methods, materials, diffusion-controlled processes*, Springer, Berlin, Germany (2007).
9. E. Hüger, D. Uxa, F. Yang, and H. Schmidt, *"The Lithiation Onset of Amorphous Silicon Thin-Film Electrodes,"* APL Special Topic New technologies and applications of advanced batteries, Appl. Phys.

Lett. **221**, 133901 (2022). https://doi.org/10.1063/5.0109610

10. K. Zhang, E. Hüger, Y, Li, H. Schmidt, F. Yang, "*Review and stress analysis on the lithiation onset of amorphous silicon films,*" Batteries **9**, 105 (2023). https://doi.org/10.3390/batteries9020105.
11. E. Hüger, L. Riedel, J. Zhu, J. Stahn, P. Heitjans, and H. Schmidt, "*Lithium niobate for fast cycling in Li-ion batteries: Review and new experimental data,*" Batteries **9**, 244 (2023). https://doi.org/10.3390/batteries9050244.
12. D. Uxa, E. Hüger, K. Meyer, L. Dörrer, H. Schmidt, "*Lithium-ion Diffusion in Near-stoichiometric Polycrystalline and Monocrystalline LiCoO2,*" Chem. Mater. **35**, 3307–3315 (2023). https://doi.org/10.1021/acs.chemmater.3c00359.
13. G. L. McVay and A. R. DuCharme, "*Diffusion of Ge in SiGe alloys,*" Phys. Rev. B **9**, 627-631 (1974).
14. J. Rahn, E. Hüger, L. Dörrer, B. Ruprecht, P. Heitjans, and H. Schmidt, "*A SIMS study on Li diffusion in single crystalline and amorphous LiNbO3,*" Defect Diffus. Forum **323-325**, 69-74 (2012). https://doi.org/10.4028/www.scientific.net/DDF.323-325.69.
15. J. Rahn, E. Hüger, L. Dörrer, B. Ruprecht, P, Heitjans, and H. Schmidt, "*Li Self-Diffusion in Lithium Niobate Single Crystals at Low Temperatures,*" Phys. Chem. Chem. Phys. **14**, 2427 (2012). https://doi.org/10.1039/C2CP23548J
16. J. Uhlendorf, B. Ruprecht, E. Witt, C. V. Chandran, L. Dörrer, E. Hüger, F. Strauß, P. Heitjans, and H. Schmidt, "*Slow Lithium Transport in Metal Oxides on the Nanoscale,*" Z. Phys. Chem. **231**, 1423–1442 (2017). https://doi.org/10.1515/zpch-2016-0939.
17. E. Hüger, L. Dörrer, and H. Schmidt, "*Permeation, Solubility, Diffusion and Segregation of Lithium in Amorphous Silicon Layers,*" Chem. Mater. **30**, 3254–3264 (2018). https://doi.org/10.1021/acs.chemmater.8b00186.
18. E. Hüger, and H. Schmidt, "*Li Permeability Increase in Nano-Sized Amorphous Silicon Layers,*" J. Phys. Chem. C **122**, 28528–28536 (2018).
19. J. Uhlendorf, Z. Galazka, and H. Schmidt, "*Oxygen diffusion in beta-Ga2O3 single crystals at high temperatures,*" Appl. Phys. Lett. **119**, 242106 (2021). https://doi.org/10.1063/5.0071729.
20. F. Strauß, E. Hüger, J. Julin, F. Munnik, and H. Schmidt, "*Lithium Diffusion in Ion-Beam Sputter-Deposited Lithium-Silicon Layers,*" J. Phys. Chem. C **124** 8616–8623 (2020). https://doi.org/10.1021/acs.jpcc.0c01244.
21. L. Dörrer, P. Tuchel, E. Hüger, R. Heller, and H. Schmidt, "*Hydrogen diffusion in proton-exchanged lithium niobate single crystals,*" J. Appl. Phys. **129**, 135105 (2021). https://doi.org/10.1063/5.0047606.
22. D. Uxa, E. Hüger, K. Meyer, L. Dörrer, and H. Schmidt, "*Lithium-ion Diffusion in Near-stoichiometric Polycrystalline and Monocrystalline LiCoO2,*" Chem. Mater. Submitted Manuscript ID: cm-2023-003599.R1 (2023).
23. N. R. Zangenberg, J. Lundsgaard Hansen, J. Fage-Pedersen, and A. Nylandsted Larsen, "Ge Self-Diffusion in Epitaxial Si1-xGex Layers," PRL **87**, 125901 (2001). DOI: 10.1103/PhysRevLett.87.125901
24. R. Kube, H. Bracht, J. Lundsgaard Hansen, A. Nylandsted Larsen, E. E. Haller, S. Paul, and W. Lerch, *"Composition dependence of Si and Ge diffusion in relaxed Si1−xGex alloys,"* J. Appl. Phys. **107**, 073520 (2010). https://doi.org/10.1063/1.3380853.
25. R. Kube, H. Bracht, E. Hüger, H. Schmidt, J. L. Hansen, A. N. Larsen, J. W. Ager III, E. E. Haller, T. Geue, and J. Stahn, "Contributions of vacancies and self-interstitials to self-diffusion in silicon under thermal equilibrium and nonequilibrium conditions," Phys. Rev. B **88**, 085206 (2013). https://doi.org/10.1103/PhysRevB.88.085206.
26. T. Südkamp, H. Bracht, "Self-diffusion in crystalline silicon: A single diffusion activation enthalpy down to 755 °C," Phys. Rev. B **94**, 125208 (2016). https://doi.org/10.1103/PhysRevB.94.125208.
27. J. Kirschbaum, T. Teuber, A. Donner, M. Radek, D. Bougeard, R. Böttger, J. Lundsgaard Hansen, A. Nylandsted Larsen, M. Posselt, and H. Bracht, "*Self-Diffusion in Amorphous Silicon by Local Bond Rearrangements,*" PRL **120**, 225902 (2018). DOI: 10.1103/PhysRevLett.120.225902.
28. H. Schmidt, M. Gupta, and M. Bruns, "*Nitrogen diffusion in amorphous silicon nitride isotope multilayers probed by neutron reflectometry,*" Phys. Rev. Lett. **96**, 055901 (2006). https://doi.org/10.1103/PhysRevLett.96.055901.
29. H. Schmidt, W. Gruber, T. Gutberlet, M. Ay, J. Stahn, U. Geckle, and M. Bruns, "*Structural relaxation and self-diffusion in covalent amorphous solids: Silicon nitride as a model system,*" J. Appl. Phys. **102**, 043516 (2007). https://doi.org/10.1063/1.2770821
30. H. Schmidt, M. Gupta, J. Stahn, T. Gutberlet, and M. Bruns, "*How to Measure Self-Diffusivities in the Sub-Nanometer Range,*" Acta Mater. **56**, 464-470 (2008). https://doi.org/10.1016/j.actamat.2007.10.005.
31. E. Hüger, H. Schmidt, J. Stahn, B. Braunschweig, U. Geckle, M. Bruns, "*Atomic Transport in Metastable Compounds: A Case Study on Si-C-N,*" Phys. Rev. B. **80**, R220101 (2009). https://doi.org/10.1103/PhysRevB.80.220101.
32. E. Hüger, H. Schmidt, T. Geue, J. Stahn, U. Tietze, D. Lott, A. Markwitz, U. Geckle, and M. Bruns, "*Nitrogen Self-Diffusion in Magnetron Sputtered Si-C-N Films,*" J. Appl. Phys. **109**, 093522 (2011). https://doi.org/10.1063/1.3585780.
33. E. Hüger, J. Rahn, J. Stahn, T. Geue, and H. Schmidt, "*Diffusivity determination in bulk*

materials on nanometric length scales," Phys. Rev. B **85**, 214102 (2012). https://doi.org/10.1103/PhysRevB.85.214102.

34. E. Hüger, J. Rahn, T. Geue, J. Stahn, P. Heitjans, and H. Schmidt, "*Lithium Diffusion in Congruent LiNbO3 Single Crystals at Low Temperatures Probed by Neutron Reflectometry*," PCCP **16**, 3670-3674 (2014). https://doi.org/10.1039/C3CP54939A.
35. S. Chakravarty, E. Hüger, H. Schmidt, J. Stahn, M. Horisberger, and N. P. Lalla, "*Self-diffusivities in Ultra-Fine Grained Metals Using Neutron Reflectometry*," Scripta Mater. **61**, 1117 (2009). https://doi.org/10.1016/j.scriptamat.2009.08.035.
36. H. Schmidt, S. Chakravarty, M. Jiang, E. Hüger, P. K. Parida, T. Geue, J. Stahn, U. Tietze, D. Lott, "*Grain Boundary Self-diffusion in Fe Films with a Stable Nanostructure*," J. Mater. Sci. **47**, 4087–4092 (2012). https://doi.org/10.1007/s10853-012-6262-0.
37. E. Hüger, R. Kube, H. Bracht, J. Stahn, T. Geue, and H. Schmidt, "*A neutron reflectometry study on silicon self-diffusion at 900 °C*," Phys. Status Solidi B **11**, 2108-2112 (2012). https://doi.org/10.1002/pssb.201248330.
38. [38=NR11] E. Hüger, U. Tietze, D. Lott, H. Bracht, E. E. Haller, D. Bougeard, and H. Schmidt, "*Self-diffusion in Germanium Isotope Multilayers at Low Temperatures*," Appl. Phys. Lett. **93**, 162104 (2008). https://doi.org/10.1063/1.3002294.
39. F. Strauß, L. Dörrer, T. Geue, J. Stahn, A. Koutsioubas, S. Mattauch, and H. Schmidt, "*Self-diffusion in amorphous silicon*," Phys. Rev. Lett. **116**, 025901 (2016). https://doi.org/10.1103/PhysRevLett.116.025901.
40. E. Hüger, J. Stahn, and H. Schmidt, "*Activation energy of diffusion determined from a single in-situ neutron reflectometry experiment*," Mat. Res. Lett. **11**, 53 (2023). https://doi.org/10.1080/21663831.2022.2114814.
41. E. Hüger, L. Dörrer, J. Rahn, T. Panzner, J. Stahn, G. Lilienkamp, and H. Schmidt, "*Lithium Transport through Nanosized Amorphous Silicon Layers*," Nano Lett. **13**, 1237–1244 (2013). https://doi.org/10.1021/nl304736t.
42. E. Hüger, J. Stahn, and H. Schmidt, "*Neutron reflectometry to measure in-situ Li permeation through ultrathin silicon layers and interfaces*," J. Electrochem. Soc. 162, A7104 (2015). DOI 10.1149/2.0131513jes.
43. E. Hüger, J. Stahn, P. Heitjans, and H. Schmidt, "*Neutron Reflectometry to Measure in-situ the Rate Determining Step of Lithium Ion Transport through Thin Silicon Layers and Interfaces*," PCCP **21**, 16444-16450 (2019). https://doi.org/10.1039/C9CP01222B.
44. F. Strauß, E. Hüger, P. Heitjans, T. Geue, J. Stahn, and H. Schmidt, "*Li Permeation through Thin Lithium-Silicon Films for Battery Applications Investigated by Neutron Reflectometry*," Energy Technol. **4**, 1582-1587 (2017). https://doi.org/10.1002/ente.201600209.
45. E. Hüger, L. Dörrer, R. Yimnirun, J. Jutimoosik, J. Stahn, and A. Paul, "Lithium permeation within lithium niobate multilayers with ultrathin chromium, silicon and carbon spacer layers," PCCP 20, 23233-23243 (2018). https://doi.org/10.1039/C8CP03345E.
46. B. Jerliu, L. Dörrer, E. Hüger, G. Borchardt, R. Steitz, U. Geckle, V. Oberst, M. Bruns, O. Schneider, and H. Schmidt, "*Neutron Reflectometry Studies on the Lithiation of Amorphous Silicon Electrodes in Lithium-Ion Batteries*," PCCP **15**, 7777-7784 (2013). DOI: 10.1039/c3cp44438d.
47. B. Jerliu, L. Dörrer, E. Hüger, B.-K. Seidlhofer, R. Steitz, U. Geckle, V. Oberst, M. Bruns, and H. Schmidt, "*Volume Expansion during Lithiation of Amorphous Silicon Thin Film Electrodes Studied by In-Operando Neutron Reflectometry*," J. Phys. Chem. C **118**, 9395–9399 (2014). https://doi.org/10.1021/jp502261t.
48. B.-K. Seidlhofer, B. Jerliu, M. Trapp, E. Hüger, S. Risse, R. Cubitt, H. Schmidt, R. Steitz, M. Ballauff, "*Lithiation of Crystalline Silicon As Analyzed by Operando Neutron Reflectivity*," ACS Nano **10**, 7458–7466 (2016). https://doi.org/10.1021/acsnano.6b02032.
49. B. Jerliu, E. Hüger, M. Horisberger, J. Stahn, and H. Schmidt, "*Irreversible lithium storage during lithiation of amorphous silicon thin film electrodes studied by in-situ neutron reflectometry*," J. Power Sources 359, 415-421 (2017). https://doi.org/10.1016/j.jpowsour.2017.05.095.
50. B. Jerliu, E. Hüger, L. Dörrer, B.-K. Seidlhofer, R. Steitz, M. Horisberger, and H. Schmidt, "*Lithium Insertion into Silicon Electrodes Studied by Cyclic Voltammetry and Operando Neutron Reflectometry*," PCCP **20**, 23480-23491 (2018). https://doi.org/10.1039/C8CP03540G.
51. D. Uxa, B. Jerilu, E. Hüger, L. Dörrer, M. Horisberger, J. Stahn, and H. Schmidt, "*On the Lithiation Mechanism of Amorphous Silicon Electrodes in Li-Ion Batteries*," J. Phys. Chem. C **123**, 22027–22039 (2019). https://doi.org/10.1021/acs.jpcc.9b06011.
52. H. Schmidt, B. Jerliu, E. Hüger, and J. Stahn, "*Volume Expansion of Amorphous Silicon Electrodes during Potentiostatic Lithiation of Li-Ion Batteries*," Electrochem. Commun. **115**, 106738 (2020). https://doi.org/10.1016/j.elecom.2020.106738.
53. Available online: http://www.ncnrs.nist.gov/resources/sldcalc.html (accessed on 22 December 2022).
54. A. Pedersen, L. Pizzagalli, H. Jonsson, "*Atomic and electronic structures of a vacancy in amorphous silicon*," Phys. Rev. B **101**, 054204 (2020). DOI: 10.1103/PhysRevB.101.054204
55. M. W. Cleveland, M. J. Demkowicz, "*Persistence of negative vacancy and self-interstitial formation*

energies in atomistic models of amorphous silicon," Phys. Rev. Mat. **6**, 013611 (2022). https://doi.org/10.1103/PhysRevMaterials.6.013611.

56. J.R. Weber, A. Janotti, C. G. Vanm de Walle, "*Dangling bonds and vacancies in germanium*," Phys. Rev. B **87**, 035203 (2013). DOI: 10.1103/PhysRevB.87.035203.
57. M. Posselt, H. Bracht, M. Ghorbani-Asl, D. Radic, "*Atomic mechanisms of self-diffusion in amorphous silicon*," AIP Advances **12**, 115325 (2022). doi: 10.1063/5.0111037.
58. E. Hüger, C. Jin, K. Meyer, D. Uxa, and F. Yang, "*Invited: Investigation of Carbon/Copper Multilayer to Examine the Influence of Copper Coating on the Li-Storage Performance of Carbon*," Energies **16**, 2740 (2023). https://doi.org/10.3390/en16062740.
59. J. Stahn, A. Glavic. "*Focusing neutron reflectometry*," Nucl. Instrum. Methods Phys. Res. Sect. A. **821**, 44–54 (2016).
60. C. Braun, Parratt32 or the reflectometry tool, HMI, Berlin, [http://www.helmholtz-berlin.de].
61. L.W. Veldhuizen, C.H.M. van der Werf, Y. Kuanga, N.J. Bakker, S.J. Yun, R.E.I. Schropp, "*Optimization of hydrogenated amorphous silicon germanium thin films and solar cells deposited by hot wire chemical vapor deposition*," Thin Solid Films **595**, 226–230 (2015). http://dx.doi.org/10.1016/j.tsf.2015.05.055.
62. A. G. Hernandez, A. E. Escobosa-Echavarría, Y. Kudriavtsev, "White luminescence emission from silicon implanted germanium," Appl. Surf. Sci. **428**, 1098–1105 (2018). https://doi.org/10.1016/j.apsusc.2017.09.234.
63. T. de Vrijer, B. Bouazzata, A. H. M. Smets, "Spectroscopic review of hydrogenated, carbonated and oxygenated group IV alloys," Vibrational Spectroscopy **121**, 103387 (2022). https://doi.org/10.1016/j.vibspec.2022.103387.

 https://doi.org/10.1051/epjconf/202328605003

Development and first results of a magnetic sample environment for polarized neutron imaging of thin metal sheets

Alex Backs[1,2,*], *Simon* Sebold[3], *Matteo* Busi[4], *Wai Tung* Lee[2], *Markus* Strobl[4,5], and *Dmytro* Orlov[1]

[1]Mechanical Engineering Sciences, Lund University (LU), 221 00 Lund, Sweden
[2]Science Directorate, European Spallation Source (ESS), 224 84 Lund, Sweden
[3]Heinz Maier-Leibnitz Zentrum (MLZ), Technische Universität München (TUM), 85748 Garching, Germany
[4]Laboratory for Neutron Scattering and Imaging, Paul Scherrer Institut (PSI), 5232 Villigen, Switzerland
[5]Nils Bohr Institute, University of Copenhagen, 2100 Copenhagen, Denmark

Abstract. Polarized neutron imaging brings the great advantage of analyzing bulk magnetic properties with good spatial resolution. The technique is based on the interaction of the neutron spin with magnetic samples or free magnetic fields and observing the changes to a spin-polarized neutron beam. The high sensitivity to even small magnetic fields is a benefit in obtaining magnetization information but simultaneously a challenge in instrumentation, since magnetic environments for the polarized neutron beam and for the sample, as well as the fringe field from the magnetic sample itself all affect the measurement and can give rise to unwanted effects. We have used finite element simulations and ray tracing simulations, to design and analyze a magnetic sample environment devised for the measurement of ferromagnetic metal sheets. Here we show an analysis of performance of the experimental setup based on the simulation results and compare them to first experimental results on a grain oriented silicon steel sample.

1 Introduction

Neutrons provide an excellent tool for the investigation of magnetic properties due to their inherent magnetic moment and therefore direct magnetic interaction. In experiments using unpolarized neutrons, this magnetic interaction can already provide information about the magnetic properties of the sample, however, to access the full potential, polarized neutrons are required. In these experiments, neutrons are filtered by their spin state, either spin-up or spin-down, which can be achieved by using equipment such as polarizing mirrors or polarized ^{3}He. However, these measurements require significantly more time, as spin-filters reduce the neutron flux and several separate exposures are necessary per measurement.

In scattering techniques, the use of polarized neutrons is well established and allow a distinction between magnetic and non-magnetic scattering contributions. For imaging applications, early experiments with polarized neutrons only started about 15 years ago [1, 2] but there is an ongoing development and great advances have been achieved in recent years [3, 4]. Polarized neutron imaging (PNI) has been used, e.g., to measure the magnetic domain structure in large grain silicon steel [5, 6], distinguish magnetic phases in steel samples through combination with conventional transmission data [7] or determine the Curie-temperature in a sample with spatial resolution [8, 9]. It is also possible to obtain images of free magnetic fields, such as the stray field of magnetic flux frozen in a superconducting sample [1, 10]. More complex measurements even allow the reconstruction of a complete magnetic vector field. This, however requires a full tomography with eighteen separate measurements per projection and is therefore a time consuming technique [11, 12].

Generally, special attention has to be given to the magnetic field setup of any PNI experiment. Once the beam is polarized, an external magnetic guide field is required to retain the neutron spin state. Hence, for any sample to be investigated as well as for additional sample environment, the interplay with this guide field has to be assessed. Samples with high permeability at low fields may be magnetized by this field. On the other hand, the guide field might be distorted by samples with high magnetic susceptibility or a high remanent magnetization and cause problems with proper neutron spin transport. Therefore, it is important to make sure the sample and the surrounding field have no unintended effects on each other, which could jeopardize the entire experiment.

Here, we report on the development of our own custom sample environment intended for the measurements of ferromagnetic metal sheets. Silicon steel laminations are routinely used as magnetic core for both transformers and electric engines and a deeper understanding of their magnetic properties and behavior can be extremely beneficial to increase their energy efficiency. For such investigations, PNI distinguishes itself from other methods by supplying spatially resolved data of the bulk material. The sample environment devised for this experiment is meant to avoid the problems mentioned above, by providing a

*e-mail: alexander.backs@ess.eu

small volume around the sample where we have high precision control over the magnetic field. This is achieved with two additional flat coils surrounding the sample; a compensation coil cancels the external guide field and a sample coil magnetizes the sample in the desired direction. The design and optimization of this setup was done using a combination of the finite element magnetic field computation software COMSOL [13] and the neutron ray tracing Monte Carlo software McStas [14, 15] to obtain the magnetic field in the experimental setup and analyze its effect on the neutron polarization, respectively.

2 Larmor Spin Precession

For PNI, a semi-classical approach to spin interactions is taken, where the spin properties of a neutron beam are described by a polarization vector $\vec{P}$, which reflects the expectation value of the neutron ensemble in density matrix formalism. Magnetic interactions affect the magnitude and direction of $\vec{P}$ and in this work, we are mainly interested in the effects due to Larmor precession. When exposed to a magnetic field $\vec{B}_{\text{sample}}$ that is not collinear with $\vec{P}$, $\vec{P}$ rotates around the field axis. In a static field, the time evolution of $\vec{P}$ can be calculated using the rotation matrix $\widehat{R}$ around $\vec{B}_{\text{sample}}$

$$\vec{P}(t) = \widehat{R}_{\vec{B}_{\text{sample}}}(\theta) \times \vec{P}(t=0) \tag{1}$$

where $\theta = \omega t = \gamma_N B_{\text{sample}} t$ is the angle of rotation, which depends on the flux density B_{sample}, the dwelling time t and the gyro-magnetic ratio of the neutron γ_N.

In conventional uniaxial polarization analysis, only a projection P_y of $\vec{P}$ on the guide field $\vec{B}_{\text{guide}}$ is measured experimentally, so that the spin precession described above is not accessible. We use the subscript 'y' since, throughout this work, the projection axis is always the vertical y-axis. P_y is calculated from two separate transmission measurements as

$$P_y = \frac{I_+ - I_-}{I_+ + I_-} \tag{2}$$

where I_+ and I_- are intensities measured with the incoming polarization direction parallel and anti-parallel to the guide field $\vec{B}_{\text{guide}}$, respectively. Ideally, our setup provides an abrupt transition from the guide field $\vec{B}_{\text{guide}}$ to the field causing spin precessions $\vec{B}_{\text{sample}}$. Abrupt, in this case, means that the neutron polarization can not follow the change from $\vec{B}_{\text{guide}}$ to $\vec{B}_{\text{sample}}$ adiabatically. As a result, the spin precession is observed as an oscillation of P_y

$$P_y(t) = A \cdot \cos(\omega \cdot t + \psi) + O \tag{3}$$

with amplitude A, angular frequency ω, phase shift ψ and offset (or mean polarization) O. The parameters depend on the orientations and values of $\vec{P}$ and $\vec{B}_{\text{sample}}$ as well as the projection axis defined by $\vec{B}_{\text{guide}}$. Experimentally, only the final state of a spin precession is accessible after the neutrons have passed through any given field. However, since the dwelling time in the field is directly proportional to the neutron wavelength $t \propto \lambda_N$, the time dependence can be recovered by scanning the wavelength. The oscillation

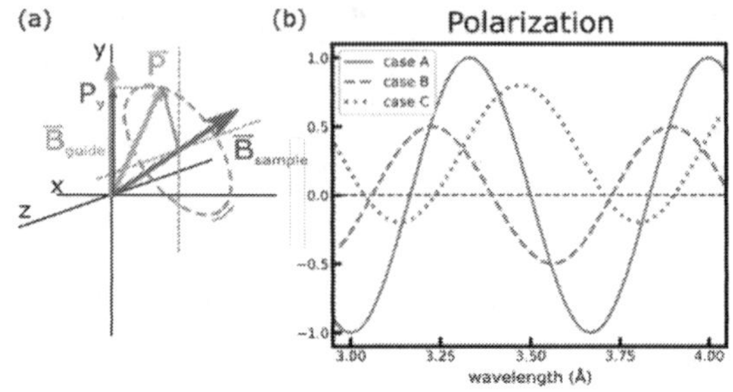

Figure 1. Inside the sample, $\vec{P}$ precesses around the field $\vec{B}_{\text{sample}}$. Experimentally, only the projection of $\vec{P}$ along the guide field $\vec{B}_{\text{guide}}$ is measured. (a) shows an arbitrary setup of the relevant vectors and indicates the Larmor precession. (b) shows three possible P_y-oscillations with the one corresponding to (a) included as magenta dotted line (case 3).

$P_y(\lambda_N)$ is equivalent to equation (3) with a new frequency $\nu \propto \omega$ with units [rad/Å].

$$P_y(\lambda_N) = A \cdot \cos(\nu \cdot \lambda_N + \psi) + O \tag{4}$$

Figure (1) illustrates how the spin rotation in a constant field results in the measured polarization oscillation. The sketch in panel (a) shows arbitrary orientations of the sample field ($\vec{B}_{\text{sample}}$, dark green) and the initial polarization vector ($\vec{P}_0$, orange), which follows the orange dashed circle during the rotation. P_y (purple) is the polarization projection on the guide field ($\vec{B}_{\text{sample}}$, light green). Its wavelength dependence, along with two additional cases, is shown in panel (b): case 1 (blue, solid) is ideal for observing the polarization oscillation with $\vec{B}_{\text{sample}} \perp \vec{P}$ and $\vec{B}_{\text{sample}} \perp \vec{y}$. The oscillation has the maximum amplitude of $A = 1$ and oscillates around zero ($O = 0$). For case 2 (red, dashed), $\vec{B}_{\text{sample}} \perp \vec{y}$ but $\vec{P}$ is inclined away from the y-axis. The result is a reduced amplitude while the offset remains zero. Finally in case 3 (purple, dotted), additionally $\vec{B}_{\text{sample}}$ is inclined towards the y-axis. This results in a reduced amplitude and in a non-zero offset. The frequency ν is the same in all cases and chosen to reflect the properties of a steel sample, while the phase angle ψ was chosen arbitrarily.

In the more general case of a field varying in space or time, equation 1 has to be integrated with respect to an infinitesimal rotation angle $\delta\phi$ or infinitesimal time δt. If the changes in the field direction are much slower than the Larmor precession, the spin evolution is adiabatic, meaning that the polarization component initially oriented along the field is retained and follows the orientation of B. If the field direction changes too rapidly, the spin evolution becomes non-adiabatic, which will result in a wavelength dependent reorientation of $\vec{P}$ with respect to $\vec{B}_{\text{sample}}$.

A second common effect important for PNI is depolarization. For spin precession, e.g. inside a single magnetic domain, only the direction of $\vec{P}$ changes while its magnitude and its component along the field are retained. In contrast, spin depolarization reduces the magnitude of $\vec{P}$.

https://doi.org/10.1051/epjconf/202328605003

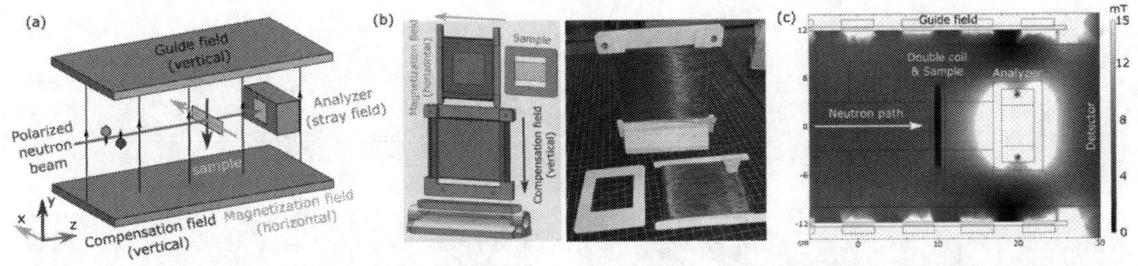

Figure 2. (a) Sketch of the magnetically relevant components in the experimental setup. For graphical reasons the double-coil-setup has been omitted, showing only the additional fields created by it. (b) Sketch and photo of the double-coil-setup. The sample, inner coil and outer coil form a nested compact sample environment that can be quickly disassembled for easy access to the sample. (c) Magnetic field distribution as calculated using COMSOL. Note that the coordinate system in panel (a) is valid for all simulations and data presented in this work.

Depolarization is generally caused by magnetic disorder. Examples are ferromagnetic materials with a large number of randomly oriented domains and domain walls between magnetic domains [16, 17]. Depolarization can also be due to a wavelength spread or geometric divergence, effectively imposing a resolution limitation on the polarization analysis.

3 Magnetic Environment

Since polarized neutrons are sensitive to to the magnetic environment, PNI requires a well defined experimental setup to avoid spurious effects from field inhomogeneities. Figure (2) (a) illustrates the key magnetic components relevant for our experiments and simulations. Arrays of permanent magnets provide the vertical guide field ($B_{\text{guide}} \approx 5.2\,\text{mT}$) along the neutron flight path to maintain high beam polarization. The analyzer, a reflection solid state bender, requires a strong magnetic field ($\approx 50\,\text{mT}$) to work properly, which is provided by permanent magnets in its housing. This, however, creates a considerable stray field. The sample is affected by the guide field and the analyzer fringe field due to its high magnetic susceptibility. Without proper counter measures the sample is magnetized, which in turn distorts the surrounding field. Therefore, a custom double-coil-setup is used to generate two additional fields (called "compensation"- and "magnetization" field) at the sample position. The compensation field is oriented anti-parallel to the guide field and is used to cancel the unwanted field at the sample position. Nested into the compensation field is the magnetization field, in perpendicular direction, which is used to magnetize the sample. A rendering and photo of the double-coil-setup are shown in panel (b) of figure (2) illustrating the nested structure of the coils with the sample on the inside. The coils are designed to accommodate thin sheets-like samples with a maximum size of 100 mm × 100 mm × 2 mm, though the samples used in this work (simulation and experiment) were smaller as shown in figure (2).

To calculate the magnetic field, we used COMSOL for finite element magnetic field calculations. The dimensions of all components were taken from the actual devices, as were the material properties to the best of our knowledge. For the guide field and the analyzer, the remanent fields of the permanent magnets were adjusted to reflect previous measurements with a Hall probe. The sample, which is a large-grain silicon-steel and therefore very inhomogeneous, was simplified as a homogeneous and isotropic material. In this way, the macroscopic effect of the sample can still be reproduced sufficiently well, while deviations due to the crystal or domain structure are interpreted as sample effects to be evaluated from the experimental data. Magnetic fields generated by the coils match reasonably well with measured values for the same driving current. Panel (c) gives an overview of the simulated magnetic flux density in the central vertical cut plane. The far-reaching stray field of the analyzer is clearly visible, which necessitates placing the sample and the double-coil-setup 8 cm upstream. In this simulation, the sample and double-coil-setup were neglected. Their effects are discussed in the following section.

As explained above, the outer coil of our setup generates a vertical field to compensate the guide field and analyzer stray field (see figure (2)). The coil has two layers of wire around a frame with dimensions 120 mm × 110 mm × 6 mm. The inner coil can be used to apply an additional field along the x-axis (perpendicular to the beam and guide field) to magnetize the sample in this direction. It is slightly smaller, fitting into the outer coil, consisting of one layer of wire with inner dimensions 120 mm × 100 mm × 3 mm. 1 mm diameter anodized aluminium wires were used to minimize neutron absorption and material activation, wound on a 3D printed plastic frame. Mounted at the center of the coils is the sample, which is a horizontally mounted stripe of silicon-steel with dimensions 50 mm × 30 mm × 0.4 mm, such that the inner coil field is aligned along the longest direction.

The impact of both coils is illustrated in figure (3), which shows COMSOL calculation results in the vicinity of the sample under four conditions. Panel (a) shows again the external field generated by the guide field and analyzer. (b) With the sample included, there is a strong distortion of the field. In panel (c), the field at the sample position ($B_{ext} \approx 6.7$ to $7.1\,\text{mT}$) is compensated by the outer coil

($B_{remaining}$ ≈ 0.5 to 0.8 mT), which completely removes the previous distortion effects. Finally, in panel (d) the inner coil is active as well with a current of I_{inner} = 2 A, providing a horizontal (along the x-axis) field of 3.1 mT. The images illustrate how the coils can be used to provide a good zero field condition and to magnetize the sample without strong field distortions.

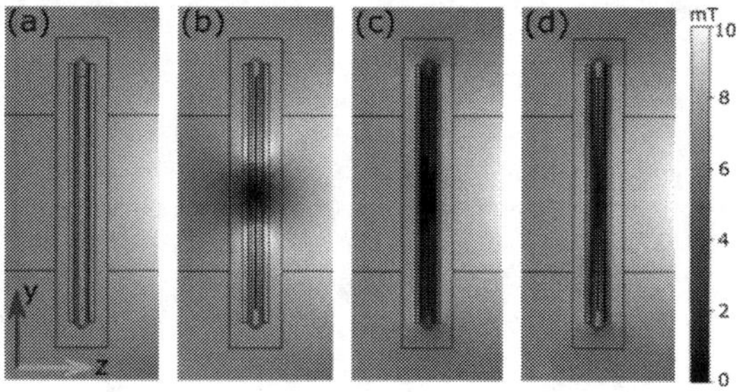

Figure 3. Magnetic flux density at the sample position, side view. (a) shows the field generated by the guide field and analyzer. (b) includes the sample which distorts the field. In (c) the vertical field has been compensated by using the outer coil of our setup. For (d) an additional horizontal field is applied with the inner coil, to magnetize the sample.

4 Spin Precession Simulations

For a full analysis of the functionality of our setup, we have used the McStas ray tracing Monte Carlo simulation tool to calculate the neutron spin evolution. In McStas, the basic pinhole geometry of an imaging beamline is modeled to include geometric effects, based on the the BOA beamline at PSI, where our experiments were conducted. As a neutron source, a "Source_simple" component was used, with a wavelength spread of 0.03 Å, representing a double crystal monochromator with $\Delta\lambda/\lambda = 1\%$. The source had a rectangular size of 15 mm × 15 mm, representing the pinhole used at BOA, and the distance between neutron source and detector was 5.27 m. As detector, a 2D position sensitive monitor recording the Cartesian projections of the polarization was used with an area of 50 mm × 50 mm and 100 × 100 pixel. Two additional such monitors were placed at the front and rear surface of the sample in the beam direction. Magnetic field values calculated in COMSOL were included in McStas as "Pol_Bfield"-components in the final 20 cm in beam direction of the instrument with a cross section of 5.2 cm × 5.2 cm. Due to technical difficulties the COMSOL results had to be included as several separate components along the beam direction. Simulations were performed with $2 \cdot 10^6$ neutron rays each.

The magnetic field region included in McStas covers the distance between the center of the guide field unit up to the center of the analyzer in the COMSOL calculations. At both positions, the field is aligned perfectly vertical (parallel to the incident polarization direction and the analyzer direction) such that no spurious effects due to abrupt boundaries are introduced. Additionally, no impairments were observed due to the additional boundaries (all parallel to the x-y-plane) introduced by splitting the field region. The sample itself, with a volume of 50 mm × 30 mm × 0.4 mm was excluded from the imported field profiles. While the field distortion outside the sample is very relevant for the spin evolution, the internal field of the sample has little direct correspondence to the real sample. Instead, values in the sample volume were replaced with a homogeneous magnetic field, representing a specific domain orientation and internal magnetic flux density. As a result, the simulations show how the chosen kind of domain would affect the measured polarization in any position in the sample. The most relevant choice for this was a field of 1.5 T oriented horizontally in x-axis, which corresponds to the experimentally determined field integral in the sample and the predominant domain orientation.

4.1 Results: Field Compensation

First, we focus on the performance of the outer coil and necessity of guide field compensation, by evaluating the polarization in the two cases of no field compensation and full field compensation (c.f. cases (b) & (c) illustrated in figure (3)). For this analysis, the Cartesian projections obtained in McStas are not optimal and instead more suitable quantities are used. i) P_{abs}: the absolute value of the polarization vector. ii) $P_{y\text{-}z}$: the magnitude of the y-z-projection of the polarization vector. iii) $\beta_{y\text{-}z}$: the angle between the y-z-projection of the polarization vector and the y-axis (the initial polarization direction). P_{abs} is a measure of the depolarization of the neutron beam. With the internal field of the sample pointing in along the x-axis, $P_{y\text{-}z}$ corresponds to the part of the polarization contributing to the spin precession. $\beta_{y\text{-}z}$ is needed for complete orientation information of the polarization and is optimal to show the spin precession occurring in the sample.

Figure (4) shows images illustrating the beam polarization just before reaching the sample using the quantities just defined. The uncompensated case is shown in the upper row (both coils inactive) and the compensated case in the lower row (active compensation coil). P_{abs} indicates that no depolarization occurs up to the sample, apart from slight artifacts at the sample edges due to the split simulation. In the uncompensated case, a pattern of horizontal stripes is visible in $P_{y\text{-}z}$, with values going down to 0.85. Fitting the sample shape, the field distortion is more pronounced in the vertical direction. $\beta_{y\text{-}z}$ is entirely inhomogeneous, changing smoothly from 90° to −90° from top to bottom. In contrast, the compensated case results in completely homogeneous images, retaining the initial polarization up to the sample. The presented results were calculated at a wavelength λ_N = 3.2 Å. In the full range used for the simulations between 3.00 to 3.40 Å, the uncompensated case shows a slight wavelength dependence which indicates a non-adiabatic spin evolution.

Inside the sample, the polarization precesses due to the chosen uniform field (B_{sample} = 1.5 T along the x-axis) according to equation (1) without unexpected side effects. In

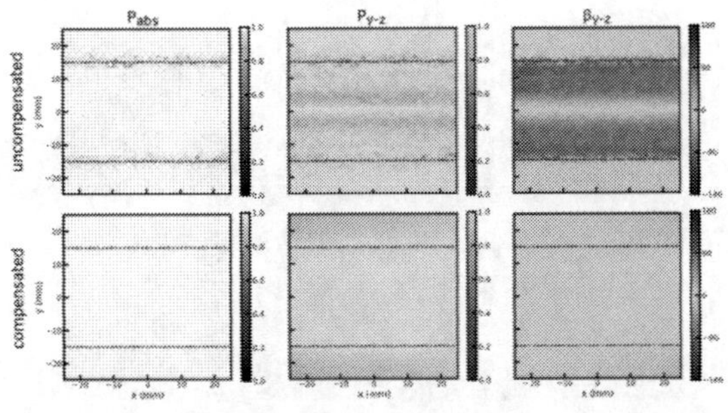

Figure 4. Orientation of the polarization vector $\vec{P}$ just before entering the sample, described by the values P_{abs}, $P_{y\text{-}z}$ and $\beta_{y\text{-}z}$ as defined in the text. The upper row shows results for an uncompensated setup and the lower row the compensated case with an active outer coil, corresponding to cases (B) and (C) in figure (3). In the uncompensated case a significant reorientation of the polarization orientation occured over the whole sample area, which is avoided by compensating the guide field. In this and the following images, the sample position is marked by dashed lines.

the 400 μm thick sample, the polarization precesses 10 to 11 times in the chosen wavelength range. Due to the wavelength spread of 0.03 Å, the spin precession angle spreads as well which reduces the measured polarization to ≈ 0.9.

In the space between the sample and the analyzer the spin evolves further, partially in a non-adiabatic way comparable to the one observed before the sample though not entirely symmetric due to the analyzer fringe field. In addition, precession around the strong guide field, combined with the wavelength spread, leads to a complete loss of polarization in the horizontal plane, such that only the polarization projection P_y is retained.

For evaluating the full impact of the setup we have simulated a wavelength scan in the range of 3.00 to 3.40 Å in 21 steps of size 0.02 Å. These simulations were analyzed analogous to the experimental data, by pixel-wise fitting of the wavelength dependence.

Ideally, we expect $P_y(\lambda_N)$ to perform sinusoidal oscillations, as described by equation (4) and illustrated in figure (1). In figure (5), three of the fitting parameters are shown, the oscillation amplitude A, the mean polarization O and instead of the oscillation frequency ν the corresponding internal field $B_{int} = 3956 \cdot \nu/\gamma_N/t_{sample}$ with the thickness of the sample $t_{sample} = 400\,\mu$m and $[\nu] = 1/$Å. The phase shift ψ does not contribute to a better understanding of the data and has been omitted.

Note, that the presented fitting parameters do not reflect the quality of the fit. If the measured oscillation deviated from a a simple sinusoidal function, this may not be reflected in the fitting result. However, since the same evaluation method is used for experimental results, a reasonable comparison can be drawn.

Again, the uncompensated case is in the upper row and the compensated case in the lower one. Starting with the oscillation amplitude, both cases have a similar range of values with $A \approx 0.65$ to 0.88 and $A \approx 0.61$ to 0.89 for the uncompensated and compensated case, respectively. The pattern, however is very different. In the uncompensated case we observe a pattern of horizontal stripes reminiscent of the one seen in figure (4). In contrast, the compensated case shows a much clearer pattern along the vertical direction, with a maximum in the center and decreasing to the edges at the top and bottom. Outside the sample, the amplitude is effectively zero as no spin precession occurs.

The mean polarization shows a similar pattern to the amplitude in the first case, with mostly positive values in the range of -0.06 to 0.22. In the second case, no pattern is observed and the values are essentially scattered around zero (-0.04 to 0.02), leaning slightly towards the negative. Outside the sample, however, the compensated case shows strongly reduced mean values $O <= 0.6$ with a vertical gradient, while the uncompensated case remains high ($O \approx 0.95$).

The internal field, again, shows the striped pattern in the uncompensated case, with an average value of 1.50 T and a rather rectangular distribution with FWHM of 0.050 T. In the compensated case, the average value is identical (1.50 T) but the distribution is closer to Gaussian with a narrower FWHM of 0.018 T. Outside the sample, no useful data is obtained as there is no oscillation.

In summary, in the uncompensated case we observe a clear fingerprint of the field distortion by the sample. Corresponding to the sample shape, the field inhomogeneity is pronounced primarily in vertical direction, leading to the horizontal stripe pattern. While a clear polarization oscillation is observed across the whole sample with high oscillation amplitude, the non zero mean polarization and the deviations in the internal field are sub-optimal. In the compensated case, a larger scale pattern seems to be present with an increasingly depolarizing effect going from the horizontal center line towards the top and bottom of the images. This can be seen in the sample area from the amplitude and in the open beam regions in the mean polarization. However, in contrast to the first case, the mean value and internal field determined in the sample are essentially homogeneous in the limit of the simulation statistics. Generally, the values obtained in the sample are fluctuating more in the uncompensated case.

While both cases show that the setup is not optimal, these are strongest in the area of the sample for the uncompensated case while they mostly affect the open beam region for the compensated case. It shows that using the active guide field compensation suppresses an unwanted magnetization of the sample and significantly improves the data quality over the uncompensated case.

4.2 Results: Magnetizing Field

As described, the additional inner coil can be used to magnetize a sample along the x-axis. The field direction here is important since the magnetic domains in the sample become (or stay) oriented in the same direction, allowing us to still observe spin precession due to the internal field. In contrast, a magnetization by the guide field in y-direction co-aligns the internal field and the initial neutron polarization, suppressing any spin precession.

 https://doi.org/10.1051/epjconf/202328605003

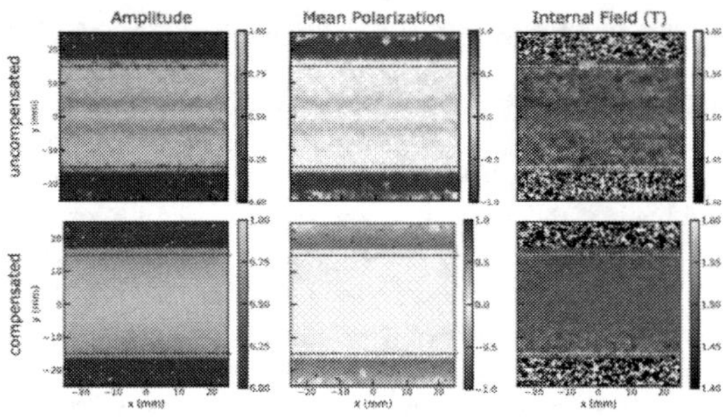

Figure 5. Fitting results of the polarization oscillation observed at the final monitor position for the two cases of an uncompensated (upper row) and compensated (lower row) setup. The uncompensated case shows a horizontal stripe pattern in the sample area, while for the compensated case, a smooth vertiacl variation is observed over the full image area. Generally, the compensated case results show less fluctuations and introduce less artifacts into the measurement.

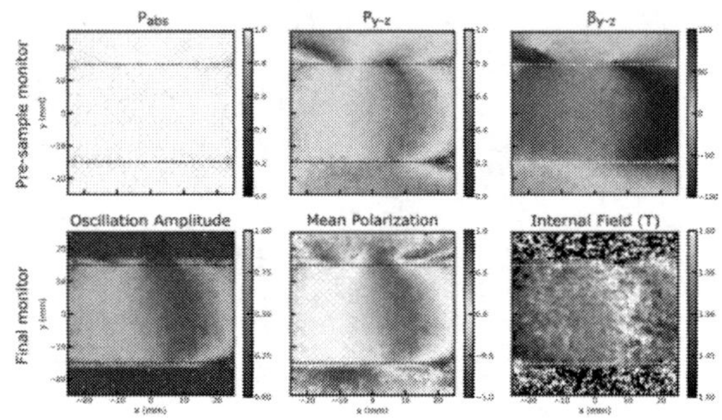

Figure 6. Simulation results with field compensation and applied magnetizing field. The upper row shows the polarization state just before entering the sample. compared to the previous cases the polarization is heavily affected by the fringe field surrounding the sample. The lower row shows pitting results of the polarization oscillation at the final monitor position. In a pattern similar to the upper row, the fitting results deviate from the perfect case.

Figure (6) shows data analogous to figures (4 & 5) for a simulation with field compensation ($I_{\text{comp}} = 2.7\,\text{A}$) and an additional magnetizing field from the inner coil ($I_{\text{mag}} = 3\,\text{A}$). The field generated by the inner coil is $B_{\text{mag}} \approx 4.6\,\text{mT}$.

In the upper row the state of the incoming polarization is shown, just before the sample, which is affected much stronger than the previous two cases. While there are still no depolarization effects, the other two parameters indicate a strong reorientation of $\vec{P}$. Especially in a crescent shaped region on the right side of the sample, $P_{y\text{-}z}$ drops below 0.3. Additionally, the open beam regions are almost entirely affected. Again, the polarization is slightly wavelength dependent indicating non-adiabatic effects.

In the polarization oscillation fitting shown in the bottom row of figure (6), the crescent shape is visible as well, though slightly distorted. On the left side of the sample the spin oscillation is observed clearly, with a high amplitude (≈ 0.85) and close to zero mean polarization. However, the internal field value obtained in this area is not reliably reproducing B_{sample} with a visible gradient and values of 1.45 to 1.63 T.

The problems observed in this example are representative for driving currents of the inner coil $I_{\text{mag}} = 1$ to 4 A, though the pattern and severity of the effects do change. The interpretation is that the polarization is to some degree reoriented towards the sample field before sample but also oriented back towards the guide field behind the sample. Depolarization in the sample, as in domain walls, can therefore still be observed, but the asymmetric spin evolution on either side of the sample and the non-adiabatic nature of this effect are challenging to isolate from the sample effect.

5 Experimental Results & Comparison

First experiments using the double-coil-setup were conducted at the BOA beamline at PSI [18]. The sample was a grain-oriented silicon-steel sheet with dimensions $70\,\text{mm} \times 30\,\text{mm} \times 0.25\,\text{mm}$. As the simulations were based on this experiment, the instrumental details are virtually the same.

A rectangular pinhole ($15\,\text{mm} \times 15\,\text{mm}$) was used, followed by a double crystal monochromator with wavelength spread of 1 % and an adiabatic RF spin flipper [19]. BOA provides a polarized neutron beam via a reflection multi-channel bender and a compact reflection solid-state bender was used as spin analyzer. As detector, a typical imaging setup was used with a $200\,\mu\text{m}$ LiFl scintillator, a mirror box and an Andor Ikon-L camera. A rectangular pinhole ($15\,\text{mm} \times 15\,\text{mm}$) was used, with the detector placed 5.65 m downstream. Close after the pinhole, a double crystal monochromator with wavelength spread of 1 % and an adiabatic RF spin flipper [19] were placed. BOA has a reflection multi-channel bender for polarizing the neutron beam and a compact reflection solid-state bender was used as spin analyzer. With a pinhole to detector distance of 5.65 m and a sample to detector distance of 17 cm, we had an $L/D = 370$ and a geometric resolution of $\approx 0.5\,\text{mm}$. Wavelength scans were performed in the range between 3.0 to 3.36 Å in 13 steps of 0.03 Å.

Preliminary analysis of some of the data is presented here to compare with the simulation results. Detailed data analysis is ongoing and not in the scope of this article. In figure (7) we present results from two wavelength scans using the same fitting procedure described for the simulated data. Corresponding to the two cases in figure (6), the upper row shows data in a fully compensated field, i.e. of the remanent state of the sample. In the bottom row, the sample was magnetized using the inner coil with a driving current of $I_{\text{inner}} = 4\,\text{A}$. According to our COMSOL calculations, this corresponds to a field of $B_{\text{inner}} = 6.2\,\text{mT}$, however, due to comparison with the simulation results we suspect a slightly lower actual field. Measurements of the field were not possible as the available Hall probe did not fit into the narrow coil. Starting with the remanent state,

 https://doi.org/10.1051/epjconf/202328605003

we first note that the open beam regions at the top and bottom of the images are uniform, with zero amplitude and high mean polarization while the internal field shows pure noise. The mean polarization measured in these regions is ≈ 0.84 without a clear wavelength dependence. It appears that the sample environment introduces a general depolarization of the beam, but it cannot be ruled out, that this is due to a slight adiabatic turning of the polarization instead. Inside the sample, a wealth of features are visible in the amplitude. The sharp boundaries indicate that they are not due to the magnetic environment but inherent to the sample. The mean polarization is very close to zero (< 0.07) almost everywhere and the internal field shows a well defined value of $B_{\text{int}} \approx 1.7\,\text{T}$ except for a large noisy area left of the center. B_{int} is an especially important parameter, as it shows that our sample has a well defined internal field in a large area leading to a sinusoidal polarization oscillation. We can therefore infer that the sample has magnetic domains extending over its full thickness. The low mean polarization indicates, that the magnetic orientation of those domains lies in the horizontal plane. From the amplitude we obtain more details about the domain structure. In the right side of the sample are two large discernable areas, in the upper and lower half, divided by a sharp ragged line. Without definite proof, we interpret this as two crystallographic grains, where the respective lattice orientations induce different domain structures. The lower region shows a very well ordered stripe pattern oriented almost horizontally, where the darker stripes mark the domain walls which locally depolarize the neutron beam. The width of these domains is $\approx 1.0\,\text{mm}$ with lengths reaching well above 10 mm. The amplitude inside the domains is ≈ 0.5 to 0.6, though, due to the geometric smearing with the depolarizing domain walls, the true value is probably above 0.6. Still, even considering the additional depolarization due to wavelength spread observed in the simulations (by a factor of 0.9) and the reduced polarization in the open beam regions, the low amplitude in the magnetic domains indicates an additional depolarization effect. In the upper region on the right side, the domain pattern appears much less well-ordered and has an overall lower amplitude of 0.25 to 0.35 in the brighter regions. As stated above, we suspect different crystallographic and therefore magnetic orientations to be the cause of the two different regions. A plausible case would be to have the magnetic domains in the lower region oriented along the x-axis (in the sample plane as well as in the horizontal plane) while the upper region is oriented slightly out of the sample plane. On the left edge of the image, similar regions to the right half are visible, separated by a large area which completely depolarized the beam. At this preliminary stage of the analysis we are unable to speculate about the internal structure of this region but it appears to consist of substantially smaller magnetic domains, especially in the direction of the neutron beam.

In the second measurement, the field of the inner coil is putting the sample in a highly magnetized state and the magnetic structure has changed drastically. The large disordered region has almost disappeared and clear stripe domains are only visible in a small area at the right bottom of the sample. We believe that most of the larger features visible in the amplitude and mean polarization are indicating crystallographic features, most likely grain boundaries. In addition to sample features, a broad roughly crescent shaped region on the left side of the sample has emerged where the amplitude is low, the mean polarization elevated and the internal field deviates strongly from values in the remaining sample up to being pure noise. We identify this region with the similar effect observed in our simulations and can therefore conclude, that this is no sample effect. Especially in the mean polarization, it is clear that some features of the sample properties are still observed such as the grain boundary between the top and bottom part or the diagonal domains at the bottom edge to the left of the sample. It is also noteworthy that the pattern of the mean polarization in the open beam region is very similar in simulation and experiment.

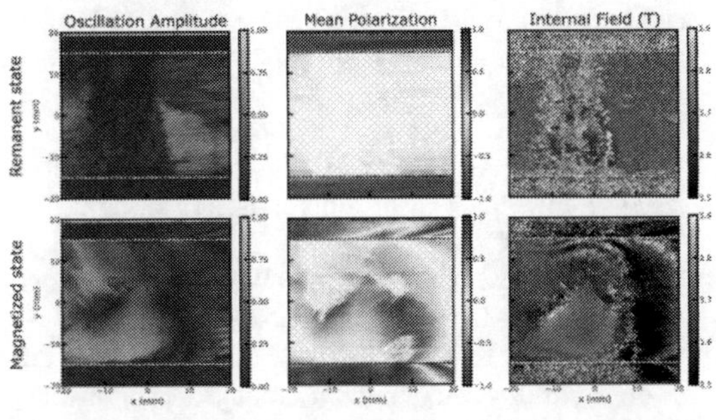

Figure 7. Fitting results of two wavelength scans. Upper row: remanent state of the sample, using only the outer coil for field compensation. Lower row: magnetized state of the sample, additionally using the inner coil. The upper measurement is free from artifacts caused by the magnetic setup, while the lower measurement shows large scale features in all fitting parameters. These results correspond with the simulated results shown in figure (6).

6 Summary

Polarized neutron imaging is a technique very sensitive to small magnetic fields, making it an ideal tool to investigate changes is the magnetic structure of a sample, but likewise very susceptible to variations in the magnetic environment. This makes a very detailed planning of the experimental setup crucial in order to obtain usable and credible data. We have shown that the combination of finite element magnetic field calculations and neutron ray tracing Monte Carlo simulations provides a powerful method to investigate the performance of a new magnetic sample environment. A double-coil-setup was designed using field calculations with COMSOL Subsequently, McStas simulations were used as well to analyze the experimental results. With the excellent agreement between simulation and experiment we are currently updating the double-coils to avoid the strong unwanted effect on the polarization in the vicinity of the sample. At the time of writing, the most promising solution appears to be the inclusion of 'window'-like iron yokes between the inner and outer coil, effectively diverting the fringe fields from the direct neutron flight path. We want to emphasize that magnetic field setups similar to the presented double-coils could be adapted easily and individually to a wide range of PNI experiments and samples. The need for a low magnetic background for a sample in a very limited volume and the necessity to apply a variable field are common prerequisites. The possibility to quickly and inexpensively create a custom sample environment by 3D printing may enable a wide range of experiments in the future.

We would like to thank Peter Willendrup and Mads Bertelsen for their help with the McStas simulations. Polarized neutron imaging experiments were conducted at the BOA beamline at PSI as part of proposal 20221166. We also acknowledge financial support through the Tillväxtverket European Regional Development Fund project SREss3 (2020-2023).

References

[1] N. Kardjilov, I. Manke, M. Strobl, A. Hilger, W. Treimer, M. Meissner, T. Krist, J. Banhart, Nat. Phys. **4**, 399 (2008)

[2] M. Schulz, P. Böni, C. Franz, A. Neubauer, E. Calzada, M. Mühlbauer, B. Schillinger, C. Pfleiderer, A. Hilger, N. Kardjilov, J. Phys. Conf. Ser. **251**, 012068 (2010)

[3] N. Kardjilov, I. Manke, R. Woracek, A. Hilger, J. Banhart, Mater. Today **21**, 652 (2018)

[4] M. Strobl, H. Heimonen, S. Schmidt, M. Sales, N. Kardjilov, A. Hilger, I. Manke, T. Shinohara, J. Valsecchi, J. Phys. D: Appl. Phys **52**, 123001 (2019)

[5] K. Hiroi, T. Shinohara, H. Hayashida, J. Parker, Y. Su, K. Oikawa, T. Kai, Y. Kiyanagi, Physica B: Cond. Matt. **551**, 146 (2018)

[6] I. Dhiman, R. Ziesche, L. Riik, I. Manke, A. Hilger, B. Radhakrishnan, T. Burress, W. Treimer, N. Kardjilov, Mater. Lett. **259**, 126816 (2020)

[7] M. Busi, E. Polatidis, C. Sofras, P. Boillat, A. Ruffo, C. Leinenbach, M. Strobl, Mater. Today Adv. **16**, 100302 (2022)

[8] P. Jorba, M. Schulz, D.S. Hussey, M. Abir, M. Seifert, V. Tsurkan, A. Loidl, C. Pfleiderer, B. Khaykovich, J. Magn. Magn. Mater. **475**, 176 (2019)

[9] M. Seifert, P. Schmakat, M. Schulz, P. Jorba, V. Hutanu, C. Geibel, M. Deppe, C. Pfleiderer, Phys. Rev. Res. **4**, 043029 (2022)

[10] I. Dhiman, R. Ziesche, V. Anand, L. Riik, G. Song, A. Islam, I. Tanaka, W. Treimer, Phys. Rev. B **96**, 104517 (2017)

[11] M. Sales, M. Strobl, T. Shinohara, A. Tremsin, L.T. Kuhn, W.R. Lionheart, N.M. Desai, A.B. Dahl, S. Schmidt, Sci. Rep. **8**, 2214 (2018)

[12] M. Sales, T. Shinohara, M.K. Sørensen, E.B. Knudsen, A. Tremsin, M. Strobl, S. Schmidt, J. Phys. D: Appl. Phys **52**, 205001 (2019)

[13] COMSOL Multiphysics® v. 6.1., *www.comsol.com*, COMSOL AB, Stockholm, Sweden (2023)

[14] P.K. Willendrup, K. Lefmann, J. Neutron Res. **22**, 1 (2020)

[15] P.K. Willendrup, K. Lefmann, J. Neutron Res. **23**, 7 (2021)

[16] O. Halpern, T. Holstein, Phys. Rev. **59**, 960 (1941)

[17] M. Burgy, D. Hughes, J. Wallace, R. Heller, W. Woolf, Phys. Rev. **80**, 953 (1950)

[18] M. Morgano, S. Peetermans, E. Lehmann, T. Panzner, U. Filges, Nucl. Instrum. Methods Phys. Res. A **754**, 46 (2014)

[19] V.I. Luschikov, Y.V. Taran, Nucl. Instrum. Methods Phys. Res. A **228**, 159 (1984)

 https://doi.org/10.1051/epjconf/202328605004

Deuterated Clathrate Hydrates as a Novel Moderator Material for Very Cold Neutrons. Project Aperçu and First Results.

Valentin Czamler[1,2,*], *Thomas C.* Hansen[2], *Michael Marek* Koza[2], *Richard* Wagner[2], and *Oliver* Zimmer[2]

[1]École Doctorale de Physique, Université Grenoble Alpes, 38402 Saint Martin d'Hères, France
[2]Institut Laue-Langevin (ILL), 71 Avenue des Martyrs, 38042 Grenoble Cedex 9, France

Abstract. Clathrate hydrates are water-based solids with large crystallographic unit cells that show promise as potential moderators for use in new, more intense sources of very cold neutrons (VCN), which would enhance neutron scattering techniques and increase sensitivity of particle physics experiments. These so-called inclusion compounds seem particularly suitable for this application due to the low-energy modes of guest molecules encaged in nano-voids formed by a crystalline network of hydrogen bonded water molecules. In this article we present first results of an ongoing, extensive experimental campaign with the aim of characterizing the scattering properties of deuterated clathrate hydrates relevant for moderator applications. Experiments include neutron diffraction (carried out at Institut Laue-Langevin (ILL) using the instrument D20), as well as measurements of the temperature-dependent dynamical structure factor $S(q, \omega)$ in absolute units (carried out on ILL's Panther and IN5). These measurements will serve as a benchmark for development of new scattering kernels.

1 Introduction

Very cold neutrons (VCN) cover a wide spectral range within the long-wavelength tail of typical sources for cold neutrons[1], with energies below, say, 1 meV (9 Å) down to few hundreds of neV (> several 100 Å), the domain of ultracold neutrons (UCN). Notably wavelengths of up to several tens of Å are in the focus of current interest of several research communities. In condensed matter physics, they serve as a sensitive probe and could improve multiple neutron scattering techniques in terms of spacial resolution as e.g. for small angle neutron scattering (SANS), or energy resolution as e.g. in time of flight (TOF) or neutron spin echo (NSE) spectrometers [1]. In particle physics, higher VCN intensities would increase the sensitivity of experiments that employ beams of slow neutrons. Examples are the search for neutron anti-neutron oscillations [2], where the figure of merit (FOM) $\propto \lambda^2$, in-beam searches for a static neutron electric dipole moment (EDM), and further searches for new fundamental forces where FOM $\propto \lambda$ [3] [4]. Furthermore, an enhancement of the flux of neutrons with wavelengths around 9 Å would be a valuable asset for in-beam UCN-sources based on super-fluid helium [5], including variants with in-situ UCN production and detection approaches, as proposed in [6]. Despite the scientific potentials of VCN beams, at present the only existing VCN beamline is the instrument PF2 located at the Institut Laue-Langevin (ILL) in Grenoble, France. At this instrument, the vertical extraction by a strongly curved neutron guide from one of ILL's cold sources transmits only the long-wavelength tail of the liquid-deuterium moderator spectrum. In order to realize higher fluxes of VCN, additional moderation to lower temperatures seems inevitable, which strongly motivates the search for appropriate, new moderator materials.

1.1 Required properties of novel cold and very cold moderator materials

The moderating properties of a material are practically determined by its excitation spectrum and its coupling strength to the neutron field via the cross section. Thus, promising materials for moderation to VCN need to possess low-energy modes that enable the incremental deposition of small quantities of the neutron's energy. This can hardly be provided by collective excitations such as phonons since their cross section, for a neutron of energies $E < k_B T_D$, in a cold medium with the Debye temperature T_D is proportional to $(\frac{E}{k_B T_D})^3$ (for single phonon emission[2]) [7]. Localized, dispersion-free excitations, such as displacement of confined molecules, often referred to as Einstein modes, molecular rotations, librations or paramagnetic excitations, on the other hand, do not suffer from this limitation and allow for an efficient neutron slowdown even at lowest neutron temperatures. Besides suitable localized excitations, the material needs to exhibit low absorption and large scattering cross sections in the corresponding energy and temperature region. In addition, a moderator material must also fulfill more technical requirements. If its thermal conductivity is low, the material

*e-mail: vczamler@ill.fr

[1]The standard cold moderator materials for both spallation and fission sources are liquid or solid deuterium (D_2), liquid hydrogen (H_2) and liquid or solid hydrocarbons (e.g. methane (CH_4)).

[2]The cross sections for processes of higher order drop even faster.

must be arranged in a manner that enables efficient cooling. On top of that it must withstand radiational damage to a degree that prevents the formation of radiolytic hydrogen as well as the accumulation of free radicals. This could result in a self-sustaining reaction of their recombination leading to a fast heating of the moderator, often referred to as "burp" [8] [9] [10]. In this article we present parts of an extensive inelastic neutron scattering study on different clathrate hydrate samples carried out on various neutron scattering instruments at the ILL.

1.2 A promising material class: Fully deuterated clathrate hydrates

Promising moderator materials investigated so far are different hydrocarbons, such as triphenylmethane [11], mesitylene and toluene [12], as well as clathrate hydrates [13] hosting methane or tetrahydrofuran (THF) [14]. Mesitylene is already in routine usage in form of a pelletized cold moderator at the IBR-2M reactor in Dubna [15]. This article focuses on the latter material class, where the encaged guest molecules provides a range of localized, low-energy excitations that can be exploited for neutron slowdown. Given their very large crystallographic unit cells (as in the case of the clathrate structure II (CS-II) 17.1 Å – 17.33 Å, see table 1), which allows for coherent scattering up to neutron wavelengths of $\lambda \sim 20$ Å, these materials promise extraordinarily large albedos for the entire cold neutron range. This allows neutrons that did not scatter down to the desired energy range to be "trapped" within the moderator material by diffuse transport, thus largely increasing the chance for further, energy decreasing interactions.
In this work, we present a study of clathrate hydrates hosting tetrahydrofuran as a guest molecule (sum formula: $17H_2O{:}C_4H_8O$), from here on referred to as THF-hydrates. A great advantage of this type of hydrate is its very reliable and reproducible manufacturing from a stochiometric mixture under ambient pressure (see section 2.1). This would even allow for an in-situ annealing, thus re-establishing the clathrate structure within the moderator volume, in order to prevent a "burp" of the moderator due to radiolysis. To some extent, the low-energy excitations of THF-hydrates have already been studied [14] [16].We have extended those studies to include deuterated materials, and to cover $S(q,\omega)$ in a large energy-momentum phase space in absolute units, which is essential for a quantitative analysis of the moderation properties in realistic scenarios.
While fully deuterated THF-hydrates ($17D_2O{:}C_4D_8O$) by themselves show very promising properties as a moderator material for VCN, they are also a good starting material for the formation of binary clathrate hydrates with both THF and molecular oxygen as guest molecules ($17D_2O{:}C_4D_8O{:}{\sim}2O_2$).[3] Therein, O_2 occupies the small cages, which are twice as abundant as the bigger cages occupied by the TDF. The magnetic triplet ground state of molecular oxygen with its zero-field splitting of 0.4 meV allows for moderation via a cooling cascade mechanism in fully deuterated O_2-clathrate hydrates, as described in [17]. It is this low excitation energy (as opposed to modes of several meV of the lattice and the THF) that will significantly enhance the moderation efficiency towards VCN, with the inelastic magnetic scattering providing a final cooling decrement.

[3]The oxygen abundance is, unlike the C_4D_8O (TDF), non-stoichiometric and strongly depends on the sample preparation. This is represented by "~".

1.3 Development of new scattering kernels in the HighNESS project

The study of novel VCN moderator materials is part of a collaborative effort within the European project HighNESS [18], committed to expanding the scientific potential of the European Spallation Source (ESS), by exploring new possible sources of slow neutrons. One of the goals is to establish a database of the dynamic structure factor $S(q,\omega)$ for various promising clathrate hydrates and to develop novel scattering kernels that allow computation of neutron spectra in realistic moderator and reflector geometries in Monte Carlo (MC) simulations (see also Ref. [19]). To this end, the experimental campaign is accompanied by the development of simulation software to describe neutron interactions in materials that we investigate. The approach is to benchmark the excitation spectrum from ab initio molecular dynamics (MD) calculations, with the experimental data presented in this article. In combination with processing codes, such as NJOY [20] and NCrystal [21], this allows the creation of scattering kernels that are callable from McStas [22], OpenMC [23], MCNP [24] or PHITS [25]. This provides the tools to simulate novel moderator and reflector materials in realistic geometries, and compare simulations with results of experiments to be performed on a later stage using a large bulk of material as a demonstrator of a VCN moderator.

2 Experimental Programm

A starting point of the experiments described here is establishing a reliable and scalable technique, that allows production of relatively large quantities of hydrates with minimal amounts of residual ice. This technique is described for THF-hydrates in section 2.1. The obtained structures were studied by neutron diffraction at the high-intensity two-axis diffractometer D20. An exemplary powder pattern and details of the associated Rietveld-refinement are given in section 2.2. The structure analysis is followed by a study of the low-energy dynamics of different THF-hydrate samples, as described in 2.3. The data presented in this article were collected during cycles N° 189, N° 190 and N° 191 at the ILL.

2.1 Manufacturing procedures

A great advantage of THF-hydrates, compared to most other clathrate hydrate compounds, is that they form at ambient conditions from a stoichiometric mixture of its two liquid components, water (H_2O) and THF (C_4H_8O),

https://doi.org/10.1051/epjconf/202328605004

in a ratio of 17 to 1. After carefully weighing and mixing the two liquids with a teflon-coated magnetic stirrer, a cool-down of the solution results in solidification in the CS-II structure (see table 1). The CS-II structure is the most common among clathrate hydrates and its ideal unit cell contains 136 (H_2O) molecules forming 16 small and 8 large cages. This structure has been extensively studied before (see e.g. [26]). The study presented here focuses on the yields of the CS-II structure and quantifying the residual ice, for two different methods to cool down the sample. A great advantage of the manufacturing technique is, that

Crystal system	Cubic
Space group	Fd3m (N°227)
Lattice description	Face centered
F Lattice parameters	a = 17.1 Å-17.33 Å $\alpha = \beta = \gamma = 90°$
Number of cages	8 large ($5^{12}\,6^4$), 16 small (5^{12})
Ideal unit cell formula	$8(5^{12}\,6^4) \cdot 16(5^{12}) \cdot 136H_2O$

Table 1: Characteristics of the CS-II hydrate crystal cell structure. In the case of THF-hydrates the unit cell formula reduces to 8 THF:136 H_2O, with the small cages remaining empty. Table adapted from [27, p. 60].

it allows to perform a contrast variation between the host and the guest contribution in spectroscopy experiments. By simply substituting one of the liquid components with its deuterated counterpart (water (H_2O) with heavy water (D_2O) and THF (C_4H_8O) with TDF (C_4D_8O)) one obtains four different samples which are summarized in table 2.

	Abbrev.	Host	Guest
Fully protonated	THF · H_2O	136 H_2O	8 C_4H_8O
Deuterated guest	TDF · H_2O	136 H_2O	8 C_4D_8O
Deuterated cage	THF · D_2O	136 D_2O	8 C_4H_8O
Fully deuterated	TDF · D_2O	136 D_2O	8 C_4D_8O

Table 2: Differently deuterated and protonated samples prepared for spectroscopy experiments. The deuteration of either the guest molecule or the host lattice allows to highlight the different parts of the sample in the scattering signal. Note that the fully or partial deuteration is able to slightly change the dynamics of the sample, while the structure remains the one described in table 1.

2.2 Neutron diffraction on fully deuterated THF-hydrates

The diffractometer was operated using the (115) reflection of a Germanium(113) monochromator at a wavelength of $\lambda = 1.546$ Å, at a take off angle of 90° resulting in a resolution of $\Delta d/d = 3 \cdot 10^{-3}$. The cylindrical sample container made from vanadium (with a diameter of 6 mm and a wall thickness of 0.1 mm) was filled with the liquid stochiometric mixture and quenched in liquid nitrogen before being loaded into the cryogenic sample environment. After equilibration to a set temperature value, it took about 30 minutes to obtain a diffraction pattern within a scattering angle range of 4° to 150° with satisfactory statistics. This allowed measurements within a temperature range of 2 K - 230 K. Afterwards, the sample was heated up above the melting point (~ 277.5 K) and slowly (about 7 K per minute) cooled down again below 200 K. We observed that the CS-II structure is formed through both rapid quenching and slower cooling of the liquid sample. Effects of texture of the polycrystalline samples were observed as a dependence of diffraction patterns on sample orientation (which for an ideal powder would be absent). They are inherent for an uncontrolled in-situ crystallization, which can lead to a small number of crystallites and ultimately to a deviation from continuous powder lines (towards single points on the Debye-Scherer cones). Mitigation of this effect was attempted by rotating the sample in increments of 20° and adding up the resulting 9 diffraction patterns, for better approximation of a powder average. Figure 1 shows an exemplary diffraction pattern obtained with this procedure at a temperature of 230 K.

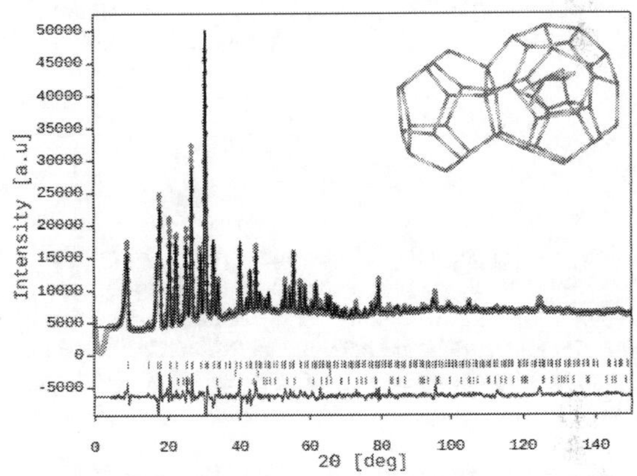

Figure 1: Diffraction pattern of fully deuterated THF-hydrates (TDF · D_2O) formed in situ taken on D20 with $\lambda = 1.546$ Å at a temperature of 230 K (red) and the corresponding multi-phase Rietveld-refinement (black). See text for experimental conditions. The experimental data is available in Ref. [28]. The blue line depicts the difference between the data and the fit. The calculated peak positions (green ticks) correspond to phases considered for the refinement, which are, besides the TDF-hydrate, the aluminum sample environment and residual hexagonal ice (see section 2.2.1) for details on the refinement). The upper right corner shows 1 of the 8 large cages ($5^{12}\,6^4$) within the unit cell hosting a TDF molecule, as well as 1 of the 16 empty small cages (5^{12}).

2.2.1 Structural Details

Data analysis and structure refinement were performed in the program Fullprof [29] [30]. A multi-phase Rietveld-refinement (see e.g. [31]) yields the lattice constants, the positional and thermal parameters, as well as the weight

Phase	$TDF \cdot D_2O$	Ice I_h
Crystal System	cubic	hexagonal
Space Group	F d -3 m	P 63/m m c
Cell parameters	a = 17.22(7)	a = b = 4.50(7) c = 7.35(4)
R_B	10.49	27.92
R_{wp}	4.89	4.89

Table 3: Refinement details of the diffraction pattern depicted in figure 1. The Bragg R-factor (R_B) and the weighted profile R-factor (R_{wp}) indicate an acceptable fit.

percentage for each crystalline phase present in the sample. In a full refinement procedure the instrumental and background parameters are determined concurrently to the structure, the respective scale factors and the phase fractions. The TDF molecule was treated as a rigid body, with its position and the orientation within the large cage as free parameters. Aside from the phase under investigation (TDF-hydrates), the refinement includes the aluminum of the sample environment and residual hexagonal ice. The weight percentage W_j for the phase j can be calculated as in [32]:

$$W_j = \frac{S_j Z_j M_j V_{cj}}{\sum_i^N S_i Z_i M_i V_{ci}}, \quad (1)$$

with S, Z, M and V being the scale factor, the number of formula units per unit cell, the mass of one formula unit and the unit-cell volume, for each phase j and i, respectively. Fullprof also accounts for the multiplicities of each site, for occupation numbers unequal to 1, via the factor f_j, and the micro absorption of neutrons (Brindley factor t_j) of each phase (see Ref. [33] for details):

$$W_j = \frac{S_j Z_j f_j^2 M_j V_j / t_j}{\sum_i^N S_i Z_i f_i^2 M_i V_i / t_i} = \frac{S_j \cdot \mathrm{ATZ} \cdot V_j}{\sum_i^N S_i \cdot \mathrm{ATZ}_i \cdot V_i}, \quad \text{with } \mathrm{ATZ} = Z_j f_j^2 M_j / t_j. \quad (2)$$

Since our hydrate is stochiometric ($f_j = 1$) and the two phases in question have very similar absorption ($t_j \sim 1$), equation 2 reduces to equation 1.

Depending on the sample, a clathrate weight percentage of 95.1 ± 1.5 % to 98.4 ± 1.6 % was reached for the samples investigated, with the uncertainty being dominated by the quality of the fit. In that context, it is important to note that deviations derived in these kinds of least-squares methods only partly reflect systematic errors which might be introduced by the instrument or sample. The details of the refinement are given in table 3. The fit suffers substantially from the size of the crystallites due to texture resulting from the in situ formation. Nevertheless, it provides evidence that the clathrate structure is formed with reliable purity from the liquid stochiometric mixture.

2.3 Time of flight spectroscopy

After having confirmed the CS-II structure of the samples solidified by quenching or slower cooldown within the cryostat we proceeded to study their low-energy excitations. To this end, experiments on the ILL time of flight (TOF) spectrometers IN5 [34] and Panther [35] were conducted. The goal of these experiments was to measure the dynamic structure function $S(q, \omega)$ over a large fraction of (q, ω)-space, which is made possible by the wide and complementary kinematic range of those two instruments. The analyzed data of these measurements will finally allow bench-marking density functional theory (DFT) and molecular dynamics (MD) simulations (elaborated in Ref. [19]). The four samples described in table 2 were measured under the configurations summarized in table 4. Additional measurements of each empty sample holder were conducted in order to subtract their contribution from the respective scattering signal. Vanadium standards were used to normalize our data to absolute units. Particular emphasis was put on selecting the geometries of the sample and the standard as identical as possible. By choosing thin hollow cylinders (with a wall thickness of $d = 0.05$ mm for the partly or fully protonated samples and $d = 0.1$ mm for the fully deuterated one) as sample geometry the multiple scattering was reduced to a minimum [36]. Note that unlike at D20, the sample containers are made from aluminum. The respective vanadium standard was cut from a sheet of the same thickness and placed inside the sample container without overlap. The vertical intensity profile of the beam was mapped out prior to the measurements. This allowed us to place the samples and the vanadium standards in the intensity distribution such as to minimize potential systematic effects due to possible small differences in their vertical extensions.

Instrument	Panther	IN5
Incident wavelength λ_i [Å]	1, 2	2, 3
Incident energies E_i [meV]	76, 19	19, 9
Temperature [K]	1.5	1.5

Table 4: Configurations for measurements conducted on ILL's TOF spectrometers. The overlap between incident wavelength allows crosschecking of the absolute units calibration.

Data reduction, analysis and visualization was done with the software package Mantid [37] [38]. Self-attenuation effects in the sample and the vanadium standard were calculated using a Monte Carlo simulation as described in [39]. The scattering of the sample environment was taken care of by subtracting measurements of the empty sample container in the cryostat. In order to normalize $S(q, \omega)$ to absolute units, the data is multiplied by a factor f, derived from the vanadium standard:

$$f = \frac{N_V \sigma_V}{N_S}. \quad (3)$$

Therein, N_V is the number density and σ_V the total scattering cross section of vanadium and N_S the number density of the sample. N_V and σ_V are well known tabulated values, while N_S can be calculated from the structure given above. Note that it is essential for this method to know exactly the

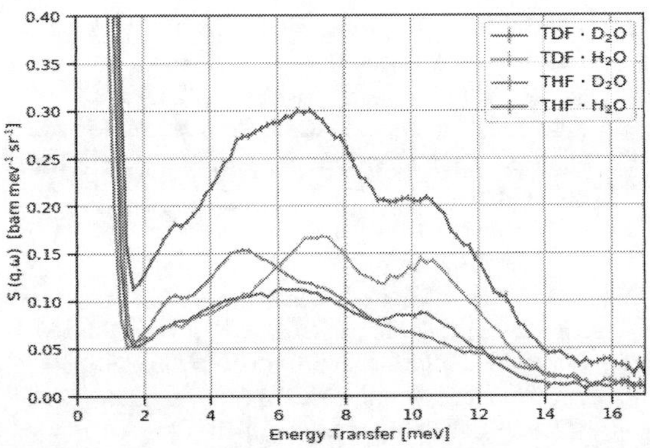

Figure 2: Constant q-slice at $q = (4 \pm 1)$ Å^{-1} through $S(q,\omega)$ for different deuterations and protonations of the THF-hydrate measured at **Panther** with an incident energy $E_i = 19\,\text{meV}$ at a temperature of $T = 1.5\,\text{K}$. The characteristic peaks at 7 meV and 10.5 meV of CS-II can be well discerned. Preliminary results, data is available under [40].

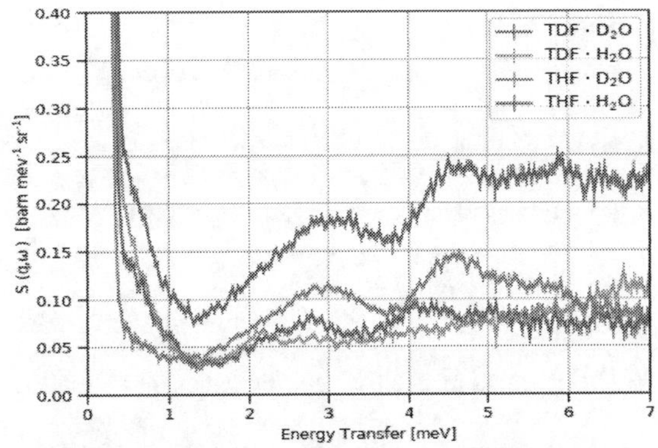

Figure 3: Constant q-slice at $q = (3.5 \pm 0.75)$ Å^{-1} through $S(q,\omega)$ for different deuterations and protonations of the THF-hydrate measured at **IN5** with an incident energy $E_i = 9\,\text{meV}$ at a temperature of $T = 1.5\,\text{K}$. The observed peaks are due to localized excitations of the THF molecules (see text). The shoulder of the elastic peak is not a feature of the sample but back scattering of the sample environment. Preliminary result, data is available under Ref. [28].

purity and geometry of both the sample and the vanadium standard.

2.3.1 *Low-energy excitations of THF-hydrates*

Figure 3 and figure 2 show two spectra for different THF hydrate samples measured at IN5 and Panther. These spectra are obtained by integration over a given q-range, providing an average of the coherent signal in this range.

In the investigated energy region the dynamics of the host structure are dominated by the translational modes of the H_2O or D_2O molecule. This leads to two characteristic peaks at about 7 meV and 10.5 meV of every CS-II hydrate structure (see e.g. Ref. [41],[42],[14],[16],[43]), which can also be observed in figure 2 for both the protonated and the deuterated host structures. The most pronounced peaks are visible in the TDF $\cdot$ H_2O sample (orange), as the contribution of the deuterated guest molecule is suppressed. The excitations of the guest molecule are located at lower energy and show very distinct peaks at about 2.9 meV and 4.7 meV, best visible in the THF $\cdot$ D_2O sample (green) in figure 3. These localized exctiations are particularly promising for moderation to the VCN range. As expected, they are shifted to slightly lower energies when substituting hydrogen with deuterium due to the increased mass and thus moment of inertia. This is the case for both translational modes of the host lattice and the excitations of the THF molecule, and consistent with phonon density of states (PDOS) computed by our collaborators [19].

3 Conclusions

VCN sources that deliver higher fluxes of very slow neutrons than current "standard" liquid deuterium sources could provide a powerful new tool for the neutron scattering community in the investigation of mesoscopic structures (e.g. in SANS), as well as tiny energy transfers (e.g. in NSE). This holds potential for bringing valuable advantages to the fields of soft condensed matter and structural biology. Furthermore they could play a significant role in increasing the sensitivity of particle physics experiments that employ beams of slow neutrons.

This work presented preliminary experimental results necessary for the development of VCN moderators using fully deuterated clathrate hydrates as a moderation material. In particular, a characterization of the structure and low-energy dynamics of THF-hydrates is elaborated. We have demonstrated that producing these hydrates is straightforward, and we were able to investigate localized, low-energy excitations using a method that involves contrast variation through partial or complete deuteration of the sample's constituents. The experimental determination of their dynamic structure factor $S(q,\omega)$ in absolute units provides an important benchmark for scattering kernels developed within the HighNESS project. Together, these works shall enable reliable simulations of future neutron moderators using fully deuterated clathrate hydrates as a moderating medium. Additionally, these experiments serve as a baseline for future experiments on binary clathrate hydrates with both THF and molecular oxygen as guest molecules ($17D_2O$:C_4D_8O:~$2O_2$).

4 Acknowledgement

We would like to thank the HighNESS collaboration, particularly Shuqi Xu from the European Spallation Source and Sara Isaline Laporte from the University of Milano-Bicocca for many inspiring discussions. This work was funded by the HighNESS project within the European Framework for Research and Innovation Horizon 2020, under grant agreement 951782.

References

[1] F. Mezei, Journal of Neutron Research **24**, 205 (2023)

[2] D.G. Phillips II, W.M. Snow et al., Physics Reports **612**, 1 (2016)

[3] F.M. Piegsa, Phys. Rev. C **88**, 045502 (2013)

[4] F.M. Piegsa, G. Pignol, Physical Review Letters **108**, 181801 (2012)

[5] R. Golub, J. Pendlebury, Physics Letters A **53**, 133 (1975)

[6] S. Degenkolb, P. Fierlinger, O. Zimmer, Journal of Neutron Research **24**, 123 (2023)

[7] I.I. Gurevich, *Low-energy neutron physics* (North-Holland, Amsterdam, 1968)

[8] D. Evans, Cryogenics **35**, 763 (1995), Nonmetallic Materials and Composites at Low Temperature-VII

[9] E. Kulagin, S. Kulikov, V. Melikhov, E. Shabalin, Nuclear Instruments and Methods in Physics Research B **215**, 181 (2004)

[10] M.H. Parajon, E. Abad, F. Bermejo, Physics Procedia **60**, 74 (2014)

[11] G. Škoro, G. Romanelli, S. Rudić, S. Lilley, F. Fernandez-Alonso, EPJ Web of Conferences **239**, 17008 (2020)

[12] I. Natkaniec, K. Holderna-Natkaniec, J. Kalus, Physica B: Condensed Matter **350**, E651 (2004)

[13] A.K. Sum, C.A. Koh, E.D. Sloan, Industrial & Engineering Chemistry Research **48**, 7457 (2009)

[14] H. Conrad, W.F. Kuhs, K. Nünighoff, C. Pohl, M. Prager, W. Schweika, Physica B: Condensed Matter **350**, E647 (2004)

[15] V. Ananiev, et al., Nuclear Instruments and Methods in Physics Research Section B: Beam Interactions with Materials and Atoms **320**, 70 (2014)

[16] M. Celli, D. Colognesi, L. Ulivi, M. Zoppi, A.J. Ramirez-Cuesta, Journal of Physics: Conference Series **340**, 012051 (2012)

[17] O. Zimmer, Physical Review C **93**, 035503 (2016)

[18] V. Santoro et al., *Development of a high intensity neutron source at the european spallation source: The highness project* (2022), `https://doi.org/10.48550/arXiv.2204.04051`

[19] S. Xu, S.I. Laporte et al., *Theoretical calculations of neutron scattering cross sections for tetrahydrofuran-containing clathrate hydrates at low temperature* (Garching, Germany, 2023)

[20] R. Macfarlane, D.W. Muir, R.M. Boicourt, A.C. Kahler, III, J.L. Conlin (2017)

[21] X.X. Cai, T. Kittelmann, Computer Physics Communications **246**, 106851 (2020)

[22] P. Willendrup, E. Farhi, K. Lefmann, Physica B: Condensed Matter **350**, E735 (2004)

[23] P.K. Romano, N.E. Horelik, B.R. Herman, A.G. Nelson, B. Forget, K. Smith, Annals of Nuclear Energy **82**, 90 (2015), Joint International Conference on Supercomputing in Nuclear Applications and Monte Carlo 2013

[24] J.T. Goorley, M.R. James et al., Tech. Rep. LA-UR-13-22934, 1086758 (2013)

[25] T. Sato, Y. Iwamoto, S. Hashimoto, T. Ogawa, T. Furuta, S. Ichiro Abe, T. Kai, P.E. Tsai, N. Matsuda, H. Iwase et al., Journal of Nuclear Science and Technology **55**, 684 (2018)

[26] S.R. Gough, D.W. Davidson, Canadian Journal of Chemistry **49**, 2691 (1971)

[27] E.D. Sloan, C.A. Koh, *Clathrate Hydrates of Natural Gases.*, 3rd edn. (CRC Press, Taylor & Francis, Berkeley, California, 2008), ISBN 978-0-8493-9078

[28] O. Zimmer et al., `https://doi.ill.fr/10.5291/ILL-DATA.1-10-42` (2020)

[29] J. Rodríguez-Carvajal, Physica B: Condensed Matter **192**, 55 (1993)

[30] J. Rodriguez-Carvajal, IUCr Newsl. **26** (2001)

[31] R. Young, *The Rietveld method*, International Union of Crystallography. Monographs on Crystallography 5 (Oxford Univ. Press, Oxford [u.a.], 1993), ISBN 0198555776

[32] R.J. Hill, C.J. Howard, Journal of Applied Crystallography **20**, 467 (1987)

[33] J. Rodriguez-Carvajal, *An introduction into the program fullprof 2000 (version july 2001)* (2001)

[34] J. Ollivier, M. Zbiri, J. Halbwachs, *Instrument layout: Disk chopper time-of-flight spectrometer IN5*, `https://www.ill.eu/users/instruments/instruments-list/in5`

[35] B. Fak, M.M. Koza, O. Meulien, *Instrument layout: Thermal neutron time-of-flight spectrometer PANTHER*, `https://www.ill.eu/users/instruments/instruments-list/panther`

[36] J. Wuttke, Physica B: Condensed Matter **266**, 112 (1999)

[37] Mantid, *Manipulation and analysis toolkit for instrument data. mantid project*, `http://dx.doi.org/10.5286/SOFTWARE/MANTID` (2013)

[38] O. Arnold, J. Bilheux, J. Borreguero, A. Buts, S. Campbell, L. Chapon, M. Doucet, N. Draper, R.F. Leal, M. Gigg et al., Nuclear Instruments and Methods in Physics Research A **764**, 156 (2014)

[39] Mantid, *Mantid documentation. algorithm contents*, `https://docs.mantidproject.org/nightly/algorithms/index.html` (2023)

[40] O. Zimmer et al., `https://doi.ill.fr/10.5291/ILL-DATA.1-10-49` (2021)

[41] B. Chazallon, H. Itoh, M. Koza, W.F. Kuhs, H. Schober, Phys. Chem. Chem. Phys. **4**, 4809 (2002)

[42] H. Schober, H. Itoh, A. Klapproth, V. Chihaia, W. Kuhs, The European Physical Journal E, Soft matter (Print) **12**, 41 (2003)

[43] D. Broseta, L. Ruffine, A. Desmedt, eds., *Gas Hydrates 1: Fundamentals, Characterization and Modeling*, 1st edn. (John Wiley & Sons, Inc., Hoboken, NJ, USA, 2017), ISBN 978-1-119-33268-8

Magnetic properties of highly ordered single crystals with layered $YBaCuFeO_5$ structure

Arnau Romaguera[1], *Xiaodong* Zhang[1], *Ruyong* Li[1], *Oscar* Fabelo[2] and *José Luis* García-Muñoz[1*]

[1]Institut de Ciència de Materials de Barcelona, ICMAB-CSIC, Campus UAB, 08193 Bellaterra, Spain
[2]Institut Laue-Langevin, 71 Avenue des Martyrs, 38042 Grenoble, France

Abstract. In the layered perovskites $YBaCuFeO_5$ (YBCFO) magnetic frustration stabilizes incommensurate spiral order up to unexpectedly high temperatures (T_S). The level of frustration and hence T_S depend directly on the fraction (n_d) of B-site (Cu/Fe) chemical disorder in the samples. We report a neutron investigation on the magnetic transitions in two $YBaCuFeO_5$ single crystals (one pure and the second slightly doped with Mn), grown by floating-zone methods, that present very low Fe/Cu cation disorder ($n_d \to 0$). The low density of antiferromagnetic Fe_2O_9 bipyramid units in these crystals precludes the stabilization of a frustrated topological network, which is at the origin of the non-collinear (helicoidal) magnetic order. Instead, the ground state is substituted by two competing commensurate collinear magnetic phases, the majority one ($\mathbf{k}_1$) attributed to the perfectly ordered YBCFO perovskite ($n_d=0$). A second type of collinear magnetic domains ($\mathbf{k}_3$) develop for samples with disorder below a critical threshold ($0<n_d<n_0$). Only in YBCFO samples with Cu/Fe disorder above the critical threshold a long-range helicoidal or spiral magnetic state can be formed.

1 Introduction

The materials presenting magnetic chirality form a rare class of frustrated compounds that attract a lot of attention owing to their unique technologically important properties. Among the properties realized in systems with chiral magnetic orders the multiferroicity is one of the most attractive. In frustrated magnets ordered spiral phases are particularly promising because the noncollinear spiral magnetic order breaks space-inversion symmetry and can induce spontaneous polarization through inverse Dzyaloshinsky-Moriya or spin-current mechanisms [1]. A common origin of polarization and spiral order guarantees a substantial coupling "by construction" in cycloidal multiferroic materials. Unfortunately, their low magnetic ordering temperatures (typically <50 K) critically restrict their potential use in spintronic and low-power magnetoelectric devices.

Exceptionally, the layered perovskites $RBaCuFeO_5$ have been reported to display spiral magnetic order (T_S) at unexpectedly high temperatures [2-6]. Surprisingly, there is no any apparent source of frustration in an ordered A- and B- site layered $YBaCuFeO_5$ structure. B-site metals form $CuFeO_9$ bilayers of corner-sharing $Cu^{2+}O_5$ and $Fe^{3+}O_5$ square pyramids that extend parallel to the *ab* plane. The bilayers formed by bipyramids are separated by Y^{3+} layers, and Ba^{2+} ions site within the bilayer spacing. Samples are prepared with partial Fe/Cu order in the structure, and the absence of perfect B-site cation order is at the origin of magnetic frustration [3-6]. In addition YBCFO exhibits an extraordinary

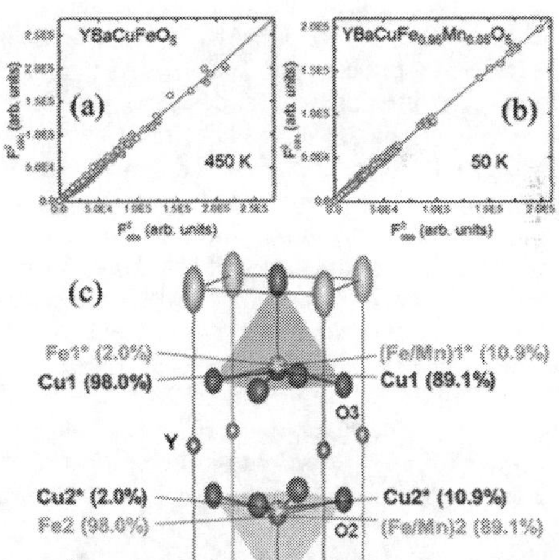

Fig. 1. Agreement plots of the nuclear structure refinements (calculated structure factors plotted against the integrated SCND nuclear reflections collected in D9@ILL) on the two single crystals of composition (a) $YBaCuFeO_5$ (450 K) and (b) $YBaCuFe_{0.95}Mn_{0.05}O_5$ (50 K). (c) Scheme of the refined *P4mm* unit cell of the YBCFO crystal showing the refined amount of Fe/Cu disorder. The color of each pyramid corresponds to that of the dominant cation in it (blue: Cu; brown: Fe).

* Corresponding author: garcia.munoz@icmab.es

tunability of its spiral ordering temperature [4,5] (T_S=T_{N2}), which can be increased by 180 K (up to ~366 K [5]), by increasing the Cu/Fe chemical disorder in the bipyramids.

A collinear commensurate (**k**$_1$=[1/2, 1/2, 1/2]) antiferromagnetic (AF) phase below T_{N1} = 440 K is followed by an incommensurate (ICM, **k**$_2$=[1/2, 1/2, 1/2±q]) spiral magnetic order (T_S=T_{N2} ~ 230 K). Coinciding with the incommensurate magnetic order, the reported electrical polarization (*P*) reached 0.4 $\mu C/cm^2$ in some polycrystalline specimens [2,3].

The need to extend these studies to single crystals led us to grow $YBaCuFeO_5$ crystals by floating zone. In this work the magnetic properties of two different crystals of compositions $YBaCuFeO_5$ and $YBaCuFe_{0.95}Mn_{0.05}O_5$ are here reported. Remarkable differences are found when comparing the crystals with their polycrystalline counterparts. Typically, B-site disorder in single crystals and powder samples is very different, and confirm the relevance of Cu/Fe disorder for the properties of these layered perovskites. This study has allowed extending the investigation on this layered structure to the regime of low Fe/Cu disorder.

2 Experimental

Single crystalline samples were grown in our Crystal Growth Lab. at the ICMAB by a modified traveling solvent floating zone (TSFZ) method from previously sintered polycrystalline rods of $YBaCuFeO_5$ and $YBaCuFe_{0.95}Mn_{0.05}O_5$. Hereafter named as YBCFO and YBCFO-5%Mn, respectively. The powders to make the rods were previously prepared as described in refs. [5,9]. A solvent consisting of CuO with 2 wt% B_2O_3 was used for the crystal growth of these incongruently melting compositions [7]. Further details on the growing conditions can be found in refs. [8,9]. In the ingot the c-axis of the YBCFO structure was perpendicular to the FZ growth direction (parallel to (110)). The elemental composition of the crystals was assessed by Inductively Coupled Plasma Optical Emission Spectroscopy (ICP-OES) that confirmed no deviations from the nominal molar ratios. Their quality and orientation was characterized by neutron Laue images and their stoichiometry, including the oxygen content, was further confirmed by single crystal neutron diffraction.

Neutron diffraction experiments on both crystals were performed at the high-flux reactor of the Institut Laue Langevin (ILL, Grenoble, France). Three different single-crystal instruments were used: Orient Express (to prove the crystal orientation), CYCLOPS (Laue single-crystal diffractometer) and D9 (hot-neutrons four-circle diffractometer, using λ=0.837 Å). The program RACER was used to integrate the omega- and omega-2theta-scans and to correct them for the Lorentz factor [11]. Data reduction and neutron refinements were made using the Fullprof package of programs [12]. Additionally the two-axis D20 diffractometer was also used (λ=2.41 Å) to characterized a powdered sample.

Table 1. Crystal structure of the single crystals and agreement factors obtained from the refinement of the D9@ILL single crystal neutron intensities (λ=0.837 Å). (*: minority fraction; n_d (disorder=Occ (Fe1*/Cu2*)). The coordinates of each metal (M: Cu or Fe) are constrained by z(M1)+ z(M2)=1).

P4mm		$YBaCuFeO_5$ (450 K)	$YBaCuFe_{0.95}Mn_{0.05}O_5$ (50 K)
$a = b$ (Å)		3.8815 (6)	3.8699 (3)
c (Å)		7.6839 (1)	7.6359 (3)
z (Y)	1*a* (00z)	0.4979 (3)	0.4979 (3)
z (Ba)	1*a* (00z)	0	0
z (Cu1)	1*b* (½ ½ z)	0.7221 (3)	0.7326 (8)
z (Fe2)	1*b* (½ ½ z)	0.2570 (3)	0.2657 (8)
z (Fe1*)	1*b* (½ ½ z)	0.7430 (3)	0.7343 (8)
z (Cu2*)	1*b* (½ ½ z)	0.2779 (3)	0.2674 (8)
z (O1)	1*b* (½ ½ z)	0.0019 (8)	-0.002 (5)
z (O2)	1*b* (½ 0 z)	0.3149 (6)	0.312 (4)
z (O3)	1*b* (½ 0 z)	0.6814 (7)	0.682 (4)
Chemical order			
n_d = Occ (Fe1*/Cu2*)		0.020 (22)	0.109 (26)
1- n_d = Occ (Cu1/Fe2)		0.980 (22)	0.891 (26)
χ^2		2.27	*1.12*
R_F		7.83	*2.95*
$R_F{}^2$		8.32	*4.36*

3 Results

3.1 Structural characterization and Fe/Cu disorder

The structure of the two single domain crystals was studied using hot-neutrons (λ=0.836 Å) at the D9 four-circle neutron diffractometer. From data collected in D9 we determined the atomic positions in the non-centrosymmetric *P4mm* structure and partial Cu/Fe occupancies in the two pyramids of the unit cell. Extinction corrections were applied following the Becker-Coppens model. Fig. 1 displays the observed and calculated squared structure factors for the nuclear structure of the two crystals, refined from 537 nuclear reflections (295 independent). Among the structural information shown in Table 1 for both crystals, the degree of Fe/Cu cationic disorder (n_d) is crucial for the properties of this system. The magnetic *3d* metals (M) occupy the two pyramids of the unit cell [sites 1*b* (1/2 1/2 z) with z ≈ 0.74 and z ≈ 0.27 for, respectively, the upper and the lower pyramids in the Fig.1)]. The z-coordinates of the two positions that a given metal M (either Fe or Cu) can occupy in the cell (M1 and M2) were constrained in the form z(M1)+z(M2)=1. The upper (lower) pyramid in Fig. 1c is mainly occupied by Cu1 (Fe2). Therefore, the improper occupancy associated to the positions Fe1* and Cu2* is a measure of the disorder. The most important observation is the low level of Fe/Cu cation disorder (n_d) in the crystals: n_d=0.020(22) and 0.109(26) for YBCFO and YBCFO-5%Mn, respectively. It should be noted that the cationic disorder found in powder samples is always much higher (above 30 %). So, the very low disorder in our single crystals (which is also disclosed in Fig. 1c) need to be compared with $n_d \approx 0.32$, the typical cationic disorder found in powder samples cooled at 300°C/h after the last annealing [5]. A first central finding is, hence, that the complex TSFZ technique used to fabricate the crystals clearly favors the ordering of B-site cations. On one hand, this entails a serious drawback to dispose of disordered crystals with cycloidal order at ambient conditions. On the other hand, the study of these crystals allows us to explore highly ordered zones of the T-n_d phase diagram that are very difficult to access with samples prepared by ceramic methods in polycrystalline form. As an example, the lowest disorder reported in Ref. [5] in powder samples was $n_d \approx 0.29(3)$ applying a very slow controlled cooling rate of 10 °C/h.

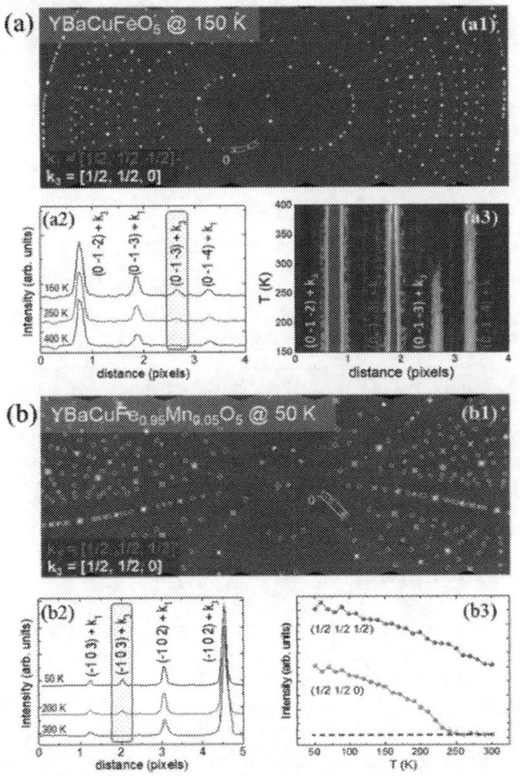

Fig. 2. Sections of the neutron Laue images collected at CYCLOPS@ILL at low temperature on the two single crystals, showing the indexation of nuclear reflections (in white) and the magnetic ones with k_1=(1/2,1/2,1/2) and k_3=(1/2,1/2,0) (in red and yellow, respectively; some magnetic reflections have also nuclear contribution). (a1) $YBaCuFeO_5$ (150 K) and (b1) $YBaCuFe_{0.95}Mn_{0.05}O_5$ (50 K). Panels (a2) and (b2) show a comparison of the integrated profiles along selected regions of the area detectors at three characteristic temperatures. The evolution with temperature of the magnetic phases is depicted in (a3) as a 2D contour map of the integrated Laue profiles, and in (b3) by the evolution of the integrated intensity of (1/2,1/2,1/2) and (1/2,1/2,0) reflections collected in D9@ILL.

3.2 Magnetic characterization by single-crystal neutron diffraction

With regards to the magnetic properties, the central observation is the lack of incommensurate spiral order in the crystals. Fig. 2 shows the temperature dependent magnetic information obtained on the neutron Laue diffractometer CYCLOPS (ILL), from measurements in the range 50-400 K. As shown in the indexed Laue patterns (Figs. 2a1 and 2b1) the magnetic phases present in the crystals are all commensurate. The antiferromagnetic phases with propagation vectors $\mathbf{k}_1$=(½, ½, ½) (AF1) and $\mathbf{k}_3$=(½, ½, 0) (AF3) are found to coexist at low temperatures in both crystals. Warming the crystals from the base temperature, the evolution of the integrated profiles for some selected characteristic magnetic reflections are displayed in Figs. 2a (YBCFO) and 2b (YBCFO-5%Mn). From the observed intensities, the collinear AF1 order is in both crystals the majority magnetic phase and the AF3 order is the minority one. Furthermore we observed that, under heating, the former (AF1) is always more stable than the AF3 configuration. The respective transition temperatures are T_{N1}=435 K and T_{N3}=290 K for YBCFO, and T_{N1}=420 K and T_{N3}=242 K for YBCFO-5%Mn.

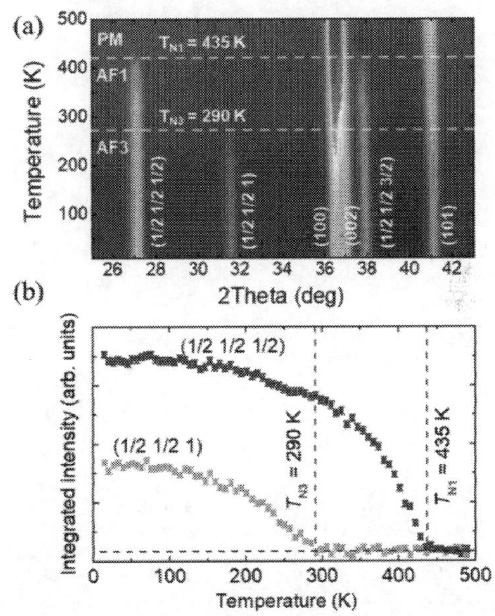

Fig. 3. Evolution of magnetic order in a powdered sample of the highly ordered YBCFO single-crystal from neutron diffraction (D20@ILL). (a) Contour map showing the 2*Theta*-T projection of the temperature dependence for the NPD intensities at the low Q-range. (b) Thermal evolution of the integrated intensities of the (1/2,1/2,1/2) and (1/2,1/2,1) magnetic reflections associated to the AF1 ($\mathbf{k}_1$) and AF3 ($\mathbf{k}_3$) phases, respectively.

 https://doi.org/10.1051/epjconf/202328605005

Additionally, in YBCFO we also performed neutron powder diffraction (NPD) measurements on a powdered sample that we obtained by crushing a portion of our large ingot. We used D20 to collect NPD patterns (λ=2.41 Å) within the temperature range 10 to 500 K. The evolution in this interval of the main magnetic reflections in YBCFO obtained by NPD measurements is illustrated in Figs. 3a,b. The neutron Rietveld refinement at 10 K for the powdered YBCFO sample is exposed in Fig. 4a.

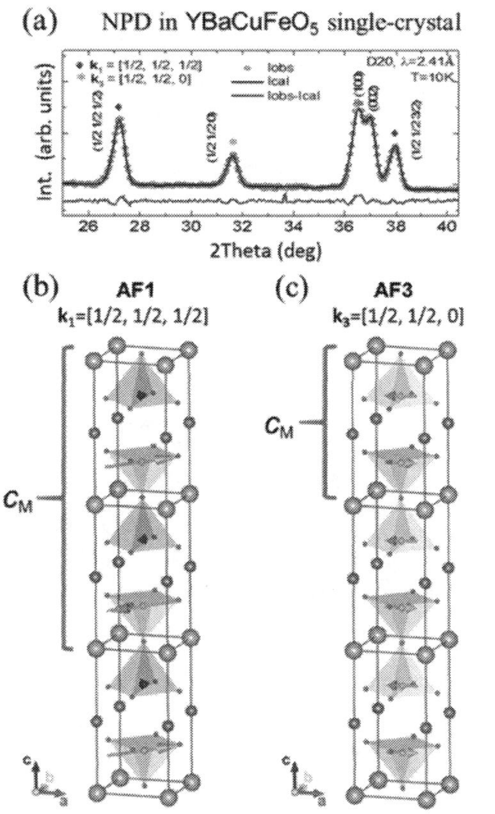

Fig. 4. (a) Rietveld refinement (black curve) of the D20@ILL (λ=2.41 Å) neutron diffraction pattern (red circles) at 10 K for the powdered YBCFO single crystal with very high Fe/Cu order (blue diamonds and orange stars indicate AF1 and AF3 magnetic reflections, respectively). Projections of the refined collinear magnetic orders with propagation vectors (b) $\mathbf{k}_1$=(1/2,1/2,1/2) and (c) $\mathbf{k}_3$=(1/2,1/2,0). The different periodicity along-c is illustrated with the help of the c_M parameter for each magnetic unit cell. For the sake of clarity, only the majoritarian magnetic atom (M) is shown in the pyramids (blue: Cu; brown: Fe)

In the T-n_d phase diagram showed in Fig. 5 the magnetic transition temperatures are plotted as a function of Cu/Fe disorder (n_d). In this phase diagram we have gathered the information from our two highly ordered crystals and the magnetic phase transitions reported in Ref. [5] for a set of more disordered YBCFO polycrystalline samples with increasing cationic disorder that were originally prepared in powder form by applying different cooling rates. To be highlighted is the existence of three different regimes in the $YBaCuFeO_5$ phase diagram versus disorder and the fact that the non-collinear (spiral) order does not form if the level of Fe/Cu disorder in the structure is too high (beyond the triple point) or too low (below a critical threshold, n_0).

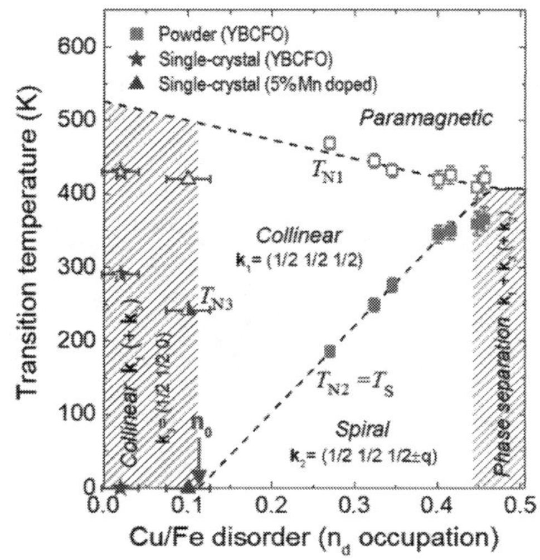

Fig. 5. Magnetic phase diagram versus Fe/Cu cation disorder for YBCFO. Dashed lines describe the phase boundaries from the onset of the magnetic phases according to neutron diffraction data, obtained on YBCFO powder samples (red squares, from Ref. [5]). The n_d-T values from the two highly ordered single crystals described in this work are also shown (blue symbols). The two crystals fall within the $0<n_d<n_0$ region, below the critical threshold of disorder for developing spiral magnetic order in this structure. Empty and filled points correspond to T_{N1} and T_{N2}=T_S, respectively, and the half-filled ones correspond to T_{N3}.

Fig. 5 reveals that the two crystals investigated in this work fall into the first phase separated region (low disorder), and consequently must be regarded as representative of the system when $n_d<n_0$. The dashed lines in Fig. 5 are lineal fits to the experimental points. A direct comparison between the ordering temperatures in the two crystals is difficult due to their slightly different composition. However the magnetic evolutions depicted in Figs. 2 and 3 for both samples strongly suggest that the AF1 and AF3 collinear magnetic phases appear in distinct regions of the crystal. The AF3 regions only order more than 100 K below the formation of AF1 magnetic domains. But the magnetic regions with $\mathbf{k}_3$=(1/2, 1/2, 0) do not substitute to the AF1 phase but rather coexists with it. It is of interest to note that, as shown in Fig. 4c, the coupling between the two spins inside the bipyramids is antiferromagnetic in the AF3 regions. In contrast to the AF1 domains, where the magnetic atoms in $FeCuO_9$ bipyramids exhibit the expected ferromagnetic coupling dictated by the positive sign of the Fe^{3+}-O-Cu^{2+} exchange (see Fig. 4b). Therefore, the second type of domains (with collinear AF3 order and $\mathbf{k}_3$=(1/2 1/2 0) propagation vector) are

very likely associated to zones in the crystal with higher cationic disorder where the Fe-Cu ferromagnetic coupling inside the bipyramids has been substituted by antiferromagnetic Fe_2O_9 and Cu_2O_9 bipyramids. From our refinement of the NPD pattern for YBCFO shown in Fig. 4a equal average ordered moments were found in the AF3 bipyramids ((~43%) m(M1)≈m(M2)≈1.12(5) μ_B at 10 K, parallel to *ab* plane). In the AF1 phase (~57%) we found m(M2)=2.65(3) μ_B antiferromagnetically coupled to m(M1)=0.32(4) μ_B, these moments separated 11(7)° from the *ab* plane (χ^2=4.6 , R_B=1.6, R_M(AF1)=4.3, R_M(AF3)=5.1).

4 Conclusions

In the layered perovskites with YBCFO structure the level of magnetic frustration and hence T_S depend directly on the fraction n_d of Cu/Fe cation disorder in the samples. Two single crystalline samples have been grown by TSFZ methods that present very low levels of Fe/Cu disorder: n_d=0.020(22) and 0.109(26). Their study by neutron diffraction has allowed extending our comprehension of the magnetic properties of these systems to the low-disorder range.

From the study of the two crystals several relevant conclusions can be drawn. First, the lack of (noncollinear) helicoidal or spiral order in these layered perovskites with low cationic disorder at the B-site. Second, although the study gives support to the "*order by disorder*" theoretical framework in this system [13,14], it also reveals that there is a critical threshold of disorder (n_0), below which a long-range spiral magnetic state cannot be formed. In this structure the critical threshold must be greater than ~10(3)%, the disorder (n_d) in our less ordered crystal. Fig. 5 confirms that this conclusion (at odd with some previous assumptions in this system) is in agreement with the linear extrapolation of the experimental (n_d,T_S) points (filled red points in Fig. 5) corresponding to YBCFO powder samples reported earlier in ref. [5]. Dashed lines in Fig. 5 are simply guides to the eye. Third, a phase separated or inhomogeneous region emerges in the phase diagram when disorder is too low. Below the threshold n_0 an insufficient density of strongly antiferromagnetic Fe_2O_9 bipyramids (named "impurity bonds" in Ref. [13] and having a central role for generating frustration) cannot stabilize long-ranged non-collinear domains (spiral order). Nevertheless in the small fraction of non-mixed M_2O_9 bipyramid units (M_2=Fe_2 or Cu_2) the M-O-M exchange is antiferromagnetic and this favors the emergence of magnetic domains with wave vector $\mathbf{k}_3$=(1/2, 1/2, 0) inside the AF1 matrix ($\mathbf{k}_1$) dominated by ferromagnetic $FeCuO_9$ bipyramids.

Due to the difficulties preparing powder YBCFO samples with a very high Fe/Cu cationic order, the high-order region of the phase diagram was until now almost unexplored. The tendency in single crystals to adopt highly ordered configurations allows overcome such limitations.

Acknowledgments.- Financial support from the Ministerio de Ciencia, Innovación y Universidades (MINCIU), Projects No. PID2021-124734OB-C22 and RTI2018-098537-B-C21, cofunded by ERDF from EU, and "Severo Ochoa" Programme for Centres of Excellence in R&D [FUNFUTURE (CEX2019-000917-S)]. R.L's work was done as a part of the Ph.D program in Materials Science at Universitat Autònoma de Barcelona. We also acknowledge the Institut Laue-Langevin (ILL) for provision of beam time (Easy-445 and Easy-447, Test-2882 and dois:10.5291/ILL-DATA.5-14-273, 10.5291/ILL-DATA.5-14-277 and 10.5291/ILL-DATA.5-31-2680 5-41-1153).

References

1. I. A. Sergienko and E. Dagotto, Phys. Rev. B **73**, 094434 (2006)
2. B. Kundys, A. Maignan, and C. Simon, Appl. Phys. Lett. **94**, 072506 (2009)
3. M. Morin *et al.*, Phys. Rev. B **91**, 064408 (2015)
4. M. Morin, E. Canévet, A. Raynaud, M. Bartkowiak, D. Sheptyakov, V. Ban, M. Kenzelmann, E. Pomjakushina, K. Conder, and M. Medarde, Nat. Commun. **7**, 13758 (2016)
5. A. Romaguera, X. Zhang, O. Fabelo, F. Fauth, J. Blasco and J. L. García-Muñoz, Phys. Rev. Research **4**, 043188 (2022)
6. T. Shang, E. Canévet, M. Morin, D. Sheptyakov, M. Teresa Fernández-Díaz, E. Pomjakushina, M. Medarde, M. T. Fernández-Díaz, E. Pomjakushina, and M. Medarde, Sci. Adv. **4**, eaau6386 (2018)
7. Y.C. Lai *et al.*, J. Phys. Condens. Matter **29**, 145801 (2017)
8. A. Romaguera Camps, PhD Thesis. Universitat Autónoma de Barcelona. 2023
9. X. Zhang, A. Romaguera, F. Sandiumenge, O. Fabelo, J. Blasco, J. Herrero-Martín, and J. L. García-Muñoz, J. Magn. Magn. Mater. **551**, 169165 (2022).
10. X. Zhang, A. Romaguera, O. Fabelo, F. Fauth, J. Herrero-Martín, and J. L. García-Muñoz, Acta Mater. **206**, 116608 (2021)
11. M. S. Lehmann, Acta Cryst. A. **30**, 580 (1974)
12. J. Rodríguez-Carvajal, Phys. B Phys. Condens. Matter **192**, 55 (1993)
13. A. Scaramucci, H. Shinaoka, M. V. Mostovoy, M. Müller, C. Mudry, M. Troyer, and N. A. Spaldin, Phys. Rev. X **8**, 011005 (2018).
14. A. Scaramucci, H. Shinaoka, M. V. Mostovoy, R. Lin, C. Mudry, and M. Müller, Phys. Rev. Res. **2**, 013273 (2020).

Influence of nonmagnetic cation substitution on magnetic order temperature in Y-type hexaferrites: $Ba_{0.5}Sr_{1.5}Zn_2Fe_{12}O_{22}$ and $Ba_{0.5}Sr_{1.5}Zn_2Al_{0.08}Fe_{11.92}O_{22}$

Kiril Krezhov[1*], *Tatyana* Koutzarova[1], *Borislava* Georgieva[1], *Svetoslav* Kolev[1,2], *Daniela* Kovacheva[3], *Benedicte* Vertruyen[4], *Raphael* Closset[4], *Lan Maria* Tran[5], *Michał* Babij[5] and *Anatoliy* Senyshyn[6]

1Institute of Electronics, Bulgarian Academy of Sciences, 72 Tsarigradsko Chaussee, 1784 Sofia, Bulgaria

2Neofit Rilski South-Western University, 66 Ivan Mihailov Str.,2700 Blagoevgrad, Bulgaria

3Institute of General and Inorganic Chemistry, Bulgarian Academy of Sciences, Acad. Georgi Bonchev str., 1113 Sofia, Bulgaria

4Greenmat, Chemistry Department, University of Liege, 11 allée du 6 août, 4000 Liège, Belgium

5Institute of Low Temperature and Structure Research, Polish Academy of Sciences, Ul. Okólna 2, 50-422 Wroclaw, Poland

6TUZEFRM Forschungs-Neutronenquelle, Heinz Maier-Leibnitz (FRM II) 85748 Garching b. München, Lichtenbergstr. 1, Germany

Abstract. With a view to the proven multiferroic properties of the title representatives of the Y-type hexaferrites as single crystals we present here a comparative study of their microstructure and magnetic properties for powders. The polycrystalline materials of $Ba_{0.5}Sr_{1.5}Zn_2Fe_{12}O_{22}$ (S1) and its doped with aluminium derivative $Ba_{0.5}Sr_{1.5}Zn_2Al_{0.08}Fe_{11.92}O_{22}$ (S2) were synthesized by two different techniques: citric-acid sol-gel auto-combustion and sonochemical co-precipitation. Their atomic-level structure determination was checked by X-ray and neutron diffraction, electron diffraction and imaging methods, and their physical characterization was carried out by magnetometry. SEM images show that the microstructure strongly depends on the method of synthesis: non-uniform hexagonal grans are formed in case of auto-combustion whereas particles of nearly perfect hexagonal shape are observed in case of sonochemical co-precipitation. With lowering the temperature to 5 K several magnetic phase transitions were observed, more clearly pronounced in S2 powders. Neutron diffraction data in vicinity and above room temperature help in revealing the effect of magnetic dilution on the observed magnetic properties.

*Corresponding author: kiril.krezhov@gmail.com

1 Introduction

Hexaferrites are an important class of magnetic oxides with diverse applications, the most attractive of which have been in permanent magnets [1], data storage and energy conversion [2,3]. Particular performance of their properties was emphasized for devices operating at microwave frequencies to mitigate electromagnetic pollution, and to enrich the stealth technology [4]. Their rich crystal chemistry and a complex structural topology made them the subject of interest in solid state science since their discovery in the 1950s. The search for multiferroics, materials in which at least two of the ferroic order parameters, including ferroelectricity, ferromagnetism and ferroelasticity, coexist and interact has added momentum and extended the scope because of high potential for new applications [5,6]. For instance, if the coupling between the electric polarization P and magnetization M is sufficiently strong one might control the magnetic phase by applying an external electric field (E), as well as to influence the electric phase by applying an external magnetic field (H). This cross-coupling effect (magneto-electric effect) gives additional functionality to materials and opens up new possibilities in spintronics [6-9]. However, in most multiferroics, the sizeable magneto-electric (ME) effect usually occurs in the low temperature range and at applied external fields that are too high to be of practical use. Also, from the practical point of view of using multiferroics in multifunctional devices, it is important to control the electrical polarization or dielectric constant using weak magnetic fields or to manipulate the magnetization by an E field at temperatures near and above room temperature.

The discovery of multiferroic behaviour of particular single-phase representatives of nearly all main types of the hexaferrite family have substantially deepened the research efforts in the study of properties of single crystals with M-, Y -, Z- [10-17], and U -type structure [18]. The measured value of magnetic field driven electric polarization was not very large, amounting generally up to several tens of $\mu C/m^2$. Kitagawa et al. [19] presented the low-field ME effect in $Sr_3Co_2Fe_{24}O_{41}$ polycrystalline ceramics (representative of Z-type hexaferrites) sintered in oxygen atmosphere, where the electric polarization was $P_s \sim 1\times10^{-3}$ $\mu C/cm^2$ at 0.25 T. In the Y-type hexaferrite $Ba_2Mg_2Fe_{12}O_{22}$, Ishiwata et al. [15] demonstrated that a low magnetic field of 300 Oe is sufficient to reverse P indicating a well pronounced ME effect. With proven significantly large ME coupling in the near-ambient temperature range are the Y-type hexaferrites $Ba_{0.5}Sr_{1.5}Zn_2(Fe_{1-x}Al_x)_{12}O_{22}$ (x=0 and 0. 08). First, Kimura et al. [14] reported H-induced ferroelectric (FE) polarization in the undoped compound x=0 in a magnetic field of 0.1 T. Later, Chun et al. [20] measured in the magnetically diluted compound x=0.08 the largest ME coefficient in single-phase hexaferrites: α_H = 20,000 ps/m. It should be noted that a large ME coefficient is observed in $Ba_{0.4}Sr_{1.6}Mg_2Fe_{12}O_{22}$ single crystal with a value of 33,000 ps/m at 10 K [21], which is believed to be due to a successful tuning of the spin cone symmetry in Y-type hexaferrites.

The atomic and magnetic order of the Y-type hexaferrite multiferroics has been well documented for single crystals. The results were elaborated by magnetic, diffraction, and Mössbauer studies. As commented in the recent review [22], due to the competition of exchange interactions and magnetic anisotropies several specific magnetic phases were determined to stabilize in particular temperature ranges, which can further be modified by applied magnetic field. Some of these phases exhibit room temperature spin-driven ferroelectricity [23] that according to the classification of Khomskii [24] corresponds to a type-II magnetoelectric coupling originated from complex magnetic orders.

Xu et al [25] investigated the substitution effects for polycrystalline $Ba_{0.5}Sr_{1.5}Zn_2(Fe_{1-x}Al_x)_{12}O_{22}$ with x=0, 0.04, 0.08 and 0.12 obtained by the conventional ceramic method, but their results were affected by the clear presence of Z and M-type phases in addition to the main Y- phase. To a large extent, these by-products of the synthesis with complex structure and magnetic behaviour accompany the very narrow temperature range for successful Y-phase formation, starting at 900°C and ending at 1200°C [2].

The subject of the present study are powder materials with x=0 and 0.08 prepared by soft chemistry routes. We summarize here the main results from our comparative studies of the structural and magnetic properties complemented with range-limited powder neutron diffraction measurements in the near and above room temperature range.

2 Experimental part

We prepared the $Ba_{0.5}Sr_{1.5}Zn_2Fe_{12}O_{22}$ (S1) and $Ba_{0.5}Sr_{1.5}Zn_2Al_{0.08}Fe_{11.92}O_{22}$ (S2) by citrate sol-gel auto-combustion method and sonochemical co-precipitation followed by thermal annealing. The resulting powder materials were characterized by X-ray diffraction (XRD) and neutron diffraction (ND) complemented by scanning electron microscopy (SEM) and magnetic measurements.

The metal nitrates ($Ba(NO_3)_2$ (99%), $Sr(NO_3)_2$ (99%), $Zn(NO_3)_{2x}6H_2O$ (98%), $Al(NO_3)_{3x}9H_2O$ (98%) and $Fe(NO_3)_{3x}9H_2O$ (ACS, 98.0-101.0%), Alfa Aesar) were used as starting materials for the sol-gel auto-combustion. Deionised water was used as solvent for preparing solutions, and citric acid acted as combustion agent. The citric acid solution was slowly added to the mixed solution of nitrates as a chelator to form stable complexes with the metal cations. The solution was slowly evaporated to form a gel. This gel was dehydrated at 120°C to obtain the barium-strontium-zinc-(aluminium)-iron citrate precursor. During the dehydration process, the gel turned into a fluffy mass and was burnt in a self-propagating combustion manner. The auto-combusted powders were subsequently calcined for 7 hours at 1170°C in air.

Alternatively, stoichiometric amounts of the corresponding metal nitrates were dissolved in deionized water and the co-precipitation process of metal cations was initiated by adding NaOH at pH = 12. High power ultrasound generated by a Sonics VCX 750 ultrasonic processor was applied for 15 minutes at a wave amplitude of 40% to assist the coprecipitation process. The obtained precursors were calcined at temperature of 1170°C in air for 7 hours.

In both routes of sample preparation, thermal annealing was completed by cooling the furnace to room temperature. To determine the morphology of the samples and the size and shape of the particles, SEM images were taken using an electron microscope system FEI XL30 FEG-ESEM.

The phase purity of the final materials as well as their crystal structures was analysed from room-temperature powder X-ray diffraction (XRD) data by Topas 4.2. The XRD patterns were collected in a Bruker D8 Advanced (filtered Cu radiation; 40 kV, 40 mA) diffractometer in Bragg-Brentano geometry equipped with a LynxEye PSD detector. Polycrystalline samples in batches of 3 g were used for the powder diffraction experiments on the SPODI diffractometer at the FRM II research reactor. The data were acquired at wavelengths of λ = 1.5482 Å and 2.537 Å down to d-values of 0.81 Å and 1.33 Å, respectively. The magnetic order evolution in the samples contained in thin-walled niobium cans was followed by successive data sets collected at 298 K, 300 K, 500 K, 900 K and 1000 K. The reduced data were treated by the Rietveld method [26] realized in the FullProf software package [27].

The M -H and M -T measurements without electric (E) field were carried out in a commercial Magnetic Property Measurement System (MPMS, Quantum Design). Isothermal magnetization curves were taken at selected temperatures. Zero-field-cooled (ZFC) and field-cooled (FC) magnetization curves vs. temperature (4.2 – 300 K) were measured at magnetic fields of 50 Oe, 100 Oe and 500 Oe. In the ZFC protocol, the sample was cooled from room temperature to 4.2 K in zero magnetic field and the magnetization was measured during warming from 4.2 - 300 K at a heating rate of 3 K/min in applied magnetic field. The FC curve was measured while warming after cooling the sample from 300 K to 4.2 K in applied magnetic field.

3 Results and discussion

The investigation of the crystal structure and morphology of samples, obtained via a sol-gel auto combustion method, was the subject of our recent work [28]. Also, some results on the magnetic properties of the compositions x=0 and x=0.08 have been reported previously [29, 30]. We found that even at the small studied level of aluminium doping, the long range atomic and magnetic order depend on the particular soft chemistry synthesis route as well as on the temperature and duration of the annealing. Here, only the most important magnetic and structural characteristics related and relevant to this work are presented.

The $Ba_{0.5}Sr_{1.5}Zn_2Fe_{12-x}Al_xO_{22}$ (x=0 and 0.08) compositions have a large structural unit cell (centric space group R-3m) with hexagonal lattice constants a = a′ ≈ 5.85 Å and c ≈ 43.46 Å, which is typical for the Y-type hexaferrite structure depicted in **Fig. 1(a)**. The unit cell contains 3 formula units (Z=3). The crystal structure is highly modular consisting of a superposition of hexagonal (T) and spinel (S) blocks along the *c*-axis direction. The unit cell is composed of a sequence of TST'ST"S" blocks, with the primes indicating rotation by 120 degrees about c-axis **[2]**. It has an easy axis of magnetization lying in a plane normal to the *c*-axis direction. The complete magnetic structure is shown to be greatly simplified by employing block-spin approximation; it is constructed by two magnetic sublattice blocks, one with a large net magnetization, L, and the other, S, with a small net magnetization, stacked alternatively along the [001] axis. These two magnetic blocks have opposite magnetizations, causing ferrimagnetism of the compounds above room temperature, and in their zero-field-cooled state host the non-polar alternating longitudinal conical (ALC) and proper-screw (PS) phases.

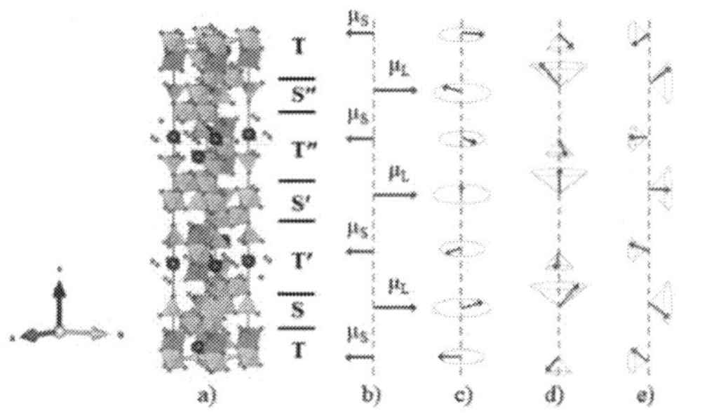

Fig. 1. (a) Schematic crystal structure of Y-type hexaferrite; (b) the collinear ferrimagnetic structure consists of alternate stacks of two spin blocks with μ_L (larger magnetic moments) and μ_S (smaller magnetic moments) along the *c* axis. (c) and (d) represent the non-polar spin order phases with propagation vector $\mathbf{k_0}$ parallel to c-axis: proper screw (PS) and alternate longitudinal conical (ALC) structure, respectively, exhibited at lower temperatures, (e) the transverse conical structure sensitive to external magnetic field.

The XRD diagrams of S1 and S2 powders showed the presence of the well crystalized Y-type hexaferrite structure as a main phase. The sonochemistry method yielded patterns of a hexaferrite structure with sharper and stronger peaks than the sol-gel auto-combustion.

Fig. 2 illustrates the X-ray diffractograms of undoped composition. The refined lattice parameters of the main phase in the R-3m space group with hexagonal settings were a=5.849(2)Å, c=43.555(3) Å and a=5.842(3)Å, c=43.681(3) Å for the auto-combustion and sonochemistry route, respectively. The few extra peaks labelled as a minor phase were identified as belonging to the spinel $ZnFe_2O_4$ amounting to less than 2.7 wt.% (standard card ICSD-04-006-1956). The detailed analysis by TOPAS using FWHM and integral breadths yielded a volume-weighted mean crystallite size L = 200 ± 30 nm and a spinel lattice parameter of 8.4332(4) Å.

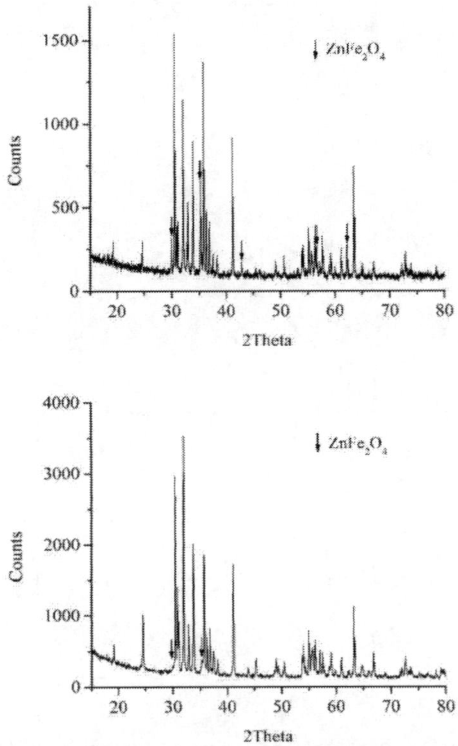

Fig. 2. XRD spectra of S1: (a) sol-gel auto-combustion, (b) sonochemistry [30].

In general, hexagonal grains were observed in the SEM images of both S1 and S2 samples. SEM images (Fig. 3a) of the S1 sample obtained by auto-combustion show particles of a very non-uniform shape and well-agglomerated to form clusters of different sizes and shapes. The average particle size varied between 200 nm and 1 μm with a mean 548 nm.

a) b)

Fig. 3. SEM images of S1 obtained by auto-combustion (a) and by sonochemical co-precipitation (b) [29].

In contrast, the particles in the S1 sample obtained by sonochemical co-precipitation had the perfect hexagonal shape typical for hexaferrites (Fig. 3b), with a particle size of 996 nm and an average thickness of 168 nm. One observes that a three-dimensional grain growth process had taken place, together with co-growth of the particles resulting in plate-like grains which stack on each other.

As seen in Fig.4a, the Al-substitution led to a melting process at 1170 °C for the S2 sample obtained by auto-combustion. Most of the particles are larger than 2 μm with different shape. Smaller particles of about 450 nm around the larger ones is also observed. On the other hand, the S2 material obtained by sonochemical co-precipitation (Fig.4b) consists of both particles of various size and shape, and of particles with a complete hexagonal shape. The hexagonal particles are in the range 1-2 μm long and are thin with a thickness of 50-70 nm. The plate-like hexagonal particles are well stuck together along the *c* axis to form an ensemble with a thickness well in excess of 100 nm.

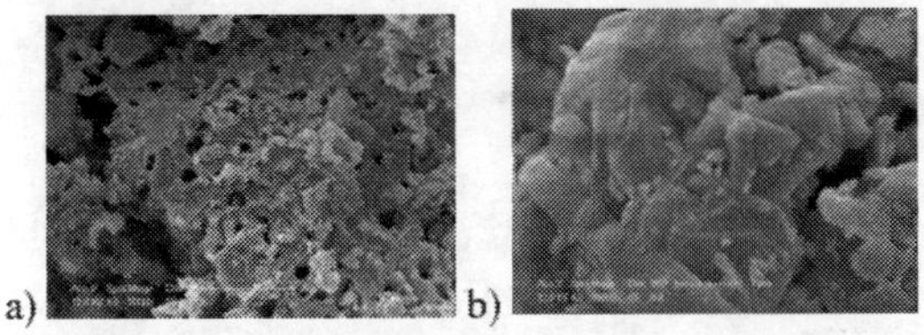

a) b)

Fig. 4. SEM images of S2 obtained by auto-combustion (a) [29] and by sonochemical co-precipitation (b) [28].

More specifically, sol–gel auto-combustion and sonochemical synthesis yielded polydomain S1 and S2 particles, with an incomplete or rather plate-like elongated hexagonal shape, respectively; however, their thickness is sufficient to not feel the possible influence of surface defects (via a significant imbalance of the long-range spin ordering) on the magnetic phase transitions.

Figure 5 shows the magnetic hysteresis loops of undoped compound measured at 4.2 K and 300 K. For each preparation route, the saturation of magnetization is achieved above 30 kOe. The sonochemical material manifests a higher saturation magnetization M_s and coercive field H_c reflecting the dominant presence of particles with a superior hexagonal shape.

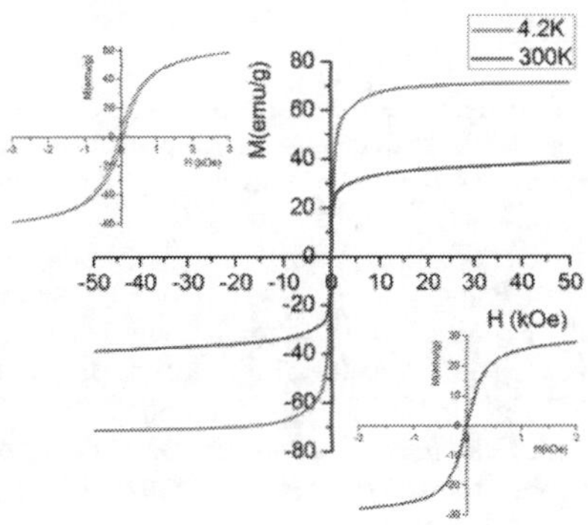

a)

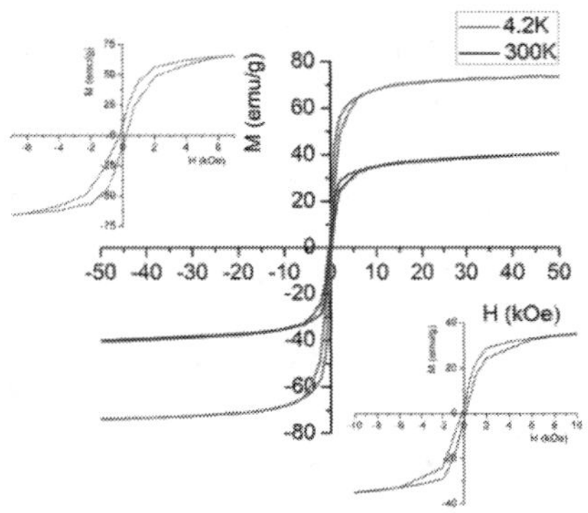

b)

Fig. 5. Hysteresis loops of S1 powders obtained by (a) sol-gel auto-combustion and (b) sonochemistry.

Unlike typical ferromagnetic substances, Figure 6 shows a triple hysteresis loop that indicates the presence of two kinds of magnetic phases in S2 by sonochemical co-precipitation. We associate this type of hysteresis at 300K at low magnetic fields to the intermediate spin arrangement between the PS and the ferrimagnetic spin order presented schematically in Fig. 1.

It should be noted that, in contrast, a triple hysteresis loop was recorded by Xu et al [25] for their ceramic sample $Ba_{0.5}Sr_{1.5}Zn_2(Fe_{1-x}Al_x)_{12}O_{22}$ with x=0.08 in magnetic fields from 0 to 0.5 T but at 5 K.

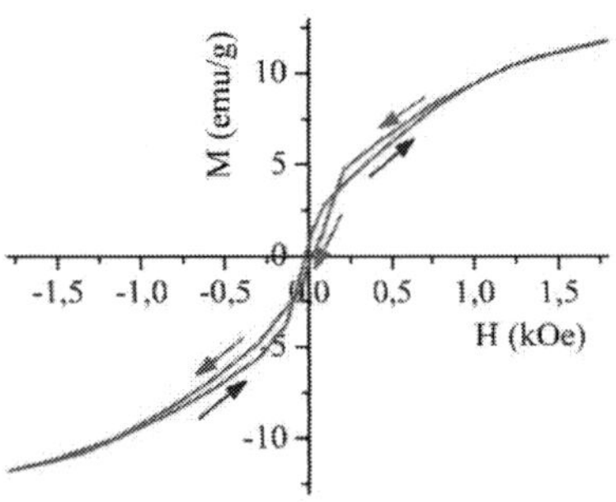

Fig. 6. Expanded view of the triple hysteresis curve at 300 K of S2 obtained by ultrasonically-assisted co-precipitation in the low magnetic field range disclosing the presence of two magnetic states [28].

Below the temperature of 326 K, the ferrimagnetic order in $Ba_{0.5}Sr_{1.5}Zn_2Fe_{12}O_{22}$ single crystals transforms into the helical spin order (see Fig. 1, c), in which the spin moments lie and rotate in the *ab*-plane [5, 9,13]. In a magnetic field applied perpendicular to the *c* axis, several magnetic transitions were found to take place in different temperature ranges, among which the so-called intermediate III phase shows a finite value of P.

Several anomalies of the magnetic behavior in the temperature range from 4.2 K to 300 K were observed for both S1 and S2 powders as presented in Figs 7-10 which can be understood in terms of the findings in single crystals.

Figure 7 shows the temperature variation of magnetization revealing the main regions of magnetic phase transitions occurring in similar temperature ranges as described for single crystals by Kimura et al [5,14] and Chun et al [17, 20].

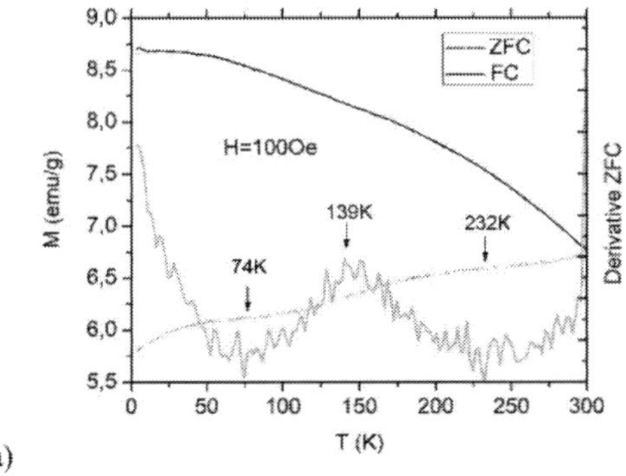

a)

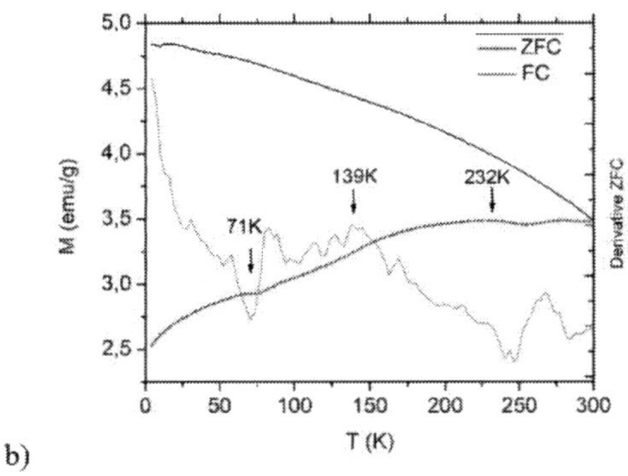

b)

Fig. 7. ZFC- and FC-magnetization vs temperature at 100 Oe for S1 obtained by (a) sol-gel auto-combustion and (b) sonochemistry [30]

Figure 8 illustrates the variation of the ac differential magnetization of S2 synthesized by sonochemical co-precipitation. By going down with temperature, the anomaly at about 285 K is to be associated with changes of magnetic arrangement from the collinear ferrimagnetic to the proper-screw spin phase while the behaviour between 85 K and 120 K is to be related with setting in the alternating longitudinal conical phase.

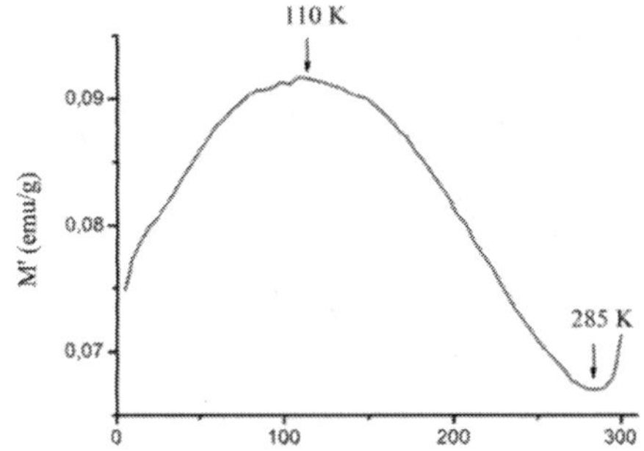

 https://doi.org/10.1051/epjconf/202328605006

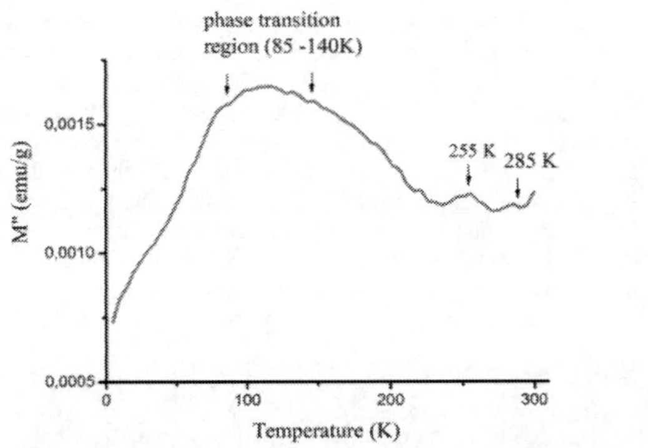

Fig. 8. Real part M' and imaginary part M" of ac differential magnetization of S2 obtained by sonochemistry vs temperature at H=10 Oe, f=1 kHz [28]

a)

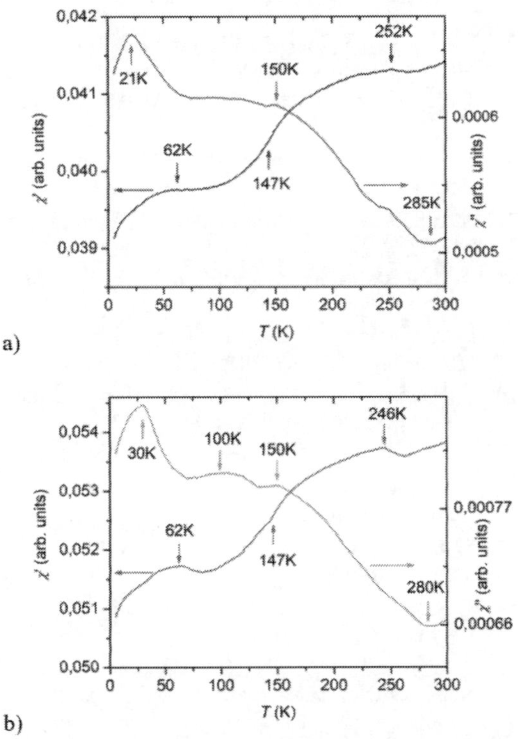

b)

Fig. 9. The differential ac susceptibility of S1 obtained by (a) sol-gel auto-combustion and (b) sonochemistry in ac magnetic field of 10 Oe and frequency of 1 kHz. [30]

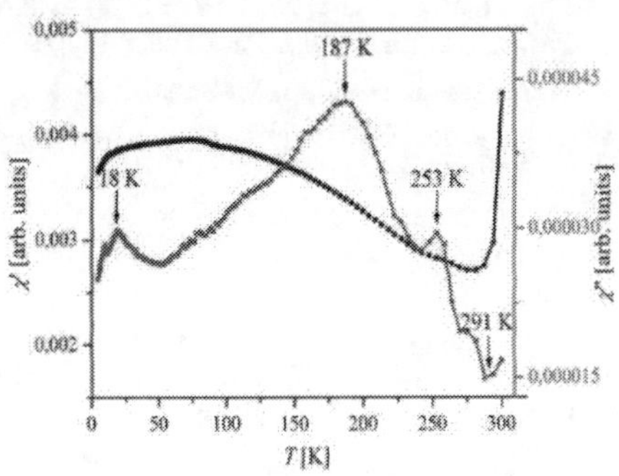

Fig. 10. The differential ac susceptibility of S2 obtained by sol-gel auto-combustion in ac magnetic field of 10 Oe and frequency of 1 kHz.

Expectedly, since the properties of the polycrystalline sample are an average of those of individual particles (taking into account the particles size distribution and their random or preferred orientation in the polycrystalline samples), the magnetic phase transitions are smoother compared to those of single crystals.

The results support the conclusion that the differences in the shape and size of the particles of the investigated soft chemistry products do not affect substantially the critical temperature of the various magnetic transitions observed in single crystals. But for the sonochemical samples, the transitions are much better pronounced due to the regular hexagonal shape of the constituent particles characteristic of hexaferrites, and the detection of metamagnetic transitions is facilitated.

The three-dimensional magnetic ordering in the Al substituted for Fe^{3+} Zn_2Y-type hexaferrite prepared by the two wet chemistry routes is clearly noticeable by the presence of additional peaks at 298 K, which rapidly lose intensity at higher temperatures as exemplified by Fig. 11.

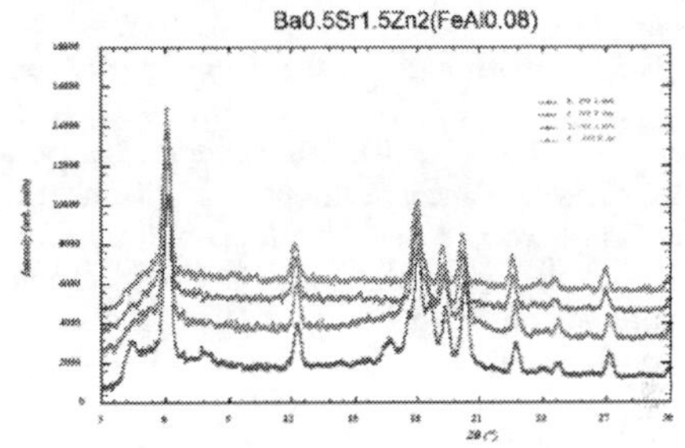

Fig. 11. Low angular part of neutron diffraction patterns of S2 collected on SPODI (MLZ) at selected temperatures. The presence of additional reflections at 298 K, which are purely magnetic in nature, is evident; they were attributed to the helicoidal arrangement formed by block-spins in the FE phase.

Due to the weak magnetic peak intensities, uncertainties in the propagation vector **k** at this stage introduce ambiguity in the associated spin ordering. Nevertheless, this finding is consistent with the low-field magnetization measurements for both S1 and S2 where the ZFC curve is observed to split from the FC curve above 300 K. For single crystal the transition from commensurate magnetic order to an incommensurate planar helix has been reported to occur at 310 K [12]. However, above 300 K we have ND measurements at temperatures of 500 K, 900 K, and 1000 K, where no satellite was observed but only Bragg peaks at wave vectors satisfying the condition -$h+k+l = 3n$, n integer. Therefore, from the present data we can confidently say that the helicoidal configuration disappears, but we cannot assert at what temperature in the range 320-290 K this magnetic transition takes place, although the temperature range is seemingly extended to a higher temperature as compared with the results of Xu et al. [25,31] for their x=0.08 ceramic material and Chang et al. [32] for single crystals $Ba_{0.7}Sr_{1.3}Zn_2(Fe_{1-x}Al_x)_{12}O_{12}$ ($0 \leq x \leq 0.12$). Going up with temperature we observe the magnetic contribution to persist at 500 K indicating the expected ferrimagnetic order, $k_0=(0,0,0)$, and at 900 K and 1000 K there are diffraction intensities of nuclear Bragg peaks only.

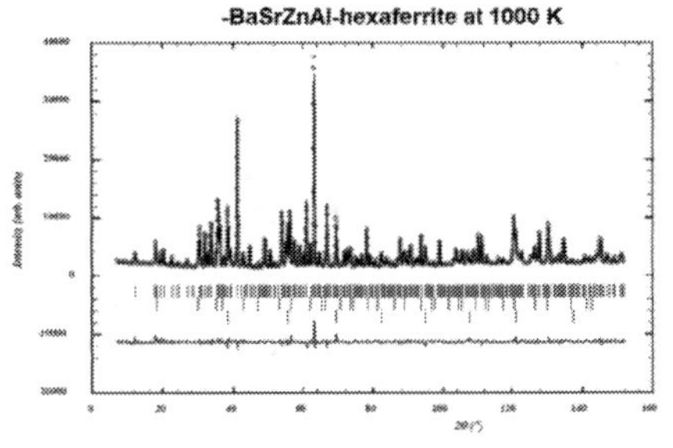

Fig. 12. Rietveld refined neutron powder diffraction pattern of S2 at 1000 K (λ = 1.5482 Å). Traces of zinc ferrite and Nb container were also considered. The series of tick marks correspond to the allowed Brag reflections.

Fig.12 illustrates the Rietveld refined neutron diffraction pattern of S2 in the paramagnetic region at 1000 K. No regions were excluded from the refinement. As variable parameters, the structural model takes into account the partial substitution of Fe occupying tetrahedral sites with Zn and Fe occupying octahedral sites with Al. The plate-like morphology of the particles observed in the SEM images is reflected strongly in the neutron diffraction lines.

4 Summary

Zn_2Y-type hexaferrite polycrystalline material of high crystallinity $Ba_{0.5}Sr_{1.5}Zn_2Fe_{12-x}Al_xO_{22}$ (x=0 and 0.08) was produced by two wet chemistry routes followed by thermal annealing. Sonochemistry yielded samples consisting of particles characterized predominantly by the hexagonal shape typical of hexaferrites. Accordingly, M_s and H_c were higher for those powders. The differential AC susceptibility shows specific peaks indicating magnetic phase transformations, and this feature is more pronounced for the powder produced by sonochemistry. Four kinds of hysteresis loops can be observed in $Ba_{0.5}Sr_{1.5}Zn_2Al_{0.08}Fe_{11.92}O_{22}$ revealing the occurrence of magnetization processes of different magnetic spin structures in given temperature ranges. The magnetic structures of both compositions studied are qualitatively similar and differ mainly in the non-collinear arrangement of iron spins especially those in octahedral sites of the Y-type hexaferrite structure. Neutron diffraction data present clear evidence for the destruction of the helical spin structure between 298 K and 300 K.

This research was supported in part by the Bulgarian National Science Fund under contract KP-06-N48/5 and research agreements between WBI, Belgium and the Bulgarian Academy of Sciences (BAS), and by a joint research project between the Institute of Low Temperature and Structure Research, Polish Academy of Sciences and the Institute of Electronics-BAS. B. G. was financially supported by the Bulgarian Ministry of Education and Science under the National Research Programme "Young scientists and postdoctoral students-2" approved by DCM 206/07.04.2022.

5 References

1. C. de Julián Fernández, C. Sangregorio, J. de la Figuera, B. Belec, D. Makovec, A. Quesada, J. Phys. D: Appl. Phys. **54**, 153001 (2021)
2. R.C. Pullar, Prog. Mater. Sci. **57**, 1191 (2012)
3. N.A. Spaldin, Proc. R. Soc. A **476**, 20190542 (2020)
4. P. Sahoo, L. Saini, A. Dixit, Oxford Open Materials Science **3**, itac012 (2023)
5. T. Kimura, Annu. Rev. Condens. Matter Phys. **3**, 93 (2012)
6. N.A. Spaldin, R. Ramesh, Nature Mater **18**, 203–2 (2019)
7. A. Hoffmann, S.D. Bader, Phys. Rev. Appl. **4**, 047001 (2015)
8. S. Fusil, V. Garcia, A. Barthélémy, M. Bibes, Ann. Rev. Mater. Res. **44**, 91 (2014)
9. F. Matsukura, Y. Tokura, H. Ohno, Nat. Nanotechnol. **10**, 209 (2015)
10. N. Momozava, Y. Yamaguchi, H. Takei, M. Mita, J. Phys. Soc. Jpn. **54**, 771 (1985)
11. N. Momozava, J. Phys. Soc. Jpn. **54**, 4007 (1986)
12. N. Momozava, Y. Yamaguchi, H. Takei, M. Mita, J. Phys. Soc. Jpn. **54**, 3895 (1985)
13. N. Momozawa, Y. Yamaguchi, J. Phys. Soc. Jpn **62**, 1292 (1993)
14. T. Kimura, G. Lawes, and A. P. Ramirez, Phys. Rev. Lett. **94**, 137201 (2005).
15. S. Ishiwata, Y. Taguchi, H. Murakawa, Y. Onose, and Y. Tokura, Science **319**, 1643 (2008)
16. M. Soda, T. Ishikura, H. Nakamura, Y. Wakabayashi, and T. Kimura, Phys. Rev. Lett. **106**, 087201 (2011)
17. S. H. Chun, Y. S. Chai, B.-G. Jeon, H. J. Kim, Y. S. Oh, I. Kim, H. Kim, B. J. Jeon, S. Y. Haam, J.-Y. Park et al., Phys. Rev. Lett. **108**, 177201 (2012)
18. K. Okumura, T. Ishikura, M. Soda, T. Asaka, H. Nakamura, Y. Wakabayashi, and T. Kimura, Appl. Phys. Lett. **98**, 212504 (2011)
19. Y. Kitagawa, Y. Hiraoka, T. Honda, T. Ishikura, H. Nakamura, and T. Kimura, Nat. Mater. 9, 797 (2010).
20. S. Chun, Y. Chai, Y. Oh, D. Jaiswal-Nagar, S. Haam, I. Kim, B. Lee, D. Nam, K-T Ko, J-H Park, J-H Chung, K. Kim, Phys. Rev. Lett. **104**, 037204 (2010)

21. K. Zhai, Y. Wu, S. Shen, W. Tian, H. Cao, Y. Chai, B. C. Chakoumakos, D. Shang, L. Yan, F. Wang, Y. Sun, Nat Commun. **12**, 519 (2017)
22. T. Koutzarova, S. Kolev, K. Krezhov, Ch. Ghelev, B. Vertruyen, L. M. Tran, A. Zaleski, *Preparation and phase transitions in Y-type magneto-electric hexaferrites*, Chapter 3, in book: *Advances in Materials Science Research*, Editor: Maryann C. Wythers, Publisher: Nova Science Publishers, Inc., **41**, 105 (2020)
23. V. Kocsis, T. Nakajima, M. Matsuda, A. Kikkawa, Y. Kaneko, J. Takashima, K. Kakurai, T. Arima, Y. Tokunaga, Y. Tokura, Y. Taguchi, Phys. Rev. B **101**, 075136 (2020)
24. D. Khomskii, Physics **2**, 20 (2009)
25. W. Xu, J. Yang, W. Bai, Y. Zhang, K. Tang, Chun-gang Duan, X. Tang, J. Chu, J. Appl. Phys. **117**, 17D909 (2015)
26. H. Rietveld, J. Appl. Crystallogr. **2**, 65 (1969)
27. J. Rodríguez-Carvajal, Physica B **192**, 55 (1993)
28. B. Georgieva, T. Koutzarova, S. Kolev, Ch. Ghelev, B. Vertruyen, R. Closset, R. Cloots, A. Zaleski, Bulg. Chem. Commun., **48 (G)**, 147 (2016)
29. B. Georgieva, S. Kolev, Ch. Ghelev, D. Kovacheva, B. Vertruyen, R. Closset, T. Koutzarova. *A Comparative study of the morphology of Y-type hexaferrite powders obtained by sol-gel auto-combustion and ultrasonic co-precipitation*. In Advanced Nanotechnologies for Detection and Defence against CBRN Agents, Petkov, P., Tsiulyanu, D., Popov, C., Kulisch, W., Ed.; NATO Science for Peace and Security Series B: Physics and Biophysics, Dordrecht, Springer, 2018; pp 31–36.
30. B. Georgieva, S. Kolev, K. Krezhov, Ch. Ghelev, D. Kovacheva, B. Vertruyen, R. Closset, L- M. Tran, M. Babij, A. J. Zaleski, T. Koutzarova, J. Magn. Magn. Mater. **477**, 131 (2019).
31. W. Xu, J. Yang, Y. Shen, W. Bei, Y. Zhang, J. Liu, K. Tang, Z. Wang, Ch.-g. Duan, X. Tang, J. Chu, AIP Advances 4, 067122 (2014.
32. H. Chang, H. B. Lee, Y.-S. Song, J.-H. Chung, S.A. Kim, I.H. Oh, J. Reehus, J. Schefer – Phys. Rev. B85, 064402 (2012)

Quasi Elastic Neutron Scattering model library

Céline Durniak[1,*], *Miguel Angel* González[2], *Anders* Markvardsen[3], *Sanghamitra* Mukhopadhyay[3], *Franz* Lang[3], and *Thomas Holm* Rod[1]

[1]European Spallation Source ERIC, Ole Maaløes Vej 3, DK-2200 Copenhagen, Denmark
[2]Institut Laue Langevin, 38042 Grenoble, France
[3]ISIS Neutron and Muon Source, STFC Rutherford Appleton Laboratory, Oxford, UK

Abstract. This paper reports on the development of a collection of dynamical models of one-dimensional peak profile functions used to fit dynamic structure factors $S(Q, \hbar\omega)$ of Quasi Elastic Neutron Scattering (QENS) data. The objective of this development is to create a maintainable and interoperable Python library with models reusable in other projects related to the analysis of data from Quasi Elastic Neutron Scattering experiments. The ambition is that the library also will serve as a platform where scientists can make their models available for others. We illustrate how the library can be used by newcomers to the field as well as by experts via different examples. These examples, provided as Jupyter notebooks, show how the QENS models can be integrated in the whole QENS data processing pipeline.

1 Motivation and significance

Quasi elastic Neutron Scattering (QENS) is a neutron scattering technique to probe displacements in temporal and spatial scales, typically from pico- to tens of nanoseconds and from Ångströms to nanometres, respectively. In this technique one measures the dynamic structure factor $S(Q, \hbar\omega)$ around the elastic line with $\hbar\omega = 0$. Here Q is the momentum exchange between the neutron and the investigated sample and $\hbar\omega$ is the energy transfer. The broadening of the elastic peak can be related to diffusive stochastic motions taking place in the sample, including translational diffusion (both unconstrained long-range translations and confined translation), molecular reorientations, localised motions (e.g., methyl rotations), etc. [1] Correspondingly, QENS finds applications in various scientific domains, such as biology (e.g., to explore the dynamics of proteins) [5, 6], soft-matter (e.g., to investigate different relaxation modes in polymers) [7, 8] or materials science (e.g., to probe H-diffusion in hydrogen storage materials) [9], among others. Each type of motion is characterised by a specific Q-dependence of the intensity and line-width of the quasi elastic signal for a given Q, which can be represented by an analytical model. Therefore, a large series of analytical models corresponding to the different types of motions mentioned above exist and can be used to fit the experimental data in order to extract the relevant physical parameters (e.g., self-diffusion constants, residence times, jump lengths, etc.) from the QENS spectra [1].

The QENS library makes a range of these mathematical models freely available to practitioners of QENS and developers of QENS analysis software in a format that makes them inter-operable and reusable. As an open-source library, the implementations can be scrutinised and validated by peers, and scientists are encouraged to add their own models to make them available for the QENS community at large.

The QENS library at its core provides Python models of mainly one-dimensional (1D) line profiles to fit experimental or simulated $S(Q, \hbar\omega \approx 0)$ data. This work was part of SINE2020 Work Package 10 on Data Treatment [10] to develop a comprehensive library of dynamical models in order to increase interoperability and reusability, for instance for rapid prototyping. The library provides different building blocks that users can combine, convolute and plug into different frameworks for visualizing or fitting.

Our approach is similar to that implemented by the `SasView` [11] developer community for small angle scattering with its SasView Marketplace [12].

The library is written in Python and depends on standard scientific Python modules (i.e., `scipy` [13] and `numpy` [14]). It is installed using the package installer for Python, `pip`.

2 Software description

The library is developed under an open-source license (BSD 3-Clause) and the source is stored in a GitHub repository at https://github.com/QENSlibrary/QENSmodels/. The models are written in Python facilitating integration with other packages of the large Python eco-system as well as easy integration in Python based workflows. Moreover, Python is a very popular scripting and programming

*e-mail: celine.durniak@ess.eu

[1]Refer to [1–4] for more details about the technique.

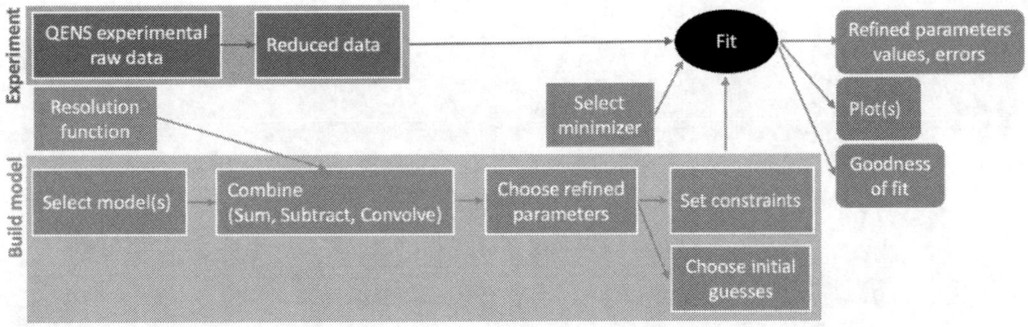

Figure 1. (Colour on-line) Workflow of QENS data processing from raw experimental data to output of fitting.

language in science and the Python interface therefore reduces the entry barrier for the many scientists already familiar with Python.

The library comes with examples of how it can be used in practise and in conjunction with existing Python modules for optimization (fitting), visualization, and for adding Graphical User Interface (GUI) elements.

Figure 1 shows a typical QENS data processing workflow. Data are collected during an experiment and then corrected (background subtraction, detector efficiency normalisation, absorption and multiple scattering corrections, etc.) in order to obtain the experimental $S(Q, \hbar\omega)$ for our sample [3]. The latter is then compared to the relevant mathematical models and typically some important parameters characterising the sample dynamics and the motion geometry are obtained by fitting the model parameters to the experimental signal. Data collected from multiple experiments can also be used in this procedure for simultaneous or sequential fittings. Different tools exist to deal with these stages with different levels of coverage of the whole workflow.

For example, MantidWorkbench [15] can deal with data reduction, fitting and plotting.

If we focus on fitting only, the available tools are more universal and not restricted to the QENS technique. There are, for example, `bumps` [16], `lmfit` [17] or `scipy` [13]. Therefore, in the present library we provide different models, (corresponding to cell 'Select models(s)' in Fig. 1) that can be combined and plugged in different frameworks. The following subsections detail the architecture of the library and its main functionalities.

2.1 Software Architecture

Figure 2 shows the structure of the repository. Each file in the QENSmodels folder contains the implementation of exactly one model in the form of a Python 3 function that returns $S(Q, \hbar\omega)$. For more complicated models, functions returning the Q-dependence of the half-width at half maximum (HWHM), the elastic incoherent structure factor (EISF) and the quasi-elastic incoherent structure factors (QISF) for the 1D peak are also provided. We refer to [8, 18] for definitions of these terms.

```
QENSmodels repository
|__QENSmodels/ ...
|  Python module files with QENS
|  models.
|__docs/
|  |__examples/ ...
|     Jupyter notebooks.
|     |__data/ ...
|     |  Reference datafiles used in
|     |  Jupyter notebooks.
|     |__using_mantid/
|__tests/ ...   Python scripts of unit tests on
|               the QENS models.
|__tools ...   Jupyter notebooks and scripts
|              to run unit tests and provide
|              additional information on QENS
|              models.
|__CONTRIBUTING.rst
|__DISCLAIMER.rst
|__LICENSE.txt
|__MANIFEST.in
|__README.rst
|__pyproject.toml
```

Figure 2. Structure of QENS models repository.

Usability in a scripting environment, such as Jupyter, is facilitated by providing rich documentation through so-called documentation strings using the Sphinx [19] standard including so-called doctest [20]. By adhering to the Sphinx standard, documentation pages can be automatically generated and updated from the code itself. Moreover, to help the interested user and to lower the entry barrier, examples using Jupyter notebooks and Python scripts are provided as well as instructions for scientists wishing to contribute their own models to the library. Online documentation is also available on `readthedocs` at https://qensmodels.readthedocs.io/.

Code quality and maintainability is facilitated by using git for version control and by implementing unit tests,

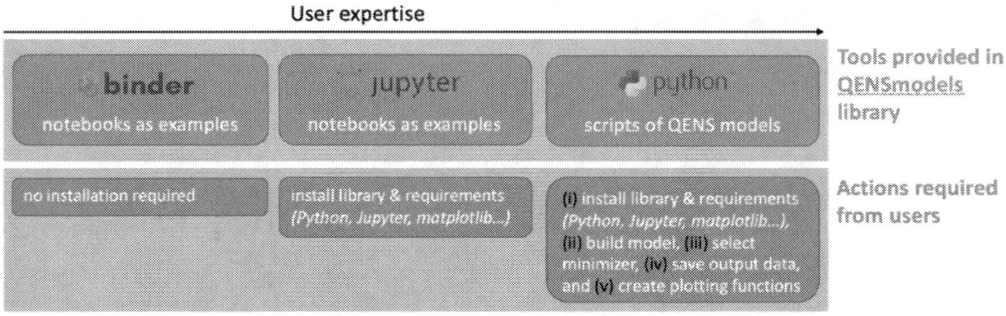

Figure 3. (Colour on-line) Diagram showing the tools provided in the QENSmodels library to help users with different levels of expertise in QENS and in programming.

including the aforementioned doctests,[2], which are used for continuous integration using GiHub Actions. This approach to software development is in alignment with the guidelines outlined in [21].

2.2 Software Functionalities

Figure 3 shows the options available for using the library. In order to give an idea of what can be done with the library without any installation, the Jupyter notebooks [22] are available through Binder [23]. Users only have to click on the link to be able to run the notebooks and check if the library can answer their needs.

On the other hand, for users only interested in the models, for example, to add them to their fitting pipelines, the installation of the library requires only a command line. In addition, they will have to perform all steps as listed in the third column of Figure 3 in order to use the model for their cases.

As mentioned in the previous section, examples are provided as Jupyter notebooks. They use a few QENS models and a few fitting engines, like `lmfit`, `scipy`, or `bumps`. The notebooks and modules required to make them run can be installed by following the guidelines on the github pages [24]. Note that the minimizers used in these notebooks were only to illustrate the specific syntaxes of the fitting engines and not because they were the most optimal engines for the investigated cases. Details about comparison of minimizers can be found at [25].

Physical units used by the models' parameters and variables are specified at the beginning of the Jupyter notebooks. A notebook, `Convert_Units.ipynb` is also provided to convert between different physical units in case the reference user's data are expressed in different units than the standards chosen for the QENS library.

[2]A unit test checks that a single component operates in the right way. It helps to isolate where something in the code is broken. Doctest provides a testing framework using code examples for documenting and testing Python code.

3 Illustrative Examples

3.1 Toy examples in the library

As tutorials, Jupyter notebooks [22] are provided showing how to (i) build a model using different functions from the QENS library, (ii) choose a fitting engine, (iii) define the settings and run the fit, (iv) extract visual and quantitative information from the outputs.

Since the syntax to build the fitting model, apply constrains and set the initial guesses depends on the chosen fitting engine, different minimizers have been used in the notebooks: `scipy` [13], `bumps` [16], `lmfit` [17]. For an easier identification, the notebooks have been named following this convention: "*[fitting engine]_[name of QENSmodels]_*fit.ipynb", e.g., "`scipy_lorentzian_fit.ipynb`".

Reference data used in the examples are either generated in the dedicated Jupyter notebook or stored in one of the files in the `data` sub-folder of the repository. We provide these reference data in binary (hdf5) and ascii formats. The resolution function is either a Gaussian profile or reduced data from a Vanadium run. And the convolution with the sample data is done using methods from `numpy` [14] or `lmfit` [17].

We are now going to detail one of these notebooks, `bumps_Brownian_Diff_fit.ipynb`. Figure 4 details its table of contents. After a short introduction describing the objective of the notebook, we specify the physical units of the refined parameters and import the required Python libraries as well as the reference data for the sample and resolution function. Figure 5 shows the evolution of these functions for different Q values. The convolution between the sample and the resolution functions is done using a method from `numpy` [14].

https://doi.org/10.1051/epjconf/202328606001

```
Brownian Translational diffusion *
Resolution with bumps
 |
 |__ Introduction
 |    |__ Physical units
 |__ Import libraries
 |__ Setting of fitting
 |    |__ Load reference data
 |    |__ Display units of input data
 |    |__ Create fitting model
 |    |__ Display initial configuration
 |    |__ Choice of minimizers for bumps
 |    |__ Setting for running bumps
 |__ Running the fit
 |__ Showing the results
      |__ Display final configuration
```

Figure 4. Table of contents of one of the Jupyter notebooks available in the library: `bumps_Brownian_Diff_fit.ipynb`

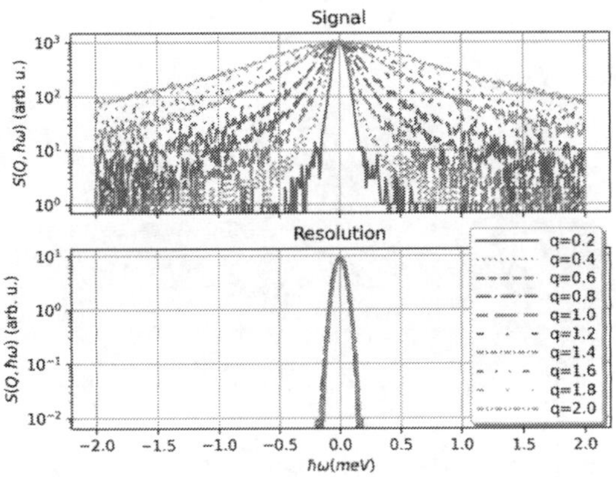

Figure 5. (Colour on-line) Plot of the reference data used in `bumps_Brownian_Diff_fit.ipynb`.

The fitting model is expressed as:

$$S(Q,\omega) = R(Q,\hbar\omega) \circledast \left[\frac{\text{scale}}{\pi} \frac{DQ^2}{(\hbar\omega - \text{center})^2 + (DQ^2)^2} \right],$$

where scale and center are the scale factor and center of the Lorentzian function describing the diffusion, of half-width at half-maximum equal to DQ^2, where D is the self-diffusion coefficient and Q the momentum transfer. Listing 1 shows how the fitting model was created following the syntax required by the fitting engine, `bumps`. In order to help less Python savvy users, several GUI widgets are provided in the notebooks, for example, to choose the minimizer, the number of iterations or for interactive plots as shown in Figure 6. After running the fit, the values and errors of the refined parameters and the fit quality are printed and a plot of the fitted model in comparison with the reference data is shown for the value of Q selected by the user via a widget.

Listing 1. Code snippet showing how to create a fitting model using 'bumps' and the QENS library

```python
import numpy as np
from QENSmodels import \
    sqwBrownianTranslationalDiffusion
import bumps.names as bmp
# Fitting model
def model_convol(x, q, scale=1, center=0, D=1,
resolution=None):
    model = sqwBrownianTranslationalDiffusion(x, q, scale,
center, D)
    return np.convolve(model,
                       resolution/resolution.sum(),
                       mode='same')
# Fit
model_all_qs = []

for i in range(len(q)):
    # Bumps fitting model
    model_q = bmp.Curve(
        model_convol,
        hw,
        sqw[:, i],
        err[:, i],
        name=f'q_{q[i]:.2f}',
        q=q[i],
        scale=1000,
        center=0.0,
        D=0.1,
        resolution=res[:,i])
    model_q.scale.range(0, 1e5)
    model_q.center.range(-0.1, 0.1)
    model_q.D.range(1e-12, 1)
    # Q-independent parameters
    if i == 0:
        D_q = model_q.D
    else:
        model_q.D = D_q
    model_all_qs.append(model_q)
problem = bmp.FitProblem(model_all_qs)
```

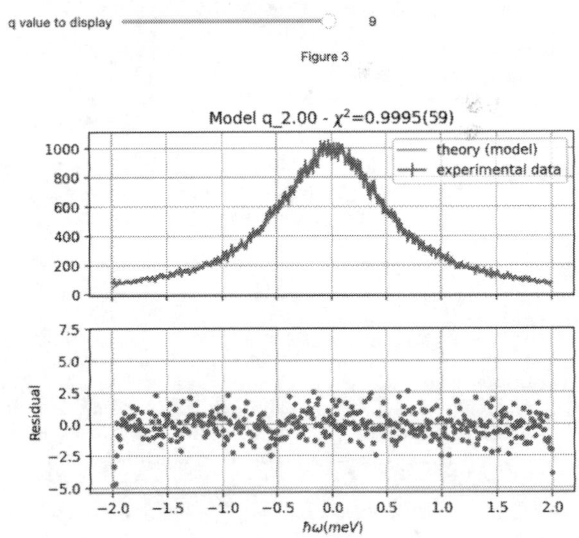

Figure 6. (Colour on-line) Widget to select the index in the list of available Q-values to plot the fitted model in comparison with the reference data as well as the residual. Here the configuration corresponding to Q = 2 Å is shown.

In addition to these notebooks, we also provide instructions to make the library importable in MantidWorkbench [15] as well as a Python script using one of the

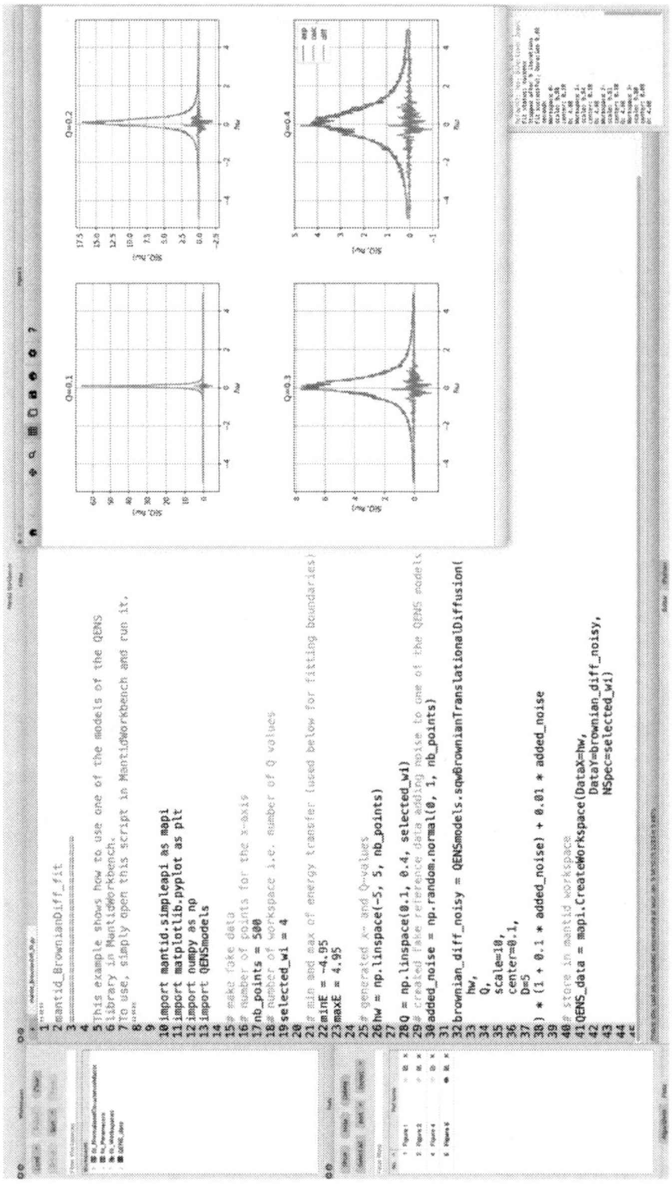

Figure 7. (Colour on-line) Graphical User Interface of MantidWorkbench [26] showing the use of the Python script `mantid_BrownianDiff_fit.py` provided in the library. The Python script is displayed in the editor. After its execution, workspaces, i.e., Mantid data structures, are available for further operations in the Graphical User Interface (left-hand side column) and a separate window displays the fitting result for the 4 Q-values considered in this example.

models to fit data and plot and display the output in MantidWorkbench (see Figure 7). Reference [27] details the QENS features available in this framework.

3.2 Benchmarking with experimental data

One of the notebooks, `lmfit_TOFTOF_delta_lorentz.ipynb` uses reduced data presented in [28].

In this publication, the authors used QENS, among other techniques, to investigate the microscopic details of the proton motions in two proton conducting, acceptor-doped, perovskites. The samples were measured at the time-of-flight spectrometer TOFTOF at MLZ at different temperatures using an incident wavelength of 2.5 Å. Nine "Q-cuts" were retained in the analysis. The reduced data files for one of the investigated samples $BaZr_{0.8}In_{0.2}O_3H_{0.2}$ at several temperatures have been provided by D. Noferini. The reduced experimental data were fitted using the following model:

$$S_{\text{meas}}(Q,\hbar\omega) = \{S(Q,\hbar\omega)[1 + n(\hbar\omega)]\hbar\omega\beta\} \circledast R(Q,\hbar\omega),$$

where $S(Q,\hbar\omega) = a_D(Q)\delta(\hbar\omega) + a_L(Q)L(Q,\hbar\omega) + \text{bkg}(Q)$, $n(\hbar\omega) = [\exp(\frac{\hbar\omega}{k_B T}) - 1]^{-1}$ is the Bose factor, and $\beta = (k_B T)^{-1}$ is the reciprocal of the thermodynamic temperature. $\delta(\hbar\omega)$ and $L(Q,\hbar\omega)$ are a Dirac and a Lorentzian peaks, respectively. $R(Q,\hbar\omega)$ is the resolution function (spectrum of a vanadium standard). bkg(Q) is the constant background, only Q-dependent and $\circledast$ is the convolution symbol.

The Jupyter notebook describes how to load the reduced data, build the model using the QENS library, select the minimizer and the range of Q-values to consider, run batch fitting over successive Q-values at different temperatures and extract, store and display results from the fit. Figure 8 shows the comparison between an experimental profile (bullet points with errorbars) and the fitted model (solid line) for $Q = 2.3$ Å at a temperature of $T = 540K$ together with the initial model (dashed curve). The good agreement between the final fit and the experimental data is clearly evident. Interested readers are encouraged to consult ref. [28] for further details.

This example could be further extended by providing loading of reduced data from different instruments or different data formats to, for example, easily compare the results collected during experiments at different institutes.

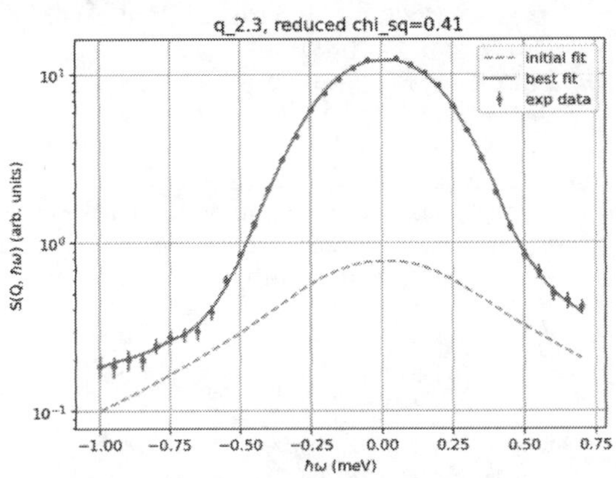

Figure 8. (Colour on-line) Experimental $S(Q,\hbar\omega)$ (bullet points with errorbars) from [28] and model with initial guesses (dashed orange line) and fitting outputs (solid red line).

4 Impact

The QENS library helps researchers and scientists by providing a list of 1D functions to fit QENS data making the analysis more FAIR, i.e., Findable Accessible Interoperable Reusable [29] by providing a public reference to fitting models allowing other researchers to reproduce the analysis.

The dynamics observed in QENS experiments are mainly related to the complexities of the samples studied. But collected signals can also contain the signature of the instrument and of the sample environment, which are getting more sophisticated. The combination of both factors may render the extraction and modelling of the experimental $S(Q,\hbar\omega)$ challenging. But this process can be made easier thanks to the flexibility of customising fitting models with the QENS library and different instrument resolution functions.

The library can be imported in different frameworks as illustrated in Section 3 with MantidWorkbench. This example shows how to use Python scripts. But the library could also be used in Graphical User Interfaces, like the QENS data analysis interface of Mantid described in [27].

Given that the library is installed using `pip`, it can be included in an automated data processing pipeline (reduction and analysis) or, for example, in a Machine Learning workflow, to build the optimal fitting model with ease, which increases the reach and usability of our development.

5 Conclusions

We have developed a Python package called QENSmodels, that provides dynamic models of 1D line profiles to fit QENS experimental or simulated data. In order to illustrate how to hook these models to the whole data processing pipeline to newcomers and experts, we provide Python scripts and Jupyter notebooks using different data and fitting engines. In the future, we plan to integrate more models and other examples using experimental data and other fitting engines. Scientists are also encouraged to contribute implementations of their work with additional models or examples.

The authors wish to thank Daria Noferini and her co-authors for providing experimental data used for testing some of the models.

This project received funding From the European Union's Horizon 2020 research and Innovation programme under grant agreement No 654000.

References

[1] M. Bée, *Quasielastic Neutron Scattering, Principles and Applications in Solid State Chemistry, Biology and Materials Science* (Taylor & Francis, 1988), ISBN 9780852743713

[2] R. Hempelmann, *Quasielastic Neutron Scattering and Solid State Diffusion*, Oxford Neutron Scattering in C (Clarendon Press, 2000), ISBN 9780198517436

[3] M. Telling, *A Practical Guide to Quasi-elastic Neutron Scattering* (Royal Society of Chemistry, 2020), ISBN 9781788019262

[4] J.P. Embs, F. Juranyi, R. Hempelmann, Zeitschrift für Physikalische Chemie **224**, 5 (2010)

[5] R.E. Lechner, S. Longeville, *Quasielastic Neutron Scattering in Biology, Part II: Applications* (Springer Berlin Heidelberg, Berlin, Heidelberg, 2006), pp. 355–397, ISBN 978-3-540-29111-4, https://doi.org/10.1007/3-540-29111-3_16

[6] D. Vural, X. Hu, B. Lindner, N. Jain, Y. Miao, X. Cheng, Z. Liu, L. Hong, J.C. Smith, Biochimica et Biophysica Acta (BBA) - General Subjects **1861**, 3638 (2017), science for Life – Recent Advances in Biochemical and Biophysical Methods

[7] V.G. Sakai, A. Arbe, Current Opinion in Colloid & Interface Science **14**, 381 (2009)

[8] Q. Berrod, K. Lagrené, J. Ollivier, J.M. Zanotti, EPJ Web of Conferences **188**, 05001 (2018)

[9] M. Karlsson, Phys. Chem. Chem. Phys. **17**, 26 (2015)

[10] *SINE2020 homepage*, https://www.sine2020.eu

[11] *sasview webpage*, www.sasview.org

[12] *sasview marketplace webpage*, http://marketplace.sasview.org/

[13] P. Virtanen, R. Gommers, T.E. Oliphant, M. Haberland, T. Reddy, D. Cournapeau, E. Burovski, P. Peterson, W. Weckesser, J. Bright et al., Nature Methods **17**, 261 (2020)

[14] C.R. Harris, K.J. Millman, S.J. van der Walt, R. Gommers, P. Virtanen, D. Cournapeau, E. Wieser, J. Taylor, S. Berg, N.J. Smith et al., Nature **585**, 357 (2020)

[15] O. Arnold, J. Bilheux, J. Borreguero, A. Buts, S. Campbell, L. Chapon, M. Doucet, N. Draper, R.F. Leal, M. Gigg et al., Nuclear Instruments and Methods in Physics Research Section A: Accelerators, Spectrometers, Detectors and Associated Equipment **764**, 156 (2014)

[16] P. Kienzle, J. Krycka, N. Patel, I. Sahin, *Bumps (version 0.8.0) [computer software]*

[17] M. Newville, T. Stensitzki, D.B. Allen, A. Ingargiola, *Lmfit: Non-linear least-square minimization and curve-fitting for python* (2014), https://doi.org/10.5281/zenodo.11813

[18] T. Matsuo, J. Peters, Life **12**, 1259 (2022)

[19] G. Brandl, URL http://sphinx-doc.org/sphinx.pdf (2021)

[20] *doctest documentation*, https://docs.python.org/3/library/doctest.html

[21] J. Wuttke, S. Cottrell, M. Gonzalez, A. Kaestner, A. Markvardsen, T. Rod, P. Rozyczko, G. Vardanyan, Journal of Neutron Research **24**, 1 (2022)

[22] *Jupyter webpage*, https://jupyter.org/

[23] Project Jupyter, Matthias Bussonnier, Jessica Forde, Jeremy Freeman, Brian Granger, Tim Head, Chris Holdgraf, Kyle Kelley, Gladys Nalvarte, Andrew Osheroff et al., *Binder 2.0 - Reproducible, interactive, sharable environments for science at scale*, in *Proceedings of the 17th Python in Science Conference*, edited by Fatih Akici, David Lippa, Dillon Niederhut, M. Pacer (2018), pp. 113 – 120

[24] *Using the jupyter notebooks in a virtual environment*, https://github.com/QENSlibrary/QENSmodels/tree/main/docs/examples#using-the-jupyter-notebooks-in-a-virtual-environment

[25] *Mantid documentation: Comparing minimizers*, https://docs.mantidproject.org/v3.8.0/concepts/FittingMinimizers.html

[26] *Mantid 6.5.0: Manipulation and analysis toolkit for instrument data.; mantid project.* (2022)

[27] S. Mukhopadhyay, B. Hewer, S. Howells, A. Markvardsen, Physica B: Condensed Matter **563**, 41 (2019)

[28] D. Noferini, M.M. Koza, S.M.H. Rahman, Z. Evenson, G.J. Nilsen, S. Eriksson, A.R. Wildes, M. Karlsson, Phys. Chem. Chem. Phys. **20**, 13697 (2018)

[29] M.D. Wilkinson, M. Dumontier, I.J. Aalbersberg, G. Appleton, M. Axton, A. Baak, N. Blomberg, J.W. Boiten, L.B. da Silva Santos, P.E. Bourne et al., Scientific Data **3**, 160018 (2016)

Neutron adiabaticity and its impact on data analysis, illustrated for polarized GISANS

Annika Stellhorn[1*], *Wai-Tung* Lee[2], *Emmanuel* Kentzinger[3], *Wangchun* Chen[4], *Jonathan* Gaudet[4,5], *Kathryn* Krycka[4] and *Elizabeth* Blackburn[1]
[1]Lund University, Division of Synchrotron Radiation Research, 22100 Lund, Sweden
[2]European Spallation Source (ESS), Instrument Technologies Division, 22484 Lund, Sweden
[3]Forschungszentrum Jülich GmbH, Jülich Centre for Neutron Science (JCNS-2) and Peter Grünberg Institut (PGI-4), JARA-FIT, 52428 Juelich, Germany
[4]NIST Center for Neutron Research, National Institute of Standards and Technology, Gaithersburg, Maryland 20899, USA
[5]Department of Materials Science and Eng., University of Maryland, College Park, MD 20742-2115

Abstract. The significance of neutron spin adiabaticity in the data analysis of polarized Grazing Incidence Small Angle Neutron Scattering (GISANS) is discussed, with the aim of minimizing the number of simulation parameters of complex magnetic models within the neutron scattering cross-section. We illustrate how an estimate of the neutron polarization direction and adiabaticity can be obtained by magnetic field and neutron ray-tracing simulations and compare the results with measurements of the magnetic field map for the beamline used in the GISANS experiment. We show how small deviations from the neutron polarization direction with respect to the magnetic field vector at the sample position significantly affect the data analysis, and how this can be considered within the magnetic model of the scattering cross section using the Distorted Wave Born Approximation.

1 Introduction and science case

In neutron scattering, the adiabaticity of neutron spin transport, A, quantifies how well the semi-classical neutron polarization vector in the density matrix formalism, $\vec{P}$, follows the direction of magnetic field, $\vec{B}$. In polarized neutron scattering, maintaining a high neutron spin-transport adiabaticity along the beam path between spin-conditioning and detection components, such as polarizer, analyzer, and spin-flippers, and the sample is crucial. If the adiabaticity drops too low in a particular region on the beam path, this can severely affect the quality of the results obtained from neutron experiments with polarization analysis. To deal with these cases of critically low adiabaticity, precise simulations of the neutron polarization evolution along the flight path are needed to evaluate the polarization at the sample position, which is then put into the simulations of the scattering pattern generated by the sample. In this article, we demonstrate the determination of the neutron adiabaticity and polarization by a combination of magnetic field modelling and neutron ray-tracing, and the effect of an imperfect neutron transport on the data analysis, using the example of polarized GISANS from the magnetic domains and domain walls in a thin film of FePd.

The adiabaticity is defined as the ratio of the Larmor frequency of the neutron spin, ω_L, and the angular frequency of the magnetic field along the neutron flight path, ω_B:

$$A = \frac{\omega_L}{\omega_B} = \frac{\gamma|\vec{B}|}{\left|\nabla\left(\frac{\vec{B}}{|\vec{B}|}\right)\cdot\vec{v_n}\right|} \tag{1}$$

where γ is the gyromagnetic ratio of the neutron, $\vec{B}(x,y,z)$, is the magnetic field vector, and $\vec{v_n}$ is the velocity of the neutrons. The gradient term is the Jacobian of the magnetic field direction vector. Multiplying the Jacobian with $\vec{v_n}$ gives the vector of the rate of change of the field direction vector along $\vec{v_n}$. It is a measure of the ability of the neutron polarization to follow the magnetic field as the field direction changes.

Two aspects have to be considered: (i) the Larmor precession of the neutron spin, i.e. the evolution of the components of $\vec{P}$ perpendicular to the magnetic field, and (ii) the component of $\vec{P}$ along the magnetic field $\vec{B}$. The evolution of the angle between $\vec{P}$ and $\vec{B}$ is of importance for the neutron adiabaticity. The adiabaticity is a measurement of part (ii). The evolution of $\vec{P}$ along the neutron flight path can be determined by a combination of 3D magnetic field simulations together with neutron ray-tracing programs, with the simulations verified by measurements of the magnetic field profile of the setup. Our experiment belongs to a class of polarized neutron experiments known as uniaxial polarization analysis [1], which measures the component of neutron polarization along $\vec{B}$. This means that it does not measure the components of $\vec{P}$ perpendicular to the magnetic field, and so the effects of

* Corresponding author: annika.stellhorn@sljus.lu.se

Larmor precession are not measured. The change of the component of $\vec{P}$ along $\vec{B}$ by the sample is one of the main focuses of uniaxial polarization analysis. Other techniques do measure the component of $\vec{P}$ perpendicular to $\vec{B}$, such as spherical polarimetry techniques using CRYOPAD [2] or MuPAD [3], or methods like spin-echo SANS [4].

To model the neutron scattering cross-section from polarized GISANS experiments, we use the Distorted Wave Born Approximation (DWBA) extended by the paracrystal theory, and with an appropriate model for the domain and domain wall pattern [5-7], which describes the lateral correlations probed by GISANS. This model of the domain wall pattern includes a number of sample-specific parameters (e.g. domain and domain wall width, the in-plane distribution and correlation length of the domain structure, and closure domain sizes). In addition, if the neutron polarization does deviate from pointing along the magnetic field axis during spin-transport between the polarizer and the sample, and between the sample and the analyzer, this will affect the result, introducing additional parameters into the simulation.

Together, the combination of magnetic field modelling, analysis of the neutron polarization by ray tracing, and simulations of the scattering cross-section, allow us to minimize the simulation parameters needed for the analysis of polarized GISANS experiments on complex magnetic structures. In this paper, we provide a detailed example of this approach, illustrating how important it is when interpreting experimental results. First, we outline the necessary information about the specific sample, and show some polarized GISANS data collected from it. We demonstrate how the magnetic field simulations combined with neutron ray-tracing can be used for data analysis on the magnetic domain structure by the DWBA. Finally, we discuss the effect of an imperfect neutron spin transport on the analysis of the magnetic domain structure by the DWBA.

2 Polarized GISANS on high-PMA FePd heterostructures

2.1 Magnetic domains and domain walls

$L1_0$-ordered FePd thin films with a high structural ordering of the Fe and Pd atoms give rise to a high perpendicular magnetic anisotropy (PMA), described by the uniaxial magnetocrystalline anisotropy constant K_u. In films with the crystalline *c* axis parallel to the surface normal, this leads to the formation of two set of domains with their magnetization respectively parallel or antiparallel to the film normal [8]. The specific forms of domains observed depends strongly on the way the film is grown, and the strength of the magneto-crystalline anisotropy energy E_{ma} and shape anisotropy energy E_{sh} can be effectively controlled [9-11]. These terms then determine the domain width, the domain wall width, the size of any closure domains, and the stray field strength associated with the magnetic domains. For materials like these FePd films, the ratio of the magnetocrystalline anisotropy constant K_u and the shape anisotropy constant K_{sh} is a measure of the strength of the PMA:

$$K = \frac{K_u}{K_{sh}} \tag{2}$$

If $K > 1$, high PMA is achieved. Further details are given in Refs. 8 and 9. For our purposes here, the sample in question develops a maze-like structure (Fig. 1(a)) and consists of no or very small expected closure domains due to the very high PMA ($K = 2 \pm 0.1$). Following Ref. [12], Bloch domain walls form between the out-of-plane oriented magnetic domains (Fig. 1(b)). The *K*-value, saturation magnetization, coercive field, measured domain periods, as well as theoretical domain and domain wall sizes for the relevant sample are listed in Table 1.

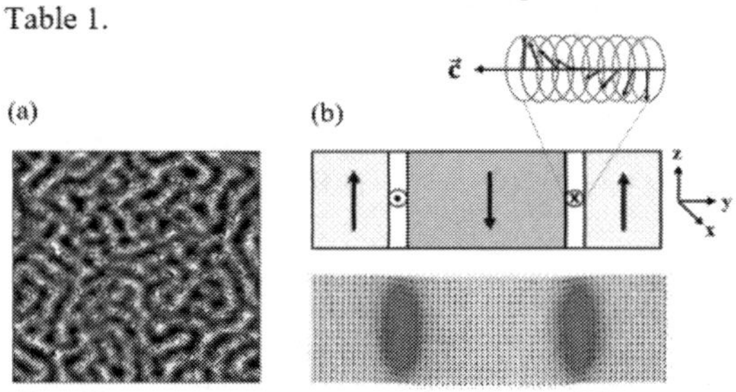

Fig. 1: (a) 3x3μm Magnetic Force Microscopy measurements on the domain pattern for a PMA FePd thin film with $K = 2 \pm 0.1$ at room temperature. (b) (top) sketch of the side-view of the magnetic domains in the out-of-plane direction (yellow and red) and the Bloch domain walls (white) containing a net chirality with chiral propagation along $\vec{C}$. (bottom) A visualization on the magnetic vector orientation by micromagnetic simulations on high-PMA FePd with $K = 1.5$ by van der Laan *et al.* [13].

Table 1. Magnetic properties of the investigated high-PMA FePd thin film [10]. The theoretical domain wall size is calculated based on the exchange constant $A_{ex} = 10^{-11}$ J/m [14].

K-value	2 ± 0.1
Saturation magnetization	$(1.09 \pm 0.05)\ 10^6$ A/m
Coercive field	$H_{c,\perp} = (6 \pm 2)$ mT; $H_{c,\parallel} = (31 \pm 1)$ mT
Measured domain width	(110 ± 5) nm
Theoretical domain wall size	≈ 8 nm

The sketch in Fig. 1(b) assumes chiral Bloch walls with a preferred orientation of chirality $\vec{C}$ parallel to the thin film surface (i.e., in the *x*-*y*-plane). Considering the centrosymmetric $L1_0$-structural phase of FePd without inversion asymmetry, no bulk-like Dzyaloshinskii–Moriya interaction (DMI) is favored in the thick FePd film investigated in this study ($d_{FePd} = 44 \pm 5$ nm). Additionally, the domain width $w_D = 110 \pm 5$ nm is large in comparison to the Bloch domain wall width $w_{DW} \approx 8$ nm (see Table 1) and so the magnetic exchange length in FePd does not support a magnetic interaction between neighboring domain walls. In summary, the formation of a net chirality from Bloch domain walls is not expected.

However, the actual sample is a heterostructure stack of Pd (cap, 1nm) / Nb (39nm) / FePd(44nm) / Pd (buffer, 60nm) / Cr (seed, 1nm) / MgO (substrate), where symmetry breaking at the interface of the FePd and the heavy metal buffer layer Pd may lead to interlayer DMI [15]. In multilayer structures composed of ultrathin films, this has been shown to lead to chiral Néel walls [15], but Pollard *et al.* [16] have shown that a combination of the interfacial symmetry breaking and the demagnetization field resulting from dipole-dipole interactions can favor the formation of chiral Bloch walls. In this experiment, we have observed asymmetric GISANS peaks in the spin flip (SF) channel that are generated by scattering from the Bloch domain walls. As will be explained in the following sections, nuclear-magnetic interactions can be ruled out as the origin of the SF asymmetry and hence it can be explained only by the existence of chiral domain walls. To be sensitive to such chiral structures, the neutron polarization direction $\vec{P}$ must have components parallel to the propagation direction of the chirality $\vec{C}$, which for Bloch domain walls lies within the surface plane (Fig. 1b). In our experiment, the expected neutron polarization is parallel to the external magnetic field applied perpendicular to the sample surface. This motivated a detailed investigation of the magnetic field vector at the sample position and the neutron adiabaticity along the beam path.

2.2 Polarized GISANS

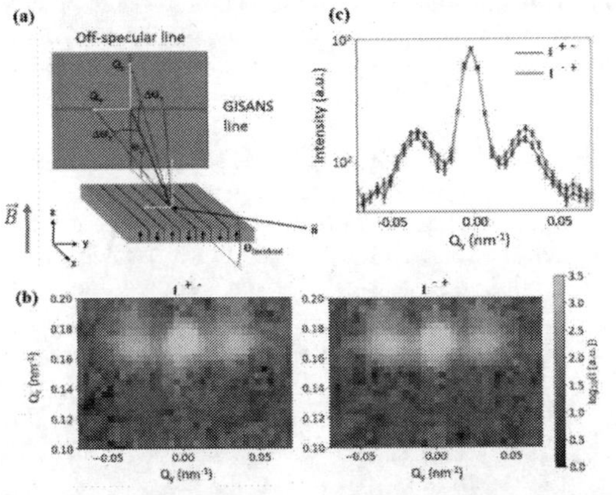

Fig. 2. (a) The scattering geometry of the polarized GISANS measurements that probe $I(Q_y,Q_z)$. The specular and GISANS lines are shown on the detector in red; the magnetic field is applied along the z-axis. (b) Measured SF intensity maps in the Q_y-Q_z plane for the channels I^{+-} (left) and I^{-+} (right) at $T = (10.10 \pm 0.02)$ K. The line at $Q_z = 0.17$ nm^{-1} corresponds to the GISANS line, *n* denotes the incident neutron beam. (c) $I(Q_y)$ integrated over 0.137 nm^{-1} < Q_z < 0.2 nm^{-1} for the data in panel (b).

In combination with polarization analysis, GISANS can provide insight into both in- and out-of-plane oriented chiral magnetic domain walls, and is therefore a logical choice for studying domain structures of the type that can be engineered into FePd thin films [11]. The neutron beam comes in with an angle close to the critical angle for total external reflection, θ_c. Using the small-angle approximation, the scattering vector $\vec{Q}$ is described by Eq. 3 [17], with the incident- and reflected angles θ_i and θ_f, as well as the off-specular scattered angles $\Delta\theta_x$ and $\Delta\theta_y$ (see Fig. 2(a)). The Q_y-line at $\theta_i = \theta_f$ is called the GISANS-line.

$$\begin{pmatrix} Q_x \\ Q_y \\ Q_z \end{pmatrix} = \frac{2\pi}{\lambda} \begin{pmatrix} \theta_i \Delta\theta_x + \frac{(\Delta\theta_x)^2}{2} + \frac{(\Delta\theta_y)^2}{2} \\ \Delta\theta_y \\ 2\theta_i + \Delta\theta_x \end{pmatrix} \quad (3)$$

Fig. 2(b) shows the SF channels of a polarized GISANS measurement on our high-PMA FePd thin film at $T = 10$ K and θ_i close to θ_c, measured at the vSANS beamline at NIST [18,19], with Q_z-integrated $I(Q_y)$ linecuts integrated over the range 0.137 nm^{-1} < Q_z <0.2 nm^{-1} shown in Fig. 2(c). The experiment was carried out with a sample-to-detector distance of 15 m and a neutron wavelength of 8 Å with a wavelength spread of 12 %. A guide field perpendicular to the thin film surface of $B_\perp = 2.6$ mT was applied. This configuration gives rise to GISANS peaks in the SF channels at $Q_y = \pm 0.033$ nm^{-1}, corresponding to a period of 96 nm. The peaks were generated by scattering from the Bloch domain walls which have a magnetic induction perpendicular to the applied field direction. A specular peak arising due to inefficiencies in the polarizer, spin-transport, spin flipper, and the analyzer. If the polarization components and spin-transport in between are perfect, there should be no specular peak at this position. The whole-system setup without the sample has a flipping ratio of 45, with the analyser efficiency previously measured to be $P_{an} = 0.988$.

The measurements show a clear asymmetry between the left and right GISANS peak (Fig. 2(c)). This arises due to the non-zero average chirality of the Bloch domain walls in the sample, which we now illustrate.

The total elastic scattering cross section for the interaction of a neutron beam with a magnetic sample is:

$$\sigma = \sigma_N + \sigma_M + \sigma_{NM} + \sigma_{M\times M} \quad (4)$$

where σ_N is the nuclear scattering contribution, σ_M is the magnetic scattering from non-chiral magnetic structure, σ_{NM} is the nuclear-magnetic interference term, and $\sigma_{M\times M}$ is an additional contribution that is only non-zero for chiral magnetic arrangements [20-22]. To describe the scattering patterns obtained from polarized measurements, the incident and final neutron polarizations need to be taken into account, such that the total elastic scattering cross section is

$$\vec{P'} I = \vec{P}\left(\tilde{N}N - \tfrac{1}{3} I_{si}\right) + \left(\vec{P} \cdot \vec{\tilde{M}}_\perp\right)\vec{M}_\perp + \left(\vec{P} \cdot \vec{M}_\perp\right)\vec{\tilde{M}}_\perp - \vec{P}\left(\vec{\tilde{M}}_\perp \cdot \vec{M}_\perp\right) + iN\left(\vec{P} \times \vec{\tilde{M}}_\perp\right) - i\tilde{N}\left(\vec{P} \times \vec{M}_\perp\right) + N\vec{\tilde{M}}_\perp + \tilde{N}\vec{M}_\perp - i\left(\vec{\tilde{M}}_\perp \times \vec{M}_\perp\right) \quad (5)$$

where $\vec{P'}$ is the outgoing polarization of the beam, $\vec{P}$ is the ingoing polarization, *N* is Fourier transform of the nuclear structure factor, with its complex conjugate $\widetilde{N}$, $\vec{M}_\perp$ is the Fourier transform of the sample magnetization component perpendicular to the scattering vector $\vec{Q}$, with complex conjugate $\vec{\widetilde{M}}_\perp$, and I_{si} the nuclear spin-incoherent scattering intensity. This shows that only magnetization components perpendicular to $\vec{Q}$ are measured.

From Eq. 5, for a specific polarization direction ν in any Cartesian direction, the asymmetric scattering contributions to σ are [22]:

$$\Delta^{SF}{}_{\nu} \approx 2i\left(\vec{\widetilde{M}}_\perp \times \vec{M}_\perp\right)_\nu \quad (6)$$

$$\Delta^{NSF}{}_{\nu} \approx 2\Re\left(N\vec{\widetilde{M}}_{\perp,\nu}\right) \quad (7)$$

for the SF and non-spin-flip (NSF) channels respectively. A contribution from any chirality with propagation vector $\vec{C}$ parallel to $\left(\vec{\widetilde{M}}_\perp \times \vec{M}_\perp\right)_\nu$ in the direction of the neutron polarization will therefore lead to asymmetric scattering peaks in the SF channels.

For our polarized GISANS measurements, we have used a setup with guide field direction perpendicular to the thin film surface and hence perpendicular to $\vec{C}$ of Bloch domain walls in high-PMA FePd as shown in Fig. 1. From Eq. 6, only components of $\vec{C}$ parallel to $\vec{P}$ will lead to asymmetric spin-flip GISANS peaks. The observed asymmetry in Fig. 2(c) could therefore arise due to a misalignment between $\vec{P}$ and the sample surface normal.

In Section 3, we discuss how the polarization direction may be affected by adiabaticity during spin-transport along the neutron flight path. First, however, we describe the magnetic model used to calculate the total scattering cross section, based on previous models developed by Toperverg and Kentzinger [5,6].

2.3 Description of the DWBA simulation of polarized GISANS measurements on high-PMA FePd

Within the Distorted Wave Born Approximation (DWBA), the neutron scattering cross section $d\sigma/d\Omega$ from a GISANS experiment can be calculated assuming a depth-dependent, laterally homogeneous magnetic potential $\hat{V}_{0,l}(z)$ inside each layer l, which is perturbed by a lateral inhomogeneous magnetic potential $\hat{V}_{pert,l}(\vec{s})$, where $\vec{s}$ is a 2D vector within the plane of the sample surface (Fig. 3).

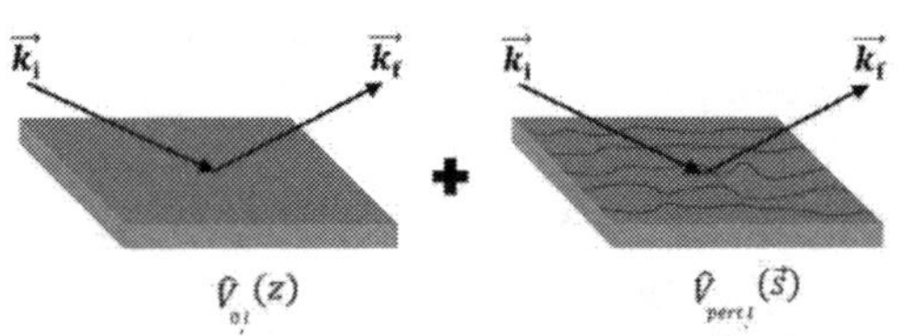

Fig. 3. Non-perturbed scattering potential $\hat{V}_{0,l}(z)$ of a homogeneous sample and perturbation potential $\hat{V}_{pert,l}(\vec{s})$ of a sample with lateral inhomogeneous structural or magnetic density [10].

This leads to off-specular scattering:

$$\frac{d\sigma}{d\Omega}(\theta_i,\theta_f,\vec{Q}_{//}) = \left|\frac{m_n}{2\pi\hbar^2}\sum_l\langle \psi_{f,l}\,|\hat{V}_{pert,l}(\vec{s})|\psi_{i,l}\rangle\right|^2 \quad (8)$$

with $\psi_{f,l}$ being the distorted wavefunction inside layer l after the scattering event, $\psi_{i,l}$ the distorted wavefunction inside l just before the scattering event, and $\vec{Q}_{//}$ the in-plane component of the scattering wave vector. Further details are given in Ref. [6].

The perturbation potential $\hat{V}_{pert,l}(\vec{s})$ is given by the sum of the nuclear and magnetic scattering length density fluctuations $\tilde{\rho}_{N,l}(\vec{s})$ and $\tilde{\rho}_{M,l}(\vec{s})$ respectively:

$$\hat{V}_{pert,l}(\vec{s}) = \frac{2\pi\hbar^2}{m_n}\left(\tilde{\rho}_{N,l}(\vec{s})\cdot\hat{1} + \tilde{\rho}_{M,l}(\vec{s})\vec{\hat{\sigma}}\cdot\vec{b}_l\right) \quad (9)$$

where $\vec{\hat{\sigma}}$ is the vector of Pauli matrices and $\vec{b}_l$ is the unit vector parallel to the magnetic field in layer l. Theoretical descriptions of $\tilde{\rho}_{M,l}(\vec{s})$ of the magnetic domain structure of a high-PMA material, like FePd, using the DWBA have been formulated by Toperverg and Kentzinger [5,6]. In a material with a maze domain structure, this model has to be extended by the paracrystal theory [7] and integration over the in-plane oriented domain orientations. The maze domain pattern has to be divided into subunits of an ordered subsystem, each consisting of two antiparallel out-of-plane oriented domains and one Bloch domain wall (Fig. 4(a)). These subunits are rotated within the surface plane; Fig. 4(b,c) shows an example of such a maze structure and division of magnetic domains into small units assuming a parallel alignment of the magnetic domains within one sub-unit. L_x represents the correlation length of one ordered subsystem parallel to the domain wall (Fig. 4(c)). The total scattering cross section can be written as sum over the scattering from all possible sub-units. The process is described in more detail by Stellhorn [10]. For the high-PMA FePd layer used here, MFM measurements show a maze distribution of magnetic domains with an average direction θ_{av} tilted away from the incident neutron beam direction of about 20°, and with a distribution of domain directions around the mean

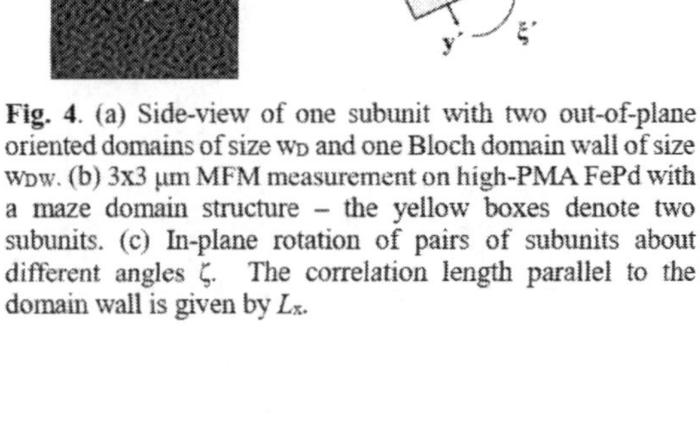

Fig. 4. (a) Side-view of one subunit with two out-of-plane oriented domains of size W_D and one Bloch domain wall of size W_{DW}. (b) 3x3 μm MFM measurement on high-PMA FePd with a maze domain structure – the yellow boxes denote two subunits. (c) In-plane rotation of pairs of subunits about different angles ζ. The correlation length parallel to the domain wall is given by L_x.

direction which can be roughly approximated with a Gaussian-type shape (Fig. 5). This distribution is used in our model. The reasons for the preferential in-plane domain orientation are not germane to this article, but are discussed elsewhere [10,11].

Within this model, the incident and outgoing polarization vectors $\vec{\boldsymbol{P}}_i$ and $\vec{\boldsymbol{P}}_f$ are included via the density matrices of the incident and reflected neutron beam $\hat{\rho}_i = \frac{1}{2}\left(\hat{1} + \vec{\boldsymbol{P}}_i\vec{\boldsymbol{\sigma}}\right)$ and $\hat{\rho}_f = \frac{1}{2}\left(\hat{1} + \vec{\boldsymbol{P}}_f\vec{\boldsymbol{\sigma}}\right)$ using:

$$\frac{d\sigma}{d\Omega}(\theta_i, \theta_f, \vec{\boldsymbol{Q}}_{//}) = tr\left(\hat{\rho}_f \hat{F}_{fi} \hat{\rho}_i \hat{F}_{fi}^{\dagger}\right) \qquad (10)$$

where $\hat{F}_{fi} = \hat{F}_{fi}(\vec{\boldsymbol{Q}}_{//})$ is the scattering amplitude operator (in the space of spin states) and $\hat{F}_{fi}^{\dagger}$ its Hermitian conjugate.

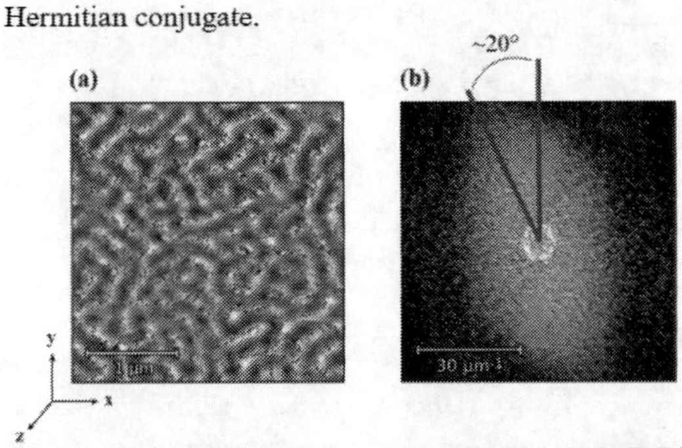

Fig. 5. (a) 3x3 µm MFM measurement on the high-PMA FePd sample measured by GISANS. A maze domain structure is seen. (b) The Fourier transform of panel (a), showing a nearly Gaussian-type domain distribution around a mean angle tilted 20° from the incident neutron beam direction.

3 Neutron spin adiabaticity

3.1 Magnetic field measurements & simulations using COMSOL

In this section we discuss how the polarization direction may have been affected by adiabaticity during spin-transport along the beam path: The setup at vSANS, NIST [17] is illustrated in Fig. 6. The neutron beam went from right to left. The incoming neutrons' polarization in the guide field was oriented in the *y*-direction, then rotated into the *z*-direction (axis parallel to the surface normal) at the sample position by the magnetic field applied from the pole-pieces of the Titan electromagnet, and was again rotated in the *x*-direction for the ³He-cell analyzer. Two guide fields just in front of and behind the pole-pieces (not shown in Fig. 6, positions marked in Fig. 7) kept the magnetic field along *y* and rotated it to the *x*-direction, respectively. A precise measurement of the 3D magnetic field configuration along the beam path has been carried out using a 3D Hall probe. The results are compared with simulations of the magnetic field configuration using finite element computational software COMSOL Multiphysics [23].

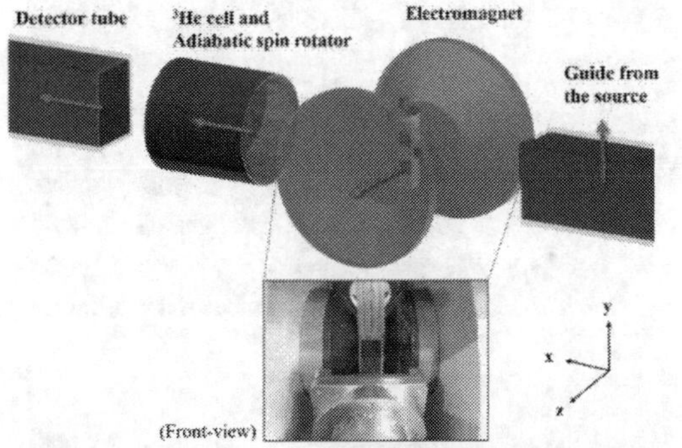

Fig. 6. (a) Beamline setup and polarization direction (red arrows) of the polarized GISANS experiments at vSANS, NIST [18], using an electromagnet with pole pieces extending close to the cryoshield containing the sample holder, visible in the inset on the bottom.

The neutron adiabaticity along the beam path may be calculated if the magnetic field configuration of the whole beamline setup is known. This can be simulated using COMSOL Multiphysics. Individual beamline components are included with their basic physical properties (such as the dielectric constant or magnetic permeability). Sources of magnetic field are included either as permanent magnets or current-carrying coils as relevant. The 3D magnetic field distribution ((B_x, B_y, B_z) in Tesla) is then calculated in a predefined area as a function of the position (x,y,z) along the neutron flight path.

Care needs to be taken when defining the mesh (shape, size, density, and number of elements within the geometry) on which the interpolation and accuracy of the solution is based on: a fine mesh that scales with the size of the object is imposed on the focused regions and the magnetic components, moderate mesh size in regions connecting them, and coarse meshing can be used everywhere else. Our input model is shown in Fig. 7. This yields a magnetic field configuration $\vec{\boldsymbol{B}}_{COMSOL}$ and adiabaticity A_{COMSOL} as given in Fig. 8.

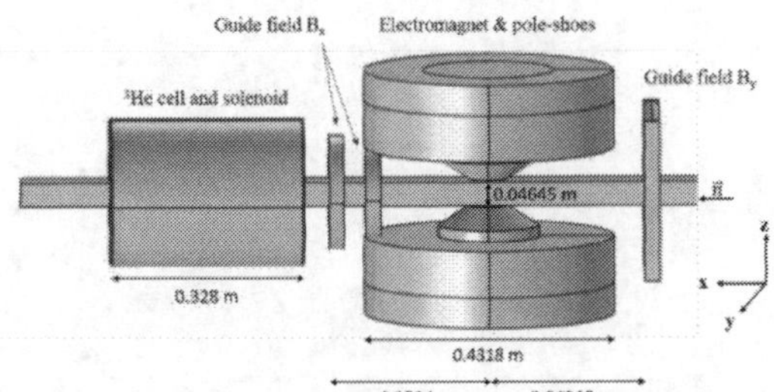

Fig. 7. The input model to COMSOL of the components of the vSANS beamline at NIST [18] used for the polarized GISANS measurements discussed here.

 https://doi.org/10.1051/epjconf/202328606002

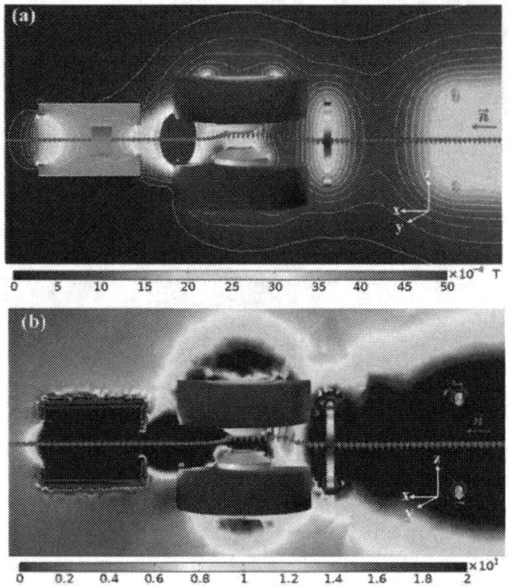

Fig. 8. (a) The calculated magnetic field map of the polarized GISANS setup at vSANS, NIST. The colour scale denotes the strength of magnetic field $B = |(B_x,B_y,B_z)|$ in (10^{-4} T). The red arrows denote the magnetic field direction along the neutron beam path. The neutrons are incident from the right moving left, as in Figures 6 and 7. Two guide fields in front of and behind the sample have been installed to adiabatically rotate the neutron spin. (b) The calculated adiabaticity map. There is a region just before the sample with a critically low adiabaticity value. The magnetic field vector along the neutron beam path is again shown by red arrows denoting the magnetic field vector.

These computational results were then compared with magnetic field measurements of the actual setup used at vSANS (NIST), made using a 3D Hall probe mounted on a fixed stage which could be slid through the setup, along the neutron beam path. The resulting measured field, $(B_x,B_y,B_z)_{meas}(x)$ is compared with $\vec{B}_{COMSOL}$ and A_{COMSOL} in Fig. 9.

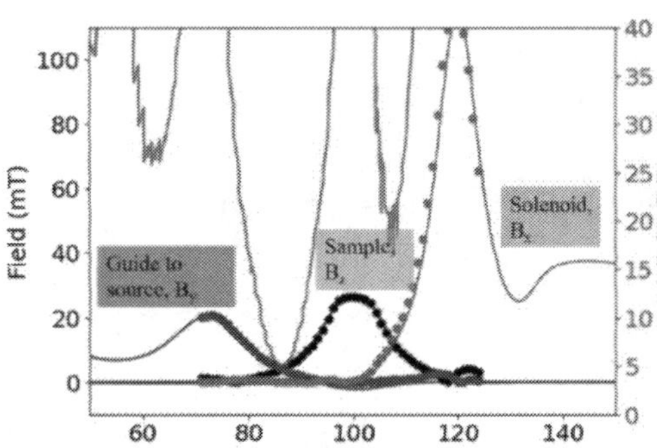

Fig. 9. The measured (circles) and simulated (lines) magnetic field components (B_x,B_y,B_z) (left axis) and the calculated adiabaticity (right axis) as a function of the position x (measured as the distance from the opening of the source guide) along the neutron flight path. The labelled guide fields are as shown in Fig. 6.

From this, two conclusions can be drawn: (i) a critically low adiabaticity $A < 5$ is reached just in front of the electromagnet during the rotation of polarization from the y to the z-direction, indicating a possible misalignment of $\vec{P}$ with respect to $\vec{B}$, and (ii) the measured and simulated field strengths coincide and so $\vec{B}_{COMSOL}$ can be used as an input to the neutron ray-tracing program McStas 2.7 [24,25].

3.2 Neutron polarization simulations using McStas

McStas is a neutron ray-tracing Monte-Carlo simulation package [24,25]. It simulates the parameters of an ensemble of neutrons (the intensity, position, velocity, and polarization) starting from a source and passing through defined beamline components. Here, we concentrate on how the neutron beam polarization $\vec{P}_{\text{McStas}}$ changes as the neutrons travel through the magnetic field map calculated using the COMSOL simulations described and validated above. The most relevant parameters defining the neutron polarization are the step-size of the magnetic field imported from the COMSOL simulations, the angular accuracy (the threshold below which two magnetic fields are considered to be parallel), the neutron wavelength and distribution, and the incident polarization after the polarizer. The resulting neutron polarization $\vec{P}_{McStas,sample}$ at the sample position of the neutrons passing through $\vec{B}_{COMSOL}$ is then compared with the applied field direction $\vec{B}_{COMSOL,sample}$ at the sample position via the angle α, defined as

$$\alpha = \cos^{-1}\left(\frac{\vec{B}_{COMSOL,sample}\cdot \vec{P}_{\text{McStas,sample}}}{|\vec{B}_{COMSOL,sample}||\vec{P}_{\text{McStas,sample}}|}\right) \quad (11)$$

The neutron polarization in out-of-plane direction (along the applied magnetic field) and the angle α between $\vec{P}_{\text{McStas,sample}}$ and $\vec{B}_{COMSOL,sample}$ as function of the transversal positions y and z around the sample position are depicted together with the sketched position and size of the investigated FePd thin film in Fig. 10. Table 2 lists

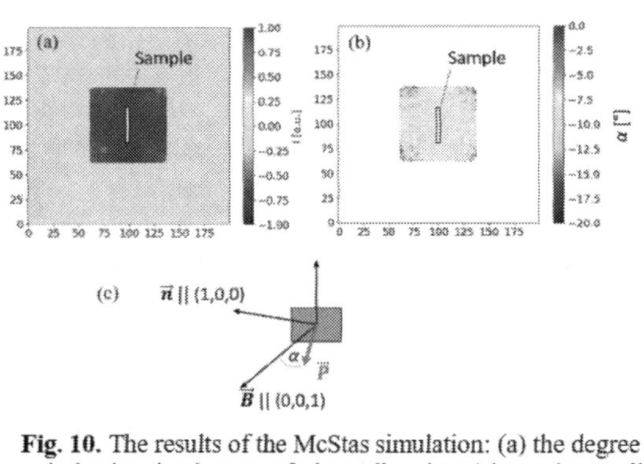

Fig. 10. The results of the McStas simulation: (a) the degree of polarization in the out-of-plane direction (along the applied magnetic field direction), and (b) the angle α between the $\vec{P}_{McStas,sample}$ and $\vec{B}_{COMSOL,sample}$. (c) A sketch of the sample geometry showing the incident neutron beam direction (x), the magnetic field vector along the surface normal (z), and the tilted polarization $\vec{P}$.

https://doi.org/10.1051/epjconf/202328606002

the relevant parameters resulting from the McStas simulations.

Table 2. Parameters and results from the McStas simulation

Step size of magnetic field	2.5 mm
Angular accuracy	1°
Wavelength + distribution	$\lambda = 8$ Å , $\Delta\lambda = 0.96$
Initial polarization	$\vec{\boldsymbol{P}}_{McStas,in} = (0, 0.97, 0)$
Resulting polarization at the sample position	$\vec{\boldsymbol{P}}_{McStas,sample} = (0.015, -0.077, -0.966)$
Magnetic field vector at the sample position	$\vec{\boldsymbol{B}}_{COMSOL,sample} = (0.1, -0.2, 2.7)$ (mT)
Angle α between $\vec{P}_{\mathrm{McStas,sample}}$ and $\vec{B}_{COMSOL,sample}$	$\alpha = 9°$

The simulation yields an angle of α = 9° between the neutron polarization and the magnetic field direction. This result can then be included in the simulations of the magnetic scattering pattern of polarized GISANS measurements described in Section 2.3 by including the value of $\vec{\boldsymbol{P}}_{McStas,sample}$. Depolarization of the neutron beam due to stray fields from the sample can lead to an additional rotation of $\vec{\boldsymbol{P}}$. We note that the electromagnet, which includes the sample stage, was tilted 4.7° away from the horizontal; this was included in the COMSOL simulations. Hence, the given $\vec{\boldsymbol{B}}_{COMSOL,sample}$ includes this tilt of the applied field at the electromagnet w.r.t to the z-direction. As the sample stage is connected to the electromagnet, the magnetic field orientation with respect to the sample was unchanged by this tilt and did not contribute to the angle between neutron polarization and the sample normal.

The strong agreement between the measured and simulated magnetic fields gives confidence in the simulated adiabaticity shown in Fig. 9. Critically low values of A for the neutron beam are reached in front of the sample position. As is clear from Eq. 5, this can have an effect on the resulting magnetic scattering pattern, as it means that scattering from the chiral magnetic moments with $\vec{\boldsymbol{M}}_{\perp} \times \vec{\boldsymbol{M}}_{\perp}$ could become visible as the asymmetry measured $I(Q_y, Q_z)$ in Fig. 2.Further improvements of the presented simulations and resulting polarization directions are still being performed, including comparisons with the neutron ray tracing software VITESS, and checking finer angular accuracies for the McStas simulation. However, the presented results show the necessity of a detailed investigation of the neutron beam adiabaticity and polarization throughout the beamline, to aid in constraining the simulation parameters required for calculations of the scattering cross section, particularly complex magnetic models.

4 Conclusions

Our results show that neutron polarization simulations are an integral part of experiments relying on polarized neutrons. They highlight that significant neutron polarization misalignments may occur even when the magnetic field direction at the sample point and the neutron polarization measurements look promising. The detailed simulations help constrain calculations of the cross section that start from the sample properties by providing detailed inputs for the neutron polarization and the magnetic field. They also provide aspect for a comprehensive data analysis of magnetic scattering cross sections, especially given that the magnetic field direction at the point of sample scattering and the neutron polarization may look good even while significant neutron polarization misalignment is occurring. Benefits are the minimization of unknown simulation parameters and a global understanding of the influence of the magnetic field vector as a function of the neutron beam path. Especially guide field conditions, with the constraint of low field values at the sample position, and a non-adiabatic neutron transport imply the need of precise scattering cross section simulations. If concurrently at the same time, complex magnetization distributions within the investigated samples have to be considered, the need of minimizing unknown parameters is even stronger.

We present a reduction of unknown parameters for the simulation of scattering cross sections from polarized GISANS measurements with polarization analysis. We have performed a comprehensive investigation of the neutron polarization orientation with respect to the applied magnetic field vector using magnetic field modelling in COMSOL and neutron ray tracing in McStas, and a comparison with magnetic field vectors obtained from measurements along the neutron beam path of the employed polarized GISANS setup. As a result, a deviation from the direction of polarization with respect to the applied magnetic field direction has been quantified, opening the possibility to be considered in simulations of the scattering cross section of the investigated sample based on the Distorted-Wave-Born-Approximation.

Acknowledgements

This research is funded in part by the Tillväxtverket grant. We acknowledge the support of the National Institute of Standards and Technology, U.S. Department of Commerce, in providing the neutron research facilities used in this work, and thank for the kind help during the measurements. Additionally, we acknowledge following support:

1. Discussions on the neutron polarization and adiabaticity at vSANS, NIST, with Julie Borchers
2. Support for the measurements and their analysis of polarized GISANS by Thomas Brückel & the JCNS

The identification of any commercial product or trade name does not imply endorsement or recommendation by the National Institute of Standards and Technology.

References

1. A. Wiedenmann, Physica B **356**, 246–253 (2005).,
2. F. Tasset, P. J. Brown, J. B. Forsyth, J. Appl. Phys. **63**, 3606-3608 (1988).
3. M. Janoschek, S. Klimko, R. Gähler, B. Roessli, P. Böni, Phys. B: Condens. Matter **397**, 125-130 (2007).
4. M. T. Rekveldt, Nucl. Instrum. Methods Phys. Res. B **114**, 366-370 (1996).
5. B. Toperverg, Phys. Met. Metallogr. **116**, 1337-1375 (2015)
6. E. Kentzinger, U. Rücker, B. Toperverg, F. Ott, T. Brückel, Phys. Rev. B **77**, 104435 (2008)
7. D. Korolkov, P. Busch, L. Willner, E. Kentzinger, U. Rücker, A. Paul, H. Frielinghaus, T. Brückel, J. Appl. Crystallogr. **45**, 245-254 (2012)
8. D. E. Laughlin, K. Srinivasan, M. Tanase, and L. Wang, Scr. Mater. **53**, 383-388 (2005)
9. J. M. D. Coey, Magnetism and Magnetic Materials, Cambridge University Press (2010)
10. A. Stellhorn, PhD Thesis "Interplay of proximity effects in superconductor/ferromagnet heterostructures, RWTH Aachen University (2021)
11. A. Stellhorn, A. Sarkar, E. Kentzinger, M. Waschk, P. Schöffmann, S. Schröder, G. Abuladze, Z. Fu, V. Pipich, T. Brückel, J. Magn. Magn. Mater. **476**, 483-486 (2019).
12. A. Hubert and R. Schäfer, "Magnetic Domains", Springer-Verlag GmbH (2008)
13. G. van der Laan, K. Chesnel, M. Belakhovsky, A. Marty, F. Livet, S. Collins, E. Dudzik, A. Haznar, J. Attané, Superlattices Microstruct. **34**, 107-126 (2003)
14. D. Ravelosona, A. Cebollada, F. Briones, C. Diaz-Paniagua, M. A. Hidalgo, F. Batallan, Phys. Rev. B **59**, 4322-4326 (1999)
15. A. Fert, V. Cros, J. Sampaio, Nat. Nanotechnol. **8**, 152-156 (2013)
16. S. D. Pollard, J. A. Garlow, K.-W. Kim, S. Cheng, K. Cai, Y. Zhu, H. Yang, Phys. Rev. Lett. **125**, 227203 (2020)
17. C. Fermon, F. Ott, B. Gilles, A. Marty, A. Menelle, Y. Samson, G. Lego, G. Francinet, Phys. B **267-268**, 162-167 (1999)
18. W. Chen, K. L. Krycka, S. M. Watson, J. G. Barker, J. Gaudet, H. Burral, J. A. Borchers, J. Phys. Conf. Ser. **2481**, 012006 (2023)
19. J.G. Barker J.J. Moyer, S.R. Kline, G.V. Jensen, J.C. Cook, C. Gagnon, E.G. Kellym J.P. Ghabot, N.C. Maliszewskyj, C. Parikh, W.C. Chen, C.J. Glinka, J. Appl. Crystallogr. **55**, 271-283 (2022)
20. M. Blume, Phys. Rev. **130**, 1670 (1963)
21. F. Tasset, P. J. Brown, E. Lelièvre-Berna, T. Roberts, S. Pujol, J. Allibon, E. Bougeat-Lami, Physica B **267-268**, 69-74 (1999)
22. W. Schweika, J. Phys. Conf. Ser. **211**, 012026 (2010)
23. COMSOL Multiphysics® v. 6.1. www.comsol.com. COMSOL AB, Stockholm, Sweden.
24. P. Willendrup, K. Lefmann, J. Neutron Res. **22**, 1-16 (2020)
25. P. Willendrup, K. Lefmann, J. Neutron Res. **23**, 7-27 (2021)

Theoretical calculations of neutron scattering cross sections for tetrahydrofuran-containing clathrate hydrates at low temperature

Shuqi Xu[1,*], *Sara Isaline* Laporte[2], *Douglas D.* DiJulio[1], *Jose Ignacio* Marquez Damian[1], *Thomas* Kittelmann[1], *Marco* Bernasconi[2], *Davide* Campi[2], *Giuseppe* Gorini[3], and *Valentina* Santoro[1]

[1]European Spallation Source ERIC, Box 176, SE-221 00 Lund, Sweden
[2]Department of Materials Science, University of Milano-Bicocca, Via R. Cozzi 55, 20125 Milano, Italy
[3]Department of Physics, University of Milano-Bicocca, Piazza della Scienza 3, 20126 Milano, Italy

Abstract. Tetrahydrofuran-containing clathrate hydrate is considered as a potential cold and very cold neutron moderator material. The fully deuterated form is more promising because of the lower neutron absorption cross section of deuterium. In this work we present theoretical calculations of neutron scattering cross sections for hydrogenated and deuterated clathrates at low temperature. The scattering cross sections are generated by using the crystalline structure and phonon density of states obtained from density functional theory calculations. The theoretical neutron scattering cross sections serve to compare against existing or future experimental data.

1 Introduction

The European Spallation Source (ESS), currently under construction in Lund, Sweden, will be the most powerful pulsed neutron source in the world [1]. The current source is located above the spallation target and designed for high cold and thermal brightness [2]. A design study project named HighNESS is dedicated to develop a second source below the target which will provide cold, very cold and ultra cold neutrons with high intensity [3, 4]. This study includes exploration of promising novel reflector and moderator materials. For the purpose of cold neutron reflectors, nano-diaomond particles and magnesium hydride have been investigated [5–7]. A promising cold neutron moderator class of materials also included in this study are clathrate hydrates, which are the focus of this work.

Clathrate hydrates are ice-like compounds having a cage structure. Small molecules, such as methane, can be enclathrated in the cage, leading to stabilisation of the structure. Clathrate hydrate hosting oxygen is considered as a promising candidate due to neutron inelastic magnetic scattering with oxygen [8]. Tetrahydrofuran (THF)-containing clathrate hydrate has also been considered as an advanced cold neutron moderator [9]. The material can be produced directly from a stoichiometric THF/water mixture at normal pressure which makes it suitable for the large volume production needed for neutron moderators/reflectors. Rattling modes of THF molecules mostly overlap with the low energy translational modes of the cage molecules at about 7 and 10.5 meV, which are common to all type II clathrates [10]. However, THF also gives rise with a lower energy peak at about 5 meV in the phonon density of states [10] that might also contribute to the inelastic neutron cross section. Both THF and fully deuterated tetrahydrofuran (TDF)-clathrates [11] have been investigated by inelastic neutron scattering measurements [9, 11]. The lower neutron absorption cross section of deuterium makes the TDF-clathrate a more promising cold moderator candidate.

The objective of this work is to provide theoretical neutron scattering cross sections and scattering kernels for THF- and TDF-clathrates to be used for simulating measurements and design studies. Therefore, we performed density functional theory (DFT) calculations using CP2k [12] and phonopy [13] in order to obtain the crystalline structure and the phonon density of states (PDOS), which captures the dynamics of the atoms around their equilibrium positions. The DFT results were used to generate scattering kernels and cross-sections using the open source software package NCrystal [14–16]. The calculated scattering kernels and cross sections can be used to compare with the neutron transmission, diffraction and inelastic neutron scattering measurements performed in another work package of this project [17].

2 Density functional theory calculations

The initial configuration of the THF-containing clathrate hydrate in structure II was taken from Ref. [18]. The unit cell of the face centered cubic lattice contains 34 H_2O molecules and two THF molecules inserted in the large cages. The unit cell was replicated to build the conventional cubic cell with 136 H_2O and 8 THF molecules (see Fig. 1). The initial position of the THF molecules was chosen by performing classical molecular dynamics (MD) simulations with GROMACS [19], the SPC/E potential for

*e-mail: shuqi.xu@ess.eu

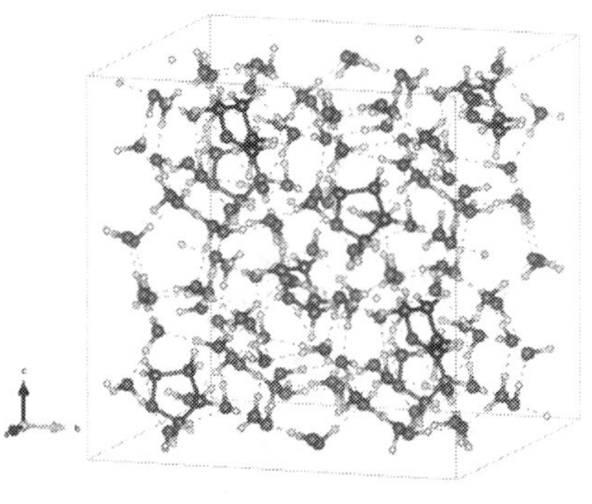

Figure 1. Crystalline structure of THF-containing clathrate hydrate (structure II) obtained from DFT calculations. THF molecules are inserted in eight large cages. TDF-clathrate shares the same crystalline structure.

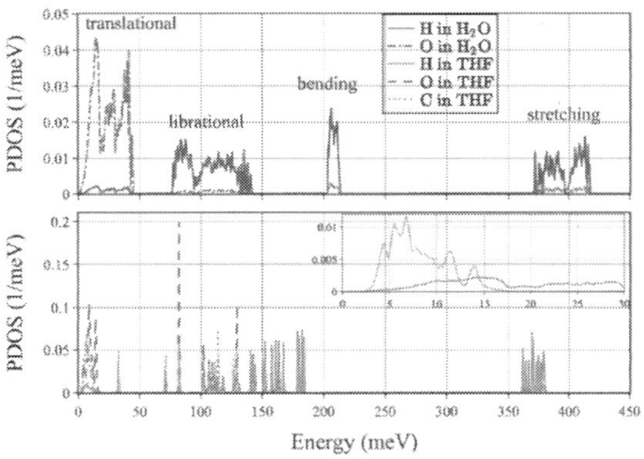

Figure 2. Normalized PDOS of THF-containing clathrate hydrate projected on different atomic species. The global PDOS can be obtained by summing the projected PDOS multiplied by the number of corresponding atoms in the unit cell.

water [20], and the general Amber force field for THF and THF-water interactions [21]. The initial structure was optimized by means of DFT calculations with CP2k [12]. We employed the Perdew-Burke-Ernzerhof approximation for the exchange and correlation functional [22] and norm conserving pseudopotentials [23]. Kohn-Sham orbitals were expanded in a triple-zeta-valence plus polarization basis set while the electronic density was expanded in plane waves up to a kinetic cutoff of 1000 Ry as implemented in CP2k. Brillouin Zone (BZ) integration for electrons was restricted to the Γ-point. Semiempirical van der Waals interactions were included according to Grimme (D3) [24].

The atomic positions were initially optimized at the experimental lattice parameter a = 17.13 Å [25] and then at different volumes. The resulting energy versus volume points were fitted by a Birch-Murnaghan equation of state which yielded a theoretical lattice parameter of 16.7 Å. As the contraction of the lattice parameter with respect to experiments leads to a stiffening of phonon frequencies, we preferred to compute the PDOS at the experimental lattice parameter.

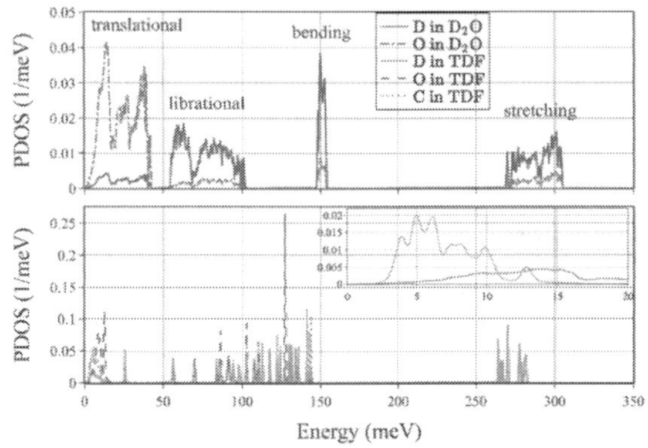

Figure 3. Normalized PDOS of fully deuterated TDF-containing clathrate hydrate projected on different atomic species. The global PDOS can be obtained by summing the projected PDOS multiplied by the number of corresponding atoms in the unit cell. The shift of the excitations to low energies is due to the increase of mass and moment of inertia of deuterium.

The phonons were computed using the CP2k code combined with phonopy [13]; the force constant matrix was computed from forces at finite atomic displacements in the conventional cubic cell (136 water molecules). The dynamical matrix, on a fine q-mesh in the BZ, was then built by phonopy. The coulombic long range contribution was computed analytically from effective charges and the dielectric constant obtained in turn from ionic forces and polarization at finite electric field within the Berry phase approach [26] implemented in the CP2k code.

The PDOS projected on different atomic species for the THF-containing clathrate hydrate are shown in Fig. 2. The upper panel of Fig. 2 shows the PDOS of hydrogen and oxygen bound in H_2O. The dynamics of the water molecules inherit the behaviours observed in ice [27]. At low energy, the translational modes of the H_2O molecules dominate. In this spectral region, the experimental neutron weighted PDOS of type II clathrate hydrates has two characteristic peaks at about 7 and 10.5 meV [11, 28]. This double peak feature is less evident in the theoretical PDOS projected on the water molecules. The position of the two peaks is also blue-shifted to about 10 and 14.5 meV. This double peak feature is instead well reproduced by classical simulations with rigid water molecules as reported for instance in Refs. [11, 29]. A possible improvement of the theoretical PDOS would then result from the use of the DFT-PDOS for the intramolecular and librational modes and the classical result for the PDOS in the translational low frequency region. The band from 70 meV to 150 meV is due to molecular librations. At higher energy, intramolecular modes are dominant with O-H bending and O-H stretching at around 200 meV and 400 meV, respectively.

The PDOS of atoms bound in the THF molecules are

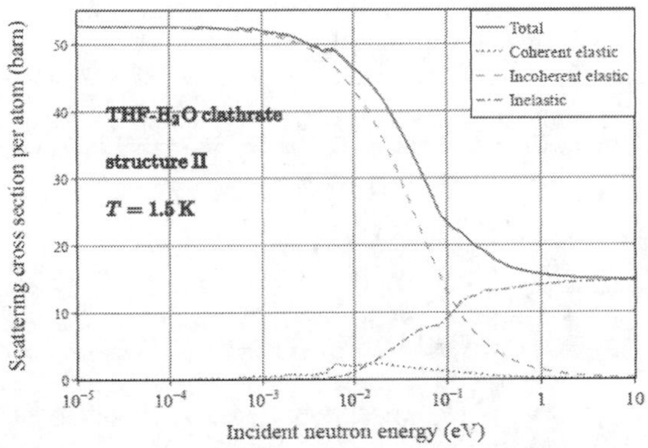

Figure 4. Neutron scattering cross sections of type II THF-containing clathrate hydrate.

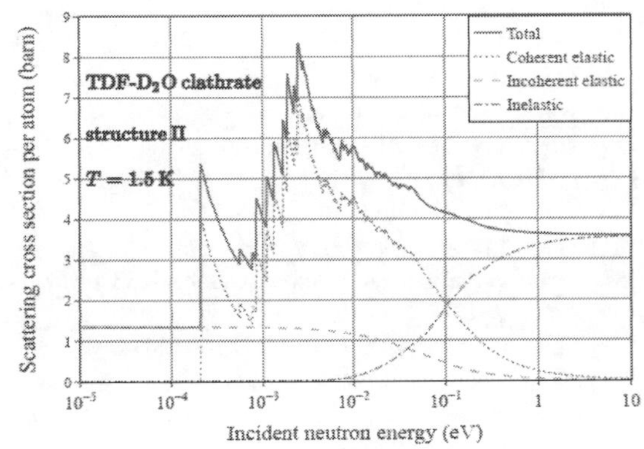

Figure 5. Neutron scattering cross sections of type II fully deuterated TDF-containing clathrate hydrate.

illustrated in the lower panel of Fig. 2. Several sharp peaks are observed in the energy range from 70 meV to 200 meV due to a weak coupling with the water molecules of the cage. Hydrogen bound in THF has excitations in a broad band from about 4 to 10 meV with contributions below the peak of the PDOS of hydrogen bound in H_2O, also considering the blueshift of the theoretical peaks of water compared to experiments that we mentioned above. This property makes THF-clathrate a better cold neutron moderator compared to ice.

The PDOS of fully deuterated TDF-clathrate projected on different atomic species are shown in Fig. 3. The PDOS are shifted to lower energies when substituting hydrogen by deuterium due to the increase of mass and moment of inertia, as in the case of ice [30]. The PDOS of the cage can be transferred to other clathrates, e.g., oxygen-clathrate, based on weak guest-host coupling assumption as discussed in Ref. [28].

3 Neutron scattering cross sections

The lattice parameters, atomic positions of the crystalline structure, and the PDOS were converted into input files for the NCrystal toolkit. Based on the Gaussian and incoherent approximations [31, 32], NCrystal is able to calculate the neutron scattering cross sections and scattering kernels including the physics of both coherent and incoherent elastic, and inelastic scattering in a wide range of materials. Fig. 4 shows the total scattering cross section of the THF-clathrate, which is the sum of coherent elastic, incoherent elastic and inelastic cross sections. Incoherent elastic scattering is dominant for neutron energy below a few meV because of the large incoherent cross section of hydrogen [33]. The coherent cross section of deuterium is more important than its incoherent part. Therefore, the coherent elastic scattering dominates in the case of TDF-clathrate as illustrated in Fig. 5. The large Bragg cutoff (around 0.2 meV or 20 Å) is an advantage for cold neutron moderation because neutrons can be reflected within the cage structure thus increasing the interactions with the guest molecules.

4 Conclusions

In this work we present current advance of the calculation of the neutron scattering cross sections of THF-containing clathrate hydrate and its fully deuterated form. The crystalline structure and phonon density of states obtained from DFT calculations have been used to compute the neutron scattering cross sections by means of the open source software package NCrystal. The resulting cross sections at different wavelength can be used to compare against existing or future measurements on THF- and TDF-clathrates. Clathrate hydrates can also be considered as potential reflector materials because of the large Bragg cutoff.

Acknowledgements

The authors wish to express their appreciation to Valentin Czamler from Institut Laue-Langevin for many helpful discussions. This work was funded by the HighNESS project at the European Spallation Source. HighNESS is funded by the European Framework for Research and Innovation Horizon 2020, under grant agreement 951782.

References

[1] R. Garoby, A. Vergara, H. Danared, I. Alonso, E. Bargallo, B. Cheymol, C. Darve, M. Eshraqi, H. Hassanzadegan, A. Jansson et al., Phys. Scr. **93**, 014001 (2017)

[2] L. Zanini et al., Nucl. Instrum and Methods in Phys. Res. A **925**, 33 (2019)

[3] V. Santoro, K.H. Andersen, D.D. DiJulio, E.B. Klinkby, T.M. Miller, D. Milstead, G. Muhrer, M. Strobl, A. Takibayev, L. Zanini et al., J. Neutron Res. **22**, 209 (2020)

[4] V. Santoro, K. Andersen, M. Bernasconi, M. Bertelsen, Y. Beßler, D. Campi, V. Czamler, D. Di Julio, E. Diane, K. Dunne et al., accepted for publication in Nucl. Sci. Eng. / arXiv preprint arXiv:2204.04051 (2023)

[5] N. Rizzi, J.I.M. Damian, T. Kittelmann, B. Lauritzen, E. Klinkby, Q. Estiez, V. Santoro, Nucl. Sci. Eng. pp. 1–9 (2023)

[6] K. Ramić, J.I. Marquez Damian, T. Kittelmann, D.D. Di Julio, D. Campi, M. Bernasconi, G. Gorini, V. Santoro, Nucl. Instrum. Methods Phys. Res. A **1027**, 166227 (2022)

[7] K. Ramić, J.I. Marquez Damian, D.D. Di Julio, T. Kittelmann, D. Campi, M. Bernasconi, A. Gosh, G. Gorini, N. Rizzi, E. Klinkby et al., Nucl. Sci. Eng. pp. 1–9 (2023)

[8] O. Zimmer, Phys. Rev. C **93**, 035503 (2016)

[9] H. Conrad, W. Kuhs, K. Nünighoff, C. Pohl, M. Prager, W. Schweika, Phys. B: Condens. Matter **350**, E647 (2004), proceedings of the Third European Conference on Neutron Scattering

[10] C. Yuan, Z. Zhang, J. Zhu, J. Zhao, L. Yang, L. Zhang, Y. Song, D. Tang, Appl. Phys. Lett. **119** (2021), 133902

[11] M. Celli, D. Colognesi, L. Ulivi, M. Zoppi, A.J. Ramirez-Cuesta, J. Phys. Conf. Ser. **340**, 012051 (2012)

[12] J. VandeVondele, M. Krack, F. Mohamed, M. Parrinello, T. Chassaing, J. Hutter, Comput. Phys. Commun. **167**, 103 (2005)

[13] A. Togo, I. Tanaka, Scr. Mater. **108**, 1 (2015)

[14] X.X. Cai, T. Kittelmann, E. Klinkby, J. Márquez Damián, J. Comput. Phys. **380**, 400 (2019)

[15] X.X. Cai, T. Kittelmann, Comput. Phys. Commun. **246**, 106851 (2020)

[16] T. Kittelmann, X.X. Cai, Comput. Phys. Commun. **267**, 108082 (2021)

[17] V. Czamler et al., *Clathrate hydrates as novel moderator material for sources of very cold neutrons*, proceedings of the 8th European Conference on Neutron Scattering, Garching, Germany (2023)

[18] A. Lenz, L. Ojamäe, J. Phys. Chem. A **115**, 6169 (2011)

[19] B. Hess, C. Kutzner, D. van der Spoel, E. Lindahl, J. Chem. Theory Comput. **4**, 435 (2008)

[20] H.J.C. Berendsen, J.R. Grigera, T.P. Straatsma, J. Phys. Chem. **91**, 6269 (1987)

[21] J. Wang, R.M. Wolf, J.W. Caldwell, P.A. Kollman, D.A. Case, J. Comp. Chem. **25**, 1157 (2004)

[22] J.P. Perdew, K. Burke, M. Ernzerhof, Phys. Rev. Lett. **77**, 3865 (1996)

[23] S. Goedecker, M. Teter, J. Hutter, Phys. Rev. B **54**, 1703 (1996)

[24] S. Grimme, J. Antony, S. Ehrlich, H. Krieg, J. Chem. Phys. **132** (2010)

[25] J.S. Tse, J. de Phys. Colloque **48**, C1 (1987)

[26] I. Souza, J. Iniguez, D. Vanderbilt, Phys. Rev. Lett. **89**, 117602 (2002)

[27] I. Morrison, S. Jenkins, Phys. B: Condens. Matter **263**, 442 (1999)

[28] H. Schober, H. Itoh, A. Klapproth, V. Chihaia, W.F. Kuhs, Eur. Phys. J. E **12**, 41 (2003)

[29] C.J. Burnham, N.J. English, J. Chem. Phys. **144**, 051101 (2015)

[30] J. Li, J. Chem. Phys. **105**, 6733 (1996)

[31] G.L. Squires, *Introduction to the Theory of Thermal Neutron Scattering*, 3rd edn. (Cambridge University Press, 2012)

[32] H. Schober, J. Neutron Res. **17**, 109 (2014)

[33] J. Dawidowski, J.R. Granada, J.R. Santisteban, F. Cantargi, L.A.R. Palomino, in *Neutron Scattering – Fundamentals*, edited by F. Fernandez-Alonso, D.L. Price (Academic Press, 2013), Vol. 44 of *Exp. Methods Phys. Sci.*, pp. 471–528

BornAgain, software for GISAS and reflectometry: releases 1.17 to 20

Ammar Nejati[1], *Mikhail* Svechnikov[1], and *Joachim* Wuttke[1,*]

[1]Forschungszentrum Jülich GmbH, Jülich Centre for Neutron Science at MLZ, Lichtenbergstraße 1, 85748 Garching, Germany

Abstract. BornAgain is a free and open source cross-platform software for simulating and fitting grazing-incidence small-angle scattering, off-specular scattering, and reflectometry. An authorative description as per release 1.16 of 2019 has been published in J. Appl. Cryst. 53, 262–276 (2020). This report explains the incremental changes from releases 1.17 to 20.

1 Introduction

The research software BornAgain was originally conceived for simulating and fitting neutron and x-ray grazing-incidence small-angle scattering (GISAS). It can also be used with off-specular scattering, reflectometry, and regular small-angle scattering. BornAgain is a cross-platform software with active support for Linux, Windows, and MacOS. BornAgain is a free and open-source project under the GNU Public License. Download links and online documentation can be reached from the project homepage [1]. Source repository and issue tracker are at [2].

The canonical literature reference for BornAgain is the comprehensive paper in J. Appl. Cryst. [3], based on software release 1.16 from 2019. The present report documents the incremental changes from releases 1.17 to 20.

2 Release history, numbering scheme

Releases 1.17 to 1.19 were published in 2020 and 2021 [4]. We then spent two full years on internal consolidation before the next release was published in March 2023. At this point the numbering scheme was changed: The major version number "1" was dropped, and the old minor promoted to major so that 1.19 was followed by version 20.

The change in the numbering scheme expresses our conviction that BornAgain has become too big for any comprehensive re-enginering or partial re-writing that would justify stepping the old major from 1 to 2. Instead, all future evolution of BornAgain will be through incremental changes in agile development cycles that should not take longer than a couple of months.

The full version names were actually ternary (1.19.0), and are now binary (20.0). In either case, the trailing ".0" expresses the patch level that is incremented only in hotfix releases, such as 20.1 from May 2023.

*e-mail: j.wuttke@fz-juelich.de

3 Source repository

The source repository [2], including the issue tracker, has been relocated from GitHub to Jugit, our institution's instance of the open-source software GitLab. External contributors can log in through a federated service [5], then self-register. Contributors from institutions that are not participating in the federation can login using their GitHub, Google, or ORCID account [6].

4 Binaries for different Python versions

We provide all necessary configuration scripts and explanations for building BornAgain from source. Most users, however, prefer binary installers, which we provide at [7] for four platforms (Linux and Windows on x64 architecture, Mac on x64 and arm). For Linux, as an alternative to the installer shell script, there exist externally maintained Debian packages [8].

For Python scripting to work, the minor version of the user's Python engine must be the same as was used in building the BornAgain binaries. In the past, version conflicts often prevented users from getting started with BornAgain scripting. Therefore, starting with release 20, we now provide binary installers for several Python minor versions (currently 3.8 to 3.10), and this for each of the four supported target platforms.

5 Python-only packages (wheels for pip)

For those who but intend to use BornAgain through the Python API only, release 20 introduced yet another, even simpler installation method: We packaged a non-GUI variant of BornAgain as a Python *wheel*. A wheel [9] is a zipped archive of Python scripts and object libraries. In our case these libraries comprise the BornAgain modules compiled from the C++ sources, including Swig-generated wrapper code, and all external library BornAgain depends on.

A wheel has a specially formatted file name that encodes software name, software version, Python language

https://doi.org/10.1051/epjconf/202328606004

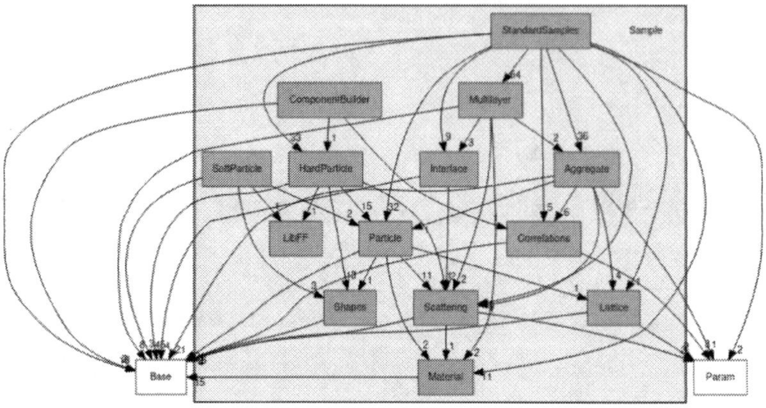

Figure 1. Include relations between code subdirectories concerned with the sample model. Graphical representation generated by Doxygen. All arrows go downwards: there are no cyclic dependences.

version, Python application binary interface version, target operating system and target processor architecture. So we prepared 12 wheels, for the four platforms and the three Python minor versions mentioned above.

We uploaded them to the standard Python package repository, for historical reasons misnomed "index": the Python Package Index (PyPI). Our wheels can be either downloaded through the "Download files" link at [10], or they can be automatically retrieved and installed using *pip* commands.

6 Internal consolidation

In the years leading to release 1.16, BornAgain has rapidly grown in scope, but also in complexity. The clarity and elegance of the initial architecture was somewhat lost in the course of unforeseen extensions. New developers found it increasingly difficult to implement new features, and most developer time was spent on analyzing how the program was working. This problem was addressed in releases 1.17 to 20 through refactoring aimed at simplifying and unifying code and data structures.

We reorganized the directory tree that holds over 1500 source files, and resolved cyclic dependence between directories [11, Sect. 5.4.1] so that include relations now form a directed graph (Fig. 1). We renamed classes, variables and functions that had evolved away from their original meaning. To facilitate code analysis, functions of same name in unrelated classes were given distinct names. In a majority of cases, C++ templates could be replaced by simpler constructs.

We removed unused code and reverted premature generalizations. To reduce the depth of call chains, numerous short functions were inlined or merged. Classes only meant to hold algorithms were dissolved, and the algorithms moved into free functions or under other classes. New abstractions helped to merge duplicate code.

Three-dimensional vector classes were moved from BornAgain into a separate library, *libheinz* [12], which in the future shall also be used by other projects at Heinz Maier-Leibnitz Zentrum. Form-factor computations [13] were also moved to a separate library, *libformfactor* [14], so that time-consuming numerical tests could be taken out of the BornAgain test suite. Regarding test coverage [11, Sect. 5.8], we removed trivial unit tests, and added functional tests, especially in form of Python scripts which have dual use as part of the example script collection.

As a result, core code could be reduced by 18 % from 38 kLoC (kilo lines of code) in release 1.16 to 31 kLoC (including 2 kLoC moved to libheinz and libformfactor) in release 20, and the GUI code by 13 %, from 60 kLoC to 52 kLoC, all while improving and extending functionality.

7 API changes

The Python Application Programming Interface (API) of BornAgain mostly consists of class constructor and class member function calls. Its ergonomy depends on the pertinence, expressivity and uniformity of function names and argument lists. If the principle of least surprise [15, 16] is well respected the user can work by analogy and develop an intuition how to use classes and functions. If the API is inconsistent the user needs to look up each function signature in the reference.

To edge out inconsistencies and make the API as uniform and unsurprising as possible, each release is currently breaking some user scripts. We are aware that it is annoying, but under the premise that BornAgain will have a lifespan of decades it is the right choice to sacrifice backward compatibility now for a more consistent API and a more maintainable code base in the future. All API changes are listed in the change log [4].

8 GUI, new sample editor

Release 20 brought two important changes to the graphical user interface (GUI): Full serialization, and a new sample editor.

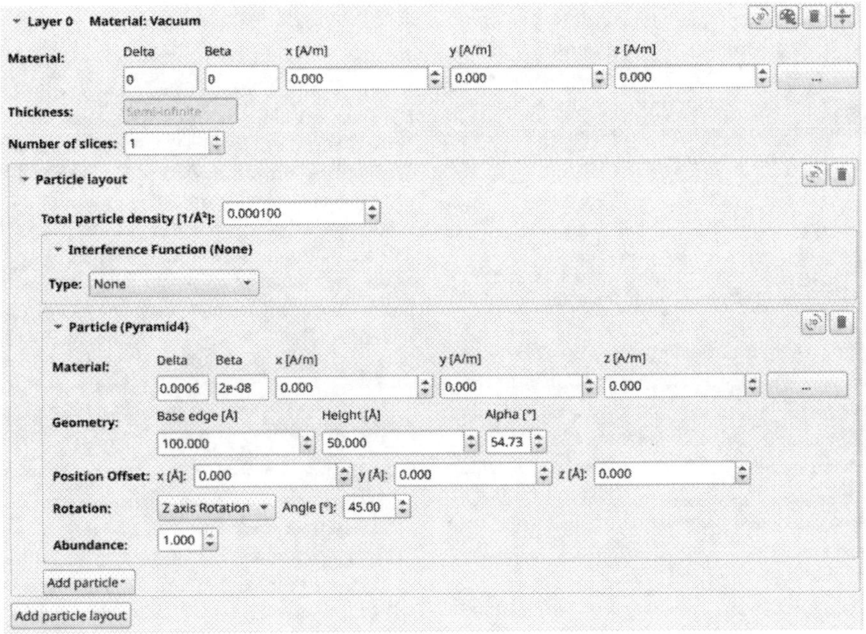

Figure 2. Screenshot of the new sample editor GUI widget, featuring a random assembly of uncorrelated pyramids in an otherwise empty surface layer.

Through serialization, the full state of a GUI session, including sample and instrument models, can be saved to and reloaded from a *project file*. Uses include resuming work after a break, reverting unsuccessful modifications of a model, sharing with collaborators, and creating fully reproducible bug reports.

The next couple of releases will likely break project file compatibility for certain model components, for the same reasons as discussed above for the Python API. In the longer run, however, we intend to ensure backward compatibility. For this purpose, each serialized feature carries a version number.

Up to release 1.19, the GUI had a graphical sample editor inspired inspired by the LabVIEW way of graphical programming. Users assembled models from components by drag and drop. This allowed to visualize the stack of layers and highlighted the internal hierarchy of layers, layouts, particles, interference functions and their modifications. Acclaimed by many novice users, this editor was aesthetically pleasing and had an almost haptic appeal. However, the block diagrams did not show model parameters, which could only be modified in a separate widget. Experienced users therefore requested a more compact sample representation.

In response to that, release 20 has a completely new sample editor, based on expandable tables. Selectable lists, parameters, their labels and editable fields are organized as grids inside nested collapsible boxes. This allows both handling layers with arbitrary internal complexity and the direct presentation of exhaustive parameter set to the user.

9 Documentation

The online documentation, accessible from the project home page [1], is now versioned: starting with 1.19, the documentation for each release will indefinitely remain readable.

Up to 1.19, the main part of the documentation was a very long tutorial on Python scripting. Usage was mostly taught by code examples, followed by explanations. In place of a reference, a documentation of the C++ application programming interface (API) was auto-generated by Doxygen.

The tutorial turned out to be cumbersome to maintain. Readability suffered from numerous repetitions. The C++ API reference was found inadequate by Python users. For these reasons, the Python scripting documentation was reorganized in release 20. It was split into three parts: a short tutorial, a collection of examples, and a human-written reference. The new, concise style is inspired by best-practice examples like the classical Unix man pages or the Qt online documentation, and is guided by the assumption of an intelligent, active reader who is able to follow links and draw obvious consequences [11, 16].

The conversion to this new style is a work in progress. Many pages in the reference section are still in old teaching-from-example style. They will be converted piece by piece in the next release cycles, preferably in conjunction with simplification and unification of the API. The "Simulation model" section is already fully converted to the new style and gives an impression how the entire reference will look like in the future.

We are aware that users of the Python API want a documentation that is more "pythonic" in style and can be queried in an interactive session. Any action on this has to wait until we have decided whether we continue to generate Python bindings through Swig (which has severe limitations especially for functional programming) or whether we switch to some alternative technology.

10 Further plans

Our work plan for the next years comprises the following categories of tasks, in roughly decreasing order of priority:

(1) User support. Questions and suggestions should preferentially be submitted to the issue tracker at [2] so that discussions are publicly visible, and will aggregate into a searchable knowledge base. Confidential requests can be sent by mail to contact@bornagainproject.org. Our biennial School and User Meeting, interrupted by the Covid-19 pandemic, shall be relaunched soon.

(2) Further code consolidation, along with API unification and simplification. In the BornAgain core, we are currently unifying the handling of scans, coordinate axes and unit conversions. In the sample model, per legacy from IsGisaxs [17, 18], inter-particle correlations are currently treated as modifiers of particles. This hierarchy shall be inverted; particles shall be attached to lattices or other structural models, which in turn are attached to coherent surface or bulk domains. In the GUI, we want to reduce the amount of repetitive ("boilerplate") code and to further simplify some data structures and class hierarchies in order to facilitate future modifications and enhancements.

(3) Improvement of usability. We need to complete the online reference, make the example collection more orthogonal, and rework parts of the GUI, taking inspiration from other reflectometry software [19–23]. Developer-oriented code instrumentation shall be withdrawn from the published examples. To support different workflows, we will probably make the GUI multi-window capable so that users can freely select and arrange the data and model views they want to access simultaneously. In a related effort, we need to support simultaneous fits of multiple data sets, for instance from different measurements of the same sample.

(4) Acceleratation of computations. Absorption shall be handled at the level of horizontally averaged sliced layers, not the level of scattering particles. This will allow us to return from complex to real valued scattering vectors, which will substantially accelerate all form factor computations. Recently obtained series expansions [24] shall be used for the efficient computation of rotationally averaged form factors.

(5) Extension of functionality, mostly in response to user requests. The modelling and computational handling of instrumental resolution and modelling of mesoparticles shall be improved, the latter needs to be fully documented. To make progress on magnetic reflectivity and scattering, we depend on users proposing theoretical models.

Acknowledgements

We thank Randolf Beerwerth, Ludwig Jäck, Tobias Knopff, Andrew McCluskey, Gennady Pospelov, Matthias Puchner and Dmitry Yurov for contributions after release 1.16. We thank Thomas Arnold, Artur Glavic, Andrew McCluskey, Karolina Mothander, Thomas Holm Rod for criticism and suggestions. The extension of BornAgain to reflectometry was partly funded under schedule NIK4#11 as part of the data analysis and modeling work package to the contribution agreement between the European Spallation Source ERIC and Forschungszentrum Jülich. J.W. thanks Andreas Meyer and Stéphane Rols for hospitality at ILL in Grenoble.

References

[1] BornAgain home page, https://bornagainproject.org.
[2] BornAgain source repository, https://jugit.fz-juelich.de/mlz/bornagain.
[3] G. Pospelov, W. Van Herck, J. Burle, J. M. Carmona Loaiza, C. Durniak, J. M. Fisher, M. Ganeva, D. Yurov and J. Wuttke, J. Appl. Cryst. **53**, 262 (2020).
[4] BornAgain change log, https://jugit.fz-juelich.de/mlz/bornagain/-/blob/main/CHANGELOG.
[5] Helmholtz Authentication and Authorisation Infrastructure, https://hifis.net/aai.
[6] How to log in to Jugit, https://computing.mlz-garching.de/tech/jugit-login.
[7] BornAgain download directory, https://bornagainproject.org/ext/files.
[8] BornAgain Debian package, https://tracker.debian.org/pkg/bornagain.
[9] D. Holth, *PEP 427 — The Wheel Binary Package Format 1.0.* https://peps.python.org/pep-0427 (2012).
[10] BornAgain without GUI as Python package, https://pypi.org/project/BornAgain.
[11] J. Wuttke, S. Cottrell, M. A. Gonzalez, A. Kæstner, A. Markvardsen, T. H. Rod and G. Vardanyan, J. Neutron Res. **24**, 33 (2022).
[12] libheinz, C++ base library of Heinz Maier-Leibnitz Zentrum, https://jugit.fz-juelich.de/mlz/libheinz.
[13] J. Wuttke, J. Appl. Cryst. **54**, 580 (2021).
[14] libformfactor, C++ library for the computation of scattering form factors, https://jugit.fz-juelich.de/mlz/libformfactor.
[15] G. James, *The Tao of Programming*, InfoBooks: Santa Monica, Calif. (1987).
[16] E. S. Raymond, *The Art of Unix Programming*, Addison-Wesley: Boston (2003).
[17] R. Lazzari, J. Appl. Cryst. **35**, 406 (2002).
[18] R. Lazzari, *IsGISAXS*. Version 2.6. http://www.insp.jussieu.fr/oxydes/IsGISAXS/isgisaxs.htm (2006).
[19] A. R. J. Nelson and S. W. Prescott, J. Appl. Cryst. **52**, 193 (2019).
[20] O. V. Penkov, I. A. Kopylets, M. Khadem and T. Qin, SoftwareX **12**, 100528 (2020).

https://doi.org/10.1051/epjconf/202328606004

[21] M. Svechnikov, J. Appl. Cryst. **53**, 244 (2020).
[22] A. Koutsioubas, J. Appl. Cryst. **54**, 1857 (2021).
[23] A. Glavic, J. Appl. Cryst. **55**, 1063 (2022).
[24] M. Wagener and S. Förster, Sci. Rep. **13**, 780 (2023).

AUTHOR INDEX

AUTHOR INDEX

AUTHOR INDEX

EDP Sciences – Web of Conferences
17, Avenue du Hoggar
Parc d'Activité de Courtabouf
BP 112
F-91944 Les Ulis Cedex A
France

ISBN 978-1-7138-8043-1

CURRAN ASSOCIATES INC.
proceedings.com